S0-CAS-501

# problems in molecular structure

editors G.J.Bullen D.J.Greenslade

pion · london

# problems in molecular structure

editors G.J.Bullen D.J.Greenslade

Pion Limited, 207 Brondesbury Park, London NW2 5JN



ISBN 0 85086 083 0

Set on IBM 72 Composers by Pion Limited, London
Printed in Great Britain by Page Bros (Norwich) Limited

Contents



Authors

Preface

1 Symmetry 1
1.1 Principles 1
A P Cracknell
1.2 Molecular symmetry 12
D J Greenslade
1.3 Crystal symmetry 37
A P Cracknell
1.4 Symmetry and spectroscopy 52
D J Greenslade, R G Jones, J R Miller

2 Diffraction 70
2.1 x-Ray diffraction 70
G J Bullen
2.2 Neutron diffraction 106
J C Speakman
2.3 Electron diffraction by gases 118
B Beagley
2.4 Electron diffraction by crystals 133
A L Mackay

3 Vibration–rotation spectroscopy 146
3.1 Infrared spectroscopy 146
N Sheppard, D B Powell
3.2 Raman spectroscopy 159
B P Stoicheff
3.3 Microwave spectroscopy 173
D H Whiffen

4 Electronic properties 182
4.1 Ultraviolet and visible spectroscopy 182
G R Eaton, J R Riter Jr, S S Eaton
4.2 Electron spin resonance 212
D H Whiffen
4.3 Magnetic moments 230
R W Jotham
4.4 Molecular electric dipole and quadrupole moments 252
A D Buckingham
4.5 Optical rotatory dispersion and circular dichroism 268
S F Mason
4.6 Photoelectron spectroscopy 294
W C Price

| | | |
|---|---|---|
| 5 | Nuclear spectroscopy | 305 |
| | 5.1 Nuclear magnetic resonance | 305 |
| | D J Greenslade, R G Jones | |
| | I Solid state n.m.r. | 305 |
| | II High-resolution n.m.r. | 312 |
| | 5.2 Nuclear quadrupole resonance | 336 |
| | E A C Lucken | |
| | 5.3 Mössbauer spectroscopy | 358 |
| | J J Zuckerman, N W G Debye | |
| 6 | Mass spectrometry | 367 |
| | 6.1 Mass spectrometry | 367 |
| | J R Gilbert | |
| 7 | Structure and energy | 378 |
| | 7.1 Bond energies and lattice energies | 378 |
| | H A Skinner | |
| | 7.2 Wave functions and bonding | 402 |
| | C A Coulson | |
| | 7.3 Arrangements of atoms in solids | 428 |
| | G J Bullen, D J Greenslade | |
| | Appendix 1 | 453 |
| | Quantities and units | 453 |
| | Table of physical quantities and conversion factors | 455 |
| | Appendix 2 | 456 |
| | Notations used in symmetry | 456 |
| | Character tables for molecular symmetry | 460 |
| | Index | 463 |

Authors

| | |
|---|---|
| **B Beagley** | University of Manchester Institute of Science and Technology, Manchester M60 1QD, England |
| **A D Buckingham** | University Chemical Laboratory, Lensfield Road, Cambridge CB2 1EW, England |
| **G J Bullen** | Department of Chemistry, University of Essex, Colchester CO4 3SQ, England |
| **C A Coulson** (deceased) | Formerly of the University of Oxford Mathematical Institute, Oxford OX1 3LB, England |
| **A P Cracknell** | Carnegie Laboratory of Physics, University of Dundee, Dundee DD1 4HN, Scotland |
| **N W G Debye** | Department of Chemistry, Towson State University, Baltimore, MD 21204, USA |
| **G R Eaton** | Department of Chemistry, University of Denver, Denver, CO 80208, USA |
| **S S Eaton** | University of Colorado at Denver, Denver, CO 80202, USA |
| **J R Gilbert** | Department of Chemistry, University of Essex, Colchester CO4 3SQ, England |
| **D J Greenslade** | Department of Chemistry, University of Essex, Colchester CO4 3SQ, England |
| **R G Jones** | Westbourne High School, Ipswich IP1 5JN, England |
| **R W Jotham** | Department of Adult Education, University of Nottingham, Nottingham NG1 4FT, England |
| **E A C Lucken** | Departement de Chimie Physique, Université de Genève, 1211 Geneva 4, Switzerland |
| **A L Mackay** | Department of Crystallography, Birkbeck College, University of London, London WC1E 7HX, England |
| **S F Mason** | Department of Chemistry, King's College, University of London, London WC2R 2LS, England |
| **J R Miller** | Department of Chemistry, University of Essex, Colchester CO4 3SQ, England |
| **D B Powell** | School of Chemistry, University of East Anglia, Norwich NR4 7TJ, England |
| **W C Price** | Department of Physics, King's College, University of London, London WC2R 2LS, England |
| **J R Riter Jr** | Department of Chemistry, University of Denver, Denver, CO 80208, USA |
| **N Sheppard** | School of Chemistry, University of East Anglia, Norwich NR4 7TJ, England |

**H W Skinner** Department of Chemistry, University of Manchester, Manchester M13 9PL, England

**J C Speakman** Department of Chemistry, University of Glasgow, Glasgow G12 8QQ, Scotland

**B P Stoicheff** Department of Physics, University of Toronto, Ontario M5S 1A1, Canada

**D H Whiffen** School of Chemistry, University of Newcastle upon Tyne, Newcastle upon Tyne NE1 7RU, England

**J J Zuckerman** Department of Chemistry, University of Oklahoma, Norman, OK 73019, USA

**Preface**

Molecular structure is of interest not only to chemists, but to many physicists, applied mathematicians, and even biologists. The techniques used in its study are varied and often complex, and a true understanding can be obtained only by working through examples of the calculations involved in application of the diverse methods. It is for this reason that the present work was planned. It should provide a useful aid for advanced undergraduate and beginning graduate students.

As may be inevitable with a work involving so many authors, the preparation has taken many years. Our apologies go to those authors who provided their manuscripts promptly, and our thanks to those who stepped in late in the day to fill gaps.

The problems have mostly been cast in terms of SI units, but where appropriate other units, often still found in published papers, are used; the student should be familiar with such units in order to read papers published in the journals—especially those prior to our SI era. We have given appendices which discuss units and symmetry notation and set out necessary group representation character tables. We have not found any way of simply and clearly distinguishing operators from algebraic quantities: the 'hat' (ˆ) used in some sources seems unduly cumbersome. Vectors are set in bold italic type and matrices in bold upright type.

We have included some purely theoretical problems and have attempted to group the various contributions into a logical framework of seven sections. The final section is nevertheless something of a miscellany, containing problems on thermochemistry, arrangements of atoms in solids (including defects), and a wave-mechanical section contributed by the late Professor C A Coulson. Even here there is a unifying theme, since the problems nearly all involve energy considerations. We have not tried to impose our style on the various authors, but have had to make firm decisions on matters such as the relative position of the problems and solutions. Each solution is placed immediately after the problem, as the same tables and diagrams are often needed in both problem and solution. Some authors have felt it worthwhile to provide a short introduction, but in general it is assumed the student will have access to the relevant textbooks and monographs. Many problems have references to appropriate journals and to the particular articles whence they are derived.

Finally we thank Ralph Brookfield of Pion, who has expended much care and effort to bring the work to press.

G J Bullen
D J Greenslade
University of Essex

# 1 Symmetry

## 1.1 Principles†

**A P Cracknell** University of Dundee

### 1.1.1

A rubber planter gave one of his labourers ten rubber-tree seedlings and told him to plant them in five rows of four seedlings in each row. An industrial dispute arose because the labourer argued that the task was impossible. Can you design a suitable plantation? If so, what symmetry elements would your plantation possess?

**Solution**

A highly symmetrical arrangement is possible, see figure 1.1.1a. (The problem did not say that the trees had to be equally spaced.) This plantation possesses a vertical fivefold rotation axis of symmetry at O and five vertical reflection planes of symmetry through AO, BO, CO, DO, and EO. These symmetry elements belong to one of the noncrystallographic molecular point groups $5m$ ($C_{5v}$).

Other less symmetrical plantations are possible, see, for example, figure 1.1.1b.

### 1.1.2

Identify the symmetry operations of an equilateral triangle and
(a) construct the group multiplication table,
(b) divide the group into classes,
(c) identify the various subgroups and state whether or not each of them is a normal (or invariant) subgroup.

**Solution**

The symmetry operations are, in the notation of figure 1.1.2,

- $E$ identity operation
- $C_3^+$ anticlockwise rotation through 120° about axis through O normal to the triangle
- $C_3^-$ clockwise rotation through 120° about axis through O normal to the triangle
- $\sigma_1$ reflection in plane through AO normal to the triangle
- $\sigma_2$ reflection in plane through BO normal to the triangle
- $\sigma_3$ reflection in plane through CO normal to the triangle.

† The character tables of the point groups, which will be needed in connection with various problems in this section, will be found in Appendix 2. The labels used for the symmetry operations in the point groups are those employed by Bradley and Cracknell (1972); see also Appendix 1.

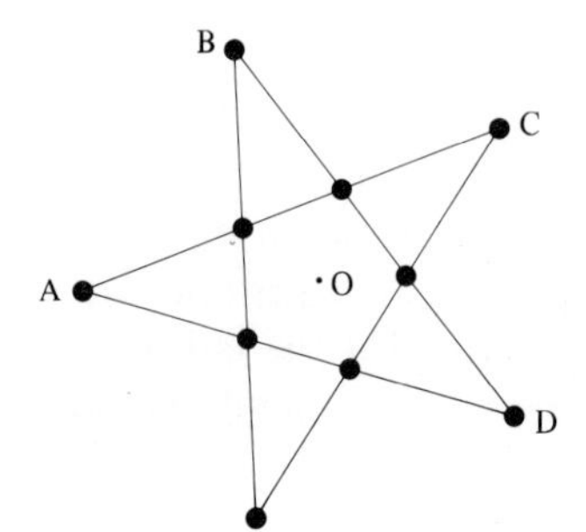


**Figure 1.1.1a.**

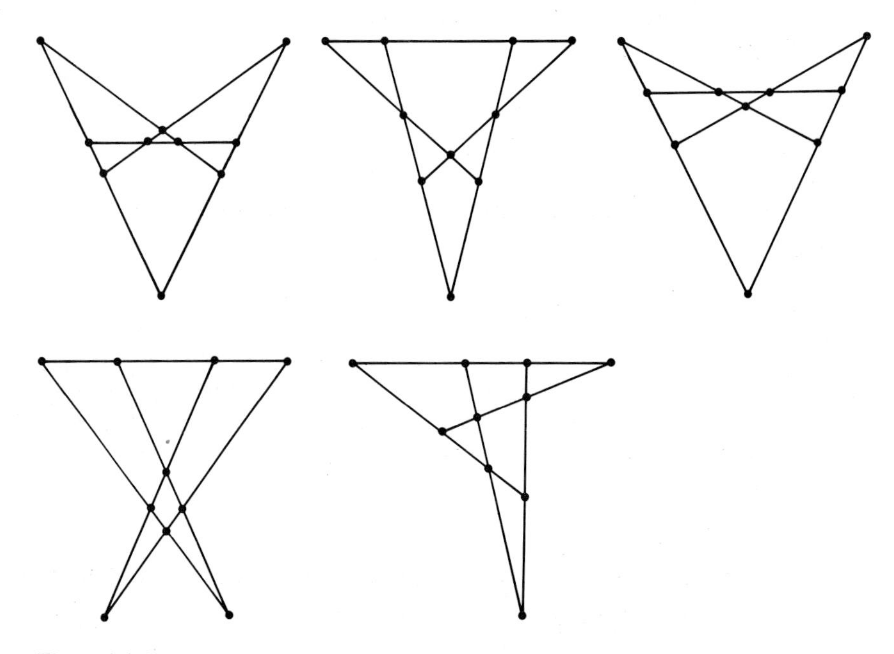

**Figure 1.1.1b.**

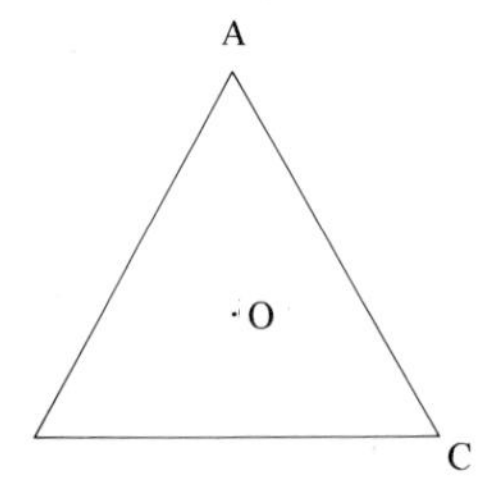


**Figure 1.1.2.**

(a)

| | | | | | |
|---|---|---|---|---|---|
| $E$ | $C_3^+$ | $C_3^-$ | $\sigma_1$ | $\sigma_2$ | $\sigma_3$ |
| $C_3^+$ | $C_3^-$ | $E$ | $\sigma_3$ | $\sigma_1$ | $\sigma_2$ |
| $C_3^-$ | $E$ | $C_3^+$ | $\sigma_2$ | $\sigma_3$ | $\sigma_1$ |
| $\sigma_1$ | $\sigma_2$ | $\sigma_3$ | $E$ | $C_3^+$ | $C_3^-$ |
| $\sigma_2$ | $\sigma_3$ | $\sigma_1$ | $C_3^-$ | $E$ | $C_3^+$ |
| $\sigma_3$ | $\sigma_1$ | $\sigma_2$ | $C_3^+$ | $C_3^-$ | $E$ |

(b) Evaluate $X^{-1}RX$ for all $X$ and for $R = E$, $C_3^+$, and $\sigma_1$.

| $X$ | $E$ | $C_3^+$ | $C_3^-$ | $\sigma_1$ | $\sigma_2$ | $\sigma_3$ |
|---|---|---|---|---|---|---|
| $X^{-1}$ | $E$ | $C_3^-$ | $C_3^+$ | $\sigma_1$ | $\sigma_2$ | $\sigma_3$ |
| $X^{-1}EX$ | $E$ | $E$ | $E$ | $E$ | $E$ | $E$ |
| $X^{-1}C_3^+X$ | $C_3^+$ | $C_3^+$ | $C_3^+$ | $C_3^-$ | $C_3^-$ | $C_3^-$ |
| $X^{-1}\sigma_1X$ | $\sigma_1$ | $\sigma_3$ | $\sigma_2$ | $\sigma_1$ | $\sigma_3$ | $\sigma_2$ . |

The classes therefore consist of

$C_1$: $E$; $C_2$: $C_3^+, C_3^-$; $C_3$: $\sigma_1, \sigma_2, \sigma_3$.

(c) Invariant subgroups: $E, C_3^+, C_3^-$; $E$.
Noninvariant subgroups: $E, \sigma_1$; $E, \sigma_2$; $E, \sigma_3$.
(An invariant subgroup consists of a sum of complete classes of the original group.)

### 1.1.3

A certain group consists of the numbers, 0, ±1, ±2, ±3. Group multiplication is defined as arithmetical addition modulo 7, that is, if a "product" goes out of the range −3 to +3 it is brought back into range by the addition or subtraction of 7. Show that this group is Abelian and hence determine its character table. Is this group isomorphic to a crystallographic point group?

**Solution**

From the rule for group multiplication of the elements the multiplication can be seen to be commutative by inspection. Therefore the group is Abelian.

The group multiplication table is:

| | | | | | | |
|---|---|---|---|---|---|---|
| 0 | 1 | 2 | 3 | −3 | −2 | −1 |
| 1 | 2 | 3 | −3 | −2 | −1 | 0 |
| 2 | 3 | −3 | −2 | −1 | 0 | 1 |
| 3 | −3 | −2 | −1 | 0 | 1 | 2 |
| −3 | −2 | −1 | 0 | 1 | 2 | 3 |
| 2 | −1 | 0 | 1 | 2 | 3 | −3 |
| −1 | 0 | 1 | 2 | 3 | −3 | −2 |

If we write $P = 1$ then we can make the identification

| $P$ | $P^2$ | $P^3$ | $P^4$ | $P^5$ | $P^6$ | $P^7$ |
|---|---|---|---|---|---|---|
| 1 | 2 | 3 | −3 | −2 | −1 | 0 . |

The conjugation of any of these elements, $P^n$, with any element $X = P^m$ of the group gives $X^{-1}P^nX = P^{-m}P^nP^m = P^n$, so that each element is in a class by itself. There are therefore seven classes so that there are seven irreducible representations which must all be one-dimensional.

The character $\chi(P)$ of $P$ is then given by

$$[\chi(P)]^7 = \chi(P^7) = 1 .$$

Therefore $\chi(P) = \exp(\frac{2}{7}\pi i m)$, where $m = 0, 1, 2, 3, 4, 5$, or 6.

The character table is therefore

| | 0 | 1 | 2 | 3 | −3 | −2 | −1 |
|---|---|---|---|---|---|---|---|
| $\Gamma_1$ | 1 | 1 | 1 | 1 | 1 | 1 | 1 |
| $\Gamma_2$ | 1 | $\alpha$ | $\alpha^2$ | $\alpha^3$ | $\alpha^4$ | $\alpha^5$ | $\alpha^6$ |
| $\Gamma_3$ | 1 | $\alpha^2$ | $\alpha^4$ | $\alpha^6$ | $\alpha$ | $\alpha^3$ | $\alpha^5$ |
| $\Gamma_4$ | 1 | $\alpha^3$ | $\alpha^6$ | $\alpha^2$ | $\alpha^5$ | $\alpha$ | $\alpha^4$ |
| $\Gamma_5$ | 1 | $\alpha^4$ | $\alpha$ | $\alpha^5$ | $\alpha^2$ | $\alpha^6$ | $\alpha^3$ |
| $\Gamma_6$ | 1 | $\alpha^5$ | $\alpha^3$ | $\alpha$ | $\alpha^6$ | $\alpha^4$ | $\alpha^2$ |
| $\Gamma_7$ | 1 | $\alpha^6$ | $\alpha^5$ | $\alpha^4$ | $\alpha^3$ | $\alpha^2$ | $\alpha$ |

where $\alpha = \exp(\frac{2}{7}\pi i)$.

There is no crystallographic point group of order 7.

**1.1.4**

The following elements form a group, **G**,

$$E, P, P^2, Q, PQ, P^2Q$$

where $P^3 = E$, $Q^2 = E$, and $QP = P^2Q$.

(a) Deduce the character table of this group, and

(b) show that there exists an isomorphism between this abstract group and the group of the symmetry operations of an equilateral triangle and also with the permutation group of three identical objects.

**Solution**

(a) Each element is its own inverse except for $P$ and $P^2$. $P^{-1} = P^2$ and $(P^2)^{-1} = P$. The group separates into three classes:

$C_1$: $E$
$C_2$: $P, P^2$
$C_3$: $Q, PQ, P^2Q$ .

The class multiplication coefficients $c_{ij,k}$, defined by $C_i \times C_j = \sum_k c_{ij,k} C_k$ (where $c_{ji,k} = c_{ij,k}$) are

$$c_{11,1} = c_{12,2} = c_{13,3} = 1 ,$$

$$c_{22,1} = 2, \quad c_{22,2} = 1, \qquad c_{23,3} = 2 ,$$

$$c_{33,1} = 3, \quad c_{33,2} = 3 ;$$

the remainder are zero. Using these values of the coefficients we solve the following equations for $\chi_i^k$:

$$h_i \chi_i^s h_j \chi_j^s = d_s \sum_{k=1}^{r} c_{ij,k} h_k \chi_k^s$$

and

$$\sum h_i \chi_i^{s*} \chi_i^t = \delta_{st} N ,$$

where

$h_i$ is the order of class $C_i$,
$\chi_i^s$ is the character of class $C_i$ in the irreducible representation $\Gamma_s$,
$\chi_i^{s*}$ is the complex conjugate of $\chi_i^s$,
$d_s$ is the dimension of $\Gamma_s$,
$r$ is the number of classes in group,
$N$ is the order of group.

The solutions give the characters of the various irreducible representations $\Gamma_s$

| | $E$ | $P, P^2$ | $Q, PQ, P^2Q$ |
|---|---|---|---|
| $\Gamma_1$ | 1 | 1 | 1 |
| $\Gamma_2$ | 1 | 1 | −1 |
| $\Gamma_3$ | 2 | −1 | 0 . |

(b) The isomorphism with the point group of the symmetry operations of the triangle may be established,

| $E$ | $P$ | $P^2$ | $Q$ | $PQ$ | $P^2Q$ |
|---|---|---|---|---|---|
| $E$ | $C_3^+$ | $C_3^-$ | $\sigma_1$ | $\sigma_3$ | $\sigma_2$ . |

If we identify the six permutations of three identical objects by

| | | | |
|---|---|---|---|
| $E$ | 1, 2, 3 → 1, 2, 3 | $F$ | 1, 2, 3 → 1, 3, 2 |
| $A$ | 1, 2, 3 → 2, 3, 1 | $G$ | 1, 2, 3 → 3, 2, 1 |
| $B$ | 1, 2, 3 → 3, 1, 2 | $H$ | 1, 2, 3 → 2, 1, 3 , |

the isomorphism with the group of permutations of three identical objects may be established,

| $E$ | $P$ | $P^2$ | $Q$ | $PQ$ | $P^2Q$ |
|---|---|---|---|---|---|
| $E$ | $A$ | $B$ | $F$ | $G$ | $H$ . |

### 1.1.5

Determine the irreducible representations that are obtained when the following representations of the point group $4mm$ ($C_{4v}$) are reduced.

| $4mm$ ($C_{4v}$) | $E$ | $C_{2z}$ | $C_{4z}^{\pm}$ | $\sigma_x, \sigma_y$ | $\sigma_{da}, \sigma_{db}$ |
|---|---|---|---|---|---|
| $X_1$ | 4 | 4 | 0 | 0 | 0 |
| $X_2$ | 5 | 1 | −1 | −3 | 1 |
| $X_3$ | 96 | 68 | 2 | −12 | 44 |
| $X_4$ | 64 | 0 | 0 | 0 | 0 . |

**Solution**

The irreducible representation labelled by $i$ appears $n_i$ times, with

$$n_i = \frac{1}{N}\sum_R \chi_i^{i*}(R)\chi(R) ,$$

where $\chi(R)$ is the character of the element $R$ in the representation that is being reduced and $N$ is the order of the group.

Therefore, for example, for $X_1$

$$n_{A_1} = \tfrac{1}{8}[(1\times 4)+(1\times 4)+0+0+0+0+0+0]$$
$$= 1 ,$$

and similarly

$$n_{A_2} = n_{B_1} = n_{B_2} = 1$$

and

$$n_E = 0 .$$

Thus, we obtain

$$X_1 = A_1+A_2+B_1+B_2 ;$$

similarly,

$$X_2 = A_2+2B_2+E$$
$$X_3 = 29A_1+13A_2+6B_1+34B_2+7E$$
$$X_4 = 8A_1+8A_2+8B_1+8B_2+16E .$$

These reductions can be verified by inspection; for example

| | $E$ | $C_{2z}$ | $C_{4z}^{\pm}$ | $\sigma_x, \sigma_y$ | $\sigma_{da}, \sigma_{db}$ |
|---|---|---|---|---|---|
| $A_2$ | 1 | 1 | 1 | −1 | −1 |
| $2B_2$ | 2 | 2 | −2 | −2 | 2 |
| $E$ | 2 | −2 | 0 | 0 | 0 |
| $A_2+2B_2+E$ | 5 | 1 | −1 | −3 | 1 . |

**1.1.6**

Assign each of the following functions to the appropriate irreducible representation of the point group $4mm$ ($C_{4v}$), the group of the symmetry operations of a square:

$x, y, z$;
$I_x, I_y, I_z$ (the infinitesimal rotation operators);
$(x^2+y^2), (x^2-y^2), z^2$;
$xy, yz, zx$.

**Solution**

Either by inspection or by use of the projection operator $\sum_R \chi^{i*}(R)R$, where $R$ ranges over the elements of the point group,

$A_1$: $(x^2+y^2), z, z^2$
$A_2$: $I_z$
$B_1$: $(x^2-y^2)$
$B_2$: $xy$
$E$: $(x, y)$; $(I_x, I_y)$; $(zx, zy)$.

1.1.7

(a) Obtain the reductions of all the Kronecker products of the irreducible representations of the point group $4mm$ $(C_{4v})$,

$$\begin{array}{lllll} A_1 \otimes A_1, & A_1 \otimes A_2, & A_1 \otimes B_1, & A_1 \otimes B_2, & A_1 \otimes E, \\ & A_2 \otimes A_2, & A_2 \otimes B_1, & A_2 \otimes B_2, & A_2 \otimes E, \\ & & B_1 \otimes B_1, & B_1 \otimes B_2, & B_1 \otimes E, \\ & & & B_2 \otimes B_2, & B_2 \otimes E, \\ & & & & E \otimes E. \end{array}$$

(b) In Landau's theory of second-order phase transitions it is necessary to evaluate the symmetrized cube $[\![\Gamma]\!]^3$ and antisymmetrized square $\{\Gamma\}^2$ of the irreducible representation $\Gamma$ of some group G, where the character of the element $R$ in $[\![\Gamma]\!]^3$ is given by

$$[\![\chi]\!]^3(R) = \tfrac{1}{3}\chi(R^3)+\tfrac{1}{2}\chi(R^2)\chi(R)+\tfrac{1}{6}\chi^3(R)$$

and in $\{\Gamma\}^2$ by

$$\{\chi\}^2(R) = \tfrac{1}{2}\chi^2(R)-\tfrac{1}{2}\chi(R^2)\,.$$

$[\![\Gamma]\!]^3$ and $\{\Gamma\}^2$ may each be either reducible or irreducible.

Identify $[\![\Gamma]\!]^3$ and $\{\Gamma\}^2$ for each of the irreducible representations of the point group $4mm$ $(C_{4v})$ in terms of the irreducible representations of this group.

**Solution**

(a)

$$\begin{array}{lllll} A_1 \otimes A_1 = A_1, & A_1 \otimes A_2 = A_2, & A_1 \otimes B_1 = B_1, & A_1 \otimes B_2 = B_2, & A_1 \otimes E = E, \\ & A_2 \otimes A_2 = A_1, & A_2 \otimes B_1 = B_2, & A_2 \otimes B_2 = B_1, & A_2 \otimes E = E, \\ & & B_1 \otimes B_1 = A_1, & B_1 \otimes B_2 = A_2, & B_1 \otimes E = E, \\ & & & B_2 \otimes B_2 = A_1, & B_2 \otimes E = E\,, \end{array}$$

$E \otimes E = A_1+A_2+B_1+B_2$.

(b) For one dimensional representations

$$[\![\chi]\!]^3(R) = \chi^3(R)$$

and

$$\{\chi\}^2(R) = 0\,.$$

Hence

$$[\![A_1]\!]^3 = A_1 \qquad \{A_1\}^2 = 0$$
$$[\![A_2]\!]^3 = A_2 \qquad \{A_2\}^2 = 0$$
$$[\![B_1]\!]^3 = B_1 \qquad \{B_1\}^2 = 0$$
$$[\![B_2]\!]^3 = B_2 \qquad \{B_2\}^2 = 0\ .$$

For $E$ the characters are

| | $E$ | $C_{2z}$ | $C_{4z}^{\pm}$ | $\sigma_x, \sigma_y$ | $\sigma_{da}, \sigma_{db}$ |
|---|---|---|---|---|---|
| $[\![E]\!]^3$ | 4 | −4 | 0 | 0 | 0 |
| $\{E\}^2$ | 1 | 1 | 1 | −1 | −1 |

so that

$$[\![E]\!]^3 = 2E\ , \qquad \{E\}^2 = A_2\ .$$

1.1.8

The quantum-mechanical angular momentum operator $\boldsymbol{L}$ is obtained by replacing $\boldsymbol{p}$ in $\boldsymbol{L} = \boldsymbol{r} \times \boldsymbol{p}$ by $-\mathrm{i}\hbar\nabla$.

(a) Show that

$$L_x L_y - L_y L_x = \mathrm{i}\hbar L_z\ ;$$
$$L_y L_z - L_z L_y = \mathrm{i}\hbar L_x\ ;$$
$$L_z L_x - L_x L_z = \mathrm{i}\hbar L_y\ ;$$
$$\boldsymbol{L}^2 L_z - L_z \boldsymbol{L}^2 = 0\ .$$

(b) Express $L_x$, $L_y$, $L_z$, and $\boldsymbol{L}^2$ in terms of spherical polar coordinates.
(c) Show that the eigenvalues of $L_z$ are $m\hbar$ ($m = -l, -l+1, ..., +l$).

**Solution**

(a)

$$\boldsymbol{L} = (-\mathrm{i}\hbar)\begin{vmatrix} \boldsymbol{i} & \boldsymbol{j} & \boldsymbol{k} \\ x & y & z \\ \dfrac{\partial}{\partial x} & \dfrac{\partial}{\partial y} & \dfrac{\partial}{\partial z} \end{vmatrix}\ ,$$

therefore
$$L_x = -\mathrm{i}\hbar\left(y\frac{\partial}{\partial z} - z\frac{\partial}{\partial y}\right),$$
$$L_y = -\mathrm{i}\hbar\left(z\frac{\partial}{\partial x} - x\frac{\partial}{\partial z}\right),$$
$$L_z = -\mathrm{i}\hbar\left(x\frac{\partial}{\partial y} - y\frac{\partial}{\partial x}\right),$$

and

$$\boldsymbol{L}^2 = L_x^2 + L_y^2 + L_z^2\ .$$

Substituting for $L_x$ and $L_y$ in $L_x L_y - L_y L_x$ we obtain

$$(-\mathrm{i}\hbar)^2\left[\left(y\frac{\partial}{\partial z}-z\frac{\partial}{\partial y}\right)\left(z\frac{\partial}{\partial x}-x\frac{\partial}{\partial z}\right)-\left(z\frac{\partial}{\partial x}-x\frac{\partial}{\partial z}\right)\left(y\frac{\partial}{\partial z}-z\frac{\partial}{\partial y}\right)\right],$$

which simplifies to

$$(-\mathrm{i}\hbar)^2\left(y\frac{\partial}{\partial x}-x\frac{\partial}{\partial y}\right) = \mathrm{i}\hbar L_z \;.$$

Hence $L_x L_y - L_y L_x = \mathrm{i}\hbar L_z$.

Similarly, the other identities follow by making the appropriate substitutions for $L_x$, $L_y$, and $L_z$.

(b) The transformations from rectangular to spherical polar coordinates are

$$x = r\sin\theta\cos\phi\,, \qquad y = r\sin\theta\sin\phi\,, \qquad z = r\cos\theta\,,$$

so that

$$r = (x^2+y^2+z^2)^{1/2}\,,$$

$$\theta = \tan^{-1}\left[\frac{(x^2+y^2)^{1/2}}{z}\right]\,,$$

$$\phi = \tan^{-1}\left(\frac{y}{x}\right)\,.$$

By using

$$\frac{\partial}{\partial x} = \frac{\partial r}{\partial x}\frac{\partial}{\partial r}+\frac{\partial\theta}{\partial x}\frac{\partial}{\partial\theta}+\frac{\partial\phi}{\partial x}\frac{\partial}{\partial\phi}\,,$$

and similar expressions for $\partial/\partial y$ and $\partial/\partial z$, we obtain

$$L_x = \mathrm{i}\hbar\left(\sin\phi\frac{\partial}{\partial\theta}+\cot\theta\cos\phi\frac{\partial}{\partial\phi}\right),$$

$$L_y = \mathrm{i}\hbar\left(-\cos\phi\frac{\partial}{\partial\theta}+\cot\theta\sin\phi\frac{\partial}{\partial\phi}\right),$$

$$L_z = -\mathrm{i}\hbar\frac{\partial}{\partial\phi}\,,$$

$$\boldsymbol{L}^2 = -\hbar^2\left[\frac{1}{\sin\theta}\frac{\partial}{\partial\theta}\left(\sin\theta\frac{\partial}{\partial\theta}\right)+\frac{1}{\sin^2\theta}\frac{\partial^2}{\partial\phi^2}\right].$$

(c) Eigenfunctions $\Phi$ and eigenvalues $\lambda$ of $L_z$ are determined by solving

$$L_z\Phi = \lambda\Phi\,.$$

Therefore

$$-\mathrm{i}\hbar\frac{\partial\Phi}{\partial\phi} = \lambda\Phi\,,$$

and hence

$$\Phi = A \exp(i\lambda\phi/\hbar) .$$

But $\Phi(\phi + 2\pi) = \Phi(\phi)$, so that $i\lambda 2\pi/\hbar = 2\pi m i$, where $m$ is an integer, and therefore $\lambda = m\hbar$. The range of $m$ is fixed by the equation in $\theta$, which only has solutions if $-l \leqslant m \leqslant +l$.
*Note*: $l$ and $m$ are the angular momentum quantum number and the magnetic quantum number respectively. Part of the significance of this problem lies in the fact that the eigenfunctions of $\boldsymbol{L}_z$ and of $\boldsymbol{L}^2$ are the spherical harmonics $Y_l^m(\theta, \phi)$ which are also bases of the irreducible representations of $O(3)$, the rotation group in three dimensions.

**1.1.9**
(a) Identify the elements of each of the five black-and-white point groups that can be derived from the point group $4/mmm$ ($D_{4h}$).
(b) Identify the four black-and-white tetragonal Bravais lattices and sketch their unit cells.

**Solution**
(a) The symmetry operations of the point group $4/mmm$ ($D_{4h}$) can be identified with the aid of figure 1.1.9a,

$$\begin{array}{llllllll} E, & C_{2z}, & C_{4z}^{+}, & C_{4z}^{-}, & C_{2x}, & C_{2y}, & C_{2a}, & C_{2b} \\ I, & \sigma_z, & S_{4z}^{-}, & S_{4z}^{+}, & \sigma_x, & \sigma_y, & \sigma_{da}, & \sigma_{db}, \end{array}$$

where the $z$ axis is perpendicular to the plane of the square and each element in the second row is the product of $I$, the inversion, with the corresponding element in the first row.

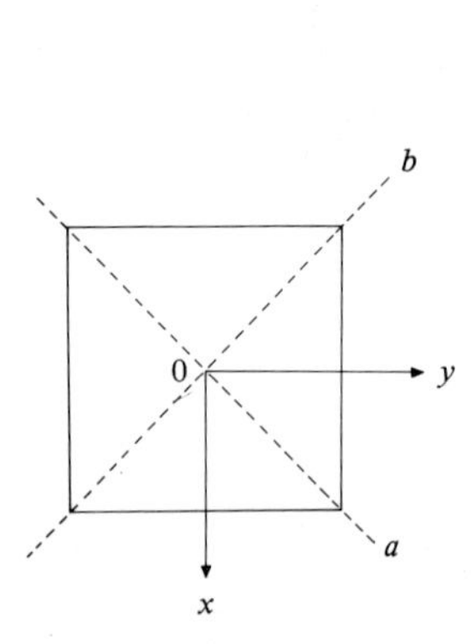

Figure 1.1.9a.

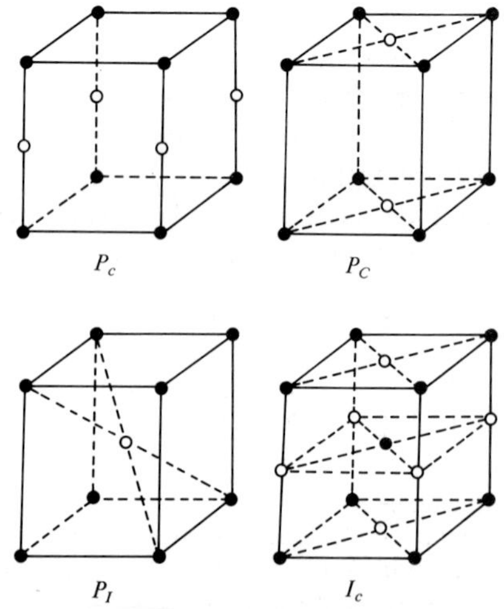

Figure 1.1.9b.

A black-and-white point group can be constructed by choosing a subgroup of index 2 and associating the remaining elements with the colour-changing operation, $\theta$. There are five choices for the subgroup of index 2, and correspondingly five magnetic point groups,

| subgroup of index 2 | black-and-white point group |
|---|---|
| 422 | $4/m'm'm'$ |
| $4mm$ | $4/m'mm$ |
| $mmm$ | $4'/mmm'$ |
| $\bar{4}2m$ | $4'/m'm'm$ |
| $4/m$ | $4/mm'm'$ |

(b) See figure 1.1.9b.

**1.1.10**

Determine the characters of the double-valued representations of the group of the symmetry operations of an equilateral triangle.

**Solution**

Each class of the point group $3m$ ($C_{3v}$) leads to two classes in the corresponding double point group. There are therefore three extra classes and three extra representations which must therefore be of dimension 1, 1, and 2. The characters must all satisfy $\chi(\bar{R}) = -\chi(R)$ for these double-valued representations. Using $(C_3^+)^3 = \bar{E}$ and $(\sigma_1)^2 = \bar{E}$ shows that, in the one-dimensional double-valued representations, $\chi(C_3^+) = -1$ and $\chi(\sigma_1) = \pm\mathrm{i}$. For the two-dimensional representation the orthogonality condition

$$\sum_R \chi^{\mathrm{i}}(R)\chi^{\mathrm{i}*}(R) = 12\,,$$

where the summation over $R$ is over all the elements of the group, determines that $\chi(\sigma_1) = 0$ and that $\chi^{\mathrm{i}}(C_3^+)\chi^{\mathrm{i}*}(C_3^+) = 1$ and, since there is only one two-dimensional representation, $\chi(C_3^+)$ must be real, hence $\chi(C_3^+) = \pm 1$. The orthogonality between this and the one-dimensional representations fixes $\chi(C_3^+) = -1$. Therefore the character table for the double-valued representations of the point group $3m$ ($C_{3v}$) of the symmetry operations of the equilateral triangle is

| | $E$ | $\bar{E}$ | $C_3^+, C_3^-$ | $\bar{C}_3^+, \bar{C}_3^-$ | $\sigma_1, \sigma_2, \sigma_3$ | $\bar{\sigma}_1, \bar{\sigma}_2, \bar{\sigma}_3$ |
|---|---|---|---|---|---|---|
| $\Gamma_4$ | 1 | −1 | −1 | 1 | i | −i |
| $\Gamma_5$ | 1 | −1 | −1 | 1 | −i | i |
| $\Gamma_6$ | 2 | −2 | 1 | −1 | 0 | 0 . |

**Reference**

Bradley, C. J., Cracknell, A. P., 1972, *The Mathematical Theory of Symmetry in Solids: Representation Theory for Point Groups and Space Groups* (Oxford University Press, Oxford).

## 1.2 Molecular symmetry

**D J Greenslade** University of Essex

### 1.2.1

Assign the following molecules to their point groups; give answers in both Schönflies and International notation: (a) methyl chloride, (b) ethane (staggered conformation), (c) chromium dibenzene, (d) ferrocene, (e) $[Ni(NH_3)_6]^{2+}$, (f) phosphorus pentachloride, (g) $[Ni(\text{ethylene diamine})_3]^{2+}$, (h) ethylene, (i) allene, (j) $W(NMe_2)_6$ (see figure 1.2.1a), (k) the *trans* and gauche conformers of 1,2-dichloroethane, and (l) $B_{12}H_{12}^{2-}$.

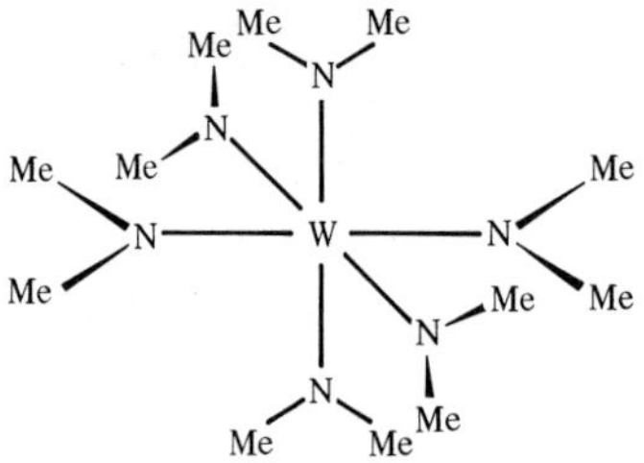

**Figure 1.2.1a.** Structure of $W(NMe_2)_6$.

**Solution**

In practice the assignment of a point group is often made by comparison with a molecule of similar structure for which the symmetry group is known. For example, methane and carbon tetrachloride have the same structure and the former belongs to the point group $T_d$, therefore so must the latter molecule. It is, nevertheless, useful to set down a systematic procedure. A number of such are possible and one is given in flowchart form in figure 1.2.1b. The procedure is based on the principle that very low or very high symmetry is identified simply. Use of the flowchart makes it straightforward to derive the answers given below; the dotted line on the flowchart pertains to ferrocene. The flowchart is given in terms of the Schönflies system, since that is most widely used for molecular systems.

(a) $C_{3v}$, $3m$; (b) $D_{3d}$, $\bar{3}m$; (c) $D_{6h}$, $6/mmm$; (d) $D_{5d}$, $\bar{5}m$;
(e) The hydrogen atoms complicate the problem but often unnecessarily as it is the site-symmetry at the nickel ion that is important in many problems. In any case, the ammonia molecules will often be freely rotating, so that the complex ion has the symmetry $O_h$ or $m3m$[(1)].
(f) In the solid this consists of two ions: $PCl_4^+$ ($T_d$ or $\bar{4}3m$) and $PCl_6^-$ ($O_h$ or $m3m$). In the gas phase the molecule has $D_{3h}$ or $\bar{6}m2$ symmetry.
(g) $D_3$, $32$; (h) $D_{2h}$, $mmm$; (i) $D_{2d}$, $\bar{4}2m$; (j) $T_h$, $m3$;
(k) *trans* conformer: $C_{2h}$, $2/m$; gauche conformer: $C_2$, $2$; (l) $I$, $532$.

[(1)] But see also problem 1.2.10.

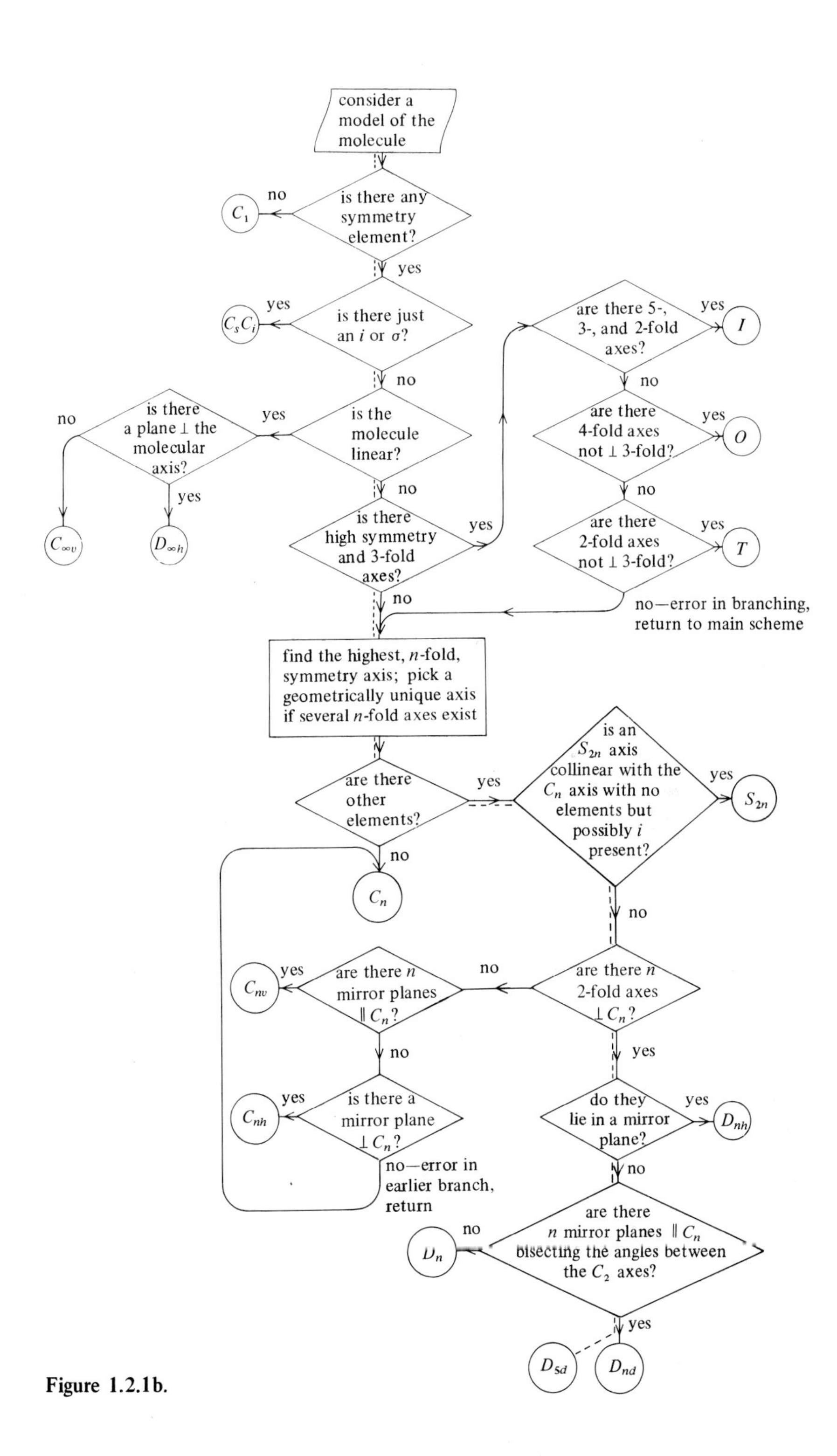

consider a model of the molecule
is there any symmetry element?
no
$C_1$
yes
is there just an $i$ or $\sigma$?
yes
$C_s C_i$
no
is the molecule linear?
yes
is there a plane ⊥ the molecular axis?
no
$C_{\infty v}$
yes
$D_{\infty h}$
no
is there high symmetry and 3-fold axes?
yes
are there 5-, 3-, and 2-fold axes?
yes
$I$
no
are there 4-fold axes not ⊥ 3-fold?
yes
$O$
no
are there 2-fold axes not ⊥ 3-fold?
yes
$T$
no—error in branching, return to main scheme
no
find the highest, $n$-fold, symmetry axis; pick a geometrically unique axis if several $n$-fold axes exist
are there other elements?
no
$C_n$
yes
is an $S_{2n}$ axis collinear with the $C_n$ axis with no elements but possibly $i$ present?
yes
$S_{2n}$
no
are there $n$ 2-fold axes ⊥ $C_n$?
no
are there $n$ mirror planes ∥ $C_n$?
yes
$C_{nv}$
no
is there a mirror plane ⊥ $C_n$?
yes
$C_{nh}$
no—error in earlier branch, return
yes
do they lie in a mirror plane?
yes
$D_{nh}$
no
are there $n$ mirror planes ∥ $C_n$ bisecting the angles between the $C_2$ axes?
no
$D_n$
yes
$D_{5d}$
$D_{nd}$


**Figure 1.2.1b.**

Further examples and stereoscopic diagrams can be found in Bernal *et al.* (1972).

**1.2.2**

(a) Deduce the combinations of s and p orbitals of X which form hybrid orbitals appropriate for a discussion of bonding in the molecule $R_3X$ having $C_{3v}$ symmetry.

(b) Similarly form the hybrid orbitals on an atom situated in a site of $C_{2v}$ symmetry.

**Solution**

(a) Suppose that the threefold axis is $0z$ (the $z$ axis), then the orbital in the $xz$ plane can be written

$$|1\rangle = a|s\rangle + b|x\rangle + c|z\rangle ,$$

where $|s\rangle$ is the atomic s orbital and $|x\rangle$ the p orbital directed along $0x$. The latter can be written in the form

$$|x\rangle = N_{np}R_{np}(r)\frac{x}{r} ,$$

where $N$ is a normalising factor, $R$ is the radial wave function, and $x/r$ is the angular function. Since the radial factors are constant under symmetry operations, $x$ is sufficient to label the wave function for our purposes. The same consideration applies to $|y\rangle$ and $|z\rangle$. We now use symmetry to deduce the symmetry-related hybrids. A rotation of $\frac{2}{3}\pi$ radians ($C_3$) about $0z$ is performed [the student unfamiliar with the transformation of $(x, y)$, should see the solution of the problem 1.2.3]; thus

$$C_3|1\rangle = |2\rangle = a|s\rangle + b(|x\rangle \cos\tfrac{2}{3}\pi + |y\rangle \sin\tfrac{2}{3}\pi) + c|z\rangle .$$

Reflection in the $xz$ plane gives

$$\sigma_{xz}|2\rangle = |3\rangle = a|s\rangle + b(|x\rangle \cos\tfrac{2}{3}\pi - |y\rangle \sin\tfrac{2}{3}\pi) + c|z\rangle .$$

This is as far as symmetry can take us, but we have normalisation as a further help:

$$|\langle x|1\rangle|^2 + |\langle x|2\rangle|^2 + |\langle x|3\rangle|^2 = 1 \qquad \text{(the total p-orbital density = 1)}$$

and so $b = (\frac{2}{3})^{1/2}$.

The same consideration gives

$$|4\rangle = (1-3a^2)^{1/2}|s\rangle - (1-3c^2)^{1/2}|z\rangle ,$$

where the minus sign is necessary to make $\langle 4|3\rangle = 0$ (orthogonality). An orbital in the $xz$ plane at an angle $\theta$ to $0z$ (that is, $|4\rangle$) has the form

$$R(r) - R'(r)(x \sin\theta - z \cos\theta) ;$$

this can be seen by applying coordinate transformation for a rotation of

$(\frac{1}{2}\pi - \theta)$ about $Oy$. We can relate $\theta$ to the interbond angle $\alpha$ by use of figure 1.2.2a:

$$l = \sin\tfrac{1}{2}\alpha\,; \qquad r = \sin\theta\,.$$

Since the base of the pyramid is an equilateral triangle,

$$l = r\cos 30^\circ = (\tfrac{3}{4})^{1/2}\sin\theta = \sin\tfrac{1}{2}\alpha\,.$$

From the above considerations, $b/c = \tan\theta$. Thus

$$\frac{b^2}{c^2} = \tan^2\theta = \frac{\sin^2\theta}{\cos^2\theta} = \frac{\frac{4}{3}\sin^2\frac{1}{2}\alpha}{1-\frac{4}{3}\sin^2\frac{1}{2}\alpha}\,.$$

Hence

$$c^2 = \frac{\frac{2}{3}(3-4\sin^2\frac{1}{2}\alpha)}{4\sin^2\frac{1}{2}\alpha} = \frac{1+2\cos\alpha}{3(1-\cos\alpha)}\,.$$

The constant $a$ is obtained by normalising $|4\rangle$:

$$1 = \langle 4|4\rangle = 1 - 3a^2 - 1 - 3c^2\,,$$

which gives, after a little algebra,

$$a^2 = -\frac{\cos\alpha}{1-\cos\alpha}\,.$$

Summarising:

$$|1\rangle = \left[-\frac{\cos\alpha}{1-\cos\alpha}\right]^{1/2}|s\rangle - (\tfrac{2}{3})^{1/2}|x\rangle + \left[\frac{1+2\cos\alpha}{3(1-\cos\alpha)}\right]^{1/2}|z\rangle\,,$$

$$|2\rangle = \left[-\frac{\cos\alpha}{1-\cos\alpha}\right]^{1/2}|s\rangle + (\tfrac{1}{6})^{1/2}|x\rangle - (\tfrac{1}{2})^{1/2}|y\rangle + \left[\frac{1+2\cos\alpha}{3(1-\cos\alpha)}\right]^{1/2}|z\rangle\,,$$

$$|3\rangle = \left[-\frac{\cos\alpha}{1-\cos\alpha}\right]^{1/2}|s\rangle + (\tfrac{1}{6})^{1/2}|x\rangle + (\tfrac{1}{2})^{1/2}|y\rangle + \left[\frac{1+2\cos\alpha}{3(1-\cos\alpha)}\right]^{1/2}|z\rangle\,,$$

$$|4\rangle = \left[\frac{1-4\cos\alpha}{1-\cos\alpha}\right]^{1/2}|s\rangle - \left[-\frac{3\cos\alpha}{1-\cos\alpha}\right]^{1/2}|z\rangle\,.$$

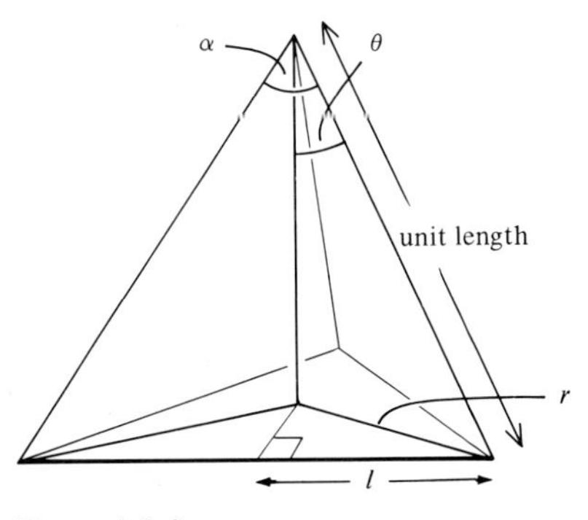


**Figure 1.2.2a.**

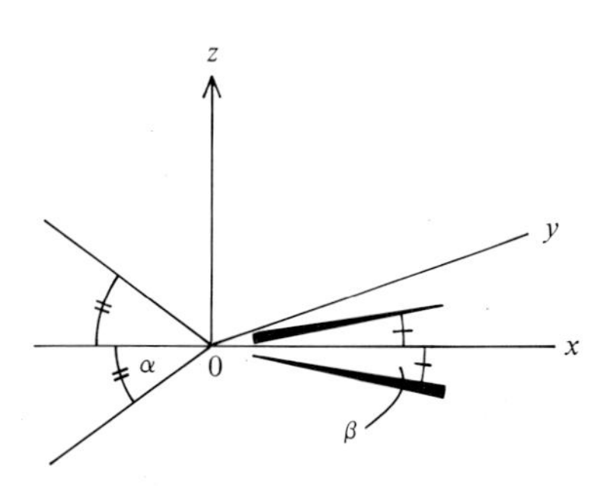


**Figure 1.2.2b.**

(b) From figure 1.2.2b, the orbitals in the $xy$ plane are

$$|1xy\rangle = a|s\rangle + b|x\rangle + (\tfrac{1}{2})^{1/2}|y\rangle ,$$

and by reflection

$$\sigma_{zx}|1xy\rangle = |2xy\rangle = a|s\rangle + b|x\rangle - (\tfrac{1}{2})^{1/2}|y\rangle .$$

Likewise, the orbitals in the $zx$ plane are

$$|3zx\rangle = c|s\rangle - d|x\rangle + (\tfrac{1}{2})^{1/2}|z\rangle ,$$

$$|4zx\rangle = c|s\rangle - d|x\rangle - (\tfrac{1}{2})^{1/2}|z\rangle .$$

The orbitals are made orthogonal,

$$\langle 1xy|2xy\rangle = 0 = a^2 + b^2 - \tfrac{1}{2} ,$$

and related, as in part (a), to their orientation to $0x$,

$$(\tfrac{1}{2})^{1/2} b^{-1} = \tan\beta ,$$

so that

$$b = (\tfrac{1}{2})^{1/2} \cot\beta .$$

Similar relations hold for $c$ and $d$. Finally, $a$ and $c$ are related since the total s-electron density must be unity:

$$2a^2 + 2c^2 = 1 , \qquad \text{or} \qquad \cot^2\alpha + \cot^2\beta = 1 .$$

Summarising:

$$\left.\begin{matrix}|1\rangle \\ |2\rangle\end{matrix}\right\} = (\tfrac{1}{2})^{1/2}[(1-\cot^2\beta)^{1/2}|s\rangle + \cot\beta|x\rangle \pm |y\rangle] ;$$

$$\left.\begin{matrix}|3\rangle \\ |4\rangle\end{matrix}\right\} = (\tfrac{1}{2})^{1/2}[(1-\cot^2\alpha)^{1/2}|s\rangle + \cot\alpha|x\rangle \pm |z\rangle] .$$

**1.2.3**

Show how the atomic p and d orbitals transform under the symmetry operations of the group $D_3(32)$. Show that each set of orbitals separates into subsets which are invariant under the operations: that is, any transformation of an orbital in the subset results in an orbital which is a linear combination of the orbitals of the subset. What significance has this result for the bonding in a molecule containing an atom in $D_3$ symmetry?

**Solution**

The atomic wave functions are given in standard texts on quantum mechanics in the form $R_{nl}(r)\Theta_{lm}(\theta)\exp(im\phi)$. Since a function of $r$ is unchanged by a symmetry transformation we will abbreviate by dropping the radial factor:

$$|\mathrm{p}, \pm 1\rangle = \mp(\tfrac{3}{4})^{1/2} \sin\theta \exp(\pm \mathrm{i}\phi)\,, \qquad |\mathrm{p}, 0\rangle = (\tfrac{3}{2})^{1/2} \cos\theta\ ;$$

$$|\mathrm{d}, \pm 2\rangle = (\tfrac{15}{16})^{1/2} \sin\theta \exp(\pm \mathrm{i}2\phi)\,, \qquad |\mathrm{d}, \pm 1\rangle = \mp(\tfrac{15}{4})^{1/2} \sin\theta \cos\theta \exp(\pm \mathrm{i}\phi)\,,$$

$$|\mathrm{d}, 0\rangle = (\tfrac{5}{8})^{1/2}(3 \cos^2\theta - 1)\,.$$

The normalising factors have been partly included to show that for the d orbitals different factors are necessary for different $m$ values.

We will consider the effect of the symmetry operations on the p orbitals in order to establish the method. We take $0z$ as the axis of quantisation and as the threefold axis. Then the effect of the threefold rotations is simply to add $\frac{2}{3}\pi$ or $\frac{4}{3}\pi$ to the coordinate $\phi$ (see figure 1.2.3). Thus, we obtain

$$C_3|\mathrm{p}, 0\rangle = C_3^2|\mathrm{p}, 0\rangle = |\mathrm{p}, 0\rangle\ ;$$

$$C_3|\mathrm{p}, \pm 1\rangle = \mp(\tfrac{3}{4})^{1/2} \sin\theta \exp[\pm \mathrm{i}(\phi + \tfrac{2}{3}\pi)] = \exp(\pm \mathrm{i}\tfrac{2}{3}\pi)\,|\mathrm{p}, \pm 1\rangle\ ;$$

likewise,

$$C_3^2|\mathrm{p}, \pm 1\rangle = \exp(\pm \mathrm{i}\tfrac{4}{3}\pi)\,|\mathrm{p}, \pm 1\rangle\,.$$

The twofold rotations change $\theta$ to $(\pi - \theta)$; $\phi$ changes to $(\pi - \phi)$ for rotation about $0y$, to $(\frac{1}{3}\pi - \phi)$ for rotation about $0l$, and to $(\frac{5}{3}\pi - \phi)$ for rotation about $0m$. We can therefore write:

$$\begin{aligned} C_{2y}|\mathrm{p}, 0\rangle = C_{2l}|\mathrm{p}, 0\rangle = C_{2m}|\mathrm{p}, 0\rangle &= (\tfrac{3}{2})^{1/2} \cos(\pi - \theta) = -(\tfrac{3}{2})^{1/2} \cos\theta \\ &= -|\mathrm{p}, 0\rangle\ ; \end{aligned}$$

$$\begin{aligned} C_{2y}|\mathrm{p}, \pm 1\rangle &= \mp(\tfrac{3}{4})^{1/2} \sin(\pi - \theta) \exp(\pm \mathrm{i}\pi) \exp(\mp \mathrm{i}\phi) \\ &= \pm(\tfrac{3}{4})^{1/2} \sin\theta \exp(\mp \mathrm{i}\phi) \\ &= |\mathrm{p}, \mp 1\rangle\ ; \end{aligned}$$

$$C_{2l}|\mathrm{p}, \pm 1\rangle = -\exp(\pm \mathrm{i}\tfrac{1}{3}\pi)\,|\mathrm{p}, \mp 1\rangle\ ;$$

$$C_{2m}|\mathrm{p}, \pm 1\rangle = -\exp(\pm \mathrm{i}\tfrac{5}{3}\pi)\,|\mathrm{p}, \mp 1\rangle\,.$$

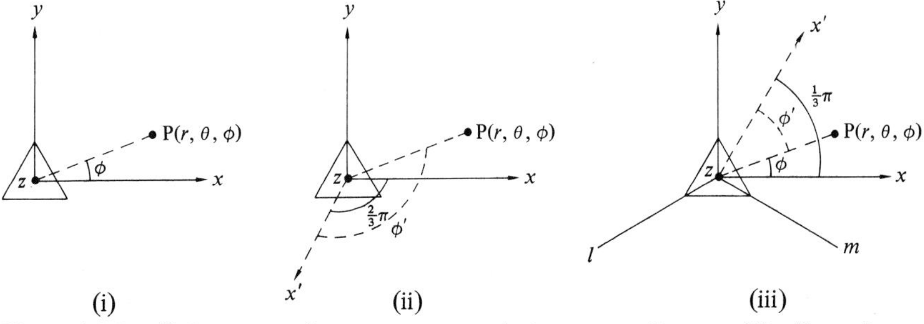


**Figure 1.2.3.** (i) Location of symmetry axes relative to coordinates; (ii) effect of $C_3$ ($\frac{2}{3}\pi$) rotation about $0z$ on coordinate $\phi$ of point p; (iii) effect of twofold rotation about $0l$. The threefold axis, $z$, is perpendicular to the plane of the page.

For bonding studies, we often use real orbitals:

$$|\mathrm{p}, x\rangle = \tfrac{1}{2}(|\mathrm{p}, -1\rangle - |\mathrm{p}, +1\rangle)\ ; \quad |\mathrm{p}, y\rangle = -\tfrac{1}{2}\mathrm{i}(|\mathrm{p}, +1\rangle + |\mathrm{p}, -1\rangle)\ ;$$
$$|\mathrm{p}, z\rangle = |\mathrm{p}, 0\rangle\ .$$

To transform these is straightforward; for example,

$$\begin{aligned} C_3|\mathrm{p}, x\rangle &= C_3[(\tfrac{3}{4})^{1/2}\sin\theta\cos\phi] = (\tfrac{3}{4})^{1/2}\sin\theta\cos(\phi + \tfrac{2}{3}\pi) \\ &= (\tfrac{3}{4})^{1/2}(\sin\theta\cos\phi\cos\tfrac{2}{3}\pi - \sin\theta\sin\phi\sin\tfrac{2}{3}\pi) \\ &= -\tfrac{1}{2}|\mathrm{p}, x\rangle - (\tfrac{3}{4})^{1/2}|\mathrm{p}, y\rangle\ . \end{aligned}$$

Clearly the operations of $D_3$ separate the orbitals into the sets $\{|\mathrm{p}, x\rangle, |\mathrm{p}, y\rangle\}$ and $\{|\mathrm{p}, z\rangle\}$, or $\{|\mathrm{p}, 0\rangle\}$ and $\{|\mathrm{p}, \pm 1\rangle\}$.

It is easily shown in a similar manner that the d orbitals separate into the sets $\{|\mathrm{d}, 3z^2 - r^2\rangle\}$, $\{|\mathrm{d}, xy\rangle, |\mathrm{d}, x^2 - y^2\rangle\}$, and $\{|\mathrm{d}, yz\rangle, |\mathrm{d}, zx\rangle\}$, or $\{|\mathrm{d}, 0\rangle\}$, $\{|\mathrm{d}, \pm 1\rangle\}$, and $\{|\mathrm{d}, \pm 2\rangle\}$, provided that one notes the different normalising factors for the different d orbitals.

In a molecule the outer electrons of any atom are strongly influenced by the other nuclei: ultimately one is not able to assign the electrons to any one nucleus. In the transition-metal complexes, for example, the d electrons were originally thought to be largely attached to the transition-metal atom but influenced by the crystal field, that is, the electrostatic forces due to the other nuclei which are only partially shielded by their own electrons. In the molecular-orbital approach the electrons are considered to be delocalised on the surrounding nuclei. In either case, the potential energy will, in the present case, have $D_3$ symmetry. The kinetic energy is spherically symmetric and so the total Hamiltonian operator will be unchanged by the operations of the group $D_3$. For any operation $T$ of the group, we have

$$T(\mathcal{H}\psi_n) = \mathcal{H}(T\psi_n) = T(E_n\psi_n) = E_n(T\psi_n)\ ,$$

where $\psi_n$ is an electronic wave function. This means that orbitals which transform into one another under the effect of operations of $D_3$ will be degenerate. This is the basis of ligand-field theory. For example, in the present case the d orbitals in a complex of a transition metal having $D_3$ symmetry will split into the three groups given above. It is clear that consideration of the symmetry does not give any information on the order of the energy levels, or indeed on the magnitudes of the splittings, if there are any.

### 1.2.4

Using (a) simple ideas of symmetry such as the effect of symmetry operations on wave functions, and (b) the methods of group representations, show that the $\pi$ orbitals of benzene form four sets of different symmetry. Deduce the algebraic form of each orbital and its energy within the Hückel approximation.

**Solution**

(a) Benzene has $D_{6h}$ symmetry, but since the p orbitals of the carbon atoms are antisymmetric with respect to reflection in the molecular plane, their combinations to form $\pi$ orbitals may be classified just by reflections perpendicular to this plane or rotations about the hexad axis, that is, by $C_{6v}$. The first linear combination of p orbitals to form a $\pi$ orbital is expected to be fully symmetric with respect to these operations [see figure 1.2.4(I)] so that it may be written

$$\pi_1 = N_1(p_1+p_2+p_3+p_4+p_5+p_6)\,.$$

$N_1$ is a normalising factor and, if the $p_i$ are assumed orthogonal (that is, are Wannier orbitals rather than pure atomic p orbitals), it takes the value $(\frac{1}{6})^{1/2}$. This function is unchanged by any of the symmetry operations of the group $C_{6v}$. Next we look for a combination which is antisymmetric with respect to reflection in one of the vertical mirror planes. Two such are to be expected. The first [figure 1.2.4(II)] has the nodal plane, across which there is antisymmetry, through two carbon atoms and is of the form

$$\pi_2 = N_2(p_2+p_3-p_5-p_6)\,, \qquad N_2 = \tfrac{1}{2}\,.$$

The second [figure 1.2.4(III)] has the nodal plane cutting bonds on opposite sides of the molecule. It can be written

$$\pi_3 = N_3(ap_1+p_2-p_3-ap_4-p_5+p_6)\,, \qquad N_3 = (4+2a^2)^{-1/2}\,.$$

The constant $a$ cannot be determined by symmetry, but this orbital must be orthogonal to all the other $\pi$ orbitals and $a$ will be chosen to ensure such orthogonality. Any other choice of symmetry plane to obtain similar orbitals is illusory: there are but two orbitals with one nodal plane, rather as there are but two p orbitals in the $xy$ plane for each atom. Any other orbital can be expressed as a linear combination of our first two choices.

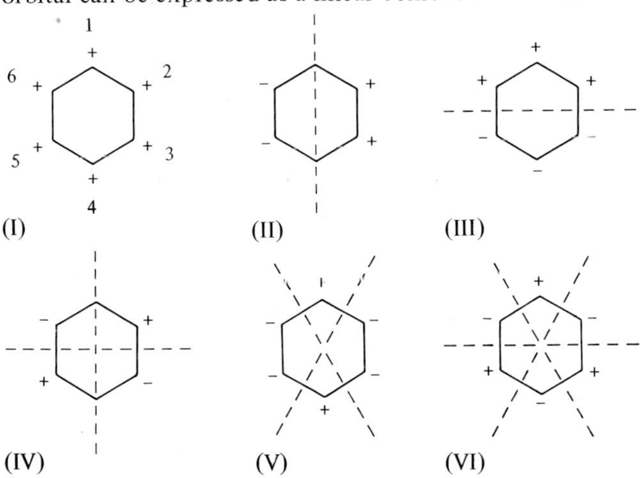


**Figure 1.2.4.** $\pi$ orbitals of benzene.

We next seek an orbital having two nodal planes (a 'd-like' orbital). Again there are two such orbitals, as shown in figure 1.2.4(IV) and (V), and we can write these as

$$\pi_4 = N_4(p_2 - p_3 + p_5 - p_6)\,, \qquad N_4 = \tfrac{1}{2}\,;$$

$$\pi_5 = N_5(bp_1 - p_2 - p_3 + bp_4 - p_5 - p_6)\,, \qquad N_5 = (4 + 2b^2)^{-1/2}\,.$$

Again the contributions of, for example, $p_2$ and $p_3$ are equal, since reflection in the $xz$ plane leaves $\pi_5$ unchanged. On the other hand no symmetry operation relates $p_1$ or $p_4$ to the other p orbitals and so orthogonality is the determining factor for $b$.

Finally we look for an orbital which is completely antisymmetric. This, shown in figure 1.2.4(VI), has an f-like appearance, being of the form

$$\pi_6 = N_6(p_1 - p_2 + p_3 - p_4 + p_5 - p_6)\,, \qquad N_6 = (\tfrac{1}{6})^{1/2}\,.$$

These orbitals must be mutually orthogonal so that

$$\langle \pi_i | \pi_j \rangle = \int \pi_i^* \pi_j \, \mathrm{d}\tau = 0\,, \qquad \text{if } i \neq j\,.$$

Thus

$$\langle \pi_6 | \pi_3 \rangle = N_6 N_3 \langle (p_1 - p_2 + p_3 - p_4 + p_5 - p_6) | (ap_1 + p_2 - p_3 - ap_4 - p_5 + p_6) \rangle$$
$$= 0\,;$$

so that, assuming

$$\langle p_i | p_j \rangle = \begin{cases} 0 & (i \neq j) \\ 1 & (i = j)\,, \end{cases}$$

and expanding the integral, we obtain

$$a - 1 - 1 + a - 1 - 1 = 0\,,$$

that is,

$$a = 2\,, \qquad N_3 = \tfrac{1}{2}(3)^{-1/2}\,.$$

Likewise, from $\langle \pi_5 | \pi_1 \rangle = 0$, we find $b = 2$, and $N_5 = \frac{1}{2}(3)^{-1/2}$.

(b) The set of carbon p orbitals will transform according to a *reducible* representation of the group $D_{6h}(6/mmm)$, but to simplify the analysis we can work in the subgroup $C_6(6)$ and correlate our result with the full group afterwards. First we find the character set of the irreducible representation. The identity operation $E$ converts the set of atomic p orbitals to the same set, so that we can write

$$E\begin{bmatrix} p_1 \\ p_2 \\ \vdots \\ p_6 \end{bmatrix} = \begin{bmatrix} 1 & & & 0 \\ & 1 & & \\ & & \ddots & \\ 0 & & & 1 \end{bmatrix} \begin{bmatrix} p_1 \\ p_2 \\ \vdots \\ p_6 \end{bmatrix}\,.$$

Adding the diagonal elements of the matrix, we see that $\chi(E) = 6$. Now every other operation of $C_6$ moves every $p_i$ to a different place in the column vector $\boldsymbol{p}$, so that the diagonal elements of the representative matrix of all the other operations are zero and so, therefore, is the character. Hence the character set for the p orbitals is (in International notation):

$$\begin{array}{lcccccc} & E & 6_z & 3_z & 2_z & 3_z^2 & 6_z^5 \\ \chi^{\mathrm{p}}(T) & 6 & 0 & 0 & 0 & 0 & 0\,. \end{array}$$

The number of molecular orbitals, $n(\Gamma)$, which can be formed from these p orbitals and which transform under the representation $\Gamma$ of $C_6$ is given by the reduction formula

$$\begin{aligned} n(\Gamma) &= \frac{1}{h}\sum_{T}\chi^{\Gamma *}(T)\chi^{\mathrm{p}}(T) \\ &= \frac{1}{h}\sum n_c\chi^{\Gamma *}(\mathrm{C})\chi^{\mathrm{p}}(\mathrm{C})\,, \end{aligned}$$

where $T$ is a symmetry operation of $C_6$, and C is a class of $C_6$ (see Heine, 1960, p.119), $h$ is of the order of the group, and $n_{\mathrm{c}}$ the number of elements in the class C.

For example, using the characters for the $A$ representation of $C_6$ (see character table in Appendix 2), we obtain

$$n(A) = \tfrac{1}{6}[(1\times 1\times 6)+(1\times 1\times 0)+\ldots] = 1\,.$$

In this case the characters are separable by inspection, and it is clear that there is one orbital of $A$ symmetry, one of $B$, two of $E_1$, and two of $E_2$.

The orbitals may be generated by use of the formula (Heine, 1960, p.119):

$$\pi_\Gamma = \sum_{\text{all } T}\chi^{\Gamma *}(T)T\phi\,. \qquad (1.2.4\text{a})$$

Here, $\phi$ can be any suitable orbital, but unless a good choice is made we may not generate all of the symmetry orbitals—it is usually obvious when this has happened. We will use $p_1$:

$$\begin{array}{lllllll} & E & 6_z & 3_z & 2_z & 3_z^2 & 6_z^5 \\ \pi_A & = p_1 & +p_2 & +p_3 & +p_4 & +p_5 & +p_6 \\ \pi_B & = p_1 & -p_2 & +p_3 & -p_4 & +p_5 & -p_6 \\ \pi_{E_{1a}} & = p_1 & -\omega p_2 & +\omega^2 p_3 & -p_4 & +\omega p_5 & -\omega^2 p_6 \\ \pi_{E_{1b}} & = p_1 & -\omega^2 p_2 & +\omega p_3 & -p_4 & +\omega^2 p_5 & -\omega p_6\,, \end{array}$$

and so forth, where $\omega = \exp(\mathrm{i}\tfrac{2}{3}\pi)$. These orbitals must be normalised.

It is clear that the first method has generated a different set of orbitals from those of the present method. In fact there are an infinite number of ways of choosing a set of degenerate orbitals, and in the former method we have derived real orbitals and in the latter complex orbitals; in fact,

$$\pi_{E_{1a}} + \pi_{E_{1b}} = \pi_3\,.$$

Surprisingly, the first method gives the proper orbitals for $D_{6h}$ symmetry: in correlating the result of method (b) with the proper symmetry group of benzene, we would have to convert to the real orbitals! It is clear that the second method is more straightforward, and relies less on mathematical intuition. On the other hand it rides on top of a large mathematical structure.

From the numbers of nodes, we expect $\pi_1$ to be the lowest orbital in energy, $\pi_2$ and $\pi_3$, possibly degenerate, next in energy, and so forth, with the first three orbitals bonding and the last three antibonding. In order to determine the energy levels within the Hückel approximation, we construct the Hückel matrix. The elements of this are given by

$$H_{ij} - ES_{ij} = \langle \pi_i | \mathcal{H} | \pi_j \rangle - ES_{ij} , \qquad [S_{ij} = 0 \ (i \neq j),\ S_{ij} = 1 \ (i = j)] ,$$

which can be expanded into terms of the form:

$$\langle p_i | \mathcal{H} | p_j \rangle = \begin{cases} \alpha & (i = j) \\ \beta & (i \text{ and } j \text{ adjacent atoms}) . \end{cases}$$

For example,

$$\begin{aligned} \langle \pi_2 | \mathcal{H} | \pi_2 \rangle &= (\tfrac{1}{2})^2 \langle (p_2 + p_3 - p_5 - p_6 | \mathcal{H} | (p_2 + p_3 - p_5 - p_6) \rangle \\ &= \tfrac{1}{4} (\langle p_2 | \mathcal{H} | p_2 \rangle + \langle p_2 | \mathcal{H} | p_3 \rangle - \ldots + \langle p_6 | \mathcal{H} | p_6 \rangle) \\ &= \alpha + \beta ; \end{aligned}$$

$$\langle \pi_2 | \mathcal{H} | \pi_4 \rangle = \tfrac{1}{4} \langle (p_2 + p_3 - p_5 - p_6 | \mathcal{H} | (p_2 - p_3 + p_5 - p_6) \rangle = 0 .$$

We find that the Hückel matrix in terms of the new symmetry-adapted orbitals is diagonal, so that its secular determinant immediately factorises to give the energies of the orbitals (note that $\beta$ is negative):

$$\begin{aligned} E(\pi_1) &= \alpha + 2\beta , \qquad & E(\pi_2, \pi_3) &= \alpha + \beta , \\ E(\pi_4, \pi_5) &= \alpha - \beta , \qquad & E(\pi_6) &= \alpha - 2\beta . \end{aligned}$$

**1.2.5**

Show that in the case of cubic symmetry the only potential which need be considered in the case of crystal-field splitting of levels involving d orbitals is of the form

$$V = D(x^4 + y^4 + z^4 - \tfrac{3}{5} r^4) .$$

**Solution**

The addition of the crystal-field potential to the Hamiltonian results in new elements in the Hamiltonian matrix of the form $\langle \Psi | V | \Psi \rangle$, where $\Psi$ is an appropriate many-electron wave function such as a Slater determinant. Since $V$ is a one-electron operator, these matrix elements, which are multiple integrals, reduce to the form

$$\langle 1 | V | 2 \rangle \times (\text{product of overlap integrals}) ,$$

where $|1\rangle$ and $|2\rangle$ are d orbitals (see Ballhausen, 1962, p.15). Since the d orbitals are, in their angular parts, second-order spherical harmonics, the product is at most fourth-order, that is of $G$ symmetry in the full rotation group. For the integral

$$\int \psi_1^{\Gamma_1 *} \psi_2^{\Gamma_2} \, d\tau$$

to be nonzero, the product $\Gamma_1 \otimes \Gamma_2$ must contain $A_1$ (in cubic symmetry). In the present case this means that we need not consider components of $V$ of order greater than four. Another way of viewing this is to note that a function of lower than $G$ symmetry will have zero overlap with a function of at least $G$ symmetry.

The potential can be expanded in terms of spherical harmonics:

$$V = \sum_i \sum_l \sum_m R_{nl}(r_i) Y_{lm}(\theta_i, \phi_i) ,$$

where $r_i$ is the radial coordinate of the $i$th electron, and $\theta_i$, $\phi_i$ are the angular coordinates. The zeroth term ($l = 0$),

$$\tfrac{1}{2}\pi^{-1/2} \sum_i R_{nl}(r_i) ,$$

is present for all states and does not lead to a splitting. The odd terms may be dropped, since the product of two d wave functions must be even and the integral of the product of an odd and an even function is zero.

Next $V$ must be invariant under the operations of the cubic group $O$ (432) so that second-order terms in $V$ must be dropped. We can then write $V = V_0 + V_4$. Clearly $V_0$ transforms as $A_1$; to see how $V_4$ transforms we must reduce the fourth-order harmonics $Y_{4m}(\theta, \phi)$. Since a rotation of $\alpha$ about the $z$ axis adds $\alpha$ to $\phi$, and so multiplies $Y_{lm}(\theta, \phi)$ by $\exp(im\alpha)$, the character for such a rotation for the set of $Y_{lm}$ is given by

$$\chi(\alpha) = \sum_{m=-l}^{l} \exp(im\alpha) = \frac{\sin(l+\frac{1}{2})\alpha}{\sin\frac{1}{2}\alpha} . \qquad (1.2.5a)$$

Thus we obtain the character set for the reducible representation of $Y_{4m}$ in the group 432:

| | $E$ | $3(8)$ | $2_z(3)$ | $2_d(6)$ | $4_z$ (6) |
|---|---|---|---|---|---|
| $\chi(Y_{4m})$ | 9 | 0 | 1 | 1 | 1 . |

By use of the character reduction formula, or simply by inspection of the character table of $O$ (Appendix 2), we see that the $Y_{4m}(\theta, \phi)$ transform as $A_1$, $E$, $T_1$, and $T_2$. The potential we require is invariant under $O$, that is, the $A_1$ potential. We can use the projection operator formula to generate this, using as a starting function

$$Y_{40} = 35z^4 - 30z^2r^2 + 3r^4 ,$$

where

$$r^2 = x^2 + y^2 + z^2 .$$

We could use the full symmetry of $O$, but this would be tedious. Instead we descend in symmetry to a lower group. If we descend too low in symmetry, we shall miss some of the potential—for example, $C_4$ only rotates $x$ and $y$ into one another and we need to mix in functions of $z$. Let us try $C_3$: then,

$$\begin{aligned} V^{A_1} &\propto (35z^4 - 30z^2r^2 + 3r^4) + (35x^4 - 30x^2r^2 + 3r^4) + (35y^4 - 30y^2r^2 + 3r^4) \\ &\propto 35(x^4 + y^4 + z^4 - \tfrac{3}{5}r^4) \,. \end{aligned}$$

**1.2.6**

Show that an atomic F term splits in the presence of a cubic crystal field of symmetry $O$ into $A_2$, $T_1$, and $T_2$ terms. Deduce the wave functions for these states in terms of the spherical symmetry eigenfunctions.

**Solution**

An F state transforms in a manner similar to that of an f orbital, that is, a spherical harmonic of third order. The character for a rotation of $\alpha$ is simply derived for the spherical harmonics $Y_{3m}$ by use of equation (1.2.5a), since the axis of quantisation, $0z$, may be arbitrarily chosen. Thus we obtain the character set for the f functions in 432:

| | $E$ | $3_z$ | $2_z$ | $2_d$ | $4_z$ |
|---|---|---|---|---|---|
| $\chi(\mathrm{f})$ | 7 | 1 | −1 | −1 | −1 . |

We now apply the character reduction formula [see solution 1.2.4(b)] to obtain the number of $A_1$ states:

$$\begin{aligned} n(A_1) &= \tfrac{1}{24}\{(1\times 7\times 1) + (8\times 1\times 1) + [3\times(-1)\times 1] + [6\times(-1)\times 1] \\ &\qquad + [6\times(-1)\times 1]\} \\ &= 0 \,. \end{aligned}$$

Similarly, $n(A_2) = 1$, $n(E) = 0$, $n(T_1) = 1$, and $n(T_2) = 1$. In summary, the F term splits into $A_2$, $T_1$, and $T_2$ terms in the presence of a cubic crystal field.

In principle, the derivation of the wave functions of these terms may be accomplished by means of the function-generator formula (1.2.4a). In practice, for a group such as $O$, having many operations, this is a lengthy and tedious process. Short cuts are possible and one such method is given here. First, note that in Cartesian form the spherical harmonics are third-order polynomials in $x$, $y$, and $z$ (divided by the symmetrical term $r^3$). The simplest of these is $xyz/r^3$. This can be written, in terms of the spherical eigenfunctions $|L, M\rangle$:

$$|xyz\rangle = (|3, 2\rangle - |3, -2\rangle) \,,$$

where the normalising factor $2^{-1/2}$ has been omitted. It is clear that all the operations of $O$ transform $xyz$ into $\pm xyz$, so that this function must belong to a one-dimensional representation of $O$ and is therefore the required $A_2$ function. Normalisation of this function gives

$$|A_2\rangle = 2^{-1/2}(|3, 2\rangle - |3, -2\rangle) .$$

The components of orbital angular momentum $\boldsymbol{L}$ transform in $O$ as $T_1$, so that an operation of any of these components on the $A_2$ function will yield a function transforming as $T_1 \otimes A_2 = T_2$. Thus the $T_2$ functions are $L_z|A_2\rangle$ and $L_\pm|A_2\rangle$, where $L_\pm = L_x \pm \mathrm{i}L_y$. Now, by setting $\hbar = 1$, we can write

$$L_z|L, M\rangle = M|L, M\rangle ,$$

and

$$L_\pm|L, M\rangle = [L(L+1) - M(M \pm 1)]^{1/2}|L, M \pm 1\rangle .$$

Thus, after normalising we obtain

$$|T_2, 0\rangle = 2^{-1/2}(|3, 2\rangle + |3, -2\rangle) ,$$

$$|T_2, 1\rangle = 8^{-1/2}(3^{1/2}|3, 3\rangle - 5^{1/2}|3, -1\rangle) ,$$

$$|T_2, -1\rangle = 8^{-1/2}(5^{1/2}|3, 1\rangle - 3^{1/2}|3, -3\rangle) .$$

To obtain the $T_1$ functions we note that $|3, 0\rangle$ has not been used so far and must therefore be a $T_1$ function, let us call it $|T_1, 0\rangle$. Finally, some of the $|3, \pm 3\rangle$ and $|3, \pm 1\rangle$ functions are left, so we write

$$|T_1, 1\rangle = a|3, 3\rangle + b|3, -1\rangle ; \qquad |T_1, -1\rangle = c|3, 1\rangle + d|3, -3\rangle .$$

Since $\langle T_2, 1|T_1, 1\rangle = 0$ and $\langle L, M|L, M'\rangle = \delta_{MM'}$,

$$3^{1/2}a + 5^{1/2}b = 0 ;$$

thus

$$|T_1, 1\rangle = 8^{-1/2}(5^{1/2}|3, 3\rangle + 3^{1/2}|3, -1\rangle) .$$

Likewise, we find that

$$|T_1, -1\rangle = 8^{-1/2}(3^{1/2}|3, 1\rangle + 5^{1/2}|3, -3\rangle) .$$

**1.2.7**
Use the method of group characters to deduce the states of an ion situated in a strong crystal field of octahedral symmetry $m3m$ ($O_h$) such that there is (a) one, (b) two, and (c) three d electrons on the ion. Outline the method by which the wave functions for these states may be deduced.

**Solution**
(a) Since the d orbitals transform gerade under the inversion operation, we may simplify by using the group 432. Since the only operation of this group which transforms any d orbital into itself is the identity operation,

we have $\chi(E) = 5$, since there are five d orbitals, and all other characters are obtained from the standard formula (1.2.5a) for rotation of spherical harmonics about an angle $\alpha$; hence

|  | $E$ | $3(8)$ | $2_z(3)$ | $2_d(6)$ | $4_z(6)$ |
|---|---|---|---|---|---|
| $\chi(\mathrm{d})$ | 5 | $-1$ | 1 | 1 | $-1$ . |

The standard character reduction method is now applied.

$$n(A_1) = \tfrac{1}{24}[(1\times5\times1)-(8\times1\times1)+(3\times1\times1)+(6\times1\times1)-(6\times1\times1)] = 0\,,$$

$$n(A_2) = \tfrac{1}{24}[(1\times5\times1)-(8\times1\times1)+(3\times1\times1)-(6\times1\times1)+(6\times1\times1)] = 0\,,$$

$$n(E) = \tfrac{1}{24}[(1\times5\times2)+(8\times1\times1)+(3\times1\times2)+0+0] = 1\,,$$

$$n(T_1) = \tfrac{1}{24}[(1\times5\times3)+0-(3\times1\times1)-(6\times1\times1)-(6\times1\times1)] = 0\,,$$

$$n(T_2) = \tfrac{1}{24}[(1\times5\times3)+0-(3\times1\times1)+(6\times1\times1)+(6\times1\times1)] = 1\,.$$

Thus the d orbitals split into two sets transforming under the $E$ and $T_2$ representations of 432, that is, the $E_g$ and $T_{2g}$ representations of $m3m$. Thus an ion containing a d electron will be in a state labelled either $\mathrm{T_{2g}}$ or $\mathrm{E_g}$ depending on the sign of the crystal field.

In order to find the symmetry-adapted d orbitals we simply apply the symmetry generation formula (1.2.4a). Thus

$$e_1 = \sum_{\text{all } T} \chi^{E*}(T)T|\mathrm{d}_0\rangle\,,$$

where $\mathrm{d}_0$ has $m_l = 0$. These orbitals may need normalising and for angular momentum considerations may need an appropriate phase factor $\exp(\mathrm{i}\phi)$, where $|\exp(\mathrm{i}\phi)| = 1$. Omitting the radial part of the wave function and unnecessary normalising factors, $\mathrm{d}_0$ has functional form

$$d_0 = (\tfrac{10}{16})^{1/2}(3\cos^2\theta - 1)\,.$$

In Cartesian form this becomes $(\frac{10}{16})^{1/2}(3z^2-r^2)$ (divided by $r^2$), which is the form most useful for symmetry transformations. Since the characters of the other operations are zero, we need only the operations $E$, 3, and $2_z$ to obtain e orbitals. Four of the threefold rotations change $z^2$ to $x^2$, and the other four rotations change $z^2$ to $y^2$; $2_z$ leaves $d_0$ unchanged. Hence, we obtain

$$e_1 = 2d_0 - 4[(\tfrac{10}{16})^{1/2}(3x^2-r^2)] - 4[(\tfrac{10}{16})^{1/2}(3y^2-r^2)] + 3(2d_0)\,;$$

but

$$3(x^2+y^2)-2r^2 = 3(r^2-z^2)-2r^2 = r^2-3z^2\,,$$

thus $e_1 = d_0$; similarly, $e_2 = 2^{-1/2}(d_2+d_{-2})$. If we choose real orbitals, we can express the $\mathrm{t_{2g}}$ orbitals as $2^{-1/2}(d_1 \pm d_{-1})$ and $2^{-1/2}(d_2-d_{-2})$.

An alternative method which is sometimes useful is the use of descent in symmetry. Suppose that the octahedral environment is stretched along the $z$ axis, the symmetry changes to $4/mmm$ ($D_{4h}$) and the representations break down:

$$E_g \rightarrow A_{1g} + B_{1g}\,, \qquad T_{1g} \rightarrow A_{2g} + E_g\,, \qquad T_{2g} \rightarrow B_{2g} + E_g\,.$$

Clearly $(3z^2 - r^2)$ transforms as $A_{1g}$ in $4/mmm$, so it must be an $e_g$ orbital in $m3m$.

(b) Naively, we might expect that since the product wave functions of the two d electrons can be written in the form $t_{2g}^2$, $t_{2g}e_g$, and $e_g^2$, the symmetry of the functions is simply given by reducing $T_2 \otimes T_2$, etc. In the present case this will work, but we must not forget the spin of the electrons and the Pauli principle. The characters of $T_2 \otimes T_2$ are simply $\chi^2(T_2)$, and the product is reduced by the usual method to give

$$T_2 \otimes T_2 = A_1 \oplus E \oplus T_1 \oplus T_2\,.$$

Since we are putting two electrons into the same orbital in some cases, some states can only be singlets. In order to deduce this we use the antisymmetrised squares and symmetrised squares of the characters (Lyubarskii, 1960). The antisymmetrised square $\{\chi\}^2(T)$, which gives the triplet state, is given by the expression

$$\{\chi\}^2(T) = \tfrac{1}{2}[\chi^2(T) - \chi(T^2)]\,.$$

The symmetrised square $[\![\chi]\!]^2(T)$ gives the singlet states and is obtained from the expression

$$[\![\chi]\!]^2(T) = \tfrac{1}{2}[\chi^2(T) + \chi(T^2)]\,.$$

Hence for $T_2$ the characters are as follows,

| | $E$ | $3(8)$ | $2_z(3)$ | $2_d(6)$ | $4_z$ (6) |
|---|---|---|---|---|---|
| $\chi^2(T) =$ | 9 | 0 | 1 | 1 | 1 |
| $\chi(T_2^2) =$ | 3 | 0 | 3 | 3 | $-1$ |
| $\{\chi\}^2(T_2) =$ | 3 | 0 | $-1$ | $-1$ | 1 |
| $[\![\chi]\!]^2(T_2) =$ | 6 | 0 | 2 | 2 | 0. |

We see that the symmetrised square corresponds to a $T_1$ state, and the antisymmetrised reduces to $A_1$, E, and $T_2$ states. We conclude that the $t_{2g}^2$ configuration in the strong-field scheme gives the states ${}^3T_{1g}$, ${}^1A_{1g}$, ${}^1E_g$, and ${}^1T_{2g}$.

The states of the configuration $t^2e$ can be obtained from the simple product representation, since we are putting two electrons into two different orbitals: there is both a singlet and a triplet from each orbital symmetry.

$$E \otimes T_2 = T_1 \oplus T_2\,,$$

so this configuration gives the states ${}^1T_1$, ${}^1T_2$, ${}^3T_1$, and ${}^3T_2$. Finally, using antisymmetrised and symmetrised squares of the characters of the $E$ representation, we see that the $e^2$ configuration gives the states $^1A_1$, $^3A_2$, and $^1E$.

In order to obtain the wave functions of these states we could start from first principles by using a product wave function as the starting point. The character-based function-generation formula is, for the high-symmetry groups, lengthy and tedious to apply (see problem 1.2.6). Fortunately, Griffith (1961, table A20) has given a table of coupling coefficients which enable the product functions to be written down immediately. These coefficients were derived by considering the operation of $C_{4z}$ and $C_{3(111)}$ $(4_z, 3_{(111)})$ on the product functions.

(c) Yet a further complication occurs in the case of the $t_{2g}^3$ configuration: the orbital and spin wave functions cannot be simply separated as a product of symmetric and antisymmetric functions for every state. The use of the antisymmetric character gives the state of maximum multiplicity (in this case $^4A_2$) since the spin functions in this case must be symmetric, which is readily seen by considering the function of maximum $M_S$, $\alpha\alpha\alpha$. The use of the methods based on the theory of the representations of the symmetric group $S_3$ enables the remaining doublet states to be deduced (see problem 1.2.8). Another method is to note that the $t_{2g}$ functions are isomorphous to the atomic p orbitals. The configuration $p^3$ gives the terms $^2P$ and $^2D$ as well as the spin quartet. Since in octahedral symmetry $\Gamma(P) \rightarrow T_1$ and $\Gamma(D) \rightarrow E \oplus T_2$, the $t_{2g}^3$ configuration will give the terms $^2E$, $^2T_1$, and $^2T_2$.

The wave functions for these states can be derived in exactly the same way as for the previous configuration. Griffith (1961) has done this, again essentially by descent in symmetry, that is, by considering just the effect of $C_{4z}$ and $C_{3(111)}$ on the simple products of the form $abc$ where $a$, $b$, and $c$ are $t_{2g}$ functions. He has listed the results in his table A24.

### 1.2.8

Give the Young tableaux for the symmetric groups $S_2$ and $S_3$ (the permutation groups of two and three objects). Deduce the number and dimension of the irreducible representations of these groups. Which of the representations correspond to the states $^1S$ and $^3P$ of the atomic configuration $p^2$, the states $^1A_1$ and $^3T_1$ of the configuration $t_2^2$ of a transition-metal ion in a strong cubic crystal field, and the states $^4S$ of $p^3$ and $^4A_2$ of $t_2^3$. Deduce the other states of $p^3$. Can you deduce the states of $p^2$ more simply?

**Solution**

The tableaux for $S_2$ are

$$\begin{array}{|c|c|}\hline 1 & 2 \\ \hline\end{array}\ ; \qquad \begin{array}{|c|}\hline 1 \\ \hline 2 \\ \hline\end{array}\ .$$

Since there are two tableaux, there are two irreducible representations of $S_2$. Each has only one standard form (numbers increasing across and down the tableau), therefore each representation is singly degenerate or of dimension one. The first corresponds to the symmetric representation and the second to the antisymmetric representation in which permutation of the indices of a second-rank tensor changes its sign.

The tableaux for $S_3$ are

$$\boxed{1}\boxed{2}\boxed{3}\ ;\ \begin{array}{c}\boxed{1}\\\boxed{2}\\\boxed{3}\end{array}\ ;\ \begin{array}{l}\boxed{1}\boxed{2}\\\boxed{3}\end{array}\ ;\ \begin{array}{l}\boxed{1}\boxed{3}\\\boxed{2}\end{array}\ .$$

Thus there are three irreducible representations. The first is the symmetric representation: permutation of the indices of a third-rank tensor in this case leaves the tensor unchanged. The second is the antisymmetric representation: a tensor transforming according to this representation will have its sign changed if the permutation of its indices is odd. The third is a two-dimensional representation, since there are two standard tableaux.

For the state $^1S$ of $p^2$ the orbital function is symmetric to permutation of the electrons and so transforms according to the representation corresponding to $\boxed{1}\boxed{2}$. The singlet spin function $2^{-1/2}[\alpha(1)\beta(2)-\alpha(2)\beta(1)]$ is antisymmetric and the corresponding tableau is $\begin{array}{c}\boxed{1}\\\boxed{2}\end{array}$. Since these are conjugate representations, the total function is antisymmetric as required by the Pauli Principle. $^1A_1$ follows the same pattern as $^1S$. The reverse is true of the states $^3P$ of $p^2$ and $^3T_1$ of $t_2^2$. Thus the spin wave function is clearly symmetric under interchange of the two electrons and so the tableau $\boxed{1}\boxed{2}$ corresponds to this spin function. The orbital function is antisymmetric by the Pauli principle, and corresponds to the tableau $\begin{array}{c}\boxed{1}\\\boxed{2}\end{array}$.

For $^4S$ of $p^3$ the spin functions are symmetric, as is always true of the maximum-multiplicity state. Thus $\alpha(1)\alpha(2)\alpha(3)$ is clearly so. The tableau is $\boxed{1}\boxed{2}\boxed{3}$ for this and the other spin functions of the $^4S$ state. The appropriate orbital function must have a conjugate diagram $\begin{array}{c}\boxed{1}\\\boxed{2}\\\boxed{3}\end{array}$. The same considerations apply to $^4A_2$ of $t_2^3$.

The other states of $p^3$ may be deduced by the use of these group theoretical methods. First we note that since an electron has only two spin states, the tableaux for spin functions can have only two rows of lengths $\lambda_1$ and $\lambda_2$ so that the spin of the state is $S = \frac{1}{2}(\lambda_1 - \lambda_2)$. For $p^3$ the spin functions correspond to the tableaux

$$\square\square\square\ ,\ \begin{array}{l}\square\square\\\square\end{array}\ ,$$

and so there are spin quartet and doublet states only.

The orbital states have diagrams which are conjugate to the spin diagrams. These orbital states are usually built up in a recursive manner, to obtain the permissible $L$ values. Thus $p^2$ is obtained symbolically:

$$\underset{p}{\square} \otimes \underset{p}{\square} = \underset{S}{\square\square} \oplus \underset{P}{\begin{array}{c}\square\\\square\end{array}}\ ;$$

and then, for $p^3$,

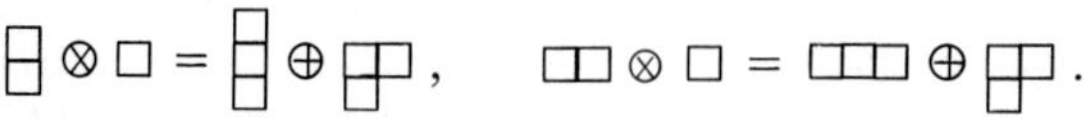

$$\ydiagram{1,1} \otimes \ydiagram{1} = \ydiagram{1,1,1} \oplus \ydiagram{2,1}\,, \qquad \ydiagram{2} \otimes \ydiagram{1} = \ydiagram{3} \oplus \ydiagram{2,1}\,.$$

By combining with the conjugate spin diagrams we see that the permissible states of $p^3$ are $^4S$, $^2P$, and $^2D$. The same method could be applied to crystal-field configurations such as $t_2^3$. This approach is discussed in Heine (1960, p.318 *et seq*) and by Hammermesh (1962, pp.266, 421–423). A simpler approach to atomic states was found by Slater (1929), and we illustrate it for $p^2$. The microstates are written, for example, in the form $\{\overset{+}{1}\bar{0}\}$, where the braces imply an antisymmetrised product starting with electron one in the p orbital of $m_l = 1$ with spin $m_s = +\frac{1}{2}$, electron two in orbital of $m_l = 0$ with spin $m_s = -\frac{1}{2}$. We write out all such products and assign them to a place in a table of permissible total $M_L$ and $M_S$ (table 1.2.8). Because of symmetry, negative $M_L$ (and, indeed $M_S$) states need not be considered. Now we note that there is a microstate $\{\overset{+}{1}\bar{1}\}$ of $M_L = 2$; it must belong to a D state, and since its highest $M_S$ is zero that state must be $^1D$. Next, the highest $M_S$ state of maximum $M_L$ is $\{\overset{+}{1}\overset{+}{0}\}$; this belongs to a triplet spin state $^3P$, as the maximum $M_L$ is 1. In the $M_L = 1$, $M_S = 0$ box there are two microstates, but two must come from the $^1D$ and $^3P$ states, so no new states are implied by this box. Likewise $\{\overset{+}{1} - \overset{+}{1}\}$ in the $M_L = 0$, $M_S = 1$ box belongs to $^3P$. Finally, the $M_L = 0$, $M_S = 0$ box has three states, two from the previous states and one new one which must be $^1S$.

**Table 1.2.8.**

| $M_L$ | $M_S$ | | |
|---|---|---|---|
| | 1 | 0 | −1 |
| 2 | | $\{\overset{+}{1}\bar{1}\}$ | |
| 1 | $\{\overset{+}{1}\overset{+}{0}\}$ | $\{\overset{+}{1}0\}, \{\bar{1}\overset{+}{0}\}$ | $\{\bar{1}\bar{0}\}$ |
| 0 | $\{\overset{+}{1} - \overset{+}{1}\}$ | $\{\overset{+}{1} - \bar{1}\}, \{\bar{1} - \overset{+}{1}\}$ | $\{\bar{1} - \bar{1}\}$ |
| | | $\{\overset{+}{0}\bar{0}\}$ | |
| ⋮ | | ⋮ | ⋮ |

## 1.2.9

Using the character table for the double group of 432′ (to be found in Appendix 2), classify the states of a transition-metal ion having a $^4F$ ground term when it is situated in a site of cubic symmetry (weak crystal-field case).

**Solution**

The $^4$F term has $L = 3$ and $S = \frac{3}{2}$, thus the $J$ value runs from $3+\frac{3}{2}$ down to $3-\frac{3}{2}$. For the $^4F_{9/2}$ state we have the character set

| $E$ | $\overline{E}$ | 3 | 3′ | $2_z, 2'_z$ | $2_d$ | $2'_d$ | $4_z$ | $4'_z$ |
|---|---|---|---|---|---|---|---|---|
| 10 | −10 | −1 | 1 | 0 | 0 | 0 | $2^{1/2}$ | $-2^{1/2}$ . |

This is obtained in a straightforward manner from the equation

$$\chi(\alpha) = \sum_{m=-j}^{j} \exp(im\alpha) = \frac{\sin(j+\frac{1}{2})\alpha}{\sin\frac{1}{2}\alpha} ,$$

together with the rules (Heine, 1960, p.141):

$$\chi(R') = -\chi(R) ,$$

unless $R$ is a twofold rotation about an axis which has a twofold axis perpendicular to it, in which case the character is zero.

The representation of the $J = \frac{9}{2}$ level is then reduced in the normal way:

$$n(A_1) = \tfrac{1}{48}[(1 \times 10 \times 1)-(1 \times 10 \times 1)-[1 \times (-1) \times 8]-(1 \times 1 \times 8)-\cdots$$
$$\cdots-(1 \times 2^{1/2} \times 6)-(1 \times 2^{1/2} \times 6)] = 0 ,$$
$$n(E'_2) = \tfrac{1}{48}(20+20-8-8+12+12) = 1 ,$$
$$n(E'_3) = \tfrac{1}{48}(20+20-8-8-12-12) = 0 ,$$
$$n(U') = \tfrac{1}{48}(40+40+8+8) = 2 .$$

Thus the $^4F_{9/2}$ state splits into an $E'_2$ and two U′ states in a cubic environment. In a similar manner we can show that the $^4F_{7/2}$ state splits into $E'_2$, $E'_3$, and U′ states, the $^4F_{5/2}$ state into $E'_3$ and U′ states, and that $^4F_{3/2}$ is a U′ state.

**1.2.10**

The molecule $CH_3NO_2$ has torsional–rotational energy levels which suggest that the molecule has $C_{6v}$ symmetry (Wilson *et al.*, 1955). What is the reason for this strange result?

**Solution**

In the classification of a molecule into its symmetry group, it is usual to assume that it has a rigid structure or that parts of it, such as methyl groups, are so free that they have full symmetry about some axis. This is, of course, an approximation for any molecule, but it does not seem to matter until one considers a molecule having different configurations separated by energy barriers. Ethane, for example, has both $D_{3d}$ and $D_{3h}$ symmetry, the former in the low-energy staggered configuration, the latter in the high-energy eclipsed configuration. For such molecules the symmetry classification given in texts and used in problem 1.2.1 must be modified.

The rules for classifying such 'nonrigid' molecules were first given by Longuet-Higgins (1963). Subsequent criticism by other authors was finally discounted (see Bunker, 1975). The rules are:
(i) The symmetry operations include all permutations, $P$, of identical nuclei (atoms) provided such a permutation might occur within the time-scale of the experiment being considered (so-called feasible permutations). Clearly the symmetry group of a molecule may be different if one is considering visible spectra rather than, say, electron paramagnetic resonance. The identity operation is included in the set of permutations.
(ii) The combination of all such permutations with inversion of the nuclei through the centre of mass of the molecule must be included. The combined operations are denoted $P^*$, and there is an exception: $E^*$ may not be a member of the symmetry group.

A clear discussion of the origin of these rules is given by Bunker (1975).

If we label the methyl protons 1, 2, and 3, and the oxygen nuclei (which can permute in an experimental timescale) 4 and 5, we can represent the molecule in an arbitrary configuration by the projection

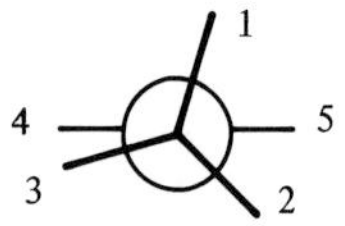

and then rotation of the $NO_2$ group about the C–C axis gives the structure

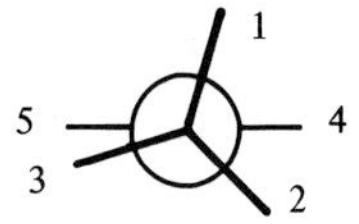

which we can represent by the usual permutation notation (45). This could be accompanied by rotation of the methyl group by 120°, that is, the permutation (123), to give a total permutation (123)(45). Each of the permutation cycles is a symmetry operation in itself, and likewise (132) and (132)(45) are symmetry operations. Next we must identify the $P^*$-type operations. Clearly $(23)^*$ is such an operation, taking the molecule into itself, but for a rotation in space:

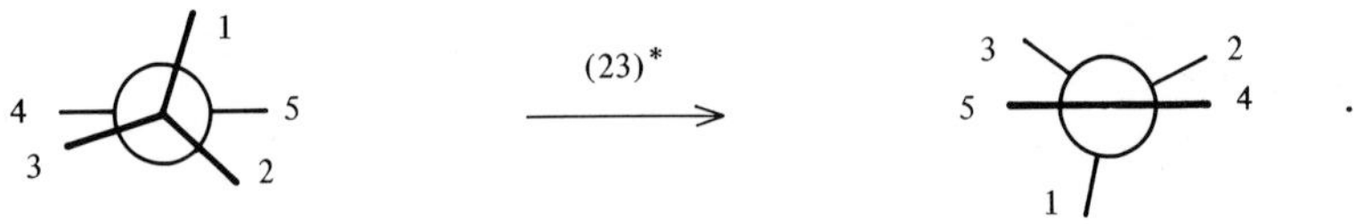

In this manner we obtain the elements of the nonrigid (perhaps 'floppy' would be a better term?) symmetry group of $CH_3NO_2$:

$E$ ; (123), (132) ; (45) ;
(123)(45), (132)(45) ; $(23)^*$, $(31)^*$, $(12)^*$ ;
$(23)(45)^*$, $(31)(45)^*$, $(12)(45)^*$ .

We have divided the elements into classes.

The division of the elements into classes and also the construction of the group multiplication table is readily carried out. For example, (123) and (45) commute, being cycles of different numbers, and so in constructing products of the form $XAX^{-1}$ in order to sort the elements into classes, we find

$$(123)(45)(123)^{-1} = (45) .$$

Indeed, the permutation nature of the group makes the construction of the group multiplication table quite simple. Thus

$$[(123)(45)](23)^* = (12)(45)^* .$$

The table so constructed is isomorphous with that of the point group $D_{3h}$. In sorting out the energy levels, however, it is the irreducible representations that are used and so any group with six classes (and therefore six irreducible representations) could be used. By a particular construction, as distinct from the general method given here, Wilson *et al.* (1955) were able to classify the torsional–rotational levels of $CH_3NO_2$ and similar molecules according to the irreducible representations of $C_{6v}$, which has six classes of elements.

**1.2.11**

(a) Consider the effect of a threefold rotation about the [111] axes and a fourfold rotation about the [001] axes on $x$, $y$, and $z$, and also on the components $L_x$, $L_y$, $L_z$ of the orbital angular momentum operator $\boldsymbol{L}$.
(b) Using your result, together with the Wigner–Eckart theorem, show how matrix elements of the form

$$\langle L, S, M_L, M_S | f(x, y, z) | L, S, M'_L, M'_S \rangle ,$$

such as those occurring in crystal-field theory, may be readily calculated.
(c) What is the effect of the fact that $[L_x, L_y] \neq 0$, whereas $[x, y] = 0$?
(d) Treat the case of the fourth-order potential of an octahedral crystal field, that is,

$$V_{\text{oct}} = D(x^4 + y^4 + z^4 - \tfrac{3}{5}r^4) ,$$

and show that this potential can be replaced, to within a constant multiplier, by

$$L_x^4 + L_y^4 + L_z^4 - \tfrac{1}{5}L(L+1)[3L(L+1)-1] .$$

**Solution**

(a) The threefold rotation rotates $x$ to $y$ to $z$ to $x$, thus,

$$L_x = -\mathrm{i}\hbar\ y\frac{\partial}{\partial z}\left(-z\frac{\partial}{\partial y}\right) \rightarrow -\mathrm{i}\hbar\left(z\frac{\partial}{\partial x} - x\frac{\partial}{\partial z}\right) = L_y .$$

Likewise, $L_y$ becomes $L_z$, and $L_z$ becomes $L_x$. Thus $x$, $y$, and $z$ transform similarly to $L_x$, $L_y$, and $L_z$. The fourfold rotation transforms $x$ to $y$ and $y$ to $-x$; $z$ is not changed; thus

$$L_x \to -\mathrm{i}\hbar\left(-x\frac{\partial}{\partial z}+z\frac{\partial}{\partial x}\right) = L_y \ .$$

Likewise, $L_y$ becomes $-L_x$ and $L_z$ is not changed. For all such symmetry operations it is found that a Cartesian coordinate and its corresponding component angular momentum operator transform in the same manner.
(b) The Wigner–Eckart theorem states that the matrix elements of operators which transform under the same irreducible representation of the symmetry group of the system are equal but for a coupling constant (see Fano and Racah, 1959; Heine, 1960, p.184, equation 20.34). Thus

$$\langle LSM_L M_S|f(x,y,z)|LSM'_L M'_S\rangle = \langle LS\|f\|LS\rangle c\langle M_L M_S|M'_L M'_S\rangle \ ,$$

where $\langle LS\|f\|LS\rangle$ is termed the reduced matrix element. From the result of part (a), we see that

$$\langle ...|f(x,y,z)|...\rangle = \mathrm{const}\langle ...|f(L_x, L_y, L_z)|...\rangle \ .$$

(c) The terms in $(x, y, z)$ which have products, explicit or implicit, of $x$, $y$, or $z$, have to be replaced by symmetrised products of $L_x$, $L_y$, $L_z$. Thus $x^2y^2$ is replaced by

$$\tfrac{1}{6}[L_x^2L_y^2+L_xL_yL_xL_y+L_xL_y^2L_x+L_y^2L_x^2+L_yL_xL_yL_x+L_yL_x^2L_y] \ .$$

This may be simplified by the use of commutation relations such as

$$[L_x, L_y] = \mathrm{i}L_z \qquad (\hbar = 1) \ .$$

(d) In $V_{\mathrm{oct}}$ we may directly replace $x^4+y^4+z^4$ by $L_x^4+L_y^4+L_z^4$. However, $r^4$ is really $(x^2+y^2+z^2)^2$, that is,

$$r^4 = x^4+y^4+z^4+2x^2y^2+2y^2z^2+2z^2x^2 \ .$$

Hence

$$x^4+y^4+z^4-\tfrac{3}{5}r^4 = x^4+y^4+z^4-\tfrac{3}{5}(x^4+y^4+z^4+2x^2y^2+2y^2z^2+2z^2x^2) \ .$$

Now $x^2y^2$ is replaced by

$$\tfrac{1}{6}[L_x^2L_y^2+L_xL_yL_xL_y+...] \ ,$$

as above. Using $[L_x, L_y] = \mathrm{i}L_z$, we simplify:

$$\begin{aligned} L_xL_yL_xL_y &= L_x(L_xL_y-\mathrm{i}L_z)L_y \\ &= L_x^2L_y^2-\mathrm{i}L_xL_zL_y \ ; \end{aligned}$$

similarly,

$$\begin{aligned} L_xL_y^2L_x &= L_xL_y[L_xL_y-\mathrm{i}L_z]L_x = L_xL_yL_xL_y-\mathrm{i}L_xL_yL_z \\ &= L_x^2L_y^2-\mathrm{i}L_xL_zL_y-\mathrm{i}L_xL_yL_z \ ; \end{aligned}$$

furthermore,

$$L_y L_x L_y L_x = L_y^2 L_x^2 + \mathrm{i} L_y L_z L_x ,$$

and

$$L_y L_x^2 L_y = L_y^2 L_x^2 + \mathrm{i} L_y L_z L_x + \mathrm{i} L_y L_x L_z .$$

Thus $6x^2y^2$ is replaced by

$$\begin{aligned} & 3(L_x^2L_y^2 + L_y^2L_x^2) - 2\mathrm{i}(L_xL_zL_y - \tfrac{1}{2}L_xL_yL_z + L_yL_zL_x + \tfrac{1}{2}L_yL_xL_z) \\ & \quad = 3(L_x^2L_y^2 + L_y^2L_x^2) - 2\mathrm{i}[L_xL_zL_y - L_yL_zL_x] - \mathrm{i}[L_x, L_y]L_z \\ & \quad = 3(L_x^2L_y^2 + L_y^2L_x^2) - 2\mathrm{i}[L_x(L_yL_z - \mathrm{i}L_x) - L_y(L_xL_z + \mathrm{i}L_y)] + L_z^2 \\ & \quad = 3(L_x^2L_y^2 + L_y^2L_x^2) - 2\boldsymbol{L}^2 + 5L_z^2 . \end{aligned}$$

Since a cyclic permutation of $x$, $y$, and $z$ is symmetrical we can write down immediately

$$6y^2z^2 \to 3(L_y^2L_z^2 + L_z^2L_y^2) - 2\boldsymbol{L}^2 + 5L_x^2 ;$$

$$6z^2x^2 \to 3(L_z^2L_x^2 + L_x^2L_z^2) - 2\boldsymbol{L}^2 + 5L_y^2 .$$

Thus $(x^4 + y^4 + z^4 - \frac{3}{5}r^4)$ is replaced by

$$\begin{aligned} L_x^4 + L_y^4 + L_z^4 - \tfrac{1}{5}[ & 3L_x^2(L_x^2 + L_y^2 + L_z^2) + 3L_y^2(L_x^2 + L_y^2 + L_z^2) \\ & + 3L_z^2(L_x^2 + L_y^2 + L_z^2) - 6\boldsymbol{L}^2 + 5(L_x^2 + L_y^2 + L_z^2)] , \end{aligned}$$

or

$$L_x^4 + L_y^4 + L_z^4 - \tfrac{1}{5}\boldsymbol{L}^2(3\boldsymbol{L}^2 - 1) ,$$

and, since

$$\boldsymbol{L}^2|L, ...\rangle = L(L+1)|L ...\rangle ,$$

finally

$$V_{\mathrm{oct}} = L_x^4 + L_y^4 + L_z^4 - \tfrac{1}{5}L(L+1)[3L(L+1) - 1] .$$

**1.2.12**

The Zeeman splitting of the atomic states ${}^{2S+1}L_J$ is calculated from the formula

$$E|L, S, J, M\rangle = E|L, S, J\rangle + g_J \mu_{\mathrm{B}} MB ,$$

where $g_J$ is the Landé $g$ factor, $\mu_{\mathrm{B}}$ the Bohr magneton, and $B$ the applied magnetic induction. The value of $g_J$ is often calculated by the vector model to be

$$g_J = 1 + \frac{J(J+1) + S(S+1) - L(L+1)}{2J(J+1)} .$$

Use the Wigner–Eckart theorem to show that this formula is approximately correct.

**Solution**

The Zeeman energy is calculated to first order in perturbation theory as

$$\langle L, S, J, M|\mathcal{H}_z|L, S, J, M\rangle = \mu_B B\langle L, S, J, M|L_z+g_S S_z|L, S, J, M\rangle ,$$

where $g_S$ is the magnetogyric ratio of the electron (2·0023). $\boldsymbol{L}$, $\boldsymbol{S}$, and $\boldsymbol{J}$ are all vector operators and so their matrix elements are related within a given $J$ (Heine, 1960, p.184):

$$\langle JM'|\boldsymbol{L}|JM\rangle = \langle J\|\boldsymbol{L}\|J\rangle\langle JM'|\boldsymbol{J}|JM\rangle ;$$

$$\langle JM'|\boldsymbol{S}|JM\rangle = \langle J\|\boldsymbol{S}\|J\rangle\langle JM'|\boldsymbol{J}|JM\rangle .$$

By adding these equations we see that

$$\langle J\|\boldsymbol{L}\|J\rangle+\langle J\|\boldsymbol{S}\|J\rangle = 1 .$$

The reduced matrix elements, $\langle J\|\boldsymbol{L}\|J\rangle$ and $\langle J\|\boldsymbol{S}\|J\rangle$ (where the latter is set equal to $\alpha$, say), are calculated by putting

$$\boldsymbol{L}^2 = (\boldsymbol{J}-\boldsymbol{S})^2 = \boldsymbol{J}^2+\boldsymbol{S}^2-2\boldsymbol{J}\cdot\boldsymbol{S} = \boldsymbol{J}^2+\boldsymbol{S}^2-2\alpha\boldsymbol{J}^2 ,$$

within constant $J$, and then calculating $\langle L, S, J, M|\boldsymbol{L}^2|L, S, J, M\rangle$. Thus,

$$L(L+1) = (1-2\alpha)J(J+1)+S(S+1) ,$$

so that,

$$\alpha = \frac{J(J+1)+S(S+1)-L(L+1)}{2J(J+1)} .$$

The Zeeman energy is then

$$\begin{aligned} g_J\mu_B BM &= \mu_B BM(\langle J\|\boldsymbol{L}\|J\rangle+g_S\langle J\|\boldsymbol{S}\|J\rangle) \\ &= \mu_B BM(1-\alpha+g_S\alpha) . \end{aligned}$$

If $g_S$ is put equal to 2, $g_J = 1+\alpha$, giving the required result.

**References**

Ballhausen, C. J., 1962, *Introduction to Ligand Field Theory* (McGraw-Hill, New York).
Bernal, I., Hamilton, W. C., Ricci, J. S., 1972, *Symmetry: A Stereoscopic Guide for Chemists* (W H Freeman, San Francisco).
Bunker, P. R., 1975, *Vib. Spectra Struct.*, **3**, 1.
Fano, U., Racah, G., 1959, *Irreducible Tensorial Sets* (Academic Press, New York).
Griffith, J. S., 1961, *The Theory of Transition-metal Ions* (Cambridge University Press, Cambridge).
Hammermesh, M., 1962, *Group Theory and Its Applications to Physical Problems* (Addison-Wesley, Reading, Mass.).
Heine, V., 1960, *Group Theory in Quantum Mechanics* (Pergamon Press, Oxford).
Longuet-Higgins, H. C., 1963, *Mol. Phys.*, **6**, 445.
Lyubarskii, G. Y., 1960, *The Application of Group Theory in Physics* (Pergamon Press, Oxford), p.71.
Slater, J. C., 1929, *Phys. Rev.*, **34**, 1293.
Wilson, E. B., Lin, C. C., Lide, D. R., 1955, *J. Chem. Phys.*, **23**, 136.

## 1.3 Crystal symmetry †

**A P Cracknell** University of Dundee

### 1.3.1

Draw a stereographic projection of the earth and mark on it the following cities:

| | |
|---|---|
| Canberra (C) | 35°S, 149°E |
| Delhi (D) | 28°N, 77°E |
| Hawaii (H) | 20°N, 156°W |
| London (L) | 52°N, 0°E |
| Moscow (M) | 56°N, 38°E |
| New York (N) | 41°N, 74°W |
| Peking (P) | 40°N, 117°E |
| Singapore (S) | 2°N, 104°E |
| Tokyo (T) | 36°N, 140°E |

**Solution**

The projection is onto the plane contained by the equator shown in figure 1.3.1a. The point $\theta$°N, $\phi$°E is on the radius of the equator $\phi$° from the axis of the Greenwich meridian. Its position on the radius is obtained by the construction shown. Points in the northern hemisphere are represented by dots, those in the southern hemisphere by circles.

### 1.3.2

Draw stereograms to represent the faces of each of the following solids:
(a) cube
(b) octahedron
(c) rhombic dodecahedron
and assign a set of Miller indices to each of the faces. Mark on each stereogram as many symmetry elements as possible without creating confusion.

**Solution**

(a) Cube, figure 1.3.2a.
(b) Octahedron, figure 1.3.2b.
(c) Rhombic dodecahedron, figure 1.3.2c.
We indicate the symmetry operations, which are the same for all three solids, on a separate stereogram, see figure 1.3.2d, where the planes drawn in the stereogram are all planes of symmetry.

† The character tables of the point groups, which will be needed in connection with various problems in this section, will be found in Appendix 2. The labels used for the symmetry operations in the point groups are those employed by Bradley and Cracknell (1972); see also Appendix 2.

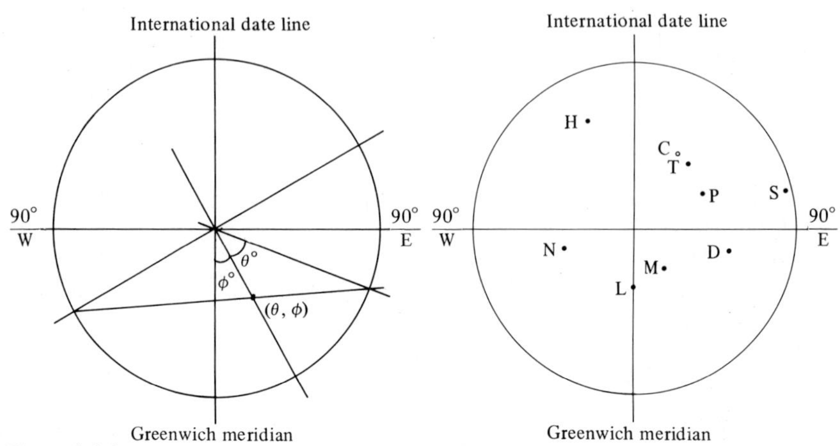


**Figure 1.3.1a.**

**Figure 1.3.1b.**

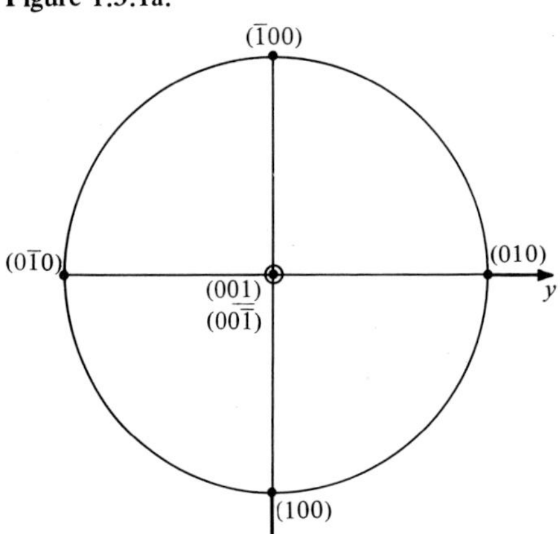


**Figure 1.3.2a.**

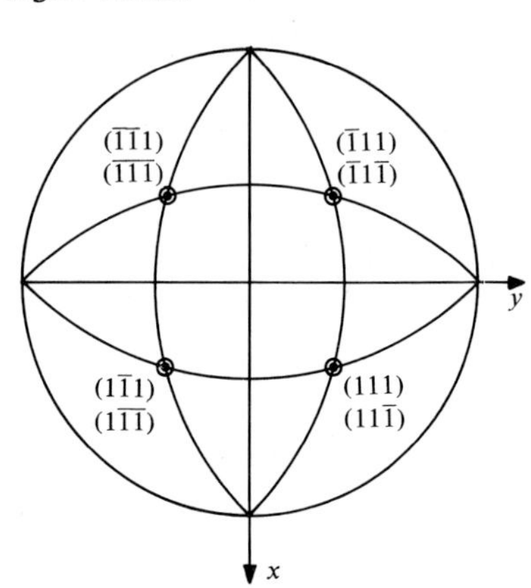


**Figure 1.3.2b.**

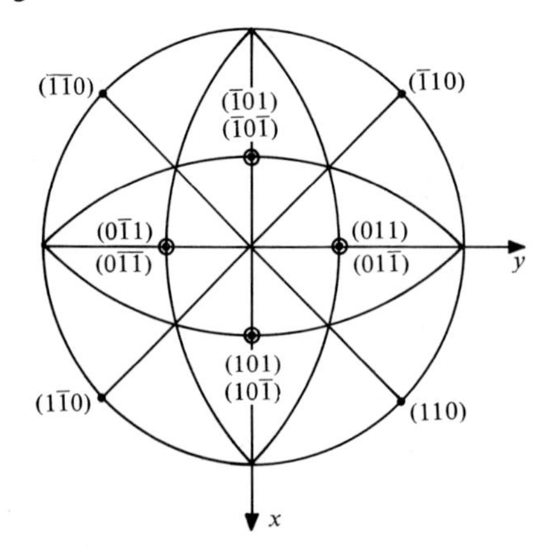


**Figure 1.3.2c.**

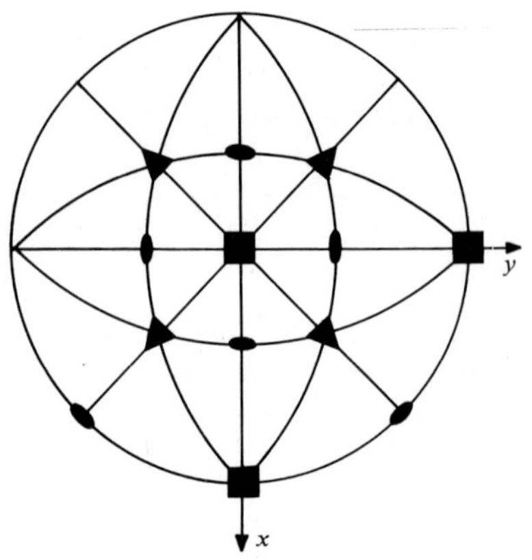


**Figure 1.3.2d.**

**1.3.3**

Show that for an orthorhombic crystal the electric polarisability tensor takes the form

$$\begin{bmatrix} \alpha_{11} & 0 & 0 \\ 0 & \alpha_{22} & 0 \\ 0 & 0 & \alpha_{33} \end{bmatrix}.$$

**Solution**

Under a general rotation, $R$, the electric polarisability tensor component $\alpha_{ij}$ is transformed into

$$\alpha'_{ij} = \sum_p \sum_q R_{ip} R_{jq} \alpha_{pq} ,$$

and if $R$ is a symmetry operation of the crystal then the tensor components are unchanged by $R$, that is, $\alpha'_{ij} = \alpha_{ij}$, and hence

$$\alpha_{ij} = \sum_p \sum_q R_{ip} R_{jq} \alpha_{pq} .$$

For an orthorhombic crystal the notations $R$ include $C_{2x}$, $C_{2y}$, $C_{2z}$. For $C_{2x}$,

$$R = \begin{bmatrix} 1 & 0 & 0 \\ 0 & -1 & 0 \\ 0 & 0 & -1 \end{bmatrix},$$

then

$$\begin{aligned} \alpha_{ij} &= \sum_p \sum_q R_{ip} R_{jq} \alpha_{pq} \\ &= \sum_p \sum_q R_{ip} \alpha_{pq} R_{qj} , \qquad \text{since } R_{jq} \text{ is symmetric.} \end{aligned}$$

Therefore

$$\begin{bmatrix} \alpha_{11} & \alpha_{12} & \alpha_{13} \\ \alpha_{21} & \alpha_{22} & \alpha_{23} \\ \alpha_{31} & \alpha_{32} & \alpha_{33} \end{bmatrix} = \begin{bmatrix} 1 & 0 & 0 \\ 0 & -1 & 0 \\ 0 & 0 & -1 \end{bmatrix} \begin{bmatrix} \alpha_{11} & \alpha_{12} & \alpha_{13} \\ \alpha_{21} & \alpha_{22} & \alpha_{23} \\ \alpha_{31} & \alpha_{32} & \alpha_{33} \end{bmatrix} \begin{bmatrix} 1 & 0 & 0 \\ 0 & -1 & 0 \\ 0 & 0 & -1 \end{bmatrix}$$

$$= \begin{bmatrix} \alpha_{11} & -\alpha_{12} & -\alpha_{13} \\ -\alpha_{21} & \alpha_{22} & \alpha_{23} \\ -\alpha_{31} & \alpha_{32} & \alpha_{33} \end{bmatrix}.$$

Equating components we therefore find that $\alpha_{12} = -\alpha_{12}$, $\alpha_{13} = -\alpha_{13}$, $\alpha_{21} = -\alpha_{21}$, and $\alpha_{31} = -\alpha_{31}$, and hence that $\alpha_{12} = \alpha_{13} = \alpha_{21} = \alpha_{31} = 0$. Similarly by using $C_{2y}$ or $C_{2z}$ we can show that $\alpha_{23} = \alpha_{32} = 0$ also. The electric polarisability tensor therefore simplifies to

$$\begin{bmatrix} \alpha_{11} & 0 & 0 \\ 0 & \alpha_{22} & 0 \\ 0 & 0 & \alpha_{33} \end{bmatrix}.$$

**1.3.4**
An atom is placed successively in crystalline electric fields with the symmetry of the point groups $m3m$, 432, 23, $\bar{4}$, and 1 ($O_h$, $O$, $T$, $S_4$, and $C_1$). Given that a $^1$P term is not split in $m3m$ and belongs to $T_1$ of 432, determine the splitting of a $^1$P term in each of these fields.

**Solution**
The $^1$P term is not split in $m3m$ ($O_h$) or in 23 ($T$). The character table of $\bar{4}$ ($S_4$) is given in Appendix 2, and on restricting to $\bar{4}$ the representation $T_{1g}$ of $m3m$ ($O_h$), to which the $^1$P term belongs, one obtains

| | $E$ | $S_{4z}^-$ | $C_{2z}$ | $S_{4z}^+$ |
|---|---|---|---|---|
| $^1$P | 3 | −1 | −1 | −1 , |

so that the $^1$P term splits into $B+E$. Hence, as long as time reversal is present, the $^1$P term splits into one nondegenerate and one twofold degenerate level. In $1(C_1)$ the $^1$P term becomes three nondegenerate levels. These levels are illustrated in figure 1.3.4, where the degeneracy of each level is given in brackets.

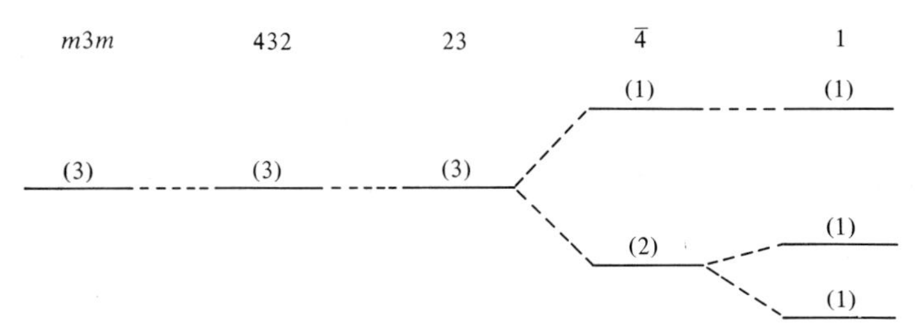


**Figure 1.3.4.**

**1.3.5**
If $\chi^L(2\pi/n)$, the character of a rotation $(2\pi/n)$ in the $(2L+1)$-dimensional irreducible representation of the rotation group, is given by

$$\chi^L\left(\frac{2\pi}{n}\right) = \frac{\sin[(L+\frac{1}{2})(2\pi/n)]}{\sin(\pi/n)}$$

$$\chi^L(E) = (2L+1)\,,$$

determine the character of each of the elements of the point group 422 ($D_4$) in the representation for which $L = 2$. Reduce this representation and thus draw a diagram showing the splitting of an atomic D term in a crystal field having the symmetry of the point group 422 ($D_4$).

**Solution**

The characters of the elements in the representation for which $L = 2$ are

| | $E$ | $C_{2z}$ | $C_{4z}^{\pm}$ | $C_{2x}, C_{2y}$ | $C_{2a}, C_{2b}$ |
|---|---|---|---|---|---|
| $(L = 2)$ | 5 | 1 | $-1$ | 1 | 1 , |

which on reduction gives $A_1+B_1+B_2+E$, see figure 1.3.5, where no significance is to be attached to the ordering of the levels.

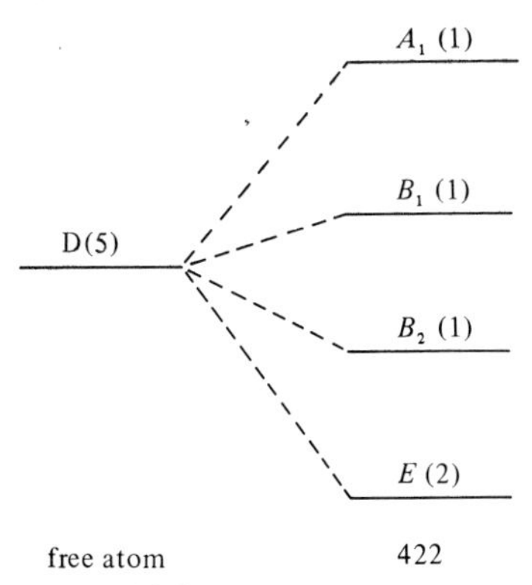


**Figure 1.3.5.**

**1.3.6**

Equivalent positions in the unit cell of the orthorhombic space group *Pnnm* ($D_{2h}^{12}$) are given by

$$\begin{array}{ll} (x, y, z) & (x+\frac{1}{2}, -y+\frac{1}{2}, -z+\frac{1}{2}) \\ (-x, -y, z) & (-x+\frac{1}{2}, y+\frac{1}{2}, -z+\frac{1}{2}) \\ (-x, -y, -z) & (-x+\frac{1}{2}, y+\frac{1}{2}, z+\frac{1}{2}) \\ (x, y, -z) & (x+\frac{1}{2}, -y+\frac{1}{2}, z+\frac{1}{2})\,. \end{array}$$

Suppose that a hypothetical material $XY_2$ is described by this structure and contains two molecules per unit cell. Suggest possible positions for the X and Y atoms in the unit cell.

**Solution**

Possible positions for the X atoms are

$$\begin{array}{lllll} (0, 0, 0) & (\frac{1}{2}, \frac{1}{2}, \frac{1}{2}) \quad \text{or} & (0, \frac{1}{2}, \frac{1}{2}) & (\frac{1}{2}, 0, 0) & \text{or} \\ (\frac{1}{2}, 0, \frac{1}{2}) & (0, \frac{1}{2}, 0) \quad \text{or} & (\frac{1}{2}, \frac{1}{2}, 0) & (0, 0, \frac{1}{2})\,, & \end{array}$$

and for the Y atoms are

$$\begin{array}{lllll} (x, y, 0) & (-x, -y, 0) & (x+\frac{1}{2}, -y+\frac{1}{2}, \frac{1}{2}) & (-x+\frac{1}{2}, y+\frac{1}{2}, \frac{1}{2}) & \text{or} \\ (0, \frac{1}{2}, z) & (0, \frac{1}{2}, -z) & (\frac{1}{2}, 0, -z+\frac{1}{2}) & (\frac{1}{2}, 0, z+\frac{1}{2}) & \text{or} \\ (0, 0, z) & (0, 0, -z) & (\frac{1}{2}, \frac{1}{2}, -z+\frac{1}{2}) & (\frac{1}{2}, \frac{1}{2}, z+\frac{1}{2})\,. & \end{array}$$

See the equivalent positions listed in Henry and Lonsdale (1965).

1.3.7

Suppose that a crystal which belongs to the primitive tetragonal space group $P4/mmm$ has an atom of element X at the point $(x, y, z)$ in the unit cell. Write down the coordinates of the equivalent positions at which there must also be other X atoms [the elements of the point group $4/mmm$ $(D_{4h})$ can be identified from the character tables in Appendix 2].

The space group of $MnF_2$ is $P4_2/mnm$ and the structure contains two molecules per unit cell. The space group of $MnF_2$ is related to $P4/mmm$ by adding $(\frac{1}{2}, \frac{1}{2}, \frac{1}{2})$ to exactly half of the equivalent positions obtained above in such a way that there are Mn atoms at $(0, 0, 0)$ and $(\frac{1}{2}, \frac{1}{2}, \frac{1}{2})$ and F atoms at $(u, u, 0)$, $(-u, -u, 0)$, $(\frac{1}{2}+u, \frac{1}{2}-u, \frac{1}{2})$, and $(\frac{1}{2}-u, \frac{1}{2}+u, \frac{1}{2})$. Hence write down the coordinates of points that are equivalent to the point $(x, y, z)$ for the space group $P4_2/mnm$.

Below 72 K, $MnF_2$ becomes antiferromagnetically ordered with the magnetic moment of the Mn atom at $(0, 0, 0)$ spontaneously aligned parallel to the $z$ axis and the magnetic moment of the Mn atom at $(\frac{1}{2}, \frac{1}{2}, \frac{1}{2})$ spontaneously aligned antiparallel to the $z$ axis; the F atoms have no magnetic moment. If $\theta$ represents the operation of time reversal, which reverses the sense of a magnetic moment, write down the Seitz space-group symbols for the black-and-white space group $P4_2'/mnm'$ of antiferromagnetic $MnF_2$.

**Solution**

The equivalent positions generated by the action of the various symmetry operations on $(x, y, z)$ are

| | | | |
|---|---|---|---|
| $E$: | $(x, y, z)$ | $I$: | $(-x, -y, -z)$ |
| $*C_{4z}^+$: | $(-y, x, z)$ | $*S_{4z}^-$: | $(y, -x, -z)$ |
| $C_{2z}$: | $(-x, -y, z)$ | $\sigma_z$: | $(x, y, -z)$ |
| $*C_{4z}^-$: | $(y, -x, z)$ | $*S_{4z}^+$: | $(-y, x, -z)$ |
| $*C_{2x}$: | $(x, -y, -z)$ | $*\sigma_x$: | $(-x, y, z)$ |
| $*C_{2y}$: | $(-x, y, -z)$ | $*\sigma_y$: | $(x, -y, z)$ |
| $C_{2a}$: | $(y, x, -z)$ | $\sigma_{da}$: | $(-y, -x, z)$ |
| $C_{2b}$: | $(-y, -x, -z)$ | $\sigma_{db}$: | $(y, x, z)$ . |

In $P4_2/mnm$ the operations indicated with an asterisk interchange the two molecules in the unit cell. The general equivalent positions are therefore:

| | |
|---|---|
| $(x, y, z)$ | $(-x, -y, -z)$ |
| $(-y+\frac{1}{2}, x+\frac{1}{2}, z+\frac{1}{2})$ | $(y+\frac{1}{2}, -x+\frac{1}{2}, -z+\frac{1}{2})$ |
| $(-x, -y, z)$ | $(x, y, -z)$ |
| $(y+\frac{1}{2}, -x+\frac{1}{2}, z+\frac{1}{2})$ | $(-y+\frac{1}{2}, x+\frac{1}{2}, -z+\frac{1}{2})$ |
| $(x+\frac{1}{2}, -y+\frac{1}{2}, -z+\frac{1}{2})$ | $(-x+\frac{1}{2}, y+\frac{1}{2}, z+\frac{1}{2})$ |
| $(-x+\frac{1}{2}, y+\frac{1}{2}, -z+\frac{1}{2})$ | $(x+\frac{1}{2}, -y+\frac{1}{2}, z+\frac{1}{2})$ |
| $(y, x, -z)$ | $(-y, -x, z)$ |
| $(-y, -x, -z)$ | $(y, x, z)$ . |

The operation of space inversion, $I$, does not reverse the direction of a magnetic moment. Hence the Seitz space-group symbols for $P4_2'/mnm'$ are,

$$
\begin{array}{ll}
\{E|0\} & \{I|0\} \\
\theta\{C_{4z}^+|\tau\} & \theta\{S_{4z}^-|\tau\} \\
\{C_{2z}|0\} & \{\sigma_z|0\} \\
\theta\{C_{4z}^-|\tau\} & \theta\{S_{4z}^+|\tau\} \\
\{C_{2x}|\tau\} & \{\sigma_x|\tau\} \\
\{C_{2y}|\tau\} & \{\sigma_y|\tau\} \\
\theta\{C_{2a}|0\} & \theta\{\sigma_{da}|0\} \\
\theta\{C_{2b}|0\} & \theta\{\sigma_{db}|0\},
\end{array}
$$

where $\tau = (\frac{1}{2}, \frac{1}{2}, \frac{1}{2})$.

**1.3.8**
The eight Seitz space-group symbols

$$
\begin{array}{llll}
\{E|000\} & \{C_{2x}|\frac{1}{2}\frac{1}{2}0\} & \{C_{2z}|000\} & \{C_{2a}|\frac{1}{2}\frac{1}{2}\frac{1}{2}\} \\
\{C_{4z}^+|00\frac{1}{2}\} & \{C_{2y}|\frac{1}{2}\frac{1}{2}0\} & \{C_{4z}^-|00\frac{1}{2}\} & \{C_{2b}|\frac{1}{2}\frac{1}{2}\frac{1}{2}\},
\end{array}
$$

where the translation vector associated with the point-group operation $R$ in the symbol $\{R|t_1t_2t_3\}$ is $t_1a\boldsymbol{i}+t_2a\boldsymbol{j}+t_3c\boldsymbol{k}$, form the space group $P4_22_12$. By forming the multiplication table show that the space group is a group of infinite order.

**Solution**

$$
\begin{aligned}
\{C_{4z}^+|00\tfrac{1}{2}\}^2 &= \{C_{2z}|001\}, \\
\{C_{4z}^+|00\tfrac{1}{2}\}^3 &= \{C_{4z}^+|00\tfrac{1}{2}\}\{C_{2z}|001\} = \{C_{4z}^-|00\tfrac{3}{2}\}.
\end{aligned}
$$

However, the inverse of $\{C_{4z}^-|00\frac{1}{2}\}$ is $\{C_{4z}^-|00-\frac{1}{2}\}$, so that we have produced two different elements with the same rotational part $C_{4z}^-$ but different translations, namely $\frac{3}{2}c\boldsymbol{k}$ and $-\frac{1}{2}c\boldsymbol{k}$. The elements cannot then form a group of order 8 but must at least form a group of order 16. By continuing the various multiplications one would generate more and more elements of the complete space group which would be found to be of infinite order.

**1.3.9**
The basic vectors of the primitive cubic Bravais lattice may be chosen as

$$\boldsymbol{t}_1 = a(1,0,0), \quad \boldsymbol{t}_2 = a(0,1,0), \quad \boldsymbol{t}_3 = a(0,0,1).$$

(a) Deduce the reciprocal lattice vectors and sketch the Brillouin zone.
(b) Identify the special points and lines of symmetry in this Brillouin zone and hence predict the essential degeneracies of energy eigenvalues all over the Brillouin zone for the space group $Pm3m$ ($O_h^1$).
(c) Using the results of (a) predict the symmetries and degeneracies of the normal modes of a CsCl crystal which has a primitive cubic structure with Cs atoms at (0, 0, 0), and Cl atoms at $(\frac{1}{2}, \frac{1}{2}, \frac{1}{2})$.

**Solution**

(a)

$$g_1 = \frac{2\pi(t_2 \wedge t_3)}{t_1 \cdot (t_2 \wedge t_3)} = 2\pi \begin{vmatrix} i & j & k \\ 0 & a & 0 \\ 0 & 0 & a \end{vmatrix} \div \begin{vmatrix} a & 0 & 0 \\ 0 & a & 0 \\ 0 & 0 & a \end{vmatrix} = \frac{2\pi}{a}(1, 0, 0).$$

Similarly

$$g_2 = \frac{2\pi}{a}(0, 1, 0) \quad \text{and} \quad g_3 = \frac{2\pi}{a}(0, 0, 1).$$

The Brillouin zone is a cube, see figure 1.3.9a.

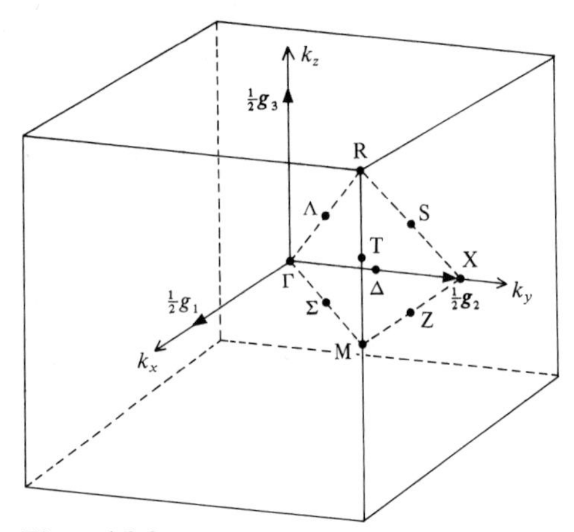


**Figure 1.3.9a.**

(b) The special points and lines of symmetry are as follows, where the coordinates are in terms of $2\pi/a$ and the symmetry operations of the wave vector $\boldsymbol{k}$ are identified:

| | | |
|---|---|---|
| Γ (0 0 0), R ($\frac{1}{2}$ $\frac{1}{2}$ $\frac{1}{2}$) | all the elements of cubic point group | $m3m$ ($O_h$) |
| M ($\frac{1}{2}$ $\frac{1}{2}$ 0) | $E, C_{4z}^{\pm}, C_{2z}, C_{2x}, C_{2y}, C_{2a}, C_{2b}$<br>$I, S_{4z}^{\mp}, \sigma_z, \sigma_x, \sigma_y, \sigma_{da}, \sigma_{db}$ | $4/mmm$ ($D_{4h}$) |
| X (0 $\frac{1}{2}$ 0) | $E, C_{4y}^{\pm}, C_{2y}, C_{2z}, C_{2x}, C_{2c}, C_{2e}$<br>$I, S_{4y}^{\mp}, \sigma_y, \sigma_z, \sigma_x, \sigma_{dc}, \sigma_{de}$ | $4/mmm$ ($D_{4h}$) |
| Σ (α α 0) | $E, C_{2a}, \sigma_z, \sigma_{db}$ | $2mm$ ($C_{2v}$) |
| Δ (0 α 0) | $E, C_{4y}^{\pm}, C_{2y}, \sigma_z, \sigma_x, \sigma_{dc}, \sigma_{de}$ | $4mm$ ($C_{4v}$) |
| Λ (α α α) | $E, C_{31}^{\pm}, \sigma_{db}, \sigma_{de}, \sigma_{df}$ | $3m$ ($C_{3v}$) |
| S (α $\frac{1}{2}$ α) | $E, C_{2c}, \sigma_y, \sigma_{de}$ | $2mm$ ($C_{2v}$) |
| T ($\frac{1}{2}$ $\frac{1}{2}$ α) | $E, C_{4z}^{\pm}, C_{2z}, \sigma_x, \sigma_y, \sigma_{da}, \sigma_{db}$ | $4mm$ ($C_{4v}$) |
| Z (α $\frac{1}{2}$ 0) | $E, C_{2x}, \sigma_y, \sigma_z$ | $2mm$ ($C_{2v}$) . |

Because the space group $Pm3m$ ($O_h^1$) is symmorphic, the energy eigenvalues of a quantum-mechanical particle or quasiparticle with one of the wave vectors given above will belong to the various representations of the appropriate point group. The possible degeneracies can therefore be determined from the character tables. It will be seen that all the eigenvalues for a given $\boldsymbol{k}$ can be expected to be nondegenerate except at M and X and along the lines $\Delta$ and T, where twofold degenerate eigenvalues can occur, and at $\Gamma$ and R, where both twofold and threefold degenerate eigenvalues can occur.

(c) If $\{R|0\}$ is an element of the space group $Pm3m$ ($O_h^1$) of the CsCl structure, the $\kappa\kappa'$ element of the 2 by 2 supermatrix $\mathbf{T}(\boldsymbol{k}; \{R|0\})$ to which the phonons at $\boldsymbol{k}$ belong is given by (Maradudin and Vosko, 1968; Warren, 1968):

$$\mathbf{T}_{\kappa\kappa'}(\boldsymbol{k}; \{R|0\}) = \delta(\kappa, \kappa'') \exp\{i\boldsymbol{k}\cdot[\boldsymbol{x}(\kappa) - R\boldsymbol{x}(\kappa')]\}\mathbf{R}$$

where $\mathbf{R}$ is the 3 by 3 matrix giving the effect of $R$ on the vector $(x, y, z)$. $\boldsymbol{x}(\kappa)$ is the position vector of the atom $\kappa$ relative to the origin of the appropriate unit cell and $\kappa''$ denotes the site to which the operation $\{R|0\}$ moves the atom at $\kappa'$. For CsCl there are two atoms per unit cell (see figure 1.3.9b), and $\kappa$ and $\kappa'$ each take the values 1 and 2, where we choose 1 for the Cs atom and 2 for the Cl atom. For the Cs atom which is at the origin of a set of Cartesian axes, all the space-group operations $\{R|0\}$ leave this atom unmoved so that $\boldsymbol{x}(\kappa') = \boldsymbol{x}(\kappa) = 0$ and the first element $\mathbf{T}_{11}(\boldsymbol{k}; \{R|0\})$, is just equal to $\mathbf{R}$. The off-diagonal elements are zero because the two atoms in the unit cell are of different elements and so

$$\mathbf{T}(\boldsymbol{k}; \{R|0\}) = \begin{bmatrix} \mathbf{R} & 0 \\ 0 & \bar{\theta}\mathbf{R} \end{bmatrix},$$

where

$$\bar{\theta} = \exp\{i\boldsymbol{k}\cdot[\boldsymbol{x}(\kappa) - R\boldsymbol{x}(\kappa')]\},$$

with $\kappa = \kappa' = 2$. $R\boldsymbol{x}(\kappa')$ can be evaluated by using the Jones (1960) symbols, which give the effect of the various operations on the vector $(x, y, z)$. We can write $\boldsymbol{x}(2) = (\frac{1}{2}a, \frac{1}{2}a, \frac{1}{2}a)$, so that, by choosing one element

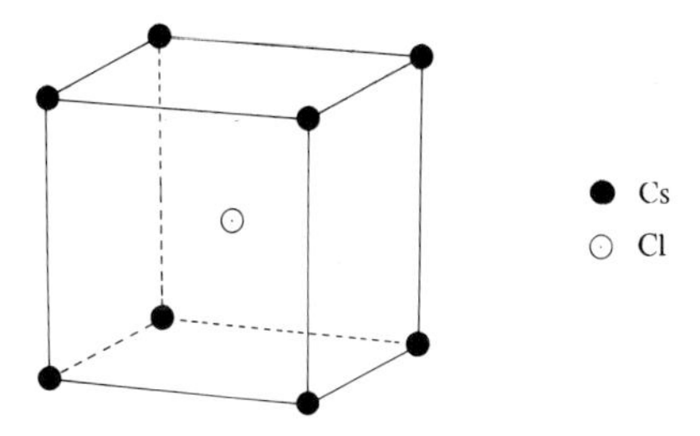



**Figure 1.3.9b.**

from each class of the point group $m3m$ $(O_h)$

$$\begin{aligned} E\boldsymbol{x}(2) &= a(\tfrac{1}{2}, \tfrac{1}{2}, \tfrac{1}{2}) & I\boldsymbol{x}(2) &= a(-\tfrac{1}{2}, -\tfrac{1}{2}, -\tfrac{1}{2}) \\ C_{2x}\boldsymbol{x}(2) &= a(\tfrac{1}{2}, -\tfrac{1}{2}, -\tfrac{1}{2}) & \sigma_x\boldsymbol{x}(2) &= a(-\tfrac{1}{2}, \tfrac{1}{2}, \tfrac{1}{2}) \\ C_{31}^{+}\boldsymbol{x}(2) &= a(\tfrac{1}{2}, \tfrac{1}{2}, \tfrac{1}{2}) & S_{61}^{-}\boldsymbol{x}(2) &= a(-\tfrac{1}{2}, -\tfrac{1}{2}, -\tfrac{1}{2}) \\ C_{2a}\boldsymbol{x}(2) &= a(\tfrac{1}{2}, \tfrac{1}{2}, -\tfrac{1}{2}) & \sigma_{da}\boldsymbol{x}(2) &= a(-\tfrac{1}{2}, -\tfrac{1}{2}, \tfrac{1}{2}) \\ C_{4x}^{+}\boldsymbol{x}(2) &= a(\tfrac{1}{2}, -\tfrac{1}{2}, \tfrac{1}{2}) & S_{4x}^{-}\boldsymbol{x}(2) &= a(-\tfrac{1}{2}, \tfrac{1}{2}, -\tfrac{1}{2}) . \end{aligned}$$

At Γ, $\boldsymbol{k} = 0$ so that $\bar{\theta} = +1$ for all $R$; therefore at Γ, $\mathbf{T}(\boldsymbol{k}; \{R|0\}) = 2T_{1u}$. At R, where $\boldsymbol{k} = (2\pi/a)(\tfrac{1}{2}, \tfrac{1}{2}, \tfrac{1}{2})$, it can be seen that $\theta = +1$ for $E$, $C_{2x}$, $C_{31}^{+}$, $\sigma_{da}$, and $S_{4x}^{-}$ [and in fact for all the elements of $\bar{4}3m$ $(T_d)$], and $\theta = -1$ for $I$, $\sigma_x$, $S_{61}^{-}$, $C_{2a}$, and $C_{4x}^{+}$ [and in fact for all the elements in $m3m$ $(O_h)$, but not in $\bar{4}3m$ $(T_d)$]. The characters of $\mathbf{T}(\boldsymbol{k}; \{R|0\})$ at R $[\boldsymbol{k} = (2\pi/a)(\tfrac{1}{2}, \tfrac{1}{2}, \tfrac{1}{2})]$ are therefore

| | $\mathbf{T}(\boldsymbol{k}; \{R\|0\})$ | $T_{2g}$ | $T_{1u}$ | | $\mathbf{T}(\boldsymbol{k}; \{R\|0\})$ | $T_{2g}$ | $T_{1u}$ |
|---|---|---|---|---|---|---|---|
| $\{E\|0\}$ | 6 | 3 | 3 | $\{I\|0\}$ | 0 | 3 | −3 |
| $\{C_{2m}\|0\}$ | −2 | −1 | −1 | $\{\sigma_m\|0\}$ | 0 | −1 | 1 |
| $\{C_{3j}^{\pm}\|0\}$ | 0 | 0 | 0 | $\{S_{6j}^{\mp}\|0\}$ | 0 | 0 | 0 |
| $\{C_{2p}\|0\}$ | 0 | 1 | −1 | $\{\sigma_{dp}\|0\}$ | 2 | 1 | 1 |
| $\{C_{4m}^{\pm}\|0\}$ | 0 | −1 | +1 | $\{S_{4m}^{\mp}\|0\}$ | −2 | −1 | −1 |

which, on reduction, give $T_{2g} + T_{1u}$. Repeating this procedure all over the Brillouin zone, we find

| | | |
|---|---|---|
| Γ | $2T_{1u}$ | $2\Gamma_{15}$ |
| R | $T_{1u} + T_{2g}$ | $\Gamma_{15} + \Gamma'_{25}$ |
| M | $B_{1u} + A_{2u} + 2E_u$ | $M'_2 + M'_4 + 2M'_5$ |
| X | $A_{1g} + E_g + A_{2u} + E_u$ | $X_1 + X_5 + X'_4 + X'_5$ |
| Σ | $2A_1 + 2B_1 + 2B_2$ | $2\Sigma_1 + 2\Sigma_4 + 2\Sigma_3$ |
| Δ | $2A_1 + 2E$ | $2\Delta_1 + 2\Delta_5$ |
| Λ | $2A_1 + 2E$ | $2\Lambda_1 + 2\Lambda_3$ |
| S | $2A_1 + A_2 + 2B_1 + B_2$ | $2S_1 + S_2 + 2S_3 + S_4$ |
| T | $A_1 + B_2 + 2E$ | $T_1 + T'_2 + 2T_5$ |
| Z | $2A_1 + A_2 + 2B_1 + B_2$ | $2Z_1 + Z_2 + 2Z_3 + Z_4$ |

where the second set of labels is that used by Bouckaert *et al.* (1936).

**1.3.10**

Determine the irreducible representations of the group of the wave vector at X $[\boldsymbol{k} = (2\pi/a)(0, \tfrac{1}{2}, 0)]$ in the nonsymmorphic primitive cubic space group $Pn3m$ $(O_h^4)$ which is generated by the elements

$$\{S_{61}^{-}|\tfrac{1}{2}\tfrac{1}{2}\tfrac{1}{2}\}, \quad \{\sigma_x|\tfrac{1}{2}\tfrac{1}{2}\tfrac{1}{2}\}, \quad \{\sigma_z|\tfrac{1}{2}\tfrac{1}{2}\tfrac{1}{2}\}, \quad \{C_{2c}|\tfrac{1}{2}\tfrac{1}{2}\tfrac{1}{2}\},$$

where the rotational parts of these operations are identified in the notation of Bradley and Cracknell (1972).

**Solution**

Let $\mathbf{T}^k$ be the group of all those Bravais lattice translations $(n_1\boldsymbol{t}_1+n_2\boldsymbol{t}_2+n_3\boldsymbol{t}_3)$ for which

$$\exp[\mathrm{i}\boldsymbol{k}\cdot(n_1\boldsymbol{t}_1+n_2\boldsymbol{t}_2+n_3\boldsymbol{t}_3)] = +1 \ ,$$

where $\boldsymbol{k}$ is the wave vector for the point X. The rotational parts of the elements in the group $\mathbf{G}^k$ of the wave vector $\boldsymbol{k}$ at X can be obtained from solution 1.3.9b, so that the elements in the factor group $\mathbf{G}^k/\mathbf{T}^k$ are

| | | | | | |
|---|---|---|---|---|---|
| $\{E\|0\}$ | $E$ | $\{C_{2c}\|\boldsymbol{\tau}\}$ | $M$ | $\{S_{4y}^-\|0\}$ | $T$ |
| $\{C_{2x}\|0\}$ | $B$ | $\{C_{2e}\|\boldsymbol{\tau}\}$ | $N$ | $\{S_{4y}^+\|0\}$ | $U$ |
| $\{C_{2y}\|0\}$ | $C$ | $\{I\|\boldsymbol{\tau}\}$ | $P$ | $\{\sigma_{dc}\|0\}$ | $V$ |
| $\{C_{2z}\|0\}$ | $D$ | $\{\sigma_x\|\boldsymbol{\tau}\}$ | $Q$ | $\{\sigma_{de}\|0\}$ | $W$ , |
| $\{C_{4y}^+\|\boldsymbol{\tau}\}$ | $K$ | $\{\sigma_y\|\boldsymbol{\tau}\}$ | $R$ | | |
| $\{C_{4y}^-\|\boldsymbol{\tau}\}$ | $L$ | $\{\sigma_z\|\boldsymbol{\tau}\}$ | $S$ | | |

together with $\{E|\boldsymbol{t}_2\}$, where $\boldsymbol{\tau} = (\frac{1}{2}\boldsymbol{t}_1+\frac{1}{2}\boldsymbol{t}_2+\frac{1}{2}\boldsymbol{t}_3)$. The letters $E, B, C, D$, etc., are used as a shorthand labelling of these operations. The group multiplication table for $\mathbf{G}^k/\mathbf{T}^k$ at X is set out in table 1.3.10a, where $B' = \{E|\boldsymbol{t}_2\}B$, $C' = \{E|\boldsymbol{t}_2\}C$, etc.

On separating the elements of this group into classes, we find

| | | | |
|---|---|---|---|
| $\mathrm{C}_1 = E$ | $\mathrm{C}_2 = E'$ | $\mathrm{C}_9 = P, P'$ | |
| $\mathrm{C}_3 = B, D, B', D'$ | | $\mathrm{C}_{10} = Q, S, Q', S'$ | |
| $\mathrm{C}_4 = C$ | $\mathrm{C}_5 = C'$ | $\mathrm{C}_{11} = R, R'$ | |
| $\mathrm{C}_6 = K, L, K', L'$ | | $\mathrm{C}_{12} = T, U, T', U'$ | |
| $\mathrm{C}_7 = M, N'$ | $\mathrm{C}_8 = M', N$ | $\mathrm{C}_{13} = W, V$ | $\mathrm{C}_{14} = W', V'$ . |

By evaluating the class multiplication coefficients $c_{ij,k}$, the equations used in solution 1.1.4 can be solved to determine the character $\chi_i^s$ of class

**Table 1.3.10a.**

| | | | | | | | | | | | | | | | |
|---|---|---|---|---|---|---|---|---|---|---|---|---|---|---|---|
| $E$ | $B$ | $C$ | $D$ | $K$ | $L$ | $M$ | $N$ | $P$ | $Q$ | $R$ | $S$ | $T$ | $U$ | $V$ | $W$ |
| $B$ | $E$ | $D$ | $C$ | $N'$ | $M'$ | $L'$ | $K'$ | $Q'$ | $P'$ | $S'$ | $R'$ | $W$ | $V$ | $U$ | $T$ |
| $C$ | $D$ | $E$ | $B$ | $L$ | $K$ | $N$ | $M$ | $R$ | $S$ | $P$ | $Q$ | $U$ | $T$ | $W$ | $V$ |
| $D$ | $C$ | $B$ | $E$ | $M'$ | $N'$ | $K'$ | $L'$ | $S'$ | $R'$ | $Q'$ | $P'$ | $V$ | $W$ | $T$ | $U$ |
| $K$ | $M$ | $L$ | $N$ | $C'$ | $E'$ | $D'$ | $B'$ | $T'$ | $V'$ | $U'$ | $W'$ | $R$ | $P$ | $S$ | $Q$ |
| $L$ | $N$ | $K$ | $M$ | $E'$ | $C'$ | $B'$ | $D'$ | $U'$ | $W'$ | $T'$ | $V'$ | $P$ | $R$ | $Q$ | $S$ |
| $M$ | $K$ | $N$ | $L$ | $B$ | $D$ | $E$ | $C$ | $V$ | $T$ | $W$ | $U$ | $Q$ | $S$ | $P$ | $R$ |
| $N$ | $L$ | $M$ | $K$ | $D$ | $B$ | $C$ | $E$ | $W$ | $U$ | $V$ | $T$ | $S$ | $Q$ | $R$ | $P$ |
| $P$ | $Q$ | $R$ | $S$ | $T$ | $U$ | $V$ | $W$ | $E$ | $B$ | $C$ | $D$ | $K$ | $L$ | $M$ | $N$ |
| $Q$ | $P$ | $S$ | $R$ | $W'$ | $V'$ | $U'$ | $T'$ | $B'$ | $E'$ | $D'$ | $C'$ | $N$ | $M$ | $L$ | $K$ |
| $R$ | $S$ | $P$ | $Q$ | $U$ | $T$ | $W$ | $V$ | $C$ | $D$ | $E$ | $B$ | $L$ | $K$ | $N$ | $M$ |
| $S$ | $R$ | $Q$ | $P$ | $V'$ | $W'$ | $T'$ | $U'$ | $D'$ | $C'$ | $B'$ | $E'$ | $M$ | $N$ | $K$ | $L$ |
| $T$ | $V$ | $U$ | $W$ | $R'$ | $P'$ | $S'$ | $Q'$ | $K'$ | $M'$ | $L'$ | $N'$ | $C$ | $E$ | $D$ | $B$ |
| $U$ | $W$ | $T$ | $V$ | $P'$ | $R'$ | $Q'$ | $S'$ | $L'$ | $N'$ | $K'$ | $M'$ | $E$ | $C$ | $B$ | $D$ |
| $V$ | $T$ | $W$ | $U$ | $Q$ | $S$ | $P$ | $R$ | $M$ | $K$ | $N$ | $L$ | $B$ | $D$ | $E$ | $C$ |
| $W$ | $U$ | $V$ | $T$ | $S$ | $Q$ | $R$ | $P$ | $N$ | $L$ | $M$ | $K$ | $D$ | $B$ | $C$ | $E$ |

$C_i$ in the irreducible representation $\Gamma_s$. Our results are shown in table 1.3.10b.

**Table 1.3.10b.**

| | $C_1$ | $C_2$ | $C_3$ | $C_4$ | $C_5$ | $C_6$ | $C_7$ | $C_8$ | $C_9$ | $C_{10}$ | $C_{11}$ | $C_{12}$ | $C_{13}$ | $C_{14}$ |
|---|---|---|---|---|---|---|---|---|---|---|---|---|---|---|
| $\Gamma_1$ | 1 | 1 | 1 | 1 | 1 | 1 | 1 | 1 | 1 | 1 | 1 | 1 | 1 | 1 |
| $\Gamma_2$ | 1 | 1 | 1 | 1 | 1 | 1 | 1 | 1 | −1 | −1 | −1 | −1 | −1 | −1 |
| $\Gamma_3$ | 1 | 1 | 1 | 1 | 1 | −1 | −1 | −1 | 1 | 1 | 1 | −1 | −1 | −1 |
| $\Gamma_4$ | 1 | 1 | 1 | 1 | 1 | −1 | −1 | −1 | −1 | −1 | −1 | 1 | 1 | 1 |
| $\Gamma_5$ | 1 | 1 | −1 | 1 | 1 | 1 | −1 | −1 | 1 | −1 | 1 | 1 | −1 | −1 |
| $\Gamma_6$ | 1 | 1 | −1 | 1 | 1 | 1 | −1 | −1 | −1 | 1 | −1 | −1 | 1 | 1 |
| $\Gamma_7$ | 1 | 1 | −1 | 1 | 1 | −1 | 1 | 1 | 1 | −1 | 1 | −1 | 1 | 1 |
| $\Gamma_8$ | 1 | 1 | −1 | 1 | 1 | −1 | 1 | 1 | −1 | 1 | −1 | 1 | −1 | −1 |
| $\Gamma_9$ | 2 | 2 | 0 | −2 | −2 | 0 | 0 | 0 | 2 | 0 | −2 | 0 | 0 | 0 |
| $\Gamma_{10}$ | 2 | 2 | 0 | −2 | −2 | 0 | 0 | 0 | −2 | 0 | 2 | 0 | 0 | 0 |
| $\Gamma_{11}$ | 2 | −2 | 0 | 2 | −2 | 0 | 0 | 0 | 0 | 0 | 0 | 0 | 2 | −2 |
| $\Gamma_{12}$ | 2 | −2 | 0 | 2 | −2 | 0 | 0 | 0 | 0 | 0 | 0 | 0 | −2 | 2 |
| $\Gamma_{13}$ | 2 | −2 | 0 | −2 | 2 | 0 | 2 | −2 | 0 | 0 | 0 | 0 | 0 | 0 |
| $\Gamma_{14}$ | 2 | −2 | 0 | 2 | 2 | 0 | −2 | 2 | 0 | 0 | 0 | 0 | 0 | 0 |

**1.3.11**

The basic vectors of the face-centred cubic Bravais lattice may be chosen as

$$\boldsymbol{t}_1 = a(0, \tfrac{1}{2}, \tfrac{1}{2}), \quad \boldsymbol{t}_2 = a(\tfrac{1}{2}, 0, \tfrac{1}{2}), \quad \boldsymbol{t}_3 = a(\tfrac{1}{2}, \tfrac{1}{2}, 0)\,.$$

(a) Deduce the reciprocal lattice vectors and sketch the Brillouin zone.
(b) Determine expressions for the eigenvalues and eigenfunctions of the Hamiltonian of an electron in a face-centred cubic metal in the free-electron (or empty-lattice) approximation.
(c) Calculate the first three free-electron eigenvalues at $\Gamma$ (the point at the centre of the Brillouin zone) and X (a point at the centre of one of the small square faces of the Brillouin zone); in each case determine the degeneracies of these eigenvalues. Describe qualitatively the splittings of these levels that would be expected when a realistic nonzero potential $V(\boldsymbol{r})$ is used.
(d) For the energies at X obtained in (c) determine the correct linear combinations of the wave functions obtained in (b) that belong to the irreducible representations of the group of the wave vector at X.

**Solution**

(a)

$$\boldsymbol{g}_1 = \frac{2\pi(\boldsymbol{t}_2 \wedge \boldsymbol{t}_3)}{\boldsymbol{t}_1\cdot(\boldsymbol{t}_2 \wedge \boldsymbol{t}_3)} = 2\pi\begin{vmatrix} \boldsymbol{i} & \boldsymbol{j} & \boldsymbol{k} \\ \tfrac{1}{2}a & 0 & \tfrac{1}{2}a \\ \tfrac{1}{2}a & \tfrac{1}{2}a & 0 \end{vmatrix} \div \begin{vmatrix} 0 & \tfrac{1}{2}a & \tfrac{1}{2}a \\ \tfrac{1}{2}a & 0 & \tfrac{1}{2}a \\ \tfrac{1}{2}a & \tfrac{1}{2}a & 0 \end{vmatrix} = \frac{2\pi}{a}(-1, 1, 1)\,.$$

Similarly

$$\boldsymbol{g}_2 = \frac{2\pi}{a}(1,-1,1), \quad \text{and} \quad \boldsymbol{g}_3 = \frac{2\pi}{a}(1,1,-1).$$

The Brillouin zone is illustrated in figure 1.3.11a.

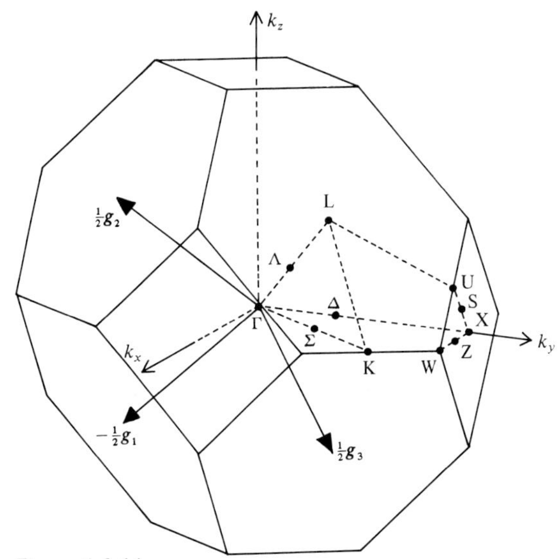


**Figure 1.3.11a.**

(b) The wave function $\psi(\boldsymbol{r})$ must satisfy the Schrödinger equation,

$$-\frac{\hbar^2}{2m}\nabla^2\psi(\boldsymbol{r}) + V(\boldsymbol{r})\psi(\boldsymbol{r}) = E\,\psi(\boldsymbol{r}),$$

and the form of $\psi(\boldsymbol{r})$ is given by Bloch's theorem

$$\psi(\boldsymbol{r}) = u_{\boldsymbol{k}}(\boldsymbol{r})\exp(\mathrm{i}\boldsymbol{k}\cdot\boldsymbol{r}),$$

where $u_{\boldsymbol{k}}(\boldsymbol{r})$ has the periodicity of the Bravais lattice of the crystal. The solutions of the Schrödinger equation then take the form (Jones, 1960):

$$E_{\boldsymbol{k}} = \frac{\hbar^2}{2m}|\boldsymbol{k} - l_1\boldsymbol{g}_1 - l_2\boldsymbol{g}_2 - l_3\boldsymbol{g}_3|^2,$$

and

$$\psi(\boldsymbol{k},\boldsymbol{r}) = \exp\{\mathrm{i}(\boldsymbol{k} - l_1\boldsymbol{g}_1 - l_2\boldsymbol{g}_2 - l_3\boldsymbol{g}_3)\cdot\boldsymbol{r}\}.$$

(c) By inspection, values of $l_1$, $l_2$, and $l_3$ and the corresponding energy eigenvalues $E_{\boldsymbol{k}}$, in units of $(\hbar^2/2m)(2\pi/a)^2$, at Γ and X are as shown in table 1.3.11a.

At Γ the first three eigenvalues are 0, 3, and 4 which have degeneracies of 1, 8, and 6, respectively, and at X the first three eigenvalues are 1, 2, and

5 which have degeneracies of 2, 4, and 8 respectively. A nonzero potential $V(\boldsymbol{r})$ will split these degenerate energy levels until the degeneracies are reduced to those of the appropriate irreducible representations of $\mathbf{G}^{k}/\mathbf{T}^{k}$.
(d) The functions $\psi(\boldsymbol{k},\boldsymbol{r}) = \exp\{\mathrm{i}(\boldsymbol{k}-l_1\boldsymbol{g}_1-l_2\boldsymbol{g}_1-l_3\boldsymbol{g}_3)\cdot\boldsymbol{r}\}$ determined in (c) will usually not belong separately to the irreducible representations of $\mathbf{G}^{k}/\mathbf{T}^{k}$, and it is necessary to use the projection operator $\sum_R \chi^{s*}(R)R$, where $\chi^{s}(R)$ is the character of the element $R$ in $\Gamma_s$.

For X the function $\psi(\boldsymbol{k},\boldsymbol{r})$ and the linear combinations, $\phi^{\mathrm{i}}$, that actually belong to irreducible representations of $\mathbf{G}^{k}/\mathbf{T}^{k}$ are given in table 1.3.11b.

Using the compatibilities between the irreducible representations along ΓΔX, and the fact that the free-electron bands are parabolic, we can construct the bands along ΓΔX, see figure 1.3.11b.

**Table 1.3.11a.**

| $l$ | $E_\Gamma$ | $E_\mathrm{X}$ | $l$ | $E_\Gamma$ | $E_\mathrm{X}$ | $l$ | $E_\Gamma$ | $E_\mathrm{X}$ |
|---|---|---|---|---|---|---|---|---|
| 000 | 0 | 1 | 011 | 4 | 5 | $\bar{1}\bar{1}\bar{1}$ | 3 | 6 |
| 100 | 3 | 2 | 101 | 4 | 1 | $01\bar{1}$ | 8 | 13 |
| 010 | 3 | 6 | 110 | 4 | 5 | $0\bar{1}1$ | 8 | 5 |
| 001 | 3 | 2 | $0\bar{1}\bar{1}$ | 4 | 5 | $10\bar{1}$ | 8 | 9 |
| $\bar{1}00$ | 3 | 6 | $\bar{1}0\bar{1}$ | 4 | 9 | $\bar{1}01$ | 8 | 9 |
| $0\bar{1}0$ | 3 | 2 | $\bar{1}\bar{1}0$ | 4 | 5 | $1\bar{1}0$ | 8 | 5 |
| $00\bar{1}$ | 3 | 6 | 111 | 3 | 2 | $\bar{1}10$ | 8 | 13 |

**Table 1.3.11b.**

| $l$ | $\psi$ | | Representation of X | $\phi^{\mathrm{i}}$ |
|---|---|---|---|---|
| $E = 1$ | | | | |
| 000 | $\exp(-\mathrm{i}y)$ | $u_1$ | $A_{1\mathrm{g}}$ | $u_1+u_2$ |
| 020 | $\exp(\mathrm{i}y)$ | $u_2$ | $A_{2\mathrm{u}}$ | $u_1-u_2$ |
| $E = 2$ | | | | |
| 100 | $\exp\{-\mathrm{i}(x-z)\}$ | $u_1$ | $A_{1\mathrm{g}}$ | $u_1+u_2+u_3+u_4$ |
| 001 | $\exp\{-\mathrm{i}(-x+z)\}$ | $u_2$ | $B_{2\mathrm{g}}$ | $u_1+u_2-u_3-u_4$ |
| $0\bar{1}0$ | $\exp\{-\mathrm{i}(x+z)\}$ | $u_3$ | $E_\mathrm{u}$ | $\{u_3-u_4$ |
| 111 | $\exp\{-\mathrm{i}(-x-z)\}$ | $u_4$ | | $\{u_1-u_2$ |
| $E = 5$ | | | | |
| 011 | $\exp\{\mathrm{i}(2x-y)\}$ | $u_1$ | $A_{1\mathrm{g}}$ | $u_1+u_2+u_3+u_4+u_5+u_6+u_7+u_8$ |
| 110 | $\exp\{\mathrm{i}(-y+2z)\}$ | $u_2$ | $A_{2\mathrm{u}}$ | $u_1+u_2+u_3+u_4-u_5-u_6-u_7-u_8$ |
| $0\bar{1}\bar{1}$ | $\exp\{\mathrm{i}(-2x-y)\}$ | $u_3$ | $B_{1\mathrm{g}}$ | $u_1-u_2+u_3-u_4-u_5+u_6+u_7-u_8$ |
| $\bar{1}\bar{1}0$ | $\exp\{\mathrm{i}(-y-2z)\}$ | $u_4$ | $B_{2\mathrm{u}}$ | $u_1-u_2+u_3-u_4+u_5-u_6-u_7+u_8$ |
| $0\bar{1}1$ | $\exp\{\mathrm{i}(y-2z)\}$ | $u_5$ | $E_\mathrm{g}$ | $\{u_1-u_2-u_3+u_4-u_5+u_6-u_7+u_8$ |
| $1\bar{1}0$ | $\exp\{\mathrm{i}(-2x+y)\}$ | $u_6$ | | $\{u_1+u_2-u_3-u_4+u_5+u_6-u_7-u_8$ |
| 112 | $\exp\{\mathrm{i}(2x+y)\}$ | $u_7$ | $E_\mathrm{u}$ | $\{u_1+u_2-u_3-u_4-u_5-u_6+u_7+u_8$ |
| 211 | $\exp\{\mathrm{i}(y+2z)\}$ | $u_8$ | | $\{u_1-u_2-u_3+u_4+u_5-u_6+u_7-u_8$ |

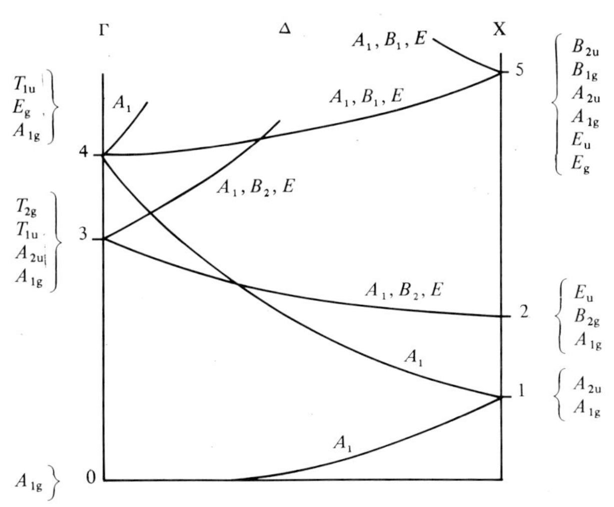


**Figure 1.3.11b.**

**References**

Bouckaert, L. P., Smoluchowski, R., Wigner, E., 1936, *Phys. Rev.*, **50**, 58.

Bradley, C. J., Cracknell, A. P., 1972, *The Mathematical Theory of Symmetry in Solids: Representation Theory for Point Groups and Space Groups* (Oxford University Press, Oxford).

Henry, N. F. M., Lonsdale, K. (Eds), 1965, *International Tables for X-ray Crystallography. Volume 1, Symmetry Groups* (Kynoch Press, Birmingham).

Jones, H., 1960, *The Theory of Brillouin Zones and Electronic States in Crystals* (North-Holland, Amsterdam).

Maradudin, A. A., Vosko, S. H., 1968, *Rev. Mod. Phys.*, **40**, 1.

Warren, J. L., 1968, *Rev. Mod. Phys.*, **40**, 38.

**General references**

Cracknell, A. P., 1968, *Applied Group Theory* (Pergamon Press, Oxford).

Heine, V., 1960, *Group Theory in Quantum Mechanics* (Pergamon Press, Oxford).

Lax, M., 1974, *Symmetry Principles in Solid State and Molecular Physics* (John Wiley, New York).

Tinkham, M., 1964, *Group Theory and Quantum Mechanics* (McGraw-Hill, New York).

## 1.4 Symmetry and spectroscopy

**D J Greenslade, R G Jones, J R Miller** University of Essex

### 1.4.1

Describe what is meant by a fundamental (normal) vibrational mode. Work out the irreducible representations of the point group $D_{4d}$ to which the normal modes of $Mn_2(CO)_{10}$ belong. State the infrared and Raman activity of each mode and comment on the infrared–Raman exclusion rule. Which modes can be described as 'C–O stretching' and which are vibrations associated solely with the linking of two $Mn(CO)_5$ groups [i.e. not internal modes of the separated $Mn(CO)_5$ halves]?

**Solution**

A fundamental mode of vibration is a vibration in which all parts of an object oscillate at the same frequency with a constant phase relationship. Applied to a molecule this means that all atoms have the same vibration frequency and all reach the extremes of their amplitudes simultaneously, and their maximum velocities at the same instant. In an overtone or combination mode these conditions do not apply.

Each fundamental mode has a frequency and a form which are properties of the force constants, the molecular geometry and the atomic masses. If a molecule has symmetry the forms (molecular shapes at the amplitude extreme) depend to some extent on the symmetry alone. Thus the tetrahedral symmetry of methane constrains one, and only one, of the nine fundamental modes to be 'fully symmetric', i.e. the hydrogen atoms must move equally along the C–H bond directions.

By assigning to each atom $i$ of the $Mn_2(CO)_{10}$ molecule three external displacement coordinates $x_i, y_i, z_i$, and determining the way in which these 66 coordinates behave under the symmetry operations of $D_{4d}$, a reducible representation of the group can be formed. The displacement coordinates $x_i, y_i, z_i$ represent a small movement of the labelled atom from its own equilibrium position, and the whole set represents the various ways—i.e. translations, rotations, and vibrations—in which the molecule can move.

The symmetry operations are listed in the character table (Appendix 2), and in investigating the behaviour of the coordinates, four points are noted.

(i) To develop a representation it is necessary to know only whether a coordinate is unaltered, inverted, or permuted with another coordinate by each symmetry operation. These effects contribute respectively +1, −1, 0 to the *character* of the representation for that symmetry operation.

(ii) All operations in the same class have the same character in any representation.

(iii) When a number of atoms are permuted by an operation, all their local displacement coordinates are permuted and contribute zero to the character. This permutation must always occur unless an atom lies *on a symmetry element*. Thus only atoms lying on an element need to be considered.

(iv) The effect of a symmetry operation on the coordinates of an atom lying on the associated element is independent of the position of the atom on the element. Thus the total effect for a set of atoms is that for one atom multiplied by the number of atoms.

For example, consider the operation $C_4$ on the coordinates of $Mn_2(CO)_{10}$. There are 22 atoms and therefore 66 coordinates. Only six of the atoms lie on the fourfold axis, the other sixteen are permuted by $C_4$ and need not be considered [rule (iii)]. The effect on $x$, $y$, $z$ for any one atom on the axis is that:
(a) $z$ is unaltered and contributes +1;
(b) $x$ and $y$ are permuted and contribute 0.
The contribution of one atom to the character is +1, and of six atoms +6 [rule (iv)]. The result for $-C_4$ is the same [rule (ii)].

The $C_2$ operation is slightly different. There are still six atoms to consider, but the effect on the coordinates of one atom is that:
(a) $z$ is unaltered and contributes +1;
(b) $x$ is inverted and contributes −1;
(c) $y$ is inverted and contributes −1.
The contribution of one atom is thus −1, and of six atoms −6.

The identity operation, $E$, leaves all the coordinates unaltered and has a character of 66.

By extending these arguments to each class of operation, the following list of characters can be derived:

| $E$ | $2C_4$ | $C_2$ | $4C_2'$ | $2S_8$ | $2S_8^3$ | $4\sigma_d$ |
|---|---|---|---|---|---|---|
| 66 | 6 | −6 | 0 | 0 | 0 | 10 . |

By the standard procedure for reduction (cf. problem 1.2.4) this list can be shown to be the following sum of irreducible representations:

$$7A_1 + 2A_2 + 2B_1 + 7B_2 + 9E_1 + 6E_2 + 9E_3 \ .$$

As the $E$ species are doubly degenerate, this implies that the total number of displacements represented is 66, a useful check on the calculation, since this should equal the number of coordinates.

The list above represents all possible movements of the molecule, including rotation and translation. Translation may be represented by three vectors $T_x$, $T_y$, $T_z$ from the centre of the molecule. If the transformation of these vectors is investigated for the symmetry operations, it can be shown that $T_z$ transforms like the representation $B_2$ ($z$ is along the fourfold axis) and the pair $T_x$, $T_y$ like $E_1$.

Rotations can be represented diagrammatically by rotating circles around the axes. It can thus be shown that $R_z$ transforms like $A_2$ and, with a little more difficulty, that the pair $R_x$, $R_y$ transforms like $E_3$.

By subtraction of these rotational and translational components from the grand total, the final vibrational list is obtained:

$$7A_1 + A_2 + 2B_1 + 6B_2 + 8E_1 + 6E_2 + 8E_3 \ .$$

The infrared activity of a mode is determined by the electric transition moment integral, $\int \psi_0^*(e\boldsymbol{r})\psi_1\,d\tau$, where $\psi_0$ is the ground-state and $\psi_1$ the excited-state wave function; $\psi_1$ has the same symmetry properties as the vibrational mode, and $\psi_0$ is fully symmetric; $\boldsymbol{r}$ is a vector and has $x$, $y$, and $z$ components. In general, $\boldsymbol{r}$ will transform according to a suitable mixture of $B_2$ (for the $z$ component) and $E_1$ for the $x$, $y$ component. These components form the direction of polarisation of the electric vector in the incident infrared light. By simple rules of multiplication the transformation properties of the integrand above can be obtained for $\boldsymbol{r} = z$ and $\boldsymbol{r} = x$ or $y$. If the integrand does not contain an $A_1$ component, it must have nodes and will always integrate to zero, hence the transition will be forbidden. The only way an $A_1$ component can occur in the above type of integrand is when the transformations of $\boldsymbol{r}$ and $\psi_1$ are the same. Thus transitions to $B_2$ modes are allowed in $z$-polarised light and to $E_1$ modes in $(x, y)$-polarised light.

In Raman spectroscopy the transition moment is of the type $\int \psi_0^*\boldsymbol{r}^2\psi_1\,d\tau$, and transitions are allowed when the polarisability of the molecule changes during vibration. In this case the integrand contains $A_1$ when $\psi_1$ transforms in the same way as quadratic functions like $x^2$, $xy$, $yz$, ... . Examination shows that these transformations are:

$$\begin{array}{ll} x^2+y^2,\ z^2, & A_1; \\ (x^2-y^2, xy), & E_2; \\ (yz, zx), & E_3. \end{array}$$

Only these modes are then Raman-allowed.

The infrared–Raman exclusion rule applies when the components of $\boldsymbol{r}$ $(x, y, z)$ do not transform in the same way as the quadratic functions. This always occurs when a centre of symmetry is present; $D_{4d}$ is an example of a noncentric group which shows exclusion. In general, exclusion will occur if the molecule has an $S_n$ axis except where $n = 1$, 3, or 4. The rotational groups $O$ and $I$ also show exclusion, owing to the presence of multiple fourfold and fivefold axes, respectively. The exclusion in $D_{4d}$ stems from the $S_8$ axis.

In metal carbonyl compounds the C–O bonds are considerably stronger and have higher force constants than other links in the molecule. As the strongest bonds are between the lightest atoms, a set of high vibration frequencies occurs, assigned as 'C–O stretching' vibrations. The details show that these vibrations involve movement of carbon and oxygen atoms along the M–C–O direction only. The M–C bond is thus contracted as the C–O bond stretches, so the above assignment is not quite accurate. However, these vibrational modes occur in a frequency range around 2000 $cm^{-1}$ which is very free of frequencies of other types, and considerable useful information can be obtained by a study of these frequencies in isolation.

The symmetry properties of C–O stretching modes are readily obtained by assigning a double-headed vector to each C–O link, and studying the behaviour of these vectors under the symmetry operations. It is helpful to separate them into two sets which are not symmetry-related to each other, i.e. two axial and eight radial groups.

The axial rotations $C_4$ and $C_2$, together with the identity operation, leave the two axial vectors unaltered and therefore have characters of 2; $C_2'$, $S_8$, and $S_8^3$ permute them and have zero characters, $\sigma_d$ leaves them unaltered. The representation is thus

| $E$ | $2C_4$ | $C_2$ | $4C_2'$ | $2S_8$ | $2S_8^3$ | $4\sigma_d$ |
|---|---|---|---|---|---|---|
| 2 | 2 | 2 | 0 | 0 | 0 | 2 , |

which reduces to $A_1+B_2$. The $A_1$ mode is Raman-allowed, the $B_2$ infrared-allowed.

Similarly, the eight radial groups produce a representation

| $E$ | $2C_4$ | $C_2$ | $4C_2'$ | $2S_8$ | $2S_8^3$ | $4\sigma_d$ |
|---|---|---|---|---|---|---|
| 8 | 0 | 0 | 0 | 0 | 0 | 2 , |

which reduces to $A_1+B_2+E_1+E_2+E_3$.

The axial and radial $A_1$ and $B_2$ modes couple together, but the $E_1$, $E_2$, and $E_3$ modes involve only the radial carbonyl groups.

A half-molecule $Mn(CO)_5$ has 11 atoms and therefore 27 vibrational modes. Therefore, of the 60 modes of $Mn_2(CO)_{10}$, six involve relative oscillations of one half-molecule against the other, rather than internal oscillation of the half-molecules. One way of finding these modes is to work out the modes of $Mn(CO)_5$ in $C_{4v}$ symmetry, form them into symmetric and antisymmetric pairs in $D_{4d}$ symmetry, and subtract from the total for $Mn_2(CO)_{10}$. A quicker method is to combine the *rotations and translations* of two $Mn(CO)_5$ halves into symmetric and antisymmetric pairs. Consider the equation

$$T_z^{(1)}+T_z^{(2)} = T_z \ ,$$

where the two halves are superscripted (1) and (2), and the whole molecule is left unsuperscripted. This equation means that when both halves translate in the $z$ direction, the whole molecule is doing just that. However, when the two halves translate in opposite directions, the whole molecule is involved in what might be called the Mn–Mn stretching mode:

$$T_z^{(1)}-T_z^{(2)} = A_1 \qquad \text{(Mn–Mn stretch)} .$$

Similarly,

$$R_z^{(1)}+R_z^{(2)} = R_z(A_2) \ ,$$

and

$$R_z^{(1)}-R_z^{(2)} = B_1 \qquad \text{(torsional oscillation about Mn–Mn)} .$$

The combinations in the $x$ and $y$ directions are interesting:

$$\left.\begin{aligned} T_x^{(1)} + T_x^{(2)} &= T_x \\ T_y^{(1)} + T_y^{(2)} &= T_y \end{aligned}\right\} (E_1) .$$

However, $[T_x^{(1)} - T_x^{(2)}]$ appears to give the molecule angular momentum about the $y$ axis, and $[T_y^{(1)} - T_y^{(2)}]$ imparts angular momentum about the $x$ axis. The apparent angular momentum can be removed by rotating the whole in the opposite sense, so that each half is actually part translated, part rotated. Notice that the two rotations must be in the same sense. Thus to get a vibration it is necessary to add a component of $[R_y^{(1)} + R_y^{(2)}]$ to $[T_x^{(1)} - T_x^{(2)}]$. Similarly the pure rotation $R_x$ involves $[R_x^{(1)} + R_x^{(2)}]$ with a component of $[T_y^{(1)} - T_y^{(2)}]$. The vibrations and the pure rotations are of species $E_3$.

The *dis*-rotations $[R_x^{(1)} - R_x^{(2)}]$ and $[R_y^{(1)} - R_y^{(2)}]$ produce a different pair of vibrations of $E_1$ symmetry.

The six 'inter-half' vibrations may then be described as follows:

$A_1$ Mn–Mn stretching;
$B_1$ torsional vibration about Mn–Mn bond;
$E_1$ *dis*-rotation of $Mn(CO)_5$ units;
$E_3$ *con*-rotation of $Mn(CO)_5$ units.

**1.4.2**

Figure 1.4.2 shows the infrared spectrum of carbon dioxide recorded at 40°C under conditions of medium resolution. The Raman spectrum shows a doublet with peaks at 1285 and 1388 $cm^{-1}$. Deduce the structure of carbon dioxide, explaining the structures of the bands.

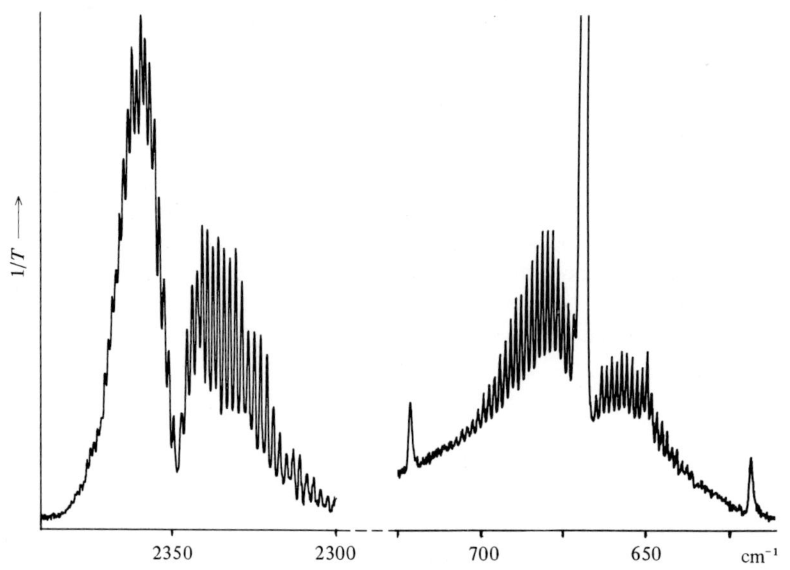


**Figure 1.4.2.** Infrared spectrum of carbon dioxide.

Assuming the frequencies to be harmonic, calculate a valence force field, stating the physical meaning of each force constant. Show what extra information would be needed to calculate the corresponding force field of the molecule SCO.

**Solution**

Carbon dioxide is triatomic and should show three fundamental modes of vibration if nonlinear and four if linear (one degenerate). The observed spectra appear to show four frequencies, but the splitting of the Raman band into a doublet is due to Fermi resonance, explained later. The molecule is linear, confirmed by the *PR* and *PQR* structures of the two infrared bands. If the molecule were bent, the fine structure would be much more complicated as the molecule would have three principal moments of inertia.

The fine structure apparent in the infrared bands arises because the transitions involve changes of rotational state as well as vibrational state. The *P* branches, which appear on the low-frequency sides of both infrared bands, correspond to rotational changes in which the rotational quantum number $J$ decreases by 1 as the vibrational quantum number $v$ increases by 1. *R* branches, on the high-frequency sides of the band centre, correspond to changes of $J$ of $+1$. The strong central *Q* branch in the band at 667 $cm^{-1}$ corresponds to zero change of $J$.

At 40°C many rotational energy states are populated and the *Q* branch at 667 $cm^{-1}$ is very strong because each populated $J$ state can be excited to a vibrational state with the same $J$ value. As there are only very minute changes of rotational energy, at low resolution these transitions appear to be superimposed, giving the very strong *Q* branch. There is a selection rule $\Delta J = 0, \pm 1$, for rotational transitions accompanying a vibrational change when the vibration concerned is a 'perpendicular' one, i.e. the vibration excited involves the molecular dipole gradient being perpendicular to the principal axis. When the dipole gradient is parallel to the principal axis the rotational selection rule is $\Delta J = \pm 1$, i.e. no *Q* branch appears. These selection rules show that the 667 $cm^{-1}$ band corresponds to the bending mode of carbon dioxide, and the 2350 $cm^{-1}$ band, with no *Q* branch, corresponds to a 'parallel' transition, i.e. is a stretching mode.

The final detail of the molecular shape, that carbon dioxide is centrosymmetric, may be deduced from the mutual exclusion of the infrared and Raman frequencies. Exclusion is caused by a $S_n$ axis ($n \neq 1, 3, 4$), and a triatomic molecule can only have such an axis if it belongs to point group $D_{\infty h}$.

We expect then three frequencies. Simple group theory shows that they belong to the species $A_{1g}$, $A_{2u}$, and $E_{1u}$ of the group $D_{\infty h}$. The $A_{1g}$ mode is Raman-allowed and the $A_{2u}$ and $E_{1u}$ modes infrared-allowed. $A_{2u}$ is a parallel mode, corresponding to asymmetric C–O stretching, and is assigned to the band centred at 2350 $cm^{-1}$. $E_{1u}$ is a doubly degenerate perpendicular mode, corresponding to $\widehat{OCO}$ bending, and is assigned to the $PQR$ band at 667 $cm^{-1}$.

The overtone of $E_{1u}$ has three components (take the *indirect* product[1] $E_{1u} \wedge E_{1u} = A_{1g} + E_{2g}$) and might be expected around 1334 $cm^{-1}$. The $A_{1g}$ component, however, enters into Fermi resonance with the $A_{1g}$ symmetric stretching mode, which by coincidence lies very near to 1334 $cm^{-1}$. The resonance is caused by the presence of cubic and higher-order anharmonic terms in the potential energy, and has the effect of mixing the two $A_{1g}$ states and pushing their energies about 100 $cm^{-1}$ apart. Hence the two $A_{1g}$ transitions appear in the Raman spectrum.

A related feature appears in the infrared spectrum. The two weak peaks at 617 and 720 $cm^{-1}$ are 'hot bands'. They correspond to $Q$ branches of transitions from the thermally populated $E_{1u}$ state to the two resonating $A_{1g}$ states. The intensities of these two peaks are very temperature-dependent. According to the Maxwell–Boltzmann distribution, the population at 40°C relative to that of the ground state is 9·3%.

A valence force field for carbon dioxide has three force constants:

$f_{CO}$, the stretching constant of each C–O bond, represents the restoring force generated when a C–O bond is stretched or compressed by a unit amount, the rest of the molecule remaining in its equilibrium position.

$\delta_{OCO}$ is the bending force constant, similarly representing the restoring couple generated when the molecule is bent through a unit angle.

$f'_{CO,CO}$ is the so-called CO–CO interaction constant. When one C–O bond is stretched, the properties of the other are affected because of electronic movements; $f'$ measures this effect.

A slightly more satisfying interpretation is that force constants are the second derivatives of the potential energy with respect to the coordinates, taken at the equilibrium geometry. They represent the curvature of the energy profile when the molecule is distorted. Note that the first derivatives must be zero at the energy minimum.

The potential energy, $V$, may be written

$$2V = f_{CO}(\Delta r_1^2 + \Delta r_2^2) + 2f'_{CO,CO}\Delta r_1 \Delta r_2 + \delta_{OCO}\Delta\theta^2$$

$$= (\Delta r_1 \Delta r_2 \Delta\theta)\begin{bmatrix} f_{CO} & f'_{CO,CO} & 0 \\ f'_{CO,CO} & f_{CO} & 0 \\ 0 & 0 & \delta_{OCO} \end{bmatrix}\begin{bmatrix} \Delta r_1 \\ \Delta r_2 \\ \Delta\theta \end{bmatrix},$$

where $\Delta r$, $\Delta\theta$ represent respectively displacements of bond lengths and the bond angle. The assumption of only harmonic forces will not hold for large displacements.

The equation above may be put conveniently in the form

$$2V = \boldsymbol{r}^{\mathrm{T}}\mathbf{f}\boldsymbol{r},$$

[1] Editors' note: obtained from the symmetrised square of the character (see problem 1.2.7).

where f is the square matrix above. A simple orthogonal transformation of the coordinates $\boldsymbol{r}$ into symmetry coordinates $\boldsymbol{S}$, such that

$$
\begin{aligned}
S_1 &= 2^{-1/2}(\Delta r_1 + \Delta r_2) , && A_{1g} ; \\
S_2 &= 2^{-1/2}(\Delta r_1 - \Delta r_2) , && A_{2u} ; \\
S_3 &= \Delta\theta , && E_{1u} ;
\end{aligned}
$$

or

$$\boldsymbol{S} = \mathbf{U}\boldsymbol{r} ;$$

leads to the equation

$$2V = \boldsymbol{r}^{\mathrm{T}}\mathbf{f}\boldsymbol{r} = \boldsymbol{S}^{\mathrm{T}}\,\mathbf{U}\mathbf{f}\mathbf{U}^{\mathrm{T}}\boldsymbol{S} = \boldsymbol{S}^{\mathrm{T}}\mathbf{F}\boldsymbol{S} .$$

The matrix $\mathbf{F} = \mathbf{U}\mathbf{f}\mathbf{U}^{\mathrm{T}}$ is the symmetry-correct force-constant matrix and has the diagonal form

$$\mathbf{F} = \begin{bmatrix} f_{\mathrm{CO}} + f'_{\mathrm{CO,CO}} & 0 & 0 \\ 0 & f_{\mathrm{CO}} - f'_{\mathrm{CO,CO}} & 0 \\ 0 & 0 & \delta_{\mathrm{OCO}} \end{bmatrix}$$

The kinetic energy, $T$, of the system may be expressed in terms of a $\mathbf{G}$ matrix such that

$$2T = \dot{\boldsymbol{S}}^{\mathrm{T}}\mathbf{G}^{-1}\dot{\boldsymbol{S}} .$$

The $\mathbf{G}$ matrix is obtained very simply as there is only one coordinate in any symmetry species. If displacements of atoms in the axial direction are denoted by $z_i$ and those perpendicular by $x_i$, and so on, then

$$\Delta r_1 = z_2 - z_1 , \qquad \Delta r_2 = z_3 - z_2 ;$$

and

$$r\Delta\theta = x_1 - 2x_2 - x_3 , \qquad \text{for small } \Delta\theta .$$

Hence, we can write

$$
\begin{aligned}
S_1 &= 2^{-1/2}(z_3 - z_1) , \\
S_2 &= 2^{-1/2}(2z_2 - z_1 - z_3) , \\
S_3 &= (x_3 - x_2)/r .
\end{aligned}
$$

To find the matrix element $G_{ii}$, square the coefficient of each $z$ or $x$ in $S_i$, divide by the mass of that atom and sum over all $z_i$:

$$G_{11} = \tfrac{1}{2}\left(\frac{1}{m_3} + \frac{1}{m_1}\right) = \mu_{\mathrm{O}} ,$$

$$G_{22} = \tfrac{1}{2}\left(\frac{4}{m_2} + \frac{1}{m_1} + \frac{1}{m_3}\right) = 2\mu_{\mathrm{C}} + \mu_{\mathrm{O}} ,$$

$$G_{33} = \frac{2\mu_{\mathrm{O}} + 4\mu_{\mathrm{C}}}{r^2} .$$

The mode frequencies are related to the eigenvalues, $\lambda_i$, of the product matrix **GF**, which is already diagonalised:

$$4\pi^2c^2\bar{\nu}_1^2 = \lambda_1 = G_{11}F_{11} = (f_{\text{CO}} + f'_{\text{CO, CO}})\mu_{\text{O}} ,$$

$$4\pi^2c^2\bar{\nu}_3^2 = \lambda_2 = G_{22}F_{22} = (f_{\text{CO}} - f'_{\text{CO, CO}})(2\mu_{\text{C}} + \mu_{\text{O}}) ,$$

$$4\pi^2c^2\bar{\nu}_2^2 = \lambda_3 = G_{33}F_{33} = \frac{\delta_{\text{OCO}}(2\mu_{\text{O}} + 4\mu_{\text{C}})}{r^2} ,$$

Hence, we obtain

$$f_{\text{CO}} = \tfrac{1}{2}\left(\frac{\lambda_1}{\mu_{\text{O}}} + \frac{\lambda_2}{2\mu_{\text{C}} + \mu_{\text{O}}}\right) ,$$

$$f'_{\text{CO, CO}} = \tfrac{1}{2}\left(\frac{\lambda_1}{\mu_{\text{O}}} - \frac{\lambda_2}{2\mu_{\text{C}} + \mu_{\text{O}}}\right) ,$$

$$\frac{\delta_{\text{OCO}}}{r^2} = \frac{\lambda_3}{4\mu_{\text{C}} + 2\mu_{\text{O}}} .$$

From the assignments suggested earlier [i.e. $\bar{\nu}_1(A_{1g}) = 1334\ \text{cm}^{-1}$; $\bar{\nu}_2(E_{1u}) = 667\ \text{cm}^{-1}$; $\bar{\nu}_3(A_{2u}) = 2350\ \text{cm}^{-1}$] the force constants are:

$$f_{\text{CO}} = 15{\cdot}48\ \text{mdyn Å}^{-1}; \qquad f'_{\text{CO, CO}} = 1{\cdot}29\ \text{mdyn Å}^{-1};$$
$$\delta_{\text{OCO}}/r^2 = 0{\cdot}630\ \text{mdyn Å}^{-1} .$$

The high value of $f_{\text{CO}}$ is almost what one would expect of a triple bond. The positive sign of $f'_{\text{CO, CO}}$ suggests that elongation of a C–O bond encourages the other C–O bond to become stiffer. This is in keeping with the $\pi$-bonding model for this molecule, as the canonical forms (II) and (III) are enhanced during vibration:

$$\underset{\text{(I)}}{\text{O=C=O}} \quad \leftrightarrow \quad \underset{\text{(II)}}{{}^{-}\text{O—C≡O}^{+}} \quad \leftrightarrow \quad \underset{\text{(III)}}{{}^{+}\text{O≡C—O}^{-}} .$$

In the case of the bending mode, a knowledge of the equilibrium bond length is necessary for complete determination of the force constant. This quantity can be estimated from the spacing of the rotational components in the infrared spectrum.

The spacings between the rotational components in the stretching mode vary across the band because the moment of inertia in the excited state is greater than that in the ground state. However, in the $E_{1u}$ mode, the moment of inertia is almost the same as that of the ground state and the spacings are more nearly even. The spacing between these levels is given by the quantity $4h/8\pi^2cI$, where $I$ can be considered to be equal to the ground-state moment of inertia, and averaging over about thirty components we obtain a mean spacing of $1{\cdot}565\ \text{cm}^{-1}$. As $I = 2m_{\text{O}}r^2$, a value $r = 1{\cdot}160$ Å is obtained, giving $\delta$ a value of $0{\cdot}848$ mdyn Å, or $0{\cdot}0148$ mdyn Å deg$^{-1}$.

The molecule SCO is linear but noncentric and the C–S and C–O bonds have different properties. Thus a total of four force constants is needed. The number of fundamental frequencies, however, remains at three, the bending mode still being doubly degenerate. If the molecular geometry is known, the bending force constant can be calculated from the bending frequency. However, no unique solution is possible for the stretching force constants. In these circumstances isotopically substituted molecules like $^{32}S^{13}C^{16}O$ or $^{34}S^{12}C^{16}O$ must be studied. The assumption is made that the isotopic substitution does not affect the geometry or the force constants. With these extra data the calculation becomes possible.

**1.4.3**

Show that the selection rule for rotational Raman scattering is

$$\Delta J = 0, \pm 1, \pm 2 \, .$$

**Solution**

In the Raman process, the incoming wave polarises the charge distribution of the molecule; the perturbation is represented by the normal Hamiltonian for the interaction of the electric vector of the electromagnetic wave with the charge distribution,

$$\mathcal{H}_{\mathrm{R}} = \Sigma q_x \nabla_x \, ,$$

where $q_x$ is proportional to an induced charge.

This Hamiltonian $\mathcal{H}_{\mathrm{R}}$ transforms as a vector. The probability of a transition from state $|i\rangle$ to state $|f\rangle$ of the molecule is proportional to a sum over products of matrix elements,

$$\Sigma \langle i|\mathcal{H}_{\mathrm{R}}|j\rangle\langle j|\mathcal{H}_{\mathrm{R}}|f\rangle \, .$$

The intermediate state is 'virtual', i.e. the sum runs over all $j$, with no one eigenstate being associated with an intermediate state of the molecule during the emission of the new photon. Thus, since all states are present in the sum

$$\sum_{\text{all } j} \ldots |j\rangle\langle j| \ldots \, ,$$

it transforms as $\mathcal{D}^{(0)}$ of the full rotation group, i.e. the fully symmetric representation. Hence the transition-probability sum transforms as

$$\mathcal{D}^{\Gamma} = \mathcal{D}^{i} \otimes \mathcal{D}^{(1)} \otimes \mathcal{D}^{(1)} \otimes \mathcal{D}^{f} \, ;$$

here $\mathcal{D}^{i}$ is the representation of $|i\rangle$, $\mathcal{D}^{f}$ that of $|f\rangle$, $i$ and $f$ being the rotational quantum numbers; $\mathcal{D}^{(1)}$ is the representation of $\mathcal{H}_{\mathrm{R}}$, since it is vectorial in form. Now

$$\mathcal{D}^{(1)} \otimes \mathcal{D}^{(1)} = \mathcal{D}^{(0)} \oplus \mathcal{D}^{(1)} \oplus \mathcal{D}^{(2)} \, .$$

Thus

$$\mathcal{D}^i \otimes \mathcal{D}^{(1)} \otimes \mathcal{D}^{(1)} = \mathcal{D}^i \otimes (\mathcal{D}^{(0)} \oplus \mathcal{D}^{(1)} \oplus \mathcal{D}^{(2)})$$
$$= 3\mathcal{D}^{(i)} \oplus 2(\mathcal{D}^{(i-1)} \oplus \mathcal{D}^{(i+1)}) \oplus \mathcal{D}^{(i+2)} \oplus \mathcal{D}^{(i-2)} .$$

For the transition from state $|i\rangle$ to $|f\rangle$ to be allowed, the representation $\mathcal{D}^P$ must contain the fully symmetric representation $\mathcal{D}^{(0)}$. But $\mathcal{D}^{(l)} \otimes \mathcal{D}^{(m)}$ cannot contain $\mathcal{D}^{(0)}$ unless $l = m$. Now the product of the first three terms in $\mathcal{D}^P$ has been decomposed to give $\mathcal{D}^{(i)}$, $\mathcal{D}^{(i+1)}$, $\mathcal{D}^{(i-1)}$, $\mathcal{D}^{(i+2)}$, $\mathcal{D}^{(i-2)}$. Thus $f$ must be $i$, $i \pm 1$, or $i \pm 2$ for an allowed transition. Since the rotational state of quantum number $J$ transforms as $\mathcal{D}^{(J)}$, as $f$ and $i$ are really the rotational quantum numbers as asserted before, we have the required selection rule $\Delta J = 0, \pm 1, \pm 2$.

**1.4.4**

Discuss the vib-rotational selection rules for $NH_3$, noting that this molecule can tunnel into an inverted conformation.

**Solution**

The symmetry of $NH_3$ is classified by means of feasible permutations of nuclear coordinates (cf. problem 1.2.10). Feasible includes, here, the tunnelling inversion. The operations of the symmetry group are

$$E;\ (123), (132);\ (12), (23), (31);\ E^*;\ (123)^*, (132)^*;\ (12)^*, (23)^*, (31)^*.$$

This group is isomorphous with $D_{3h}$ and its classes have been separated by semicolons, to correspond to the $D_{3h}$ classes

$$E;\ 2C_3;\ 3C_2;\ \sigma_h;\ 2S_3;\ 3\sigma_v .$$

The dipole moment operator $\mu_z$ transforms according to a representation $\Gamma^*$ of character $-1$ for operations involving inversion, but $+1$ otherwise, since permutation of like particles does not affect $\mu$. By inspection of the character table for $D_{3h}$ (Appendix 2), we see the required representation is $A_1''$. Further, there is the selection rule for rotational wave functions $\Delta J = 0, \pm 1$ obtained by use of the three-dimensional rotation group. The dipole moment operator along the $z$ axis, $\mu_z$, transforms according to $A_2''$ and $\mu_x$, $\mu_y$ as $E'$. For vibrational transitions to be allowed from a state $v'$ to $v''$, the product of their representations $\Gamma_{v'} \otimes \Gamma_{v''}$ must contain $A_2''$ or $E'$. Then for a vibrational transition involving $\mu_z$ the rotational transition will be connected by an operator transforming as $A_2'$, since $A_2'' \otimes A_2' = A_1''$, which is $\Gamma^*$. The resultant allowed transitions are given by Bunker (1975, p.60) who discusses breakdown of the selection rules and statistical weights of the energy levels of this molecule.

**1.4.5**

Use symmetry arguments to examine the splittings of the lowest states of the transition-metal ions $Cu^{2+}$, $V^{3+}$, $Cr^{3+}$, and $Mn^{2+}$, when they are in octahedral ($O_h$) or tetrahedral ($T_d$) symmetry. Consider the surroundings to provide a weak field, except in the case of $V^{3+}$. On symmetry grounds, what transitions might be observed in the visible or near-visible regions of the spectrum of $Cu^{2+}$ and $Cr^{3+}$ in systems expected at first sight to be of $O_h$ symmetry? What effect will a small trigonal electric field have in the case of $V^{3+}$ in $O_h$ symmetry?

**Solution**

The $Cu^{2+}$ ion has a $^2D$ ground state in spherical symmetry, and no other state near it, i.e. the next state is some tens of thousands of wavenumbers higher in energy. In octahedral symmetry this state splits into $^2T_{2g}$ and $^2E_g$ states, as can be seen by examining the characters of the d wave functions (cf. problem 1.2.7). Symmetry arguments do not give the order of these states. Tetrahedral symmetry gives identical splittings, except that the designation g is dropped, since the inversion operation does not occur in the group $T_d$. By such character methods (applied to corresponding atomic wave functions) it is readily seen that the $^3F$ ground state of $V^{3+}$ splits in $O_h$ to $^3T_1$, $^3T_2$, and $^3A_2$; and the next state, $^3P$, becomes $^3T_1$. The $^4F$ state of $Cr^{3+}$ splits into $^4A_2$, $^4T_2$, and $^4T_1$; and the next state, $^4P$, becomes $^4T_1$. All these have subscript g in $O_h$. The $^6S$ state of $Mn^{2+}$ is not split.

Apparently these results assume that the crystal field is weak compared to the electron–electron repulsion. We can see this is not so, by examining the strong-field states of $V^{3+}$ $(3d^2)$. The d orbitals split under a strong field into a $t_{2g}$ and $e_g$ set, so that three configurations are possible, namely $t_{2g}^2$, $t_{2g}e_g$, and $e_g^2$. By use of the character method (cf. problem 1.2.7), we see that the $t_{2g}^2$ configuration corresponds to a $T_{1g}$ state, the $t_{2g}e_g$ to $T_{2g}$ and $T_{1g}$ states, and $e_g^2$ to an $A_{2g}$ state. The order of these levels will depend on the relative magnitudes of crystal-field and electron–electron repulsion. Tetrahedral surroundings give rise to similar splittings, but with reversed ordering. Electric dipole transitions are, strictly speaking, forbidden in octahedral symmetry, since under inversion the d orbitals have even parity, so must electronic states derived from them, whereas the electric dipole has odd parity. Some of the vibrations of the ligands remove inversion symmetry and these give rise to weak transitions. One way of interpreting this effect is to suppose that an asymmetric vibration along a coordinate $q$ changes the crystal field $V$ by a small amount, expressed as a Taylor series:

$$V = V_0 + q\left(\frac{\partial V}{\partial q}\right)_0 + \ldots .$$

Now, since $V$ transforms as the totally symmetric representation, and $q$ as an odd (u, under inversion) representation, $(\partial V/\partial q)_0$ must also transform as an odd representation. Thus, by first-order perturbation theory

$$\psi = \psi_g - \frac{q\int \psi_g (\partial V/\partial q)_0 \psi_u \, d\tau}{E_u - E_g} \psi_u \, ,$$

that is, the asymmetric vibration mixes an odd wave function into the nominally even ground state. Now the transition moment $R$ has to include a vibrational part:

$$R = \langle A, n \,|\boldsymbol{r}|\, B, m \rangle \, ,$$

where $A$ and $B$ specify the electronic parts of the wave function, and $n$ and $m$ are vibrational states. The vibrational part separates out, being $\langle n\,|q|\,m\rangle$, which implies that $n = m \pm 1$, for an allowed transition (of a vibrational transition moment). Thus on symmetry grounds we expect the $1 \leftarrow 0$ band of the allowed electronic transitions to be seen at low temperature. The electronic transition is allowed when product of the representations of initial and final states includes that of $\boldsymbol{r}$, i.e. $T_1$ in $O$, $T_2$ in $T_d$, $E$ or $A$ in $D_4$. For $Cu^{2+}$, the ground state is $E_g$, but by the result of Jahn and Teller, a distortion will occur as this state is degenerate. Incidentally, the Jahn–Teller 'theorem' was deduced from symmetry considerations. By the character method the $^2E_g$ state splits in $D_{4h}$ symmetry into a $^2B_{1g}$ and a $^2A_{1g}$ state. The excited $^2T_{2g}$ state splits into $^2E_g$ and $^2B_{2g}$ states. By the character method, or by using representation product tables (see, e.g. Griffith, 1960, table A9), we find, for example,

$$B_{1g} \otimes B_{2g} = A_{2g} \text{ and } B_1 \otimes E = E \, .$$

The total electronic part involves the electric dipole operator $\boldsymbol{r}$ which transforms as $(A_{2u} + E_u)$; thus the total electronic part of the transition moment for the transition $B_2 \leftarrow B_1$ transforms as

$$(A_{2u} \oplus E_u) \otimes A_{2g} = A_{1u} \oplus E_u \, .$$

Thus, if the vibrational part contains these representations the total transition moment is symmetric and therefore not necessarily zero. In fact the ground-state vibrational wave function is represented $A_{1g}$, and the first excited states belong to the representations $2A_{1g}$, $B_{1g}$, $B_{2g}$, $E_g$, $2A_{2u}$, $B_{1u}$, and $3E_u$ (the method of deducing these is given in problem 1.4.1). Since $z$ transforms as $A_{2u}$, which is not contained in this list, when the electric vector of the light is aligned along the $z$ axis $0z$ a transition is induced; similarly for the electric vectors in the $xy$ plane the presence of $E_u$ implies an allowed, but weak, transition.

Assuming such vibronic coupling, then for $Cr^{3+}$ we expect transitions from the $^4A_{2g}$ ground state to $^4T_{2g}$, $^4T_{1g}(F)$, and $^4T_{1g}(P)$ excited states.

In the case of $V^{3+}$ a trigonal field implies descent in symmetry from $O_h$ to $D_{3h}$; then by the character method $^3T_{1g}$ splits to $^3E$ and $^3A_2$. Time-reversal symmetry (Kramer's theorem) does not apply, as there is an even number of electrons ($3d^2$). In fact the splitting which therefore might occur in zero magnetic field, the zero-field splitting, is often found. Such a splitting arises because of the presence both of spin–orbit coupling and of the trigonal crystal field, but consideration of this takes us beyond mere symmetry considerations.

**1.4.6**

Consider the effect of a symmetry operation which interchanges two identical molecules in a crystal on their combined wave functions in the ground and in excited states. Show that electronic absorption bands of the isolated molecules will be split.

**Solution**

If the ground-state wave functions of each molecule are $\phi_1$ and $\phi_2$ the combined function, $\psi$, can be written $\psi = A\phi_1\phi_2$, $A$ being the anti-symmetrising operator. The Hamiltonian may be written

$$\mathcal{H} = \mathcal{H}_1 + \mathcal{H}_2 + V_{12} ,$$

$\mathcal{H}_1$, $\mathcal{H}_2$ referring to the isolated molecules, and $V_{12}$ being an interaction potential. To first order in perturbation theory, the ground-state energy $E_G$ is given by

$$E_G = E_1 + E_2 + \langle \phi_1\phi_2 | V_{12} | \phi_1\phi_2 \rangle .$$

The excited state has two possibilities:

$$\psi' = \phi_1'\phi_2 \quad \text{or} \quad \phi_1\phi_2' ,$$

where $\phi'$ is the excited-state wave function for the free molecule. These $\psi'$ have energies given by solution of the determinantal secular equation

$$\begin{vmatrix} E_1' + E_2 + \langle \phi_1'\phi_2 | V_{12} | \phi_1'\phi_2 \rangle - E & L \\ L & E_1 + E_2' + \langle \phi_1\phi_2' | V_{12} | \phi_1\phi_2' \rangle - E \end{vmatrix} = 0 ,$$

where the excitation exchange integral, $L$, is equal to $\langle \phi_1'\phi_2 | V_{12} | \phi_1\phi_2' \rangle$. If a symmetry operation relates the pair of molecules in the crystal, then the secular equation factorises (cf. problem 1.2.4), to give two energies, and two possible transition energies,

$$\Delta E_{\pm} - E' + D \pm L .$$

The free-molecule excitation energy, $\Delta E_0$, is simply $E' - E$. To first order,

$$D = \langle \phi_1'\phi_2 | V_{12} | \phi_1\phi_2' \rangle - \langle \phi_1\phi_2 | V_{12} | \phi_1\phi_2 \rangle .$$

The splitting, $2L$, is closely related to Davydov splitting, for which a 'full crystal' treatment is needed (Fischer, 1974).

### 1.4.7

Show that the transitions in nuclear magnetic resonance between the symmetric and antisymmetric states of the (spin $\frac{1}{2}$ only) $AB_2$ high-resolution system are forbidden (see Harris and Lynden-Bell, 1969).

**Solution**

One example is chosen here, the transition between the symmetric state $2^{-\frac{1}{2}}\alpha(\alpha\beta+\beta\alpha)$ and the antisymmetric state $2^{-\frac{1}{2}}\beta(\alpha\beta-\beta\alpha)$.

Experimentally, a radio-frequency field $B_1 \cos\omega t$ is applied perpendicular to the static field $B_0$, say in the $x$ direction. This adds a small time-dependent term $\mathcal{H}_1$ to the Hamiltonian,

$$\mathcal{H}_1(t) = (2\pi)^{-1}\sum_i \gamma B_1 I_{xi} \cos\omega t \,,$$

where $I_{xi}$ are the $x$ components of the individual nuclear spin operators, $\boldsymbol{I}_i$, so that the probability per unit time of the transition taking place is proportional to $|\langle n|\mathcal{H}_1|m\rangle|$, where $|n\rangle$ and $|m\rangle$ are the spin states involved, here identified as $2^{-\frac{1}{2}}\alpha(\alpha\beta+\beta\alpha)$ and $2^{-\frac{1}{2}}\beta(\alpha\beta-\beta\alpha)$, respectively. Expanding the transition-moment integral, we obtain

$$\begin{aligned}\langle n|\mathcal{H}_1|m\rangle &= \tfrac{1}{2}\langle\alpha(\alpha\beta+\beta\alpha)|\mathcal{H}_1|\beta(\alpha\beta-\beta\alpha)\rangle \\ &= \tfrac{1}{2}\gamma(2\pi)^{-1}B_1\cos\omega t\,\langle\alpha(\alpha\beta+\beta\alpha)|\sum_i I_{xi}|\beta(\alpha\beta-\beta\alpha)\rangle\,.\end{aligned}$$

The action of $I_x$ in this context can be derived from the step-up and step-down operators, $I_+$ and $I_-$ respectively (Harris and Lynden-Bell, 1969, p.54). Since

$$I_+|\alpha\rangle = 0\,, \qquad I_+|\beta\rangle = |\alpha\rangle\,,$$

and

$$I_-|\alpha\rangle = |\beta\rangle\,, \qquad I_-|\beta\rangle = 0\,,$$

then, because

$$I_+ = I_x + \mathrm{i}I_y\,, \qquad I_- = I_x - \mathrm{i}I_y \qquad (\mathrm{i}^2+1 = 0)\,,$$

we can write

$$I_x|\alpha\rangle = \tfrac{1}{2}|\beta\rangle \quad \text{and} \quad I_x|\beta\rangle = \tfrac{1}{2}|\alpha\rangle\,.$$

So,

$$\langle\alpha(\alpha\beta+\beta\alpha)|\sum_i I_{xi}|\beta(\alpha\beta-\beta\alpha)\rangle$$

becomes

$$\begin{aligned}&\langle\alpha\alpha\beta|I_{x1}+I_{x2}+I_{x3}|\beta\alpha\beta\rangle - \langle\alpha\alpha\beta|I_{x1}+I_{x2}+I_{x3}|\beta\beta\alpha\rangle \\ &\quad + \langle\alpha\beta\alpha|I_{x1}+I_{x2}+I_{x3}|\beta\alpha\beta\rangle - \langle\alpha\beta\alpha|I_{x1}+I_{x2}+I_{x3}|\beta\beta\alpha\rangle\,,\end{aligned}$$

with the nuclei labelled 1, 2, and 3 for convenience, with the identities $1 \equiv A$, $2 \equiv B$, $3 \equiv B$. Expanding again, we obtain

$$
\begin{aligned}
\langle\alpha\alpha\beta|I_{x1}|\beta\alpha\beta\rangle &= \tfrac{1}{2}\langle\alpha\alpha\beta|\alpha\alpha\beta\rangle = \tfrac{1}{2}; & \langle\alpha\alpha\beta|I_{x2}|\beta\alpha\beta\rangle &= \tfrac{1}{2}\langle\alpha\alpha\beta|\beta\beta\beta\rangle = 0;\\
\langle\alpha\alpha\beta|I_{x3}|\beta\alpha\beta\rangle &= \tfrac{1}{2}\langle\alpha\alpha\beta|\beta\alpha\alpha\rangle = 0; & -\langle\alpha\alpha\beta|I_{x1}|\beta\beta\alpha\rangle &= -\tfrac{1}{2}\langle\alpha\alpha\beta|\alpha\beta\alpha\rangle = 0;\\
-\langle\alpha\alpha\beta|I_{x2}|\beta\beta\alpha\rangle &= -\tfrac{1}{2}\langle\alpha\alpha\beta|\beta\alpha\beta\rangle = 0; & -\langle\alpha\alpha\beta|I_{x3}|\beta\beta\alpha\rangle &= -\tfrac{1}{2}\langle\alpha\alpha\beta|\beta\beta\beta\rangle = 0;\\
+\langle\alpha\beta\alpha|I_{x1}|\beta\alpha\beta\rangle &= \tfrac{1}{2}\langle\alpha\beta\alpha|\alpha\alpha\beta\rangle = 0; & +\langle\alpha\beta\alpha|I_{x2}|\beta\alpha\beta\rangle &= \tfrac{1}{2}\langle\alpha\beta\alpha|\beta\beta\beta\rangle = 0;\\
+\langle\alpha\beta\alpha|I_{x3}|\beta\alpha\beta\rangle &= \tfrac{1}{2}\langle\alpha\beta\alpha|\beta\alpha\alpha\rangle = 0; & -\langle\alpha\beta\alpha|I_{x1}|\beta\beta\alpha\rangle &= -\tfrac{1}{2}\langle\alpha\beta\alpha|\alpha\beta\alpha\rangle = -\tfrac{1}{2};\\
-\langle\alpha\beta\alpha|I_{x2}|\beta\beta\alpha\rangle &= -\tfrac{1}{2}\langle\alpha\beta\alpha|\beta\alpha\alpha\rangle = 0; & -\langle\alpha\beta\alpha|I_{x3}|\beta\beta\alpha\rangle &= -\tfrac{1}{2}\langle\alpha\beta\alpha|\beta\beta\beta\rangle = 0.
\end{aligned}
$$

The net result is $(\tfrac{1}{2}-\tfrac{1}{2}) = 0$.

The same result can be derived for all symmetric–antisymmetric transitions.

A more general solution makes use of the equivalence of symmetry operators and total spin operators. Use of orthogonal sets of wave functions, each set characterised by a given value of total spin $I_T$, enables the secular determinant to be factorised.

**1.4.8**

Which symmetry groups are isomorphic with the n.m.r. permutation group in each of the following molecules (I)–(VI), where the molecule is in (a) an isotropic liquid or solution phase, and (b) an anisotropic phase (solution in liquid crystal)?
(See Jones, 1969.)

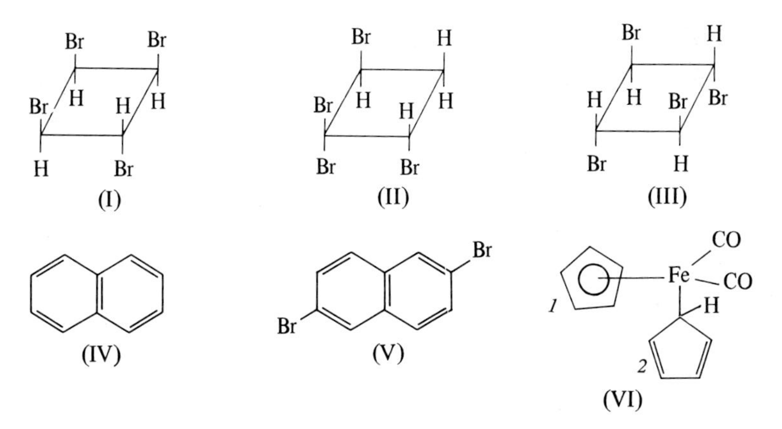


**Solution**

The n.m.r. permutation group is most satisfactorily defined as containing all the nuclear permutations which do not alter the nuclear parameters (Woodman, 1966).

*Molecule (I).* (a) and (b), $C_s$ $(E, \sigma)$. The plane of symmetry includes the two bromine atoms *trans*, diagonally, across the ring; the spin system is $AB_2C$ [or $AM_2X$ for (a) depending on the magnetic field].

*Molecule (II).* (a) and (b), $C_2$ ($E$, $C_2$). The twofold axis passes through the secondary and quaternary carbon atoms; the system can be described as $[AB]_2$ [or $[AX]_2$ for (a) at high field].

*Molecule (III).* (a) The four protons are chemically equivalent, and because there is no coupling to another group the symmetry group is isomorphic with a permutation group $P_4$ [$E$, (12), (13), (14), (23), (24), (34), (123), (132), (124), (142), (234), (243), (134), (143), (1234), (1324), (1423), (1432), (12)(34), (13)(24), (14)(23)], although the apparent symmetry is $D_{2d}$. This example emphasises the importance of the n.m.r. Hamiltonian in determining the permutation group.
(b) For the isolated group $\mathcal{H} = \sum_i \nu_i I_{zi}$, where $\nu_i$ are the chemical shifts, suffices since the spin–spin coupling is not observed in the isotropic phase spectrum. The problem is made more complicated by inclusion of the dipole–dipole coupling terms when the molecule is dissolved in an anisotropic solvent. Now the Hamiltonian reads (Diehl and Khetrapel, 1969; Jones, 1969):

$$H = \Sigma \nu_i I_{zi} + \sum_{i<j} J_{ij} \boldsymbol{I}_i \cdot \boldsymbol{I}_j + \sum_{i<j} D_{ij} [I_{zi} I_{zj} - 0{\cdot}5(I_{xi} I_{xj} + I_{yi} I_{yj})] ,$$

and the appropriate permutation group is isomorphic with $D_{2d}$ ($E$, $2S_4$, $C_2$, $2C_2'$, $2\sigma_d$).

*Molecule (IV).* (a) The n.m.r. permutation group again depends on the interactions which are considered important. One approximation is to neglect scalar coupling between the rings when the n.m.r. permutation group is isomorphic with $C_2$ or $C_s$ ($[AB]_2$). $D_{2h}$ is appropriate when the inter-ring coupling is included ($[AB]_4$).
(b) The inter-ring dipole–dipole coupling must be included when the molecule is partially oriented by interaction with a liquid-crystal solvent in a magnetic field. The appropriate symmetry group is then $D_{2h}$.

*Molecule (V).* (a) The omission of inter-ring scalar coupling leads to a trivial point group $C_1$ ($E$ only) when the system is labelled ABC or AMX, depending on the magnetic field. The appropriate group when inter-ring coupling is included is isomorphic with $C_{2h}$; this is also true in case (b), for which the notation reads $[ABC]_2$.

*Molecule (VI).* (a) In isotropic media the two ring systems can be considered separately since the dipole-dipole coupling is averaged to zero over the rapid motions. At high temperatures the spectrum consists of two sharp lines (Wilkinson and Piper, 1956), so that each cyclopentadienyl ring undergoes a rapid motion ($\pi$-bonded ring (*1*) 'rotates'; $\sigma$-bonded ring (*2*) jumps rapidly from one binding site to another); and the $P_5$ permutation group is appropriate in each case.

At 170 K the line at low field appears as a multiplet, while the high-field line is unchanged (Wilkinson and Piper, 1956). The low-field

multiplet is assigned to the $\sigma$-bonded cyclopentadienyl ring where the system is described by $[AB]_2X$ and the appropriate symmetry group is $C_2$. The $\pi$-bonded cyclopentadienyl ring is considered to rotate very quickly, even at the lower temperature, so the $P_5$ group is still valid.
(b) At the higher temperature the notation for the system will depend on the rate of 'rotation' of the cyclopentadienyl rings compared with the values of the $D_{ij}$ coupling constants between the protons in the rings. The resultant spin systems are $[AB]_2C$ ($D_5$) for slow rotation of ring *1* and fast rotation of ring *2*, and $A_5B_5$ for fast rotation of ring *1* and ring *2*.

**References**

Bunker, P. R., 1975, *Vib. Spectra Struct.*, **3**, 1.
Diehl, P., Khetrapel, C. L., 1969, in *Nuclear Magnetic Resonance, Basic Principles and Progress,* Ed. P. Diehl (Springer, Berlin), p.1.
Fischer, G., 1974, in *Transfer and Storage of Energy by Molecules,* Eds G. M. Burnett, A. M. North, J. N. Sherwood, Volume 4 (John Wiley, Chichester, Sussex), p.16.
Griffith, J. S., 1960, *The Theory of Transition-metal Ions* (Cambridge University Press, Cambridge).
Harris, R. K., Lynden-Bell, R. M., 1969, *Nuclear Magnetic Resonance Spectroscopy* (Nelson, London).
Jones, R. G., 1969, in *Nuclear Magnetic Resonance, Basic Principles and Progress,* Ed. P. Diehl (Springer, Berlin), p.97.
Wilkinson, G., Piper, J. S., 1956, *J. Inorg. Nucl. Chem.*, **3**, 104.
Woodman, C. M., 1966, *Mol. Phys.*, **11**, 109.

# 2 Diffraction

## 2.1 x-Ray diffraction

**G J Bullen** University of Essex

### 2.1.1

The first ten lines in the x-ray diffraction pattern of a crystalline powder occur at the following Bragg angles $\theta$ (°): 8·9, 12·7, 15·6, 18·1, 20·3, 22·4, 24·3, 26·1, 27·8, and 29·4. The measurements were made to the nearest 0·1° and the wavelength of the x-radiation was 1·542 Å [1].

Show that the crystal is cubic, deduce its Bravais lattice type, and find the lattice parameter $a$. Suggest how the accuracy of the parameter could be improved by making other measurements.

**Solution**

For a cubic crystal, the interplanar spacing $d$ is related to the lattice constant $a$ and the Miller indices $(hkl)$ of the crystal plane by the equation

$$\frac{1}{d^2} = \frac{h^2+k^2+l^2}{a^2} . \tag{2.1.1a}$$

$d$ is also related to the Bragg angle $\theta$ by the Bragg equation

$$\lambda = 2d\sin\theta , \tag{2.1.1b}$$

where $\lambda$ is the wavelength of the x-radiation. Combining equations (2.1.1a) and (2.1.1b), we obtain

$$\sin^2\theta = \left(\frac{\lambda}{2a}\right)^2 (h^2+k^2+l^2) \tag{2.1.1c}$$

$$= \text{constant} \times (h^2+k^2+l^2) .$$

Thus for the sequence of lines in the x-ray pattern the values of $\sin^2\theta$ should increase in proportion to $(h^2+k^2+l^2)$, which is integral since $h$, $k$, and $l$ are integers.

Provided there are no systematic absences of x-ray reflections arising from the lattice type (i.e. the lattice is primitive), the sum $(h^2+k^2+l^2)$ will take the values 1, 2, 3, 4, 5, 6, 8, 9, 10, 11 for the first ten lines and hence $\sin^2\theta$ will increase in these proportions. (There is no value 7 because there is no combination of $h$, $k$, and $l$ which will produce it.) If the lattice is either body-centred ($I$) or face-centred ($F$), certain combinations of $h$, $k$, and $l$ are forbidden (systematic absences) and the sequence will be different (table 2.1.1a).

The $\sin^2\theta$ values for the given pattern are listed in table 2.1.1b. The sequence is revealed by dividing each value by $\sin^2\theta$ for line 1. It shows the Bravais lattice to be body-centred since the seventh line has a $\sin^2\theta$ which is seven times that of the first line. Once the Miller indices have

[1] 1 Å = $10^{-10}$ m = $10^{-1}$ nm = $10^2$ pm.

been assigned to each line the lattice parameter can be calculated from equation (2.1.1c).

The accuracy of the lattice parameter can be improved by measuring lines with $\theta$ near 90° (high-order lines). The error in $a$ depends on the error in $\sin\theta$ which becomes small as $\theta$ approaches 90°, as

$$\left[\frac{d(\sin\theta)}{d\theta}\right]_{\theta \to 90^\circ} = 0\,.$$

**Table 2.1.1a.**

| Lattice type | Restriction | Sequence of $(h^2+k^2+l^2)$ | | | | | | | | | |
|---|---|---|---|---|---|---|---|---|---|---|---|
| $P$ | None | 1 | 2 | 3 | 4 | 5 | 6 | 8 | 9 | 10 | 11 |
| $I$ | $(h+k+l)$ even | 2 | 4 | 6 | 8 | 10 | 12 | 14 | 16 | 18 | 20 |
| $F$ | $h, k, l$ all even or all odd | 3 | 4 | 8 | 11 | 12 | 16 | 19 | 20 | 24 | 27 |

**Table 2.1.1b.**

| Line | $\theta$ (°) | $\sin^2\theta$ | $N$[a] | $h^2+k^2+l^2$ | $hkl$ | $a$ (Å) |
|---|---|---|---|---|---|---|
| 1 | 8·9 | 0·0239 | 1 | 2 | 110 | 7·048[b] |
| 2 | 12·7 | 0·0483 | 2·02 | 4 | 200 | 7·015 |
| 3 | 15·6 | 0·0723 | 3·03 | 6 | 211 | 7·023 |
| 4 | 18·1 | 0·0965 | 4·04 | 8 | 220 | 7·019 |
| 5 | 20·3 | 0·1203 | 5·03 | 10 | 310 | 7·028 |
| 6 | 22·4 | 0·1452 | 6·08 | 12 | 222 | 7·008 |
| 7 | 24·3 | 0·1693 | 7·08 | 14 | 321 | 7·011 |
| 8 | 26·1 | 0·1935 | 8·10 | 16 | 400 | 7·011 |
| 9 | 27·8 | 0·2175 | 9·10 | 18 | 411 | 7·013 |
| 10 | 29·4 | 0·2410 | 10·08 | 20 | 420 | 7·024 |
| | | | | | Mean | 7·017 |
| | | | | | E.s.d. | 0·007 |

[a] $N = \sin^2\theta/0{\cdot}0239$. [b] Neglected in calculating mean value for $a$.

## 2.1.2

Determine the unit-cell dimensions of an orthorhombic crystal from the following.

(a) An x-ray rotation photograph (recorded on a cylindrical film) of a single crystal, rotated about the $c$ axis, shows layer lines up to $l = 7$. The distances of the layer lines from the zero layer ($y_l$) are.

| Layer index $l$ | 1 | 2 | 3 | 4 | 5 | 6 | 7 |
|---|---|---|---|---|---|---|---|
| $y_l$ (mm) | 3·45 | 7·0 | 11·0 | 15·5 | 21·1 | 29·1 | 42·9 . |

The x-ray wavelength, $\lambda$, is 1·542 Å, and the camera radius 28·78 mm. Calculate the length of the $c$ axis of the crystal.

(b) The Bragg angles, $\theta$, of a number of high-order $hk0$ reflections, recorded with the use of Cu $K\alpha_1$ radiation ($\lambda = 1\cdot5405$ Å), are given in table 2.1.2a. The lattice parameters $a$ and $b$ have previously been found approximately as $a = 17\cdot9$, $b = 16\cdot5$ Å. Use the Bragg angles to determine $a$ and $b$ more accurately.

**Table 2.1.2a.**

| $h\ k\ l$ | $\theta$ (°) | $h\ k\ l$ | $\theta$ (°) |
|---|---|---|---|
| 20 1 0 | 59·73 | 0 20 0 | 68·59 |
| 20 2 0 | 60·18 | 4 19 0 | 64·31 |
| 20 4 0 | 62·02 | 8 19 0 | 71·73 |
| 20 7 0 | 67·33 | 10 19 0 | 79·64 |
| 22 2 0 | 72·47 | 10 18 0 | 70·53 |
| 22 4 0 | 75·23 | 12 18 0 | 80·04 |
| 22 5 0 | 77·88 | | |

**Solution**

(a) The $c$ axis length is given by the Laue condition

$$l\lambda = c \sin\phi_l \,, \tag{2.1.2a}$$

where the semiangle of the cone of rays making up the $l$th layer line is $(90° - \phi_l)$ and the x-ray wavelength is $\lambda$. $\phi_l$ is found from

$$\tan\phi_l = \frac{y_l}{r} \,, \tag{2.1.2b}$$

where $r$ is the camera radius. The evaluation of $c$ is tabulated in table 2.1.2b.

(b) The interplanar spacings $d$ are first calculated from the Bragg angles (see problem 2.1.1). For an orthorhombic crystal $d$ is related to the unit-cell parameters by

$$\frac{1}{d^2} = \frac{h^2}{a^2} + \frac{k^2}{b^2} + \frac{l^2}{c^2} \,, \tag{2.1.2c}$$

**Table 2.1.2b.**

| $l$ | $y_l$ (mm) | $\tan\phi_l$ | $\sin\phi_l$ | $c$ (Å) |
|---|---|---|---|---|
| 1 | 3·45 | 0·1199 | 0·1191 | 12·95 |
| 2 | 7·0 | 0·2432 | 0·2363 | 13·05 |
| 3 | 11·0 | 0·3822 | 0·3571 | 12·95 |
| 4 | 15·5 | 0·5386 | 0·4742 | 13·01 |
| 5 | 21·1 | 0·7331 | 0·5913 | 13·04 |
| 6 | 29·1 | 1·011 | 0·7110 | 13·01 |
| 7 | 42·9 | 1·491 | 0·8304 | 13·00 |
| | | | Mean | 13·00 |
| | | | E.s.d. | 0·04 |

which reduces to

$$\frac{1}{d^2} = \frac{h^2}{a^2} + \frac{k^2}{b^2} \tag{2.1.2d}$$

for $hk0$ reflections. We thus have a set of second-order equations (2.1.2d) in the two unknowns $a$ and $b$ which can be solved by an iterative procedure starting from approximate values of $a$ and $b$. The reflections in table 2.1.2a have been chosen so that they have a high $h$ value and low $k$ or vice versa. If we rearrange equation (2.1.2d) to give

$$a^2 = d^2\left[h^2 + k^2\left(\frac{a}{b}\right)^2\right]$$

or (2.1.2e)

$$a = d\left[h^2 + k^2\left(\frac{a}{b}\right)^2\right]^{1/2},$$

**Table 2.1.2c.**

| $h\ k\ l$ | $\theta$ (°) | $d$ (Å) | $a$ (Å) calculated with | |
|---|---|---|---|---|
| | | | $\left(\frac{a}{b}\right)^2 = 1\cdot177$[a] | $\left(\frac{a}{b}\right)^2 = 1\cdot167$[b] |
| Calculation of $a$ | | | | |
| 20 1 0 | 59·73 | 0·89184 | 17·863 | 17·863 |
| 20 2 0 | 60·18 | 0·88779 | 17·860 | 17·859 |
| 20 4 0 | 62·02 | 0·87220 | 17·850 | 17·846 |
| 20 7 0 | 67·33 | 0·83474 | 17·858 | 17·848 |
| 22 2 0 | 72·47 | 0·80776 | 17·857 | 17·856 |
| 22 4 0 | 75·23 | 0·79657 | 17·862 | 17·859 |
| 22 5 0 | 77·88 | 0·78781 | 17·851 | 17·847 |
| | | Mean | 17·86 | 17·854 |
| | | | E.s.d. | 0·007 |

| $h\ k\ l$ | $\theta$ (°) | $d$ (Å) | $b$ (Å) calculated with | | |
|---|---|---|---|---|---|
| | | | $\left(\frac{b}{a}\right)^2 = 0\cdot850$[c] | $\left(\frac{b}{a}\right)^2 = 0\cdot8566$[d] | $\left(\frac{b}{a}\right)^2 = 0\cdot8585$[e] |
| Calculation of $b$ | | | | | |
| 0 20 0 | 68·59 | 0·82734 | 16·55 | 16·547 | 16·547 |
| 4 19 0 | 64·31 | 0·85473 | 16·54 | 16·545 | 16·546 |
| 8 19 0 | 71·73 | 0·81114 | 16·53 | 16·541 | 16·543 |
| 10 19 0 | 79·64 | 0·78301 | 16·54 | 16·548 | 16·552 |
| 10 18 0 | 70·53 | 0·81697 | 16·52 | 16·536 | 16·539 |
| 12 18 0 | 80·01 | 0·78204 | 16·52 | 16·541 | 16·546 |
| | | Mean | 16·53 | 16·543 | 16·546 |
| | | | | E.s.d. | 0·004 |

[a] $\left(\frac{17\cdot9}{16\cdot5}\right)^2$; [b] $\left(\frac{17\cdot86}{16\cdot53}\right)^2$; [c] $\left(\frac{16\cdot5}{17\cdot9}\right)^2$; [d] $\left(\frac{16\cdot53}{17\cdot86}\right)^2$; [e] $\left(\frac{16\cdot543}{17\cdot854}\right)^2$.

we see that the term in $(a/b)$ is relatively unimportant if $h \gg k$ and so $a$ can be calculated directly from $d$ by inserting the known approximate value of $(a/b)$; $b$ is calculated analogously by rearranging equation (2.1.2d) to give

$$b = d\left[k^2 + h^2\left(\frac{b}{a}\right)^2\right]^{\frac{1}{2}}.$$

These new values of $a$ and $b$ give an improved value for $(a/b)$ or $(b/a)$ and the calculations are then repeated until the adjustment of the ratio $(a/b)$ has no further effect on the results for $a$ and $b$. The steps in the procedure are shown in table 2.1.2c.

After two cycles of calculation there is no change in the mean value of $a$. For $b$ three cycles are needed. This is because the condition $k \gg h$ is not well-satisfied. More suitable reflections happened to have very low intensities and could not be measured accurately.

## 2.1.3
Tetrameric phosphonitrilic chloride,

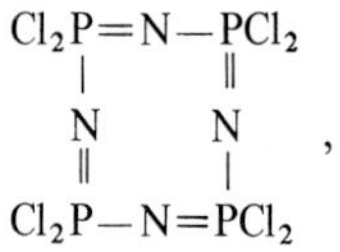

```
Cl2P=N—PCl2
  |     ||
  N     N    ,
  ||    |
Cl2P—N=PCl2
```

forms tetragonal crystals with $a = b = 10{\cdot}84$, $c = 5{\cdot}96$ Å, and two molecules in the unit cell. Its x-ray diffraction pattern shows the following symmetry: a fourfold rotation axis parallel to $c$ and a mirror plane perpendicular to $c$. No other symmetry elements are present.

x-Ray reflections are systematically absent as follows: $00l$ when $l$ is odd, and $hk0$ when $(h+k)$ is odd.

State the point-group symmetry of the diffraction pattern and deduce the space group of the crystal. Find the positions of the molecules in the unit cell and their minimum symmetry.

**Solution**

The point-group symmetry of the diffraction pattern is $4/m$. This shows that the point group of the crystal may be either 4, $\bar{4}$, or $4/m$, since a diffraction pattern is always centrosymmetric (Friedel's Law) and addition of a centre of symmetry to point group 4 or $\bar{4}$ produces $4/m$. Any other tetragonal point group would give a diffraction pattern of symmetry $4/mmm$. The space group is indicated by the systematic absences. Since these affect only the special classes of reflections $00l$ and $hk0$ the Bravais lattice is primitive ($P$). The conclusions that can be drawn are listed in table 2.1.3.

Since the diffraction symmetry is $4/m$, the screw axis must be $4_2$. The space group is therefore $P4_2/n$ (and the crystal point group is $4/m$).

The arrangement of symmetry elements in the unit cell is shown in figure 2.1.3a [the symbols follow the usual conventions, see *International Tables for x-ray Crystallography*, Volume 1 (1952)]. An atom placed at the general position $(x, y, z)$ will be repeated by the symmetry elements to give the following set of eight general equivalent positions: $(x, y, z)$, $(\bar{x}, \bar{y}, z)$, $(\bar{y}, x, \bar{z})$, $(y, \bar{x}, \bar{z})$, $(\frac{1}{2}+x, \frac{1}{2}+y, \frac{1}{2}-z)$, $(\frac{1}{2}-x, \frac{1}{2}-y, \frac{1}{2}-z)$, $(\frac{1}{2}-y, \frac{1}{2}+x, \frac{1}{2}+z)$, $(\frac{1}{2}+y, \frac{1}{2}-x, \frac{1}{2}+z)$. These positions are shown in figure 2.1.3b. Since there are only two molecules in the unit cell they must lie on a symmetry element, i.e. at special positions. There are only two possibilities: (a) molecules at (0, 0, 0) and $(\frac{1}{2}, \frac{1}{2}, \frac{1}{2})$, or (b) molecules at $(0, 0, \frac{1}{2})$ and $(\frac{1}{2}, \frac{1}{2}, 0)$. In either case the molecules lie on a $\bar{4}$ axis (see figure 2.1.3a) and the choice between set (a) and set (b) is arbitrary since the relative placing of the molecules is the same in both sets. The minimum molecular symmetry is therefore $\bar{4}$.

This is of importance with regard to the ring conformation. Known conformations for eight-membered phosphonitrilic rings are: planar, saddle, boat, chair, crown, and a hybrid of the crown and saddle (Bullen and Tucker, 1972). Only the first three of these can show symmetry $\bar{4}$. The full symmetries of these three conformations are: boat $\bar{4}$, saddle $\bar{4}2m$, and planar $4/mmm$; $\bar{4}$ is a subgroup of $\bar{4}2m$ and $4/mmm$. $(NPCl_2)_4$ in fact has the boat conformation in this crystal (Hazekamp *et al.*, 1962).

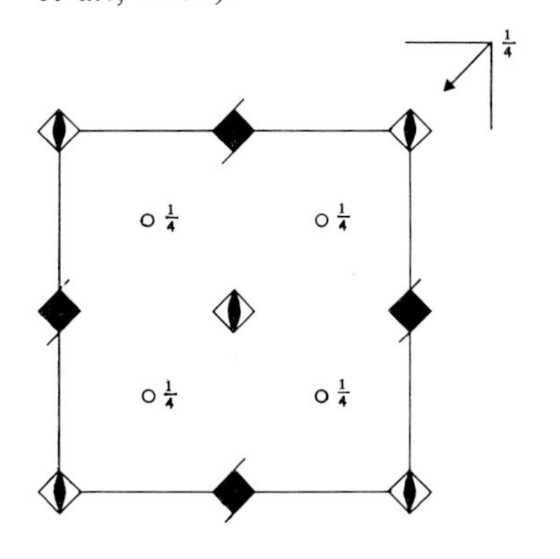


**Figure 2.1.3a.** The symmetry elements in space group $P4_2/n$.

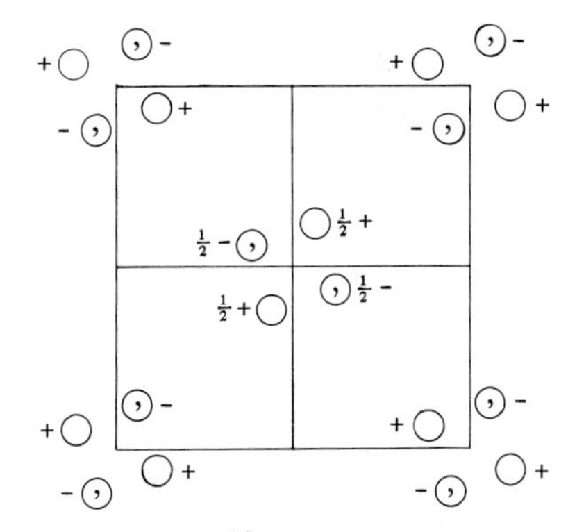


**Figure 2.1.3b.** The general equivalent positions.

**Table 2.1.3.**

| Reflections | Systematic absence | Conclusion |
|---|---|---|
| $hkl$ | None | Primitive lattice $P$ |
| $00l$ | Absent for $l$ odd | Screw axis parallel to $c$, either $2_1$, $4_2$, or $6_3$ |
| $hk0$ | Absent for $(h+k)$ odd | $n$ glide plane perpendicular to $c$ [translation $\frac{1}{2}(a+b)$] |

2.1.4

Boron phosphide, BP, crystallises in the cubic system with $a = 4\cdot538$ Å and density $2\cdot98$ g cm$^{-3}$. x-Ray reflections occur only when $h$, $k$, and $l$ are either all even or all odd. The relative intensities $I$ of some x-ray powder lines ($\lambda = 1\cdot542$ Å) are given in table 2.1.4a[(2)].

How many boron and phosphorus atoms are there in one unit cell and what is the Bravais lattice? Suggest possible arrangements for the atoms in the unit cell and use the intensity data to show which arrangement is correct. Hence calculate the shortest boron–phosphorus interatomic distance.

*Data needed.* Atomic weights B 10·81, P 30·97; Avogadro's number $6\cdot022 \times 10^{23}$ mol$^{-1}$; atomic scattering factors:

| $\sin\theta/\lambda$ (Å$^{-1}$) | 0 | 0·1 | 0·2 | 0·3 | 0·4 | 0·5 | 0·6 | 0·7 |
|---|---|---|---|---|---|---|---|---|
| $f_B$ (electrons) | 5·00 | 4·07 | 2·71 | 1·99 | 1·69 | 1·53 | 1·41 | 1·28 |
| $f_P$ (electrons) | 15·00 | 13·17 | 10·34 | 8·59 | 7·54 | 6·67 | 5·83 | 5·02 . |

**Table 2.1.4a.**

| *hkl* | *I* | *hkl* | *I* |
|---|---|---|---|
| 111 | 100 | 331 | 6 |
| 200 | 25 | 420 | 4 |
| 220 | 30 | 422 | 6 |
| 311 | 20 | 511, 333 | 12 |
| 222 | 3 | 440 | 3 |
| 400 | 3 | | |

**Solution**

The volume of the unit cell is $4\cdot538^3$ Å$^3$ = $4\cdot538^3 \times 10^{-24}$ cm$^3$; hence the mass of one unit cell is $4\cdot538^3 \times 10^{-24} \times 2\cdot98$ g.

The mass of one B atom is $10\cdot81/(6\cdot022 \times 10^{23})$ g and that of one P atom is $30\cdot97/(6\cdot022 \times 10^{23})$ g.

If there are $n$ B atoms and $n$ P atoms in the unit cell, the mass of a unit cell is $n(10\cdot81+30\cdot97)/(6\cdot022 \times 10^{23})$ g; therefore

$$n = \frac{4\cdot538^3 \times 10^{-24} \times 2\cdot98 \times 6\cdot022 \times 10^{23}}{41\cdot78} = 4\cdot01.$$

As $n$ must be an integer, there are four boron and four phosphorus atoms in the unit cell.

The Bravais lattice is indicated by the systematic absences of x-ray reflections. The occurrence of reflections only when $h$, $k$, and $l$ are all even or all odd is characteristic of an $F$ lattice, in which all faces of the unit cell are centred.

(2) Data taken from Perri *et al.* (1958).

The unit cell of an $F$ lattice contains four lattice points: (0, 0, 0), $(0, \frac{1}{2}, \frac{1}{2})$, $(\frac{1}{2}, 0, \frac{1}{2})$, and $(\frac{1}{2}, \frac{1}{2}, 0)$. Since the number of boron and phosphorus atoms in the unit cell is the same as the number of lattice points, the two types of atom must be placed so that each individually forms a face-centred arrangement. The overall structure is then composed of the two interpenetrating face-centred lattices. We may arbitrarily choose the origin so that one type of atom, say boron, lies at the points: (0, 0, 0), $(0, \frac{1}{2}, \frac{1}{2})$, $(\frac{1}{2}, 0, \frac{1}{2})$, and $(\frac{1}{2}, \frac{1}{2}, 0)$. The four phosphorus atoms then lie at $(x, y, z)$, $(x, \frac{1}{2}+y, \frac{1}{2}+z)$, $(\frac{1}{2}+x, y, \frac{1}{2}+z)$, and $(\frac{1}{2}+x, \frac{1}{2}+y, z)$, where the values of $x$, $y$, and $z$ remain to be found.

The overall structure must possess cubic symmetry, the essential feature of which is four triad symmetry axes parallel to the body diagonals of the unit cell. The only phosphorus atom positions which will satisfy the cubic symmetry are those with (a) $x = y = z = \frac{1}{2}$, or (b) $x = y = z = \frac{1}{4}$. A third valid possibility, $x = y = z = \frac{3}{4}$, need not be considered as it is stereochemically identical to solution (b). If any other positions were chosen, the triad symmetry operations combined with the $F$-lattice centring would produce more than four atoms in the unit cell. (A three-dimensional model will help one to appreciate this point.) Solution (a) is in fact the sodium chloride structure and solution (b) the zinc blende structure.

The x-ray intensities corresponding to these two solutions can be found by calculating the structure factor, the expression for which is

$$F(hkl) = \sum_n f_n \exp[2\pi i(hx_n + ky_n + lz_n)] = A + iB \, , \tag{2.1.4a}$$

with

$$A = \sum_n f_n \cos[2\pi(hx_n + ky_n + lz_n)] \, , \tag{2.1.4b}$$

$$B = \sum_n f_n \sin[2\pi(hx_n + ky_n + lz_n)] \, , \tag{2.1.4c}$$

where $(x_n, y_n, z_n)$ is the position of the $n$th atom and $f_n$ is its atomic scattering factor. Then, the intensity is

$$I \propto |F(hkl)|^2 = A^2 + B^2 \, . \tag{2.1.4d}$$

For solution (a), insertion of the proposed atomic coordinates into equation (2.1.4b) gives

$$\begin{aligned} A = {} & f_B\{\cos 0 + \cos[2\pi(\tfrac{1}{2}k + \tfrac{1}{2}l)] + \cos[2\pi(\tfrac{1}{2}h + \tfrac{1}{2}l)] + \cos[2\pi(\tfrac{1}{2}h + \tfrac{1}{2}k)]\} \\ & + f_P\{\cos[2\pi(\tfrac{1}{2}h + \tfrac{1}{2}k + \tfrac{1}{2}l)] + \cos[2\pi(\tfrac{1}{2}h + k + l)] \\ & + \cos[2\pi(h + \tfrac{1}{2}k + l)] + \cos[2\pi(h + k + \tfrac{1}{2}l)]\} \, , \end{aligned} \tag{2.1.4e}$$

where $f_B$ and $f_P$ are the atomic scattering factors for boron and phosphorus for the reflection in question. Since reflections occur only if $h$, $k$, and $l$ are all even or all odd, (i) sums like $(\frac{1}{2}k + \frac{1}{2}l)$ are integral and $\cos[2\pi(\frac{1}{2}k + \frac{1}{2}l)] = 1$, whereas (ii) sums like $(\frac{1}{2}h + \frac{1}{2}k + \frac{1}{2}l)$ or $(\frac{1}{2}h + k + l)$ are integral when the indices are even, but contain an odd half-integer when the indices are odd.

Hence for even indices

$$A = 4f_B + 4f_P , \qquad (2.1.4f)$$

while for odd indices

$$A = 4f_B - 4f_P . \qquad (2.1.4g)$$

Likewise

$$\begin{aligned} B &= f_B\{\sin 0 + \sin[2\pi(\tfrac{1}{2}k + \tfrac{1}{2}l)] + ...\} \\ &\quad + f_P\{\sin[2\pi(\tfrac{1}{2}h + \tfrac{1}{2}k + \tfrac{1}{2}l)] + \sin[2\pi(\tfrac{1}{2}h + k + l)] + ...\} \\ &= 0 \end{aligned} \qquad (2.1.4h)$$

for all reflections. [$B$ is always zero for structure (a) because it is centrosymmetric with the centre of symmetry at the origin.]

For solution (b)

$$\begin{aligned} A = 4f_B + f_P\{&\cos[2\pi(\tfrac{1}{4}h + \tfrac{1}{4}k + \tfrac{1}{4}l)] + \cos[2\pi(\tfrac{1}{4}h + \tfrac{3}{4}k + \tfrac{3}{4}l)] \\ &+ \cos[2\pi(\tfrac{3}{4}h + \tfrac{1}{4}k + \tfrac{3}{4}l)] + \cos[2\pi(\tfrac{3}{4}h + \tfrac{3}{4}k + \tfrac{1}{4}l)]\} . \end{aligned} \qquad (2.1.4j)$$

Here the result depends on whether $(h+k+l) = 4m$, $4m+2$, or $4m \pm 1$, $m$ being an integer, as follows:

| $(h+k+l)$ | $4m$ | $4m+2$ | $4m \pm 1$ |
|---|---|---|---|
| $A$ | $(4f_B + 4f_P)$ | $(4f_B - 4f_P)$ | $4f_B$ . |

Likewise

$$B = f_P\{\sin[2\pi(\tfrac{1}{4}h + \tfrac{1}{4}k + \tfrac{1}{4}l)] + \sin[2\pi(\tfrac{1}{4}h + \tfrac{3}{4}k + \tfrac{3}{4}l)] + ...\} \qquad (2.1.4k)$$

$$= 0 \quad \text{for even indices} ,$$

$$= 4f_P \quad \text{for } h+k+l = 4m+1 ,$$

$$= -4f_P \quad \text{for } h+k+l = 4m-1 .$$

A summary of the two solutions is given in table 2.1.4b.

**Table 2.1.4b.**

| $h+k+l$ | $4m$ | $4m+2$ | $4m+1$ | $4m-1$ |
|---|---|---|---|---|
| Solution (a) | | | | |
| $A$ | $4f_B+4f_P$ | $4f_B+4f_P$ | $4f_B-4f_P$ | $4f_B-4f_P$ |
| $B$ | 0 | 0 | 0 | 0 |
| $\|F\|^2$ | $16(f_B+f_P)^2$ | $16(f_B+f_P)^2$ | $16(f_B-f_P)^2$ | $16(f_B-f_P)^2$ |
| Solution (b) | | | | |
| $A$ | $4f_B+4f_P$ | $4f_B-4f_P$ | $4f_B$ | $4f_B$ |
| $B$ | 0 | 0 | $4f_P$ | $-4f_P$ |
| $\|F\|^2$ | $16(f_B+f_P)^2$ | $16(f_B-f_P)^2$ | $16(f_B^2+f_P^2)$ | $16(f_B^2+f_P^2)$ |

Clearly reflections with $h+k+l = 4m$ will not serve to distinguish between the two solutions. Note also that reflections with $h+k+l$ equal to $4m+1$ or $4m-1$ give the same results for $|F|^2$, and hence for the intensity. Before the intensity of a powder line can be compared with $|F|^2$, account must be taken of the number of superimposed reflections $g$ and of the Lorentz and polarisation factors, leading to the expression

$$I \propto I' = \frac{1+\cos^2 2\theta}{\sin^2\theta \cos\theta} g|F|^2 . \qquad (2.1.41)$$

$g$, the multiplicity, is equal to the total number of permutations of $h$, $k$, and $l$, with both positive and negative indices allowed. For example, the line 220 will include 202, 022, $2\bar{2}0$, $\bar{2}20$, etc., making $g$ equal to 12. The multiplicities are listed in table 2.1.4c, together with the values of $f_B$ and $f_P$ appropriate to each reflection (these are found graphically by plotting $f_B$ and $f_P$ against $\sin\theta/\lambda$). The calculated intensities $I'$ for the two solutions are shown in table 2.1.4d and comparison of these with the measured values in table 2.1.4a shows (b) to be the correct solution. However, although the general trend in intensities is correct, we see that the measured values fall off with respect to the calculated values as $h$, $k$, and $l$ increase. The reason for this is that we have neglected the temperature factor, which causes a steady decrease in $|F|$ as $\sin\theta$ increases (a simple form of temperature factor is used in problem 2.1.8 and a more complex form occurs in problem 2.1.10). In order to take the temperature factor into consideration in calculating the intensities it is necessary to know, or to postulate, the amplitudes of thermal vibration of the atoms in the crystal. However, even with neglect of the temperature factor, it is clear that solution (b) gives much better agreement of intensities than solution (a).

**Table 2.1.4c.**

| $hkl$ | $h^2+k^2+l^2$ | $\frac{\sin\theta}{\lambda}$ | $f_B$ | $f_P$ | $\frac{1+\cos^2 2\theta}{\sin^2\theta\cos\theta}$ | $g$ |
|---|---|---|---|---|---|---|
| 111 | 3 | 0·1908 | 2·81 | 10·58 | 20·34 | 8 |
| 200 | 4 | 0·2204 | 2·51 | 9·94 | 14·66 | 6 |
| 220 | 8 | 0·3116 | 1·94 | 8·48 | 6·368 | 12 |
| 311 | 11 | 0·3654 | 1·75 | 7·90 | 4·320 | 24 |
| 222 | 12 | 0·3817 | 1·70 | 7·73 | 3·908 | 8 |
| 400 | 16 | 0·4407 | 1·61 | 7·19 | 2·968 | 6 |
| 331 | 19 | 0·4803 | 1·56 | 6·82 | 2·739 | 24 |
| 420 | 20 | 0·4927 | 1·54 | 6·71 | 2·728 | 24 |
| 422 | 24 | 0·5398 | 1·50 | 6·30 | 2·991 | 24 |
| 511 | 27 | 0·5725 | 1·46 | 6·04 | 3·585 | 24 |
| 333 | 27 | 0·5725 | 1·46 | 6·04 | 3·585 | 8 |
| 440 | 32 | 0·6233 | 1·40 | 5·64 | 6·733 | 12 |

The shortest boron–phosphorus distance, from the B atom at (0, 0, 0) to the P atom at $(\frac{1}{4}, \frac{1}{4}, \frac{1}{4})$, is $3^{1/2} \times \frac{1}{4}a = 1 \cdot 965$ Å.

**Table 2.1.4d.** Values of $I' = \dfrac{1+\cos^2 2\theta}{\sin^2\theta \cos\theta} g|F|^2$ for solutions (a) and (b).

| *hkl* | $I'$ | | $I'$ scaled to 100 for reflection 111 | | $I$ from table 2.1.4a |
|---|---|---|---|---|---|
| | (a) | (b) | (a) | (b) | |
| 111 | 9824 | 19499 | 100 | 100 | 100 |
| 200 | 13634 | 4856 | 139 | 25 | 25 |
| 220 | 8297 | 8297 | 84 | 43 | 30 |
| 311 | 3921 | 6788 | 40 | 35 | 20 |
| 222 | 2780 | 1137 | 28 | 6 | 3 |
| 400 | 1379 | 1379 | 14 | 7 | 3 |
| 331 | 1819 | 3218 | 19 | 17 | 6 |
| 420 | 4456 | 1750 | 45 | 9 | 4 |
| 422 | 4367 | 4367 | 44 | 22 | 6 |
| 511, 333[a] | 2406 | 4430 | 24 | 23 | 12[b] |
| 440 | 4004 | 4004 | 41 | 21 | 3 |

[a] Reflections 511 and 333, having the same $\sin\theta/\lambda$, are superposed; their $I'$ values have been added together.
[b] Poor agreement of $I'$ and $I$; the measured value is too high.

## 2.1.5

Crystals of bis-(*N*-ethylsalicylaldiminato)palladium,

$$\mathrm{Pd}\left[\text{(O-C}_6\text{H}_4\text{-CH=N(C}_2\text{H}_5\text{))}\right]_2,$$

are monoclinic, $a = 8 \cdot 43$, $b = 5 \cdot 60$, $c = 17 \cdot 97$ Å, $\beta = 94°\ 42'$, with two molecules in the unit cell (Frasson *et al.*, 1964). Systematic absences of x-ray reflections are: $0k0$ when $k$ is odd, and $h0l$ when $l$ is odd.

(a) Deduce the space group, and find the positions of the palladium atoms and the minimum symmetry of the molecule. Hence comment on the coordination of the palladium atom.

(b) Explain how the heavy-atom technique could be applied to elucidate the complete crystal structure and what difficulties would arise in its application to this particular crystal.

**Solution**

(a) The space group is determined by drawing the following conclusions from the systematic absences of x-ray reflections:

| Reflections | Systematic absence | Conclusion |
|---|---|---|
| $hkl$ | None | Bravais lattice is $P$ |
| $0k0$ | When $k$ odd | $2_1$ screw axis parallel to $b$[3] |
| $h0l$ | When $l$ odd | $c$ glide plane perpendicular to $b$ |

The space group is therefore $P2_1/c$, in which the set of general equivalent positions comprises the four points $(x, y, z)$, $(x, \frac{1}{2}+y, \frac{1}{2}-z)$, $(x, \frac{1}{2}-y, \frac{1}{2}+z)$, and $(\bar{x}, \bar{y}, \bar{z})$. As there are only two palladium atoms in the unit cell they must occupy special positions which for this space group can only be centres of symmetry. These special positions can be any one of the four sets: (i) $(0, 0, 0)$ and $(0, \frac{1}{2}, \frac{1}{2})$; (ii) $(\frac{1}{2}, 0, 0)$ and $(\frac{1}{2}, \frac{1}{2}, \frac{1}{2})$; (iii) $(0, \frac{1}{2}, 0)$ and $(0, 0, \frac{1}{2})$; or (iv) $(\frac{1}{2}, \frac{1}{2}, 0)$ and $(\frac{1}{2}, 0, \frac{1}{2})$. The choice between the sets is immaterial, merely acting as a choice of origin. Let us select $(0, 0, 0)$ and $(0, \frac{1}{2}, \frac{1}{2})$. The molecule as a whole must then be centrosymmetric in order to conform to the palladium site symmetry. The coordination of the palladium atom has to be square planar with the ligands mutually *trans*.

(b) Since the space group is centrosymmetric and we have a centre of symmetry at the origin, the phases of all x-ray reflections will be either 0° or 180° ($B = 0$; see problem 2.1.4). The reflections are said to be positive or negative. The heavy-atom technique in this case involves calculation of the contribution made by the palladium atoms to the structure factor for each reflection and the assumption that the sign of each reflection will be the same as that of the palladium contribution. A Fourier synthesis calculated with the use of the signs so obtained should then reveal the positions of the lighter carbon, nitrogen, and oxygen atoms.

The structure factor equation for a centrosymmetric crystal is

$$F(hkl) = \sum_n f_n \cos 2\pi(hx_n + ky_n + lz_n) \,,$$

where $(x_n, y_n, z_n)$ are the fractional coordinates of the $n$th atom, $f_n$ is its atomic scattering factor, and the summation is made over all atoms in the unit cell. The contribution of the palladium atoms to $F(hkl)$ is

$$\begin{aligned} F_{\mathrm{Pd}} &= f_{\mathrm{Pd}}[\cos 2\pi(0+0+0) + \cos 2\pi(0+\tfrac{1}{2}k+\tfrac{1}{2}l)] \\ &= f_{\mathrm{Pd}}[1+\cos\pi(k+l)] \\ &= 2f_{\mathrm{Pd}} \quad \text{when } (k+l) \text{ is even, or} \\ &= 0 \quad \text{when } (k+l) \text{ is odd}\,. \end{aligned}$$

[3] Note that the only type of screw axis possible in monoclinic symmetry is $2_1$; compare problem 2.1.3.

Since $f_{Pd}$ is always positive, in applying the heavy-atom technique all reflections with $(k+l)$ even will be assumed to have a positive sign but nothing can be said about those reflections with $(k+l)$ odd because the palladium atoms make no contribution to them. The latter group of reflections (half of all the reflections) must therefore be omitted from the Fourier synthesis. This causes no difficulty provided we confine ourselves to calculating a *projection* of the electron-density synthesis on (010). For this projection we use only the $h0l$ reflections, all of which have $(k+l)$ even because they are present only when $l$ is even. We can therefore give a positive sign to all the $h0l$ reflections and so calculate a valid electron-density projection from which it should be possible to determine the $x$ and $z$ coordinates of the lighter atoms. This is especially likely to be successful for the given crystal because it has a short $b$ axis so that overlap of neighbouring molecules in projection on (010) is improbable.

In order to find the $y$ coordinates of the atoms we must use reflections of type $hkl$ or possibly $0kl$ or $hk0$. In each of these groups half of the reflections have $(k+l)$ odd and hence $F_{Pd} = 0$ so that no sign can be given to them. The effect of omitting them from the Fourier synthesis is to introduce false mirror planes into the synthesis at $y = 0$ and $y = \frac{1}{2}$, thus superposing a false image of the light atoms on the real image. [An alternative way of looking at the falsification of the synthesis is to say that the inclusion of only those reflections which have $(k+l)$ even is equivalent to treating the lattice as $A$ face-centred when it is in reality primitive. The resultant space group, $A2_1/c$, contains the extra mirror planes referred to above.] The heavy-atom technique can be of no further help and in order to find the $y$ coordinates either the real image must be unscrambled from the false image, or approximate values for the coordinates must be proposed on the basis of known bond lengths and the $x$ and $z$ coordinates already deduced. In either case prior stereochemical knowledge (at least approximate) will have to be used.

This situation is quite common as many *trans* complexes of metals (either square planar or octahedral) crystallise in space group $P2_1/c$ with the metal atoms located on centres of symmetry. Analogous difficulties arise with some other centrosymmetric space groups, such as *Pbca*.

2.1.6

The space group of crystals of phenyl phosphorodiamidate is *Pbca* and the unit cell contains eight molecules of $(PhO)P(:O)(NH_2)_2$. Some prominent peaks in the Patterson function which can be attributed to phosphorus–phosphorus vectors occur at the following positions:

(I) (0, 0·38, 0·5), (0, 0·62, 0·5), (0·40, 0·5, 0), (0·60, 0·5, 0);

(II) (0·40, 0·12, 0·5), (0·40, 0·88, 0·5), (0·60, 0·12, 0·5), (0·60, 0·88, 0·5), (0·10, 0·5, 0·40), (0·10, 0·5, 0·60), (0·90, 0·5, 0·40), (0·90, 0·5, 0·60).

The peaks in group (I) are markedly stronger than those in group (II).
(a) Where in the Patterson function are the Harker sections and lines for the space group *Pbca*?
(b) Deduce the coordinates of the phosphorus atoms from the Patterson peaks listed.

**Solution**

(a) The space group symbol *Pbca* is an abbreviation for $P\frac{2_1}{b}\frac{2_1}{c}\frac{2_1}{a}$. The set of general equivalent positions for this space group (with the origin placed at a centre of symmetry) is

(1) $(x, y, z)$; (2) $(\frac{1}{2}+x, \frac{1}{2}-y, \bar{z})$; (3) $(\bar{x}, \frac{1}{2}+y, \frac{1}{2}-z)$; (4) $(\frac{1}{2}-x, \bar{y}, \frac{1}{2}+z)$;
(5) $(\bar{x}, \bar{y}, \bar{z})$; (6) $(\frac{1}{2}-x, \frac{1}{2}+y, z)$; (7) $(x, \frac{1}{2}-y, \frac{1}{2}+z)$; (8) $(\frac{1}{2}+x, y, \frac{1}{2}-z)$.

Harker sections (or lines) are those sections (or lines) in the Patterson function which contain peaks arising from vectors between the atoms in this set, that is, between symmetry-related atoms. For example, the vector $(U, V, W)$ between atom (1) and atom (2), which are related by the $2_1$ screw axis parallel to $a$, will have $U = \frac{1}{2}$ irrespective of the value of $x$, and all such vectors will therefore produce peaks in the Harker section $(\frac{1}{2}, V, W)$. A glide operation, such as between atom (1) and atom (6), gives vectors with two fixed components, in this case $(U, \frac{1}{2}, 0)$, and the locus of these vectors is then a Harker line. The Harker sections and lines for the space group *Pbca* are listed in table 2.1.6.
(b) There are eight molecules, and hence eight phosphorus atoms, in the unit cell. The phosphorus atoms therefore occupy one set of general equivalent positions (the only alternative, that they lie on centres of symmetry, is precluded by the structure of the molecule). The Patterson peaks given occur in the Harker sections $(U, V, \frac{1}{2})$ and $(U, \frac{1}{2}, W)$. Let us look first at section $(U, V, \frac{1}{2})$. This section also contains the Harker line $(0, V, \frac{1}{2})$ so that the group (I) peaks $(0, 0{\cdot}38, 0{\cdot}5)$ and $(0, 0{\cdot}62, 0{\cdot}5)$ may belong to this line or they may be merely part of the section $(U, V, \frac{1}{2})$ with the $U$ coordinate accidentally equal to zero.

**Table 2.1.6.**

| Harker section or line | Symmetry element responsible |
|---|---|
| Sections | |
| $\frac{1}{2}, V, W$ | $2_1$ axis parallel to $a$ |
| $U, \frac{1}{2}, W$ | $2_1$ axis parallel to $b$ |
| $U, V, \frac{1}{2}$ | $2_1$ axis parallel to $c$ |
| Lines | |
| $U, \frac{1}{2}, 0$ | $b$ glide perpendicular to $a$ |
| $0, V, \frac{1}{2}$ | $c$ glide perpendicular to $b$ |
| $\frac{1}{2}, 0, W$ | $a$ glide perpendicular to $c$ |

The vectors between atoms in the set of general equivalent positions which produce peaks on the line $(0, V, \frac{1}{2})$ are $(0, \frac{1}{2}-2y, \frac{1}{2})$ and $(0, \frac{1}{2}+2y, \frac{1}{2})$. Each of these vectors occurs four times between the eight atoms in the set, and so they are said to have a weight of four. On the other hand, the vectors responsible for peaks in the section $(U, V, \frac{1}{2})$, namely $(\frac{1}{2}-2x, 2y, \frac{1}{2})$, $(\frac{1}{2}-2x, -2y, \frac{1}{2})$, $(\frac{1}{2}+2x, 2y, \frac{1}{2})$, and $(\frac{1}{2}+2x, -2y, \frac{1}{2})$, each occur only twice within the set of eight atoms and therefore have a weight of two. Since the group (I) peaks are markedly stronger than the group (II) peaks it is reasonable to suppose that the peaks at (0, 0·38, 0·5) and (0, 0·62, 0·5) are of type $(0, \frac{1}{2}-2y, \frac{1}{2})$ or $(0, \frac{1}{2}+2y, \frac{1}{2})$; therefore $\frac{1}{2}+2y = 0{\cdot}62$, and $y = 0{\cdot}06$.

(There are several other solutions. Thus, putting $\frac{1}{2}+2y = 0{\cdot}38$ gives $y = -0{\cdot}06$, but this can be disregarded as it merely represents a reversal of the $y$ axis. We could put $\frac{1}{2}+2y = 1{\cdot}62$, giving $y = 0{\cdot}56$, since there is a Patterson peak at $V = 1{\cdot}62$, i.e. in the next unit cell. This can again be disregarded as it is equivalent to moving the origin along the $y$ axis by $\frac{1}{2}b$. Any one solution can be chosen arbitrarily, so defining the origin and the direction of the $y$ axis.)

Knowing $y$ (and noting that $-2y = -0{\cdot}12$ which is equivalent to 0·88) we can deduce $x$ from the other peaks in the section $(U, V, \frac{1}{2})$: $\frac{1}{2}-2x = 0{\cdot}40$ and $\frac{1}{2}+2x = 0{\cdot}60$, which gives $x = 0{\cdot}05$. (Alternative solutions arise here as in the deduction of $y$.)

The peaks belonging to the Harker section $(U, \frac{1}{2}, W)$, which contains the Harker line $(U, \frac{1}{2}, 0)$, can be treated in an analogous way. The vectors giving rise to the group (I) peaks on the line $(U, \frac{1}{2}, 0)$ are $(\frac{1}{2}-2x, \frac{1}{2}, 0)$ and $(\frac{1}{2}+2x, \frac{1}{2}, 0)$; these confirm the solution $x = 0{\cdot}05$. The peaks in the section $(U, \frac{1}{2}, W)$ result from vectors $(2x, \frac{1}{2}, \frac{1}{2}-2z)$, $(2x, \frac{1}{2}, \frac{1}{2}+2z)$, $(-2x, \frac{1}{2}, \frac{1}{2}-2z)$, and $(-2x, \frac{1}{2}, \frac{1}{2}+2z)$, so that $\frac{1}{2}-2z = 0{\cdot}40$, $\frac{1}{2}+2z = 0{\cdot}60$, and $z = 0{\cdot}05$.

The phosphorus atoms therefore lie at (0·05, 0·06, 0·05) and other points related to this position by the symmetry operations of the space group.

### 2.1.7

1,8,10,9-Triazaboradecalin,

H H
N B N
$H_2C$ $CH_2$
$H_2C$ N $CH_2$ ,
C C
$H_2$ $H_2$

is monoclinic with $a = 18{\cdot}01$, $b = 5{\cdot}13$, $c = 8{\cdot}66$ Å, $\beta = 93{\cdot}6°$. The space group is $P2_1/n$ and there are four molecules in the unit cell. A projection of the sharpened Patterson function on the (010) plane is shown in figure 2.1.7a. Use the Patterson map to deduce (a) the orientation of the fused ring system, and (b) the positions of the atoms in the unit-cell projection (that is, their $x$ and $z$ fractional coordinates).

(Since boron, carbon, and nitrogen atoms do not differ greatly in x-ray scattering power, all atoms in the ring system should be treated as equivalent. The hydrogen atoms have a much lower scattering power and should be neglected.)

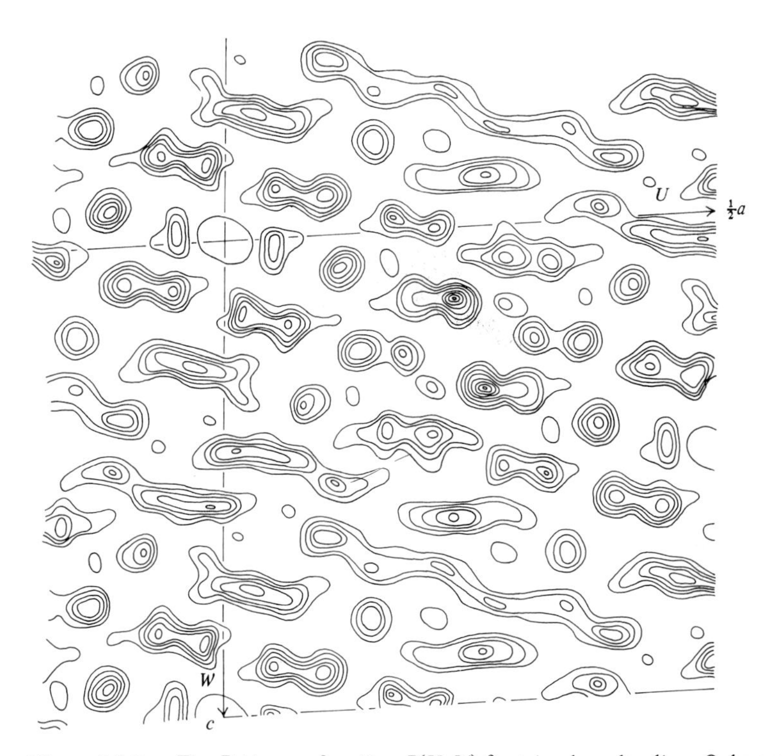


**Figure 2.1.7a.** The Patterson function $P(U, W)$ for triazaboradecalin. Only one-half of the unit cell is shown. Contours have been omitted from the origin peak; otherwise the contours are at equal intervals.

**Solution**

(a) There are four general equivalent positions in the unit cell of space group $P2_1/n$. With the origin at a centre of symmetry their coordinates are

$$(\mathrm{A})\ (x, y, z),\ (\mathrm{B})\ (\tfrac{1}{2}-x, \tfrac{1}{2}+y, \tfrac{1}{2}-z),\ (\mathrm{C})\ (\tfrac{1}{2}+x, \tfrac{1}{2}-y, \tfrac{1}{2}+z),\ (\mathrm{D})\ (\bar{x}, \bar{y}, \bar{z}).$$

The arrangement of these points and of the symmetry elements in the unit cell is shown in figure 2.1.7b. Points A and B are related by the $2_1$ screw axis, A and C by the $n$ glide plane, and A and D (or B and C) by a centre of symmetry.

The only possible *special* positions in this space group are the centres of symmetry, and since the molecule is not centrosymmetric it cannot occupy these special positions. The four molecules in the unit cell must therefore be assigned to the four general equivalent positions, with the values of $x$, $y$, and $z$ for individual atoms as yet unknown.

The length of the crystallographic $b$ axis (5·13 Å) gives some indication of the orientation of the ring system. Since the distances between molecules are usually about 3·5–4 Å, $b$ is sufficiently long to make it likely that the rings in the molecule are tilted slightly away from the (010) plane [as in figure 2.1.7c (II)] but not tilted so far that they become aligned along $b$ [as in figure 2.1.7c (III)].

In examining the Patterson function it is most profitable to look first at the region within 3 Å of the origin (marked by the circle in figure 2.1.7d). This will contain peaks arising from vectors between the atoms within one six-membered ring since such vectors will not be longer than 3 Å. Peaks 1–6 are within 1·5 Å of the origin and give the directions of the six bonds in the ring. The remaining peaks (7–14) are due to vectors across the ring between nonbonded pairs of atoms. The projected shape and orientation of the ring deduced from these peaks are shown in figure 2.1.7e. Note that the group of peaks within the circle in figure 2.1.7d

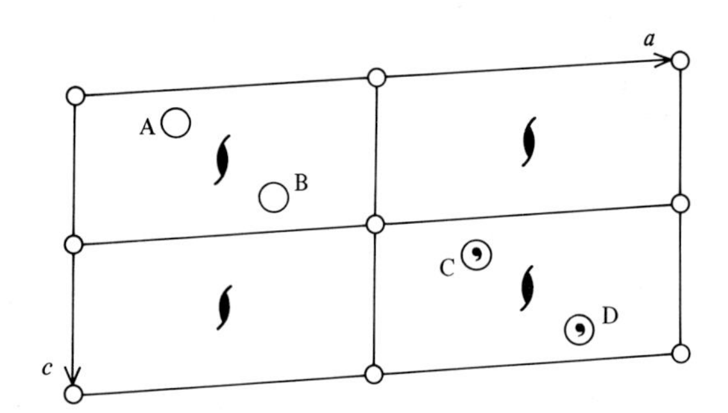


**Figure 2.1.7b.** The general equivalent positions in space group $P2_1/n$. The $y$ heights of the points are: (A) $y$, (B) $\frac{1}{2}+y$, (C) $\frac{1}{2}-y$, and (D) $\bar{y}$.

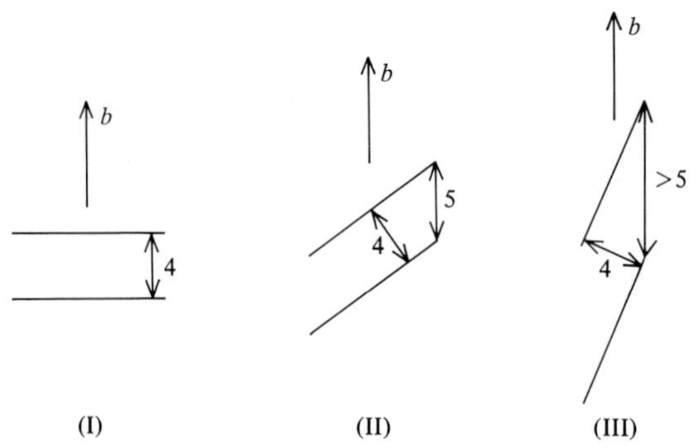


**Figure 2.1.7c.** Change in unit-cell length as molecular orientation changes; distances are in Å.

has approximate $mm2$ symmetry and that this is the approximate symmetry of the ring shape in figure 2.1.7e.

The pair of fused rings making up the molecule may be oriented in three ways (figure 2.1.7f). Orientation (I) is compatible with the peaks further away from the origin, whereas orientations (II) and (III) would require vectors to low regions of the Patterson function, e.g. the regions marked M and N in figure 2.1.7d.

(b) In order to determine the positions of the molecules from the Patterson map we must make use of *inter*molecular vectors, that is vectors between atoms in molecules related to each other by the centre of

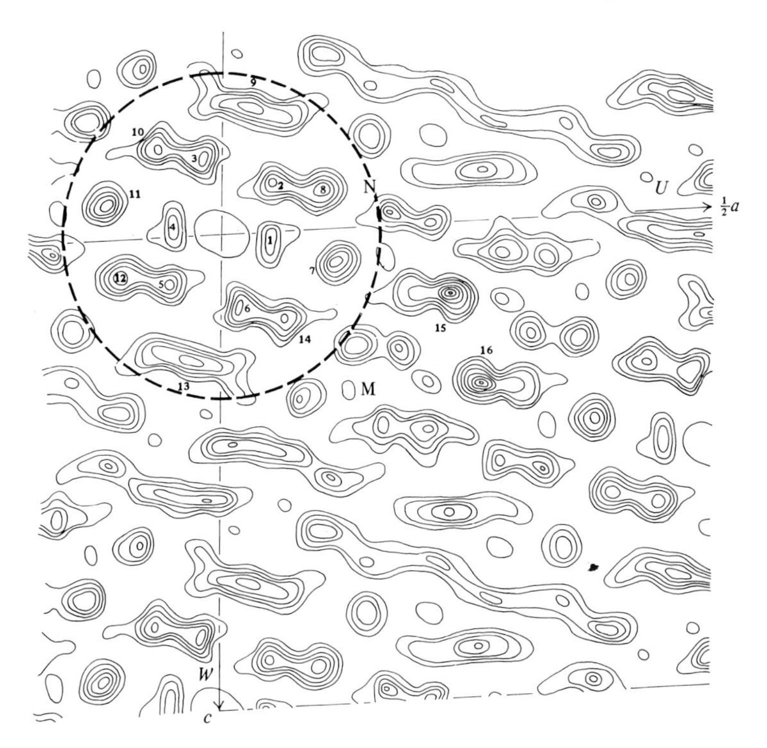


**Figure 2.1.7d.** The Patterson map with peaks numbered. The broken circle has a radius of 3 Å.

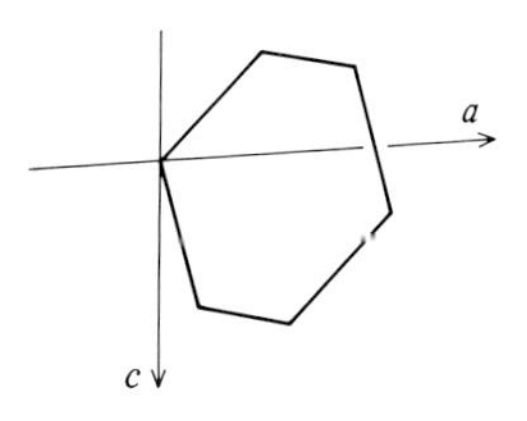


**Figure 2.1.7e.** Shape and orientation of one ring. The position of the origin of the $a$ and $c$ axes is arbitrary, the axes serving only to indicate the orientation of the ring.

symmetry or the $2_1$ screw axis. The number of these vectors will be very large and one might anticipate considerable difficulty in allocating Patterson peaks to them. However, the solution is made much easier by the molecule being approximately centrosymmetric. [This may seem to contradict the earlier statement that the molecule is not centrosymmetric. However, now that we are treating boron, carbon, and nitrogen as equivalent, the molecule as shown in figure 2.1.7f (I) is effectively centrosymmetric.] When two molecules which are themselves centrosymmetric (or nearly so) are related by a centre of symmetry (as in figure 2.1.7g) a number of vectors, such as $6' \cdots 2$ or $7' \cdots 3$, will coincide giving a high peak in the Patterson function, which should be easy to identify. In our Patterson map (figure 2.1.7d) there are in fact two very high peaks, peak 15 at (0·23, 0·15) and peak 16 at (0·27, 0·35)

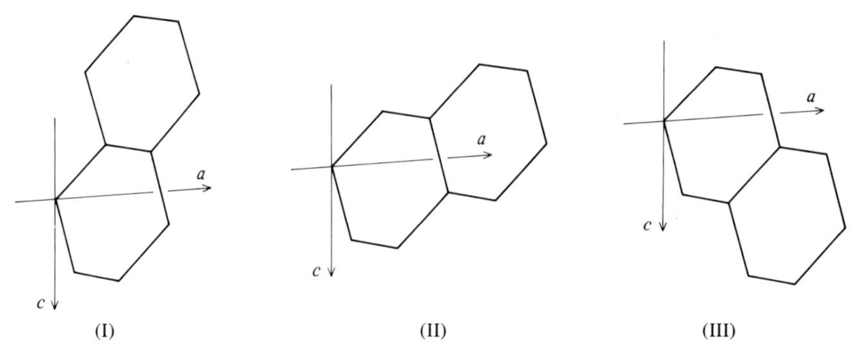


**Figure 2.1.7f.** Possible orientations of the pair of fused rings. The origin of the axes is again arbitrary.

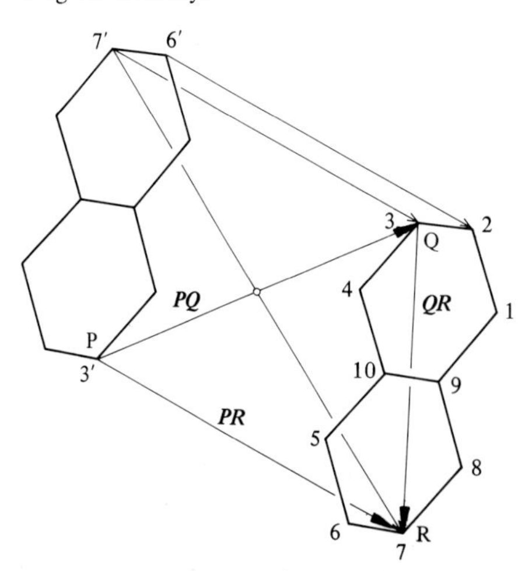


**Figure 2.1.7g.** A pair of molecules related by a centre of symmetry, with atoms numbered.

(fractional $U$ and $W$ values), which can be attributed to these coincident vectors. These peaks are 4·5 and 5·7 Å from the origin respectively. The reason for there being two peaks is as follows. In projection on (010) the screw axis produces the same effect as a centre of symmetry. We therefore have a pair of molecules related by the true centre of symmetry (as B and C in figure 2.1.7b) and a second pair related by the apparent centre caused by the screw axis (as A and B in figure 2.1.7b). Each of these pairs produces a high peak but it is not possible to tell from the Patterson function which is which. Let us use the vector (0·27, 0·35). In order to deduce the coordinates of, say, atoms 3 and 7 we measure the *intra*molecular vector 3···7 from figure 2.1.7f. We find it to be (−0·03, 0·55) (again as fractions of the unit-cell edges). Then since, from figure 2.1.7g, $\boldsymbol{PQ} = \boldsymbol{PR} - \boldsymbol{QR}$, the coordinates of atom 3 referred to the centre of symmetry as origin $(x_3, z_3)$ will be

$$x_3 = \tfrac{1}{2}[0{\cdot}27-(-0{\cdot}03)] = 0{\cdot}15\ ,$$

$$z_3 = \tfrac{1}{2}[0{\cdot}35-0{\cdot}55] = -0{\cdot}10\ ;$$

**Table 2.1.7.** Fractional coordinates of the atoms in the triazaboradecalin molecule.

| Atom | $x$ | $z$ | Atom | $x$ | $z$ |
|---|---|---|---|---|---|
| 1 | 0·21 | 0·07 | 6 | 0·07 | 0·44 |
| 2 | 0·20 | −0·09 | 7 | 0·12 | 0·45 |
| 3 | 0·15 | −0·10 | 8 | 0·17 | 0·34 |
| 4 | 0·10 | 0·02 | 9 | 0·16 | 0·18 |
| 5 | 0·06 | 0·28 | 10 | 0·11 | 0·17 |

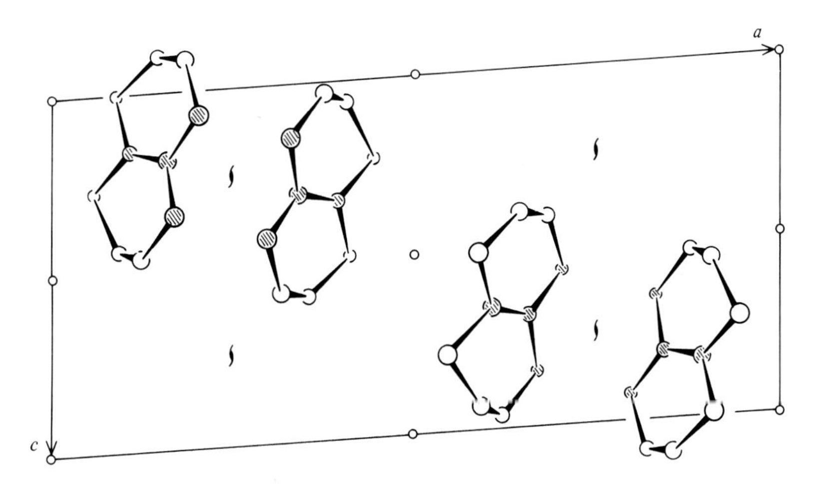


**Figure 2.1.7h.** The complete structure. This diagram, taken from Bullen and Clark (1969), shows the atomic positions as obtained by a full crystal-structure analysis, including refinement (open circles C atoms, hatched circles B and N atoms).

similarly

$$x_7 = \tfrac{1}{2}[0\cdot27+(-0\cdot03)] = 0\cdot12\,,$$

$$z_7 = \tfrac{1}{2}[0\cdot35+0\cdot55] = 0\cdot45\,.$$

By using the intramolecular vectors between the pairs of atoms on opposite sides of the molecule, namely, 1 ··· 5, 2 ··· 6, 3 ··· 7, 4 ··· 8, and 9 ··· 10 (measured from figure 2.1.7f), together with the intermolecular vector (0·27, 0·35), we obtain the coordinates shown in table 2.1.7. The complete structure is shown in figure 2.1.7h.

The alternative choice of (0·23, 0·15) for the coincident intermolecular vectors would give a different set of atomic coordinates. These would however define the same structure but with reference to the point $(\frac{1}{4}, \frac{1}{4})$ as origin.

**2.1.8**

(+)-α-Bromocamphor ($C_{10}H_{15}OBr$) and (+)-α-chlorocamphor ($C_{10}H_{15}OCl$) are isomorphous, both being monoclinic with space group $P2_1$ and having the following unit-cell dimensions:

| | $a$ (Å) | $b$ (Å) | $c$ (Å) | $\beta$ |
|---|---|---|---|---|
| Bromocamphor | 7·38 | 7·57 | 9·12 | 94° 0′ |
| Chlorocamphor | 7·25 | 7·51 | 9·04 | 93° 15′ . |

In each case there are two molecules in the unit cell.

Structure-factor amplitudes for $h0l$ reflections of the two compounds, obtained by measurement of x-ray intensities, are given in table 2.1.8a

**Table 2.1.8a.**

| $h0l$ | Structure amplitudes ($\|F\|$) for | | $f_{Br}$ [a] | $f_{Cl}$ [a] | $\sin\theta$ [b] |
|---|---|---|---|---|---|
| | bromocamphor | chlorocamphor | | | |
| 001 | 45 | 36 | 34·21 | 16·45 | 0·085 |
| 100 | 13 | 10 | 33·82 | 16·16 | 0·105 |
| $10\bar{1}$ | 15 | 35 | 33·24 | 15·75 | 0·130 |
| 103 | 32 | 34 | 28·86 | 12·74 | 0·282 |
| 300 | 7 | <4 | 27·91 | 12·03 | 0·314 |
| $20\bar{3}$ | 19 | 11 | 27·81 | 11·95 | 0·318 |
| $30\bar{1}$ | 27 | 10 | 27·76 | 11·91 | 0·320 |
| 301 | 40 | 19 | 27·44 | 11·74 | 0·331 |
| 004 | 22 | 12 | 27·22 | 11·63 | 0·339 |
| $80\bar{1}$ | 9 | 7 | 17·51 | 6·99 | 0·836 |
| 800 | 8 | 4 | 17·49 | 6·99 | 0·837 |

[a] Atomic scattering factors, in electrons.
[b] Calculated from the bromocamphor unit-cell dimensions (and $\lambda = 1\cdot542$ Å). These same values can be used for chlorocamphor without introducing appreciable error.

(data from Wiebenga and Krom, 1946). These amplitudes are on an absolute scale, that is, in electrons, and calculated for the complete contents of the unit cell.

Given that the bromine atoms in bromocamphor lie at (0·264, 0, 0·164) and (−0·264, 0·5, −0·164) (fractional coordinates), use the isomorphous replacement method to deduce the phases of the reflections in table 2.1.8a. In your calculations use a temperature factor of $\exp(-1\cdot 6\sin^2\theta)$ for both compounds. Values of $\sin\theta$ and the atomic scattering factors of bromine and chlorine for each reflection are listed in table 2.1.8a.

**Solution**

The expression for the structure factor (see problem 2.1.4) is

$$F = A + \mathrm{i}B\,, \tag{2.1.8a}$$

where

$$A = \sum_{n=1}^{N} f_n \cos 2\pi(hx_n + ky_n + lz_n)\,,$$

$$B = \sum_{n=1}^{N} f_n \sin 2\pi(hx_n + ky_n + lz_n)\,,$$

the summations being over the $N$ atoms contained in the unit cell. In the space group $P2_1$ the set of general equivalent positions is $(x, y, z)$ and $(\bar{x}, \frac{1}{2}+y, \bar{z})$. The $h0l$ reflections have their phases restricted to 0° and 180° because the summation $B$ will contain terms in $\sin 2\pi(hx+lz)$ and in $\sin 2\pi(-hx-lz)$ which will cancel, so that $B = 0$. Hence, for $h0l$ reflections

$$F = A = 2\sum_{n=1}^{\frac{1}{2}N} f_n \cos 2\pi(hx_n + lz_n)\,. \tag{2.1.8b}$$

If $A$ is positive the phase is 0° and if it is negative the phase is 180°.

The structure factor for bromocamphor, $F_b$, may be separated into the contribution of the bromine atoms $F_{Br}$ and the contribution of the remainder of the molecule $F_r$, where

$$F_b = F_{Br} + F_r\,.$$

Since chlorocamphor and bromocamphor are isomorphous we may assume, to a first approximation, that the positions of corresponding atoms in the unit cells of the two compounds are the same (there are in fact slight differences in the positions as is shown by the differences in the unit-cell dimensions). Therefore, to this approximation, $F_r$ will be the same for both compounds and we can write the structure factor for chlorocamphor, $F_c$, as

$$F_c = F_{Cl} + F_r\,,$$

where $F_{Cl}$ is the contribution of the chlorine atoms; therefore,

$$F_b - F_c = F_{Br} - F_{Cl}\,. \tag{2.1.8c}$$

Since we know $|F_b|$ and $|F_c|$ from table 2.1.8a, we can find their signs by calculating $F_{Br}$ and $F_{Cl}$.

The unit cell contains two bromine atoms, at (0·264, 0, 0·164) and (−0·264, 0·5, −0·164); therefore

$$F_{Br} = 2f_{Br}\cos 2\pi(0{\cdot}264h + 0{\cdot}164l)\,. \qquad (2.1.8d)$$

In order to compare $(F_{Br} - F_{Cl})$ with the measured $|F_b|$ and $|F_c|$ we must include the temperature factor in the calculation of $F_{Br}$, equation (2.1.8d) being modified to

$$F_{Br} = 2f_{Br}\cos 2\pi(0{\cdot}264h + 0{\cdot}164l)\exp(-1{\cdot}6\sin^2\theta)\,.$$

Since the chlorine atom in chlorocamphor is assumed to be at the same position in the unit cell as the bromine atom

$$F_{Cl} = 2f_{Cl}\cos 2\pi(0{\cdot}264h + 0{\cdot}164l)\exp(-1{\cdot}6\sin^2\theta)\,;$$

therefore

$$F_{Br} - F_{Cl} = 2(f_{Br} - f_{Cl})\cos 2\pi(0{\cdot}264h + 0{\cdot}164l)\exp(-1{\cdot}6\sin^2\theta)\,. \quad (2.1.8e)$$

The calculation of $(F_{Br} - F_{Cl})$ for the various reflections is summarised in table 2.1.8b.

Since each of $F_b$ and $F_c$ can be either positive or negative there are four possible values of $(F_b - F_c)$ for each reflection (table 2.1.8c). Comparison of these with $(F_{Br} - F_{Cl})$ leads to the signs for $F_b$ and $F_c$ shown in the final columns of table 2.1.8c. $(F_{Br} - F_{Cl})$ and $(F_b - F_c)$ do not match perfectly because of experimental error in the intensity measurements. We see that it is not possible to deduce all the signs. For reflection 103, $|F_b|$ and $|F_c|$ are so close that the sign combinations ++ and −− give similar small values for $(F_b - F_c)$ and we cannot be certain which should be matched with $(F_{Br} - F_{Cl})$. In the case of reflection 300, $(F_{Br} - F_{Cl})$ happens to fall midway between the limiting values of $(F_b - F_c)$ for the sign combinations ++ and +−. Thus although we can say with

**Table 2.1.8b.**

| $h0l$ | $f_{Br} - f_{Cl}$ | $\cos 2\pi(0{\cdot}264h + 0{\cdot}164l)$ | $\exp(-1{\cdot}6\sin^2\theta)$ | $F_{Br} - F_{Cl}$ |
|---|---|---|---|---|
| 001 | 17·76 | 0·514 | 0·99 | 18·1 |
| 100 | 17·66 | −0·088 | 0·98 | −3·0 |
| $10\bar{1}$ | 17·49 | 0·809 | 0·97 | 27·4 |
| 103 | 16·12 | 0·038 | 0·88 | 1·1 |
| 300 | 15·88 | 0·261 | 0·85 | 7·0 |
| $20\bar{3}$ | 15·86 | 0·975 | 0·85 | 26·3 |
| $30\bar{1}$ | 15·85 | −0·694 | 0·85 | −18·7 |
| 301 | 15·70 | 0·962 | 0·84 | 25·4 |
| 004 | 15·59 | −0·557 | 0·83 | −14·4 |
| $80\bar{1}$ | 10·52 | 0·947 | 0·33 | 6·6 |
| 800 | 10·50 | 0·762 | 0·33 | 5·3 |

confidence that $F_b$ is positive, no judgement can be made as to the sign of $F_c$.

Wiebenga and Krom (1946) were able by this method to determine the signs of 116 $h0l$ reflections which were then used to calculate Fourier syntheses of the two structures in projection on the (010) plane. The shape of the two camphor molecules and the atomic positions were deduced from these Fourier syntheses.

**Table 2.1.8c.**

| $h0l$ | $F_{Br}-F_{Cl}$ | $F_b-F_c$ for $F_b, F_c$ sign combinations | | | | Deduced signs for | |
|---|---|---|---|---|---|---|---|
| | | ++ | +− | −+ | −− | $F_b$ | $F_c$ |
| 001 | +18 | +9 | +81 | −81 | −9 | + | + |
| 100 | −3 | +3 | +23 | −23 | −3 | − | − |
| $10\bar{1}$ | +27 | −20 | +50 | −50 | +20 | − | − |
| 103 | +1 | −2 | +66 | −66 | +2 | | |
| 300 | +7 | $>+3$ | $<+11$ | $>-11$ | $<-3$ | + | |
| $20\bar{3}$ | +26 | +8 | +30 | −30 | −8 | + | − |
| $30\bar{1}$ | −19 | +17 | +37 | −37 | −17 | − | − |
| 301 | +25 | +21 | +59 | −59 | −21 | + | + |
| 004 | −14 | +10 | +34 | −34 | −10 | − | − |
| $80\bar{1}$ | +7 | +2 | +16 | −16 | −2 | + | + |
| 800 | +5 | +4 | +12 | −12 | −4 | + | + |

2.1.9

The absolute configuration of the molecule D(−)-isoleucine hydrobromide, $CH_3CH_2CH(CH_3)CH(NH_2)COOH.HBr$, was found by measuring differences between the intensities of symmetry-related x-ray reflections when anomalous scattering occurs. The radiation used was uranium $L_{\alpha_1}$ which has its wavelength (0·911 Å) sufficiently close to the $K$ absorption edge of bromine (0·920 Å) to excite the bromine atoms.

With the structure factor for a reflection $hkl$ expressed as $F_{hkl} = A_{hkl}+iB_{hkl}$, for isoleucine hydrobromide $A_{hkl}$ and $B_{hkl}$ can be written

$$A_{hkl} = A_s+A_{Br}$$

and

$$B_{hkl} = B_s+B_{Br} ,$$

where $A_{Br}$ and $B_{Br}$ are the contributions of the bromine atoms in the unit cell, and $A_s$ and $B_s$ are the sums of the contributions of the remaining atoms (carbon, nitrogen, oxygen, and hydrogen). Further, $A_{Br}$ and $B_{Br}$ can be written

$$A_{Br} = f_{Br}A'_{Br} ,$$

$$B_{Br} = f_{Br}B'_{Br} ,$$

where $f_{\mathrm{Br}}$ is the atomic scattering factor of bromine and $A'_{\mathrm{Br}}$ and $B'_{\mathrm{Br}}$ embody the geometric structure factor and the temperature factor for the reflection concerned.

The intensities found for pairs of reflections $hkl$ and $\bar{h}\,\bar{k}\,\bar{l}$ when $\mathrm{U}L_{\alpha_1}$ radiation is diffracted by a crystal of D(−)-isoleucine hydrobromide monohydrate (space group $P2_12_12_1$) are compared in table 2.1.9a (data from Trommel and Bijvoet, 1954). A model of the molecule with its absolute configuration chosen to conform with the Fischer convention leads to the structure-factor components $A_{\mathrm{s}}$, $B_{\mathrm{s}}$, $A'_{\mathrm{Br}}$, and $B'_{\mathrm{Br}}$ listed in tables 2.1.9b and 2.1.9c. Use these components to calculate $|F_{hkl}|^2$ for each reflection in a pair, and, by comparing them with the intensity data, show that the model chosen is the correct enantiomorph. The bromine atomic scattering factors for the various reflections are given in table 2.1.9d.

**Table 2.1.9a.** Comparison of intensities of $hkl$ and $\bar{h}\,\bar{k}\,\bar{l}$ reflections.

| $hkl$ | $I_{hkl}$ | | $I_{\bar{h}\bar{k}\bar{l}}$ | $hkl$ | $I_{hkl}$ | | $I_{\bar{h}\bar{k}\bar{l}}$ |
|---|---|---|---|---|---|---|---|
| 111 | | $<$ | | 181 | | $>$ | |
| 121 | | $>$ | | 191 | | $>$ | |
| 131 | | $<$ | | 1 10 1 | | $>$ | |
| 141 | | $<$ | | 211 | | $>$ | |
| 151 | | $<$ | | 221 | | $>$ | |
| 171 | | $>$ | | 241 | | $<$ | |

**Table 2.1.9b.** Structure-factor contributions of all atoms except bromine.

| $hkl$ | $A_{\mathrm{s}}$ | $B_{\mathrm{s}}$ | $hkl$ | $A_{\mathrm{s}}$ | $B_{\mathrm{s}}$ |
|---|---|---|---|---|---|
| 111 | 47·08 | 12·66 | 181 | 3·51 | 17·75 |
| 121 | 13·95 | 20·91 | 191 | −6·12 | −11·22 |
| 131 | −19·84 | 17·38 | 1 10 1 | −12·37 | −32·91 |
| 141 | −4·63 | 15·53 | 211 | 18·44 | −31·89 |
| 151 | −6·71 | −67·24 | 221 | −19·93 | 20·98 |
| 171 | −4·90 | 23·79 | 241 | 15·87 | −9·02 |

**Table 2.1.9c.**

| $hkl$ | $A'_{\mathrm{Br}}$ | $B'_{\mathrm{Br}}$ | $hkl$ | $A'_{\mathrm{Br}}$ | $B'_{\mathrm{Br}}$ |
|---|---|---|---|---|---|
| 111 | 0·73 | 1·33 | 181 | 2·94 | −0·15 |
| 121 | 2·89 | −0·29 | 191 | −1·20 | 0·45 |
| 131 | −1·35 | 0·17 | 1 10 1 | −1·86 | −0·46 |
| 141 | −2·49 | −0·40 | 211 | 0·62 | −1·96 |
| 151 | 0·45 | −1·44 | 221 | 1·01 | 1·05 |
| 171 | 0·91 | 1·07 | 241 | −0·87 | 1·46 |

**Table 2.1.9d.** Atomic scattering factor for bromine, in electrons: $f_0$ is the scattering factor without dispersion correction; $\Delta f' = -4{\cdot}48$ e and $\Delta f'' = 3{\cdot}39$ e for uranium $L\alpha_1$ radiation.

| *hkl* | $f_0$ | *hkl* | $f_0$ | *hkl* | $f_0$ |
|---|---|---|---|---|---|
| (000 | 35·00) | 151 | 30·55 | 1101 | 26·71 |
| 111 | 32·57 | 171 | 29·06 | 211 | 29·16 |
| 121 | 32·28 | 181 | 28·25 | 221 | 28·98 |
| 131 | 31·81 | 191 | 27·47 | 241 | 28·34 |
| 141 | 31·22 | | | | |

**Solution**

The atomic scattering factor normally used in x-ray diffraction is the real number $f_0$. Since, however, the bromine atoms in the crystal have an absorption edge at a wavelength just a little longer than the wavelength of the incident x rays, anomalous scattering occurs and the atomic scattering factor for bromine must be represented by a complex quantity

$$f_{\mathrm{Br}} = f_0 + \Delta f' + \mathrm{i}\Delta f'' \text{ , or}$$

$$f_{\mathrm{Br}} = f' + \mathrm{i}\Delta f'' \text{ .}$$

Although $f_0$ varies rapidly with $\sin\theta/\lambda$, and hence with $h$, $k$, and $l$, (see table 2.1.9d and also data in problems 2.1.4 and 2.1.8), the dispersion corrections $\Delta f'$ and $\Delta f''$ change very little and may, with sufficient accuracy, be regarded as constants for all reflections. As a result of including the imaginary correction $\Delta f''$ the contribution of the bromine atoms to the structure factor undergoes a small phase advance $\phi$, where $\phi = \tan^{-1}(\Delta f''/f')$ (figure 2.1.9). This phase *advance* applies both to *hkl*

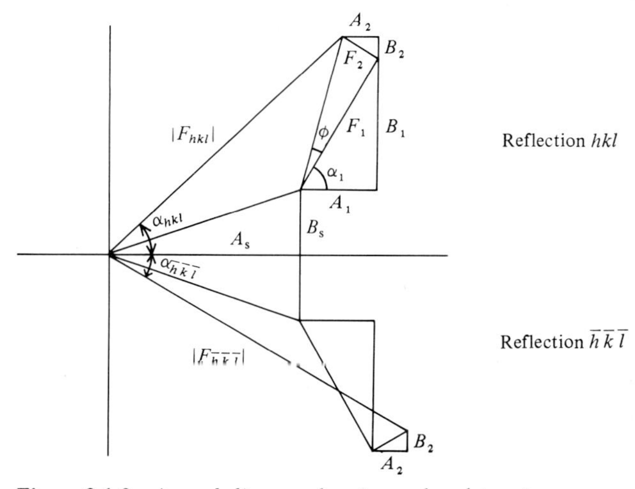


**Figure 2.1.9.** Argand diagram showing real and imaginary components of the structure factors for the reflections *hkl* and $\bar{h}\,\bar{k}\,\bar{l}$.

and to $\bar{h}\bar{k}\bar{l}$ reflections so that when the bromine atom contribution is added (vectorially) to that of the other atoms in the structure $|F_{hkl}|$ and $|F_{\bar{h}\bar{k}\bar{l}}|$ are no longer equal (figure 2.1.9). Also $\alpha_{hkl} \neq -\alpha_{\bar{h}\bar{k}\bar{l}}$.

For a reflection $hkl$ the bromine atom contribution is modified as follows. Let $A_1 = f'A'_{Br}$, $B_1 = f'B'_{Br}$, and $F_1^2 = A_1^2 + B_1^2$. Then, from figure 2.1.9,

$$F_2 = F_1 \tan\phi = F_1 \frac{\Delta f''}{f'} ,$$

$$A_2 = F_2 \sin\alpha_1 = F_1 \frac{\Delta f''}{f'} \sin\alpha_1 = B_1 \frac{\Delta f''}{f'} = B'_{Br} \Delta f'' ;$$

similarly,

$$B_2 = F_2 \cos\alpha_1 = F_1 \frac{\Delta f''}{f'} \cos\alpha_1 = A_1 \frac{\Delta f''}{f'} = A'_{Br} \Delta f'' .$$

Therefore the real part of the complete structure factor becomes

$$A_{hkl} = A_s + A_{Br} = A_s + A_1 - A_2 = A_s + f'A'_{Br} - \Delta f'' B'_{Br} , \tag{2.1.9a}$$

and the imaginary part

$$B_{hkl} = B_s + B_{Br} = B_s + B_1 + B_2 = B_s + f'B'_{Br} + \Delta f'' A'_{Br} . \tag{2.1.9b}$$

The intensity is given by

$$I_{hkl} \propto |F_{hkl}|^2 = A_{hkl}^2 + B_{hkl}^2 . \tag{2.1.9c}$$

On the other hand, for reflection $\bar{h}\bar{k}\bar{l}$ we obtain

$$A_{\bar{h}\bar{k}\bar{l}} = A_s + f'A'_{Br} + \Delta f'' B'_{Br} , \tag{2.1.9d}$$

$$B_{\bar{h}\bar{k}\bar{l}} = -B_s - f'B'_{Br} + \Delta f'' A'_{Br} = -(B_s + f'B'_{Br} - \Delta f'' A'_{Br}) \tag{2.1.9e}$$

(see figure 2.1.9), and

$$I_{\bar{h}\bar{k}\bar{l}} \propto |F_{\bar{h}\bar{k}\bar{l}}|^2 = A_{\bar{h}\bar{k}\bar{l}}^2 + B_{\bar{h}\bar{k}\bar{l}}^2 . \tag{2.1.9f}$$

The calculation of $|F_{hkl}|^2$ and $|F_{\bar{h}\bar{k}\bar{l}}|^2$ for the pairs of reflections from equations (2.1.9a)–(2.1.9f) is shown in table 2.1.9e. Wherever $I_{hkl} > I_{\bar{h}\bar{k}\bar{l}}$ (table 2.1.9a) we find $|F_{hkl}|^2 > |F_{\bar{h}\bar{k}\bar{l}}|^2$ and vice versa, showing that the model chosen was the correct enantiomorph. Had the enantiomorph been incorrect we would have found disagreement in each case instead of agreement.

By this method, and using the measurements presented in table 2.1.9a, Trommel and Bijvoet (1954) demonstrated that the Fischer convention agrees with the absolute configuration of D(−)-isoleucine as found in its hydrobromide.

Table 2.1.9e.

| $hkl$ | $A_s$ | $f'A'_{Br}$ | $\Delta f''B'_{Br}$ | $B_s$ | $f'B'_{Br}$ | $\Delta f''A'_{Br}$ |
|---|---|---|---|---|---|---|
| 111 | 47·08 | 20·51 | 4·51 | 12·66 | 37·36 | 2·47 |
| 121 | 13·95 | 80·34 | −0·98 | 20·91 | −8·06 | 9·80 |
| 131 | −19·84 | −36·90 | 0·58 | 17·38 | 4·65 | −4·58 |
| 141 | −4·63 | −66·58 | −1·36 | 15·53 | −10·70 | −8·44 |
| 151 | −6·71 | 11·73 | −4·88 | −67·24 | −37·54 | 1·53 |
| 171 | −4·90 | 22·37 | 3·63 | 23·79 | 26·30 | 3·08 |
| 181 | 3·51 | 69·88 | −0·51 | 17·75 | −3·57 | 9·97 |
| 191 | −6·12 | −27·59 | 1·53 | −11·22 | 10·35 | −4·07 |
| 1101 | −12·37 | −41·35 | −1·56 | −32·91 | −10·23 | −6·31 |
| 211 | 18·44 | 15·30 | −6·64 | −31·89 | −48·37 | 2·10 |
| 221 | −19·93 | 24·75 | 3·56 | 20·98 | 25·73 | 3·42 |
| 241 | 15·87 | −20·76 | 4·95 | −9·02 | 34·84 | −2·95 |

| | $A_{hkl}$ | $B_{hkl}$ | $A_{\bar{h}\bar{k}\bar{l}}$ | $B_{\bar{h}\bar{k}\bar{l}}$ | $\lvert F_{hkl}\rvert^2$ | $\lvert F_{\bar{h}\bar{k}\bar{l}}\rvert^2$ |
|---|---|---|---|---|---|---|
| 111 | 63·08 | 52·49 | 72·10 | −47·55 | 6734 | 7459 |
| 121 | 95·27 | 22·65 | 93·31 | −3·05 | 9589 | 8716 |
| 131 | −57·32 | 17·45 | −56·16 | −26·61 | 3590 | 3862 |
| 141 | −69·85 | −3·61 | −72·57 | −13·27 | 4892 | 5442 |
| 151 | 9·90 | −103·25 | 0·14 | 106·31 | 10759 | 11302 |
| 171 | 13·84 | 53·17 | 21·10 | −47·01 | 3019 | 2655 |
| 181 | 73·90 | 24·15 | 72·88 | −4·21 | 6044 | 5329 |
| 191 | −35·24 | −4·94 | −32·18 | −3·20 | 1266 | 1046 |
| 1101 | −52·16 | −49·45 | −55·28 | 36·83 | 5166 | 4412 |
| 211 | 40·38 | −78·16 | 27·10 | 82·36 | 7740 | 7518 |
| 221 | 1·26 | 50·13 | 8·38 | −43·29 | 2515 | 1944 |
| 241 | −9·84 | 22·87 | 0·06 | −28·77 | 620 | 828 |

## 2.1.10

Hexamethylmelamine,



,

forms hexagonal crystals with $a = 9 \cdot 99$, $c = 7 \cdot 11$ Å, space group $P6_3/m$. There are two molecules in the unit cell, with the molecular centres at $(\frac{2}{3}, \frac{1}{3}, \frac{1}{4})$ and $(\frac{1}{3}, \frac{2}{3}, \frac{3}{4})$, the molecular site-symmetry being $3/m$. The asymmetric unit thus comprises two nitrogen and three carbon atoms (one third of a molecule; hydrogen atoms are neglected), all situated on a mirror plane at $z = \frac{1}{4}c$. Their fractional coordinates and anisotropic thermal parameters are given in tables 2.1.10a and 2.1.10b (data from Bullen *et al.*, 1972).

Give the reason why $b_{13}$ and $b_{23}$ are zero for all nitrogen and carbon atoms. Calculate the elements of the thermal vibration tensor $\boldsymbol{U}$ for each atom, referred to orthogonal axes $a'$, $b$, $c$ (i.e. $a' \perp b$). Then, assuming that the molecule is undergoing rigid-body oscillations in the lattice, use the tensors $\boldsymbol{U}$ to calculate the symmetric tensors $\boldsymbol{T}$ and $\boldsymbol{\omega}$ describing the motion of the whole rigid molecule: $\boldsymbol{T}$ for translational vibrations of the molecular centre and $\boldsymbol{\omega}$ for angular oscillations about the centre.

Calculate the bond lengths from the coordinates in table 2.1.10a and then use $\boldsymbol{\omega}$ to obtain the corrections which must be applied to give the true bond lengths.

**Table 2.1.10a.**

| | $x$ | $y$ | $z$ |
|---|---|---|---|
| N(1) | 0·7915 | 0·4825 | 0·25 |
| N(2) | 0·6195 | 0·5713 | 0·25 |
| C(1) | 0·6446 | 0·4496 | 0·25 |
| C(2) | 0·7480 | 0·7299 | 0·25 |
| C(3) | 0·4634 | 0·5483 | 0·25 |

**Table 2.1.10b.** Anisotropic thermal parameters (× $10^4$), the temperature factor being in the form: $\exp[-(b_{11}h^2+b_{22}k^2+b_{33}l^2+2b_{12}hk+2b_{13}hl+2b_{23}kl)]$. For all atoms $b_{13} = b_{23} = 0$.

| | $b_{11}$ | $b_{22}$ | $b_{33}$ | $b_{12}$ |
|---|---|---|---|---|
| N(1) | 183 | 181 | 300 | 86 |
| N(2) | 255 | 200 | 439 | 139 |
| C(1) | 216 | 183 | 226 | 109 |
| C(2) | 359 | 203 | 561 | 137 |
| C(3) | 291 | 326 | 443 | 230 |

**Solution**

The anisotropic temperature factor has the form

$$\exp[-2\pi^2(U^c_{11}h^2a^{*2}+U^c_{22}k^2b^{*2}+U^c_{33}l^2c^{*2}+2U^c_{12}hka^*b^*+2U^c_{13}hla^*c^* + 2U^c_{23}klb^*c^*)] , \qquad (2.1.10a)$$

where $U^c_{11}$ etc. are the elements of a symmetric tensor

$$\boldsymbol{U}^c = \begin{pmatrix} U^c_{11} & U^c_{12} & U^c_{13} \\ U^c_{12} & U^c_{22} & U^c_{23} \\ U^c_{13} & U^c_{23} & U^c_{33} \end{pmatrix} , \qquad (2.1.10b)$$

which represents the mean square amplitude of vibration of the atom and is referred to the crystallographic axial system. For purposes of calculation and tabulation of data (as in table 2.1.10b) the temperature factor is often written in the form

$$\exp[-(b_{11}h^2+b_{22}k^2+b_{33}l^2+2b_{12}hk+2b_{13}hl+2b_{23}kl)] , \qquad (2.1.10c)$$

so that

$$b_{11} = 2\pi^2 a^{*2} U^c_{11}, \qquad b_{12} = 2\pi^2 a^* b^* U^c_{12}, \text{ and so on.} \tag{2.1.10d}$$

A tensor property such as $\boldsymbol{U}^c$ can be represented by an ellipsoid. The symmetry of the atomic position imposes restrictions on the orientation (and sometimes the shape) of the thermal vibration ellipsoid and hence on the values $U_{ij}$ or $b_{ij}$. In hexamethylmelamine the placing of all nitrogen and carbon atoms on the mirror plane perpendicular to $c$ requires the thermal ellipsoids to be oriented so that they all have a principal axis parallel to $c$. This makes $U^c_{13} = U^c_{23} = 0$ and hence $b_{13} = b_{23} = 0$ for all these atoms.

The crystallographic axial system to which $\boldsymbol{U}^c$ is referred is not orthogonal in the case of hexamethylmelamine and it is necessary to transform to the tensor $\boldsymbol{U}$ referred to orthogonal axes $a'$, $b$, $c$ before the molecular vibrations can be analysed. Transforming coordinates from hexagonal to orthogonal axes we have

$$\left.\begin{aligned} x' &= x\sin\gamma \\ y' &= x\cos\gamma + y \\ z' &= z \end{aligned}\right\}, \tag{2.1.10e}$$

so that the transformation matrix is

$$\boldsymbol{A} = \begin{pmatrix} \sin\gamma & 0 & 0 \\ \cos\gamma & 1 & 0 \\ 0 & 0 & 1 \end{pmatrix}.$$

The equation for transformation of the tensor (Lipson and Cochran, 1966) is

$$\boldsymbol{U} = (\boldsymbol{AD})\boldsymbol{U}^c(\boldsymbol{AD})^{\mathrm{T}}, \tag{2.1.10f}$$

where

$$\boldsymbol{D} = \begin{pmatrix} 1/\sin\gamma & 0 & 0 \\ 0 & 1/\sin\gamma & 0 \\ 0 & 0 & 1 \end{pmatrix},$$

$(\boldsymbol{AD})$ is obtained by post-multiplication of $\boldsymbol{A}$ by $\boldsymbol{D}$, and $(\boldsymbol{AD})^{\mathrm{T}}$ is the transpose of $(\boldsymbol{AD})$. Equation (2.1.10f) gives the following elements of $\boldsymbol{U}$

$$\left.\begin{aligned} U_{11} &= U^c_{11} \\ U_{22} &= \frac{U^c_{11}}{\tan^2\gamma} + \frac{2U^c_{12}}{\sin\gamma\tan\gamma} + \frac{U^c_{22}}{\sin^2\gamma} \\ U_{33} &= U^c_{33} \\ U_{12} &= \frac{U^c_{11}}{\tan\gamma} + \frac{U^c_{12}}{\sin\gamma} \\ U_{13} &= U_{23} = 0\,. \end{aligned}\right\} \tag{2.1.10g}$$

The $U_{ij}$ values for N(1), N(2), C(1), C(2), and C(3), calculated by use of equations (2.1.10g) and (2.1.10d), are given in table 2.1.10c.

The mean square amplitude of vibration of an atom in the direction of a unit vector $\boldsymbol{l}$ with components $l_i$ is

$$\overline{u^2} = \sum_{i=1}^{3} \sum_{j=1}^{3} U_{ij} l_i l_j \,. \tag{2.1.10h}$$

When the motion of the molecule as a rigid body is expressed in terms of the two symmetric tensors $\boldsymbol{T}$ and $\boldsymbol{\omega}$, the translational contribution to $\overline{u^2}$ is $\sum_{i=1}^{3} \sum_{j=1}^{3} T_{ij} l_i l_j$, and is the same for all atoms. The mean square amplitude of angular oscillation about an axis defined by a unit vector $\boldsymbol{t}$ through the molecular centre is $\sum_{i=1}^{3} \sum_{j=1}^{3} \omega_{ij} t_i t_j$. An atom at $\boldsymbol{r} = (X, Y, Z)$ (defined in terms of the orthogonal coordinate system and with the molecular centre as origin) will move in the direction $\boldsymbol{l}$ when the molecule turns about an axis parallel to $\boldsymbol{l} \wedge \boldsymbol{r}$ and, if the angle of oscillation $\epsilon$ is small, the displacement of the atom in direction $\boldsymbol{l}$ will be $|\boldsymbol{l} \wedge \boldsymbol{r}|\epsilon$. Therefore, provided $\epsilon$ is small, the contribution of $\omega$ to the displacement in direction $\boldsymbol{l}$ will be $\sum_{i=1}^{3} \sum_{j=1}^{3} \omega_{ij} (\boldsymbol{l} \wedge \boldsymbol{r})_i (\boldsymbol{l} \wedge \boldsymbol{r})_j$, and this will be different for different atoms. Therefore

$$\sum_{i=1}^{3} \sum_{j=1}^{3} U_{ij} l_i l_j = \sum_{i=1}^{3} \sum_{j=1}^{3} \{T_{ij} l_i l_j + \omega_{ij} (\boldsymbol{l} \wedge \boldsymbol{r})_i (\boldsymbol{l} \wedge \boldsymbol{r})_j\} \,. \tag{2.1.10j}$$

**Table 2.1.10c.** Atomic coordinates and thermal vibration tensors (Å$^2$) referred to orthogonal axes.

| | $x$ | $y$ | $X$ (Å) | $Y$ (Å) | $U_{11}$ | $U_{22}$ | $U_{33}$ | $U_{12}$ |
|---|---|---|---|---|---|---|---|---|
| N(1) | 0·7915 | 0·4825 | 1·0797 | 0·8666 | 0·0694 | 0·0712 | 0·0768 | −0·0024 |
| N(2) | 0·6195 | 0·5713 | −0·4084 | 2·6129 | 0·0967 | 0·0631 | 0·1124 | 0·0050 |
| C(1) | 0·6446 | 0·4496 | −0·1912 | 1·2717 | 0·0819 | 0·0647 | 0·0579 | 0·0004 |
| C(2) | 0·7480 | 0·7299 | 0·7034 | 3·5554 | 0·1361 | 0·0787 | 0·1437 | −0·0186 |
| C(3) | 0·4634 | 0·5483 | −1·7589 | 3·1628 | 0·1103 | 0·0853 | 0·1135 | 0·0370 |
| | | | $X'$ (Å) | $Y'$ (Å) | | | | |
| N(1′) | | | 0·2106 | −1·3683 | 0·0728 | 0·0678 | 0·0768 | 0·0002 |
| N(2′) | | | 2·4670 | −0·9528 | 0·0672 | 0·0926 | 0·1124 | 0·0120 |
| C(1′) | | | 1·1969 | −0·4703 | 0·0687 | 0·0779 | 0·0579 | 0·0072 |
| C(2′) | | | 2·7274 | −2·3869 | 0·1092 | 0·1056 | 0·1437 | 0·0342 |
| C(3′) | | | 3·6185 | −0·0581 | 0·0595 | 0·1361 | 0·1135 | −0·0077 |
| N(1″) | | | −1·2903 | 0·5017 | 0·0687 | 0·0719 | 0·0768 | 0·0022 |
| N(2″) | | | −2·0586 | −1·6602 | 0·0758 | 0·0840 | 0·1124 | −0·0170 |
| C(1″) | | | −1·0057 | −0·8014 | 0·0693 | 0·0773 | 0·0579 | −0·0076 |
| C(2″) | | | −3·4308 | −1·1685 | 0·0769 | 0·1379 | 0·1437 | −0·0156 |
| C(3″) | | | −1·8596 | −3·1047 | 0·1236 | 0·0720 | 0·1135 | −0·0293 |

There is an equation of type (2.1.10j) for each atom in the molecule, $\boldsymbol{T}$ and $\boldsymbol{\omega}$ being the same throughout but $\boldsymbol{U}$ and $\boldsymbol{r}$ differing from one atom to another.

We have already noted that for all atoms $U_{13} = U_{23} = 0$. Further, by Neumann's Principle, the molecular tensors $\boldsymbol{T}$ and $\boldsymbol{\omega}$ must possess at least the symmetry of the molecule ($3/m$) and hence

$$T_{11} = T_{22} \,,$$

$$T_{12} = T_{13} = T_{23} = 0 \,,$$

$$\omega_{11} = \omega_{22} \,,$$

and

$$\omega_{12} = \omega_{13} = \omega_{23} = 0 \,.$$

Expanding the two sides of equation (2.1.10j), we obtain

$$U_{11}l_1^2 + U_{22}l_2^2 + U_{33}l_3^2 + 2U_{12}l_1l_2 = T_{11}(l_1^2 + l_2^2) + T_{33}l_3^2 + \omega_{11}[(l_2Z - l_3Y)^2 + (l_1Z - l_3X)^2] + \omega_{33}(l_1Y - l_2X)^2 \,. \qquad (2.1.10\text{k})$$

By grouping coefficients of $l_il_j$ and equating them on the two sides of equation (2.1.10k), it can be seen that

$$\left.\begin{aligned} U_{11} &= T_{11} + Z^2\omega_{11} + Y^2\omega_{33} \,, \\ U_{22} &= T_{11} + Z^2\omega_{11} + X^2\omega_{33} \,, \\ U_{33} &= T_{33} + (X^2 + Y^2)\omega_{11} \,, \\ U_{12} &= -XY\omega_{33} \,. \end{aligned}\right\} \qquad (2.1.10\text{l})$$

Since, relative to the molecular centre, all atoms have $Z = 0$, we have

$$\left.\begin{aligned} U_{11} &= T_{11} + Y^2\omega_{33} \,, \\ U_{22} &= T_{11} + X^2\omega_{33} \,, \\ U_{33} &= T_{33} + (X^2 + Y^2)\omega_{11} \,, \\ U_{12} &= -XY\omega_{33} \,. \end{aligned}\right\} \qquad (2.1.10\text{m})$$

As there are 20 $U_{ij}$ values (derived from the 20 $b_{ij}$ in table 2.1.10b) from which only the four unknowns $T_{11}$, $T_{33}$, $\omega_{11}$, and $\omega_{33}$ are to be found, it is best to use a least-squares procedure for the solution (Cruickshank, 1956a). The normal equations comprise a third-order and a second-order equation:

$$\begin{bmatrix} n & 0 & \Sigma Y^2 \\ 0 & n & \Sigma X^2 \\ \Sigma Y^2 & \Sigma X^2 & \Sigma(X^4 + Y^4 + X^2Y^2) \end{bmatrix} \begin{bmatrix} T_{11} \\ T_{11} \\ \omega_{33} \end{bmatrix} = \begin{bmatrix} \Sigma U_{11} \\ \Sigma U_{22} \\ \Sigma(Y^2U_{11} + X^2U_{22} - XYU_{12}) \end{bmatrix} \qquad (2.1.10\text{n})$$

and

$$\begin{bmatrix} n & \Sigma(X^2+Y^2) \\ \Sigma(X^2+Y^2) & \Sigma(X^2+Y^2)^2 \end{bmatrix} \begin{bmatrix} T_{33} \\ \omega_{11} \end{bmatrix} = \begin{bmatrix} \Sigma U_{33} \\ \Sigma(X^2+Y^2)U_{33} \end{bmatrix} , \qquad (2.1.10\text{p})$$

the summations being over all ($n$) atoms in the molecule. Note that for hexamethylmelamine $n = 15$, it being necessary to include all atoms related by symmetry to those listed in table 2.1.10a.

In order to solve equations (2.1.10n) and (2.1.10p) we must first obtain the coordinates (in Å, relative to orthogonal axes with the molecular centre as origin) and the $U_{ij}$ values for the fifteen atoms from the data in tables 2.1.10a and 2.1.10c.

*Transformation of coordinates*

The coordinates in table 2.1.10a are fractional and referred to hexagonal axes. They are transformed to coordinates in Å referred to orthogonal axes with the molecular centre as origin as follows.

(a) If the origin is moved to the molecular centre, the fractional coordinates of an atom become

$$x' = x - \tfrac{2}{3} ,$$
$$y' = y - \tfrac{1}{3} ,$$
$$z' = z - \tfrac{1}{4} \quad (= 0 \text{ for all atoms}).$$

On transformation to orthogonal axes $X$ and $Y$ (see figure 2.1.10a), the coordinates (in Å) become

$$X = x'a \sin\gamma = (x - \tfrac{2}{3})a(\tfrac{3}{4})^{1/2} , \qquad (2.1.10\text{q})$$

$$Y = x'a \cos\gamma + y'b = -\tfrac{1}{2}(x - \tfrac{2}{3})a + (y - \tfrac{1}{3})a = (-\tfrac{1}{2}x + y)a . \qquad (2.1.10\text{r})$$

$X$ and $Y$ are listed in table 2.1.10c. $Z$ is zero for all atoms.

(b) The complete molecule is shown in figure 2.1.10b. Atoms N(1′) and N(1″) are related to N(1) by rotations through 120° and 240° respectively. The $X$ and $Y$ coordinates transform under rotation through an angle $\phi$ about $Z$ according to the equations

$$\left.\begin{aligned} X' &= X\cos\phi + Y\sin\phi , \\ Y' &= -X\sin\phi + Y\cos\phi . \end{aligned}\right\} \qquad (2.1.10\text{s})$$

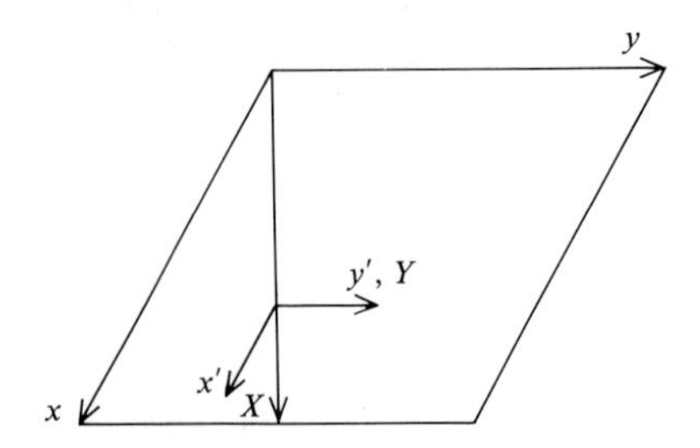


**Figure 2.1.10a.**

Substitution of $\phi = 120°$ or $240°$ in these equations leads to the coordinates for the other ten atoms in table 2.1.10c.

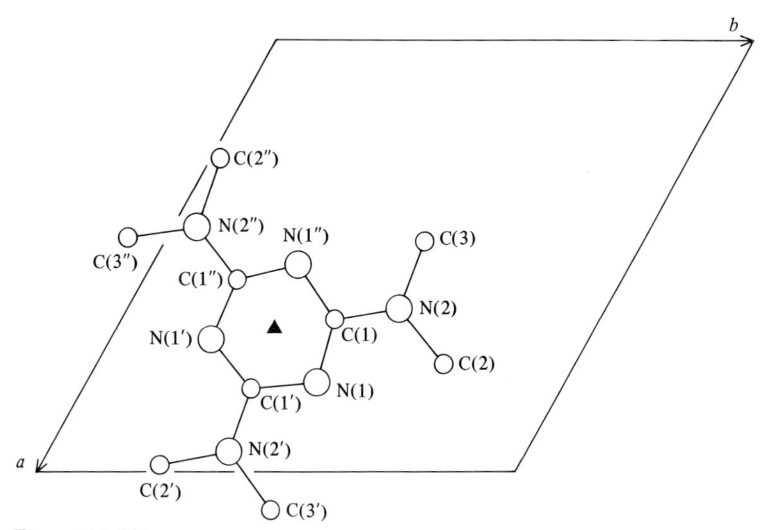


**Figure 2.1.10b.**

*Transformation of $U_{ij}$*

The $\boldsymbol{U}$ tensors for the symmetry-related atoms must be obtained by applying a threefold rotation operation to the tensors listed for N(1) etc. in table 2.1.10c. The tensor elements transform under the rotation through an angle $\phi$ according to

$$\boldsymbol{U}' = \boldsymbol{B}\begin{pmatrix} U_{11} & U_{12} & 0 \\ U_{12} & U_{22} & 0 \\ 0 & 0 & U_{33} \end{pmatrix}\boldsymbol{B}^{\mathrm{T}} , \tag{2.1.10t}$$

where

$$\boldsymbol{B} = \begin{pmatrix} \cos\phi & \sin\phi & 0 \\ -\sin\phi & \cos\phi & 0 \\ 0 & 0 & 1 \end{pmatrix} .$$

The $U_{ij}$ values, calculated for the symmetry-related atoms by putting $\phi = 120°$ or $240°$, are also given in table 2.1.10c. From the data in this table, summing over all fifteen atoms, we have

$$\begin{aligned}
&\Sigma X^2 = \Sigma Y^2 = 55\cdot 196 \text{ Å}^2 , \\
&\Sigma X^4 = \Sigma Y^4 = 449\cdot 33 \text{ Å}^4 , \\
&\Sigma X^2 Y^2 = 149\cdot 78 \text{ Å}^4 , \\
&\Sigma U_{11} = \Sigma U_{22} = 1\cdot 2861 \text{ Å}^2 , \\
&\Sigma Y^2 U_{11} = \Sigma X^2 U_{22} = 6\cdot 0705 \text{ Å}^4 , \\
&\Sigma - XYU_{12} = 0\cdot 8107 \text{ Å}^4 , \\
&\Sigma U_{33} = 1\cdot 5129 \text{ Å}^2 , \\
&\Sigma Y^2 U_{33} = \Sigma X^2 U_{33} = 6\cdot 6049 \text{ Å}^4 .
\end{aligned}$$

The equalities $\Sigma X^2 = \Sigma Y^2$, $\Sigma Y^2 U_{11} = \Sigma X^2 U_{22}$, etc., result from the high symmetry of the molecule. Substitution in equation (2.1.10n) gives

$$\left.\begin{aligned} 15T_{11}+55\cdot196\omega_{33} &= 1\cdot2861\,, \\ 15T_{11}+55\cdot196\omega_{33} &= 1\cdot2861\,, \\ 2\times55\cdot196T_{11}+1048\cdot44\omega_{33} &= 12\cdot9517\,, \end{aligned}\right\} \tag{2.1.10u}$$

one of the three equations being redundant because of the equalities noted above. Solution of equations (2.1.10u) gives

$$T_{11} = 0\cdot0658\ \text{Å}^2, \quad \text{and} \quad \omega_{33} = 0\cdot00543\ \text{rad}^2\,.$$

Similarly, use of equation (2.1.10p) gives

$$\left.\begin{aligned} 15T_{33}+110\cdot392\omega_{11} &= 1\cdot5129\,, \\ 110\cdot392T_{33}+1198\cdot22\omega_{11} &= 13\cdot2098\,, \end{aligned}\right\} \tag{2.1.10v}$$

from which $T_{33} = 0\cdot0613$ Å$^2$ and $\omega_{11} = 0\cdot00538$ rad$^2$.

Hence

$$T = \begin{pmatrix} 0\cdot0658 & 0 & 0 \\ 0 & 0\cdot0658 & 0 \\ 0 & 0 & 0\cdot0613 \end{pmatrix} \text{Å}^2\,,$$

and

$$\omega = \begin{pmatrix} 0\cdot00538 & 0 & 0 \\ 0 & 0\cdot00538 & 0 \\ 0 & 0 & 0\cdot00543 \end{pmatrix} \text{rad}^2$$

$$= \begin{pmatrix} 17\cdot66 & 0 & 0 \\ 0 & 17\cdot66 & 0 \\ 0 & 0 & 17\cdot83 \end{pmatrix} \text{deg}^2.$$

Bond lengths calculated from the atomic coordinates ($X$, $Y$) are listed in table 2.1.10d. Cruickshank (1956b, 1961) has shown that atomic positions obtained by an x-ray analysis of a crystal in which the molecules are undergoing angular oscillations are in error because there is an apparent displacement of the atomic position towards the molecular centre by an amount

$$\Delta r = \tfrac{1}{2}r(\psi_1^2+\psi_2^2)\,, \tag{2.1.10w}$$

**Table 2.1.10d.** Bond lengths (Å).

| | Uncorrected | Corrected |
|---|---|---|
| C(1)—N(1) | 1·334 | 1·341 |
| C(1)—N(1″) | 1·342 | 1·349 |
| C(1)—N(2) | 1·359 | 1·366 |
| C(2)—N(2) | 1·458 | 1·465 |
| C(3)—N(2) | 1·458 | 1·466 |

where $r$ is the distance of the atom from the molecular centre and $\psi_1^2$ and $\psi_2^2$ are the mean square amplitudes of angular oscillation of the molecule about two mutually perpendicular axes which are also perpendicular to $\boldsymbol{r}$. In the present case, because the molecule is planar and of high symmetry, $(\psi_1^2+\psi_2^2)$ is the same for all atoms and is equal to $(\omega_{33}+\omega_{11})$. This is explained as follows. As $\boldsymbol{\omega}$ has the special form

$$\begin{pmatrix} \omega_{11} & 0 & 0 \\ 0 & \omega_{11} & 0 \\ 0 & 0 & \omega_{33} \end{pmatrix},$$

the angular oscillation of the molecule can be described in terms of oscillations about three principal axes parallel to $a'$, $b$, $c$, with mean square amplitudes $\omega_{11}$, $\omega_{11}$, $\omega_{33}$, respectively. Since the vectors $\boldsymbol{r}$ are parallel to (001) for all atoms, we may always take $c$ as the axis of oscillation $\psi_1$ and hence put $\psi_1^2 = \omega_{33}$. The direction of the second axis (of oscillation $\psi_2$), although different for different atoms, is immaterial because it is always parallel to (001) and the section of the $\boldsymbol{\omega}$ ellipsoid parallel to (001) is circular. Hence $\psi_2^2$ is always equal to $\omega_{11}$, and $\Delta r = \frac{1}{2}r(\omega_{33}+\omega_{11})$. The consequent corrections $\Delta X = \frac{1}{2}X(\omega_{33}+\omega_{11})$ and $\Delta Y = \frac{1}{2}Y(\omega_{33}+\omega_{11})$ to the atomic coordinates are given in table 2.1.10e. The corrected (true) bond lengths are presented in table 2.1.10d.

**Table 2.1.10e.** Corrections to the atomic coordinates (Å).

| | $\Delta X$ | $\Delta Y$ | $X+\Delta X$ | $Y+\Delta Y$ |
|---|---|---|---|---|
| N(1) | 0·0058 | 0·0047 | 1·0855 | 0·8713 |
| N(2) | −0·0022 | 0·0141 | −0·4106 | 2·6270 |
| C(1) | −0·0010 | 0·0069 | −0·1922 | 1·2786 |
| C(2) | 0·0038 | 0·0192 | 0·7072 | 3·5746 |
| C(3) | −0·0095 | 0·0171 | −1·7684 | 3·1799 |
| N(1″) | −0·0070 | 0·0027 | −1·2973 | 0·5044 |

**References**

Bullen, G. J., Clark, N. H., 1969, *J. Chem. Soc. A.*, 404.
Bullen, G. J., Corney, D. J., Stephens, F. S., 1972, *J. Chem. Soc. Perkin Trans. 2*, 642.
Bullen, G. J., Tucker, P. A., 1972, *J. Chem. Soc. Dalton Trans.*, 2437.
Cruickshank, D. W. J., 1956a, *Acta Crystallogr.*, **9**, 754.
Cruickshank, D. W. J., 1956b, *Acta Crystallogr.*, **9**, 757.
Cruickshank, D. W. J., 1961, *Acta Crystallogr.*, **14**, 896.
Frasson, E., Panattoni, C., Sacconi, L., 1964, *Acta Crystallogr.*, **17**, 85.
Hazekamp, R., Migchelsen, T., Vos, A., 1962, *Acta Crystallogr.*, **15**, 539.
*International Tables for x-Ray Crystallography*, 1952, Volume 1, pp.49, 50 (Kynoch Press, Birmingham).
Lipson, H., Cochran, W., 1966, *The Determination of Crystal Structures* (Bell, London), p.302.
Perri, J. A., LaPlaca, S., Post, B., 1958, *Acta Crystallogr.*, **11**, 310.
Trommel, J., Bijvoet, J. M., 1954, *Acta Crystallogr.*, **7**, 703.
Wiebenga, E. H., Krom, C. J., 1946, *Rec. Trav. Chim. Pays-Bas*, **65**, 663.

## 2.2 Neutron diffraction

**J C Speakman** University of Glasgow

### 2.2.0 Introduction

According to the de Broglie equation, $\lambda = h/mv$, particles moving with a momentum $mv$ are associated with a wavelength $\lambda$. They may therefore suffer diffraction. For slow 'thermal' neutrons the wavelength is about 1 Å ($10^{-10}$ m) (see problem 2.2.1), which is of the same order of magnitude as that of the x rays used in crystal-structure analysis.

So far as its elementary mathematical formulation is concerned, the diffraction of neutrons by a crystal is identical to that of x rays. In both cases the intensity of a particular reflection, defined by the Miller indices $h$, $k$, and $l$, depends on the square of a structure factor, $F$; and, for a centrosymmetric structure, this can be calculated by the equation

$$F = \sum_i T_i f_i \cos 2\pi(hx_i + ky_i + lz_i) , \qquad (2.2.0a)$$

where $x_i$, $y_i$, and $z_i$ are the fractional coordinates of an atom $i$ within the unit cell and the summation is over all the atoms in the cell. $T_i$ is a factor which diminishes the contribution of each atom to $F$ according to the degree to which that atom is 'smeared' by its vibrations.

The difference between x rays and neutrons lies principally in $f$, the atomic scattering factor. As x rays are scattered by the cloud of electron density around the atomic nucleus, the atomic scattering factor $f^{x}$ depends on the atomic number, $Z$. In fact the maximum value of $f^{x}$ is equal to $Z$, but because the spatial spread of the cloud is comparable with the wavelength of the x rays, $f^{x}$ falls rapidly with increasing Bragg angle, $\theta$. With carbon or oxygen atoms, for example, $f^{x}$ falls to less than $\frac{1}{4}Z$ for reflections near the limits of an x-ray photograph. With hydrogen, which has no inner-shell electrons, the situation is particularly unfavourable; not only does this element have the lowest starting value of $f^{x}$ ($Z = 1$), but also the decline with $\theta$ is most severe: to about 0·02 near the photographic limit. Hence x rays are ill-adapted for locating hydrogen atoms in crystals (though with modern methods of x-ray analysis they are usually detectable unless the structure contains a number of heavy atoms).

In general, neutrons are scattered only by the atomic nuclei. The scattering factor, $f^{n}$, depends on the particular nuclide. It tends to increase with atomic number, but this trend is small and it is heavily overborne by specific interactions occurring during the collision between a neutron and the nucleus at which it is being scattered. Furthermore, the atomic nucleus is virtually a point compared with the neutron wavelength, so that $f^{n}$ does not diminish with $\theta$.

A different sort of unit is needed for $f^{n}$. The scattering power of any isotopic species is connected, though not simply connected, with the diameter of the nucleus, which is of the order of $10^{-12}$ cm.

Therefore neutron scattering factors are often expressed in units of $10^{-12}$ cm. However a new, ad hoc unit has recently been introduced and may be more convenient. This is the fermi, which is defined as 1 femtometre (fm) or $10^{-13}$ cm. Table 2.2.0 lists some representative values of $f^n$, some of which will be needed in the problems. In some cases, the value given is the effective average for the usual isotopic mixture.

For a few nuclides $f^n$ is negative. This merely means that there is an unorthodox change of phase during the scattering of the neutron from these particular nuclei. Since the proportion of deuterium present in ordinary hydrogen is negligible for this purpose, a consequence of the negative value of $f^n$ for H is that hydrogen atoms appear as negative peaks—or, less deviously, as hollows—in 'Fourier maps' based on neutron diffraction measurements. This is in contrast to x-ray analysis: in an electron-density synthesis all significant peaks are positive.

Hydrogen atoms are much easier to detect with neutrons than with x rays: $f^n_H$, though negative, is comparable in magnitude to other scattering factors; moreover, it does not decrease with $\theta$. Deuterium has a larger and positive factor. If we use a deuterated crystal, hydrogen atoms can be located as precisely as can carbon atoms. (Problems 2,2.2, 2.2.3, and 2.2.4 are concerned with various aspects of hydrogen atom location.)

Another application of neutron diffraction in chemistry is to discriminate between atoms of nearly equal atomic number, which might be hard to distinguish with x rays. Cobalt and nickel, with atomic numbers 27 and 28, would be rather difficult to distinguish even by modern x-ray methods. They are easily distinguished by neutrons, as can be surmised from table 2.2.0. A classic example is the spinel, $MgAl_2O_4$, in which we can determine the distribution of the three Mg/Al cations among the single four-coordinated, and the two six-coordinated, sites. Problem 2.2.5 illustrates a more sophisticated application of this principle.

Neutron diffraction is almost always used as an adjunct to x-ray analysis. A crystal structure is normally determined by x rays first; neutrons are then used only if the problems needing further elucidation

**Table 2.2.0.** Some neutron scattering factors.

| Nuclide | $f^n$ (fm) | Nuclide | $f^n$ (fm) |
|---|---|---|---|
| H | −3·72 | Al | 3·45 |
| D | 6·70 | K | 3·7 |
| Li | −1·94 | Cl | 9·58 |
| C | 6·63 | Co | 2·5 |
| N | 9·40 | Ni | 1·03 |
| O | 5·75 | Gd | ~15 |
| Na | 3·51 | Th | 9·9 |
| Mg | 5·2 | U | 8·4 |

seem important enough to justify the expense. Neutron diffraction is a more difficult technique. The primary reason is that the flux of neutrons in the monochromated beam is several powers of ten lower than the flux of photons in an ordinary x-ray beam. Only the largest research reactors produce an adequate flux of neutrons; this is the chief reason for high cost. A further disadvantage is that, to get reasonable counts in the various reflected beams, a wide beam of neutrons and a large crystal must be used. Whereas the single crystals used in x-ray work have linear dimensions in tenths of a millimetre, dimensions of millimetres are needed for neutron diffraction. It is by no means always possible to grow good-quality crystals of such a size.

The commonest use of neutron diffraction by chemists is for the better determination of hydrogen atom positions. A subtler advantage is that the vibrations of any atoms are more accurately measured. The factor $T_i$ in equation (2.2.0a) is virtually the only cause of the 'smearing' of the position of the atom as found by neutron diffraction. With x rays there is also the inherent spatial spread of the electron-density cloud. Moreover, to avoid severe computational difficulties, it is almost always assumed in x-ray studies that the atom is of spherical shape. This assumption works surprisingly well, but it cannot be strictly true when, for instance, there is covalent bonding. Hence the apparent vibrational parameters found for an atom by x rays may include distortion of the electron cloud by bonding. The vibrational parameters found with neutrons are more trustworthy. Indeed, by a combination of x-ray and neutron analyses, provided that each is done with high precision, it may be possible to explore the genuine distortion of the spherical atom when it becomes involved in bonding.

**2.2.1 Production of a beam of monochromatic neutrons from the thermal neutrons issuing from the core of the reactor**

(a) Take the temperature of the reactor core to be 300 K. The most probable velocity, $v_{\mathrm{mp}}$, for the molecules of any gas is given by the equation

$$v_{\mathrm{mp}} = \left(\frac{2kT}{m}\right)^{1/2}, \qquad (2.2.1a)$$

where $k$ is the Boltzmann constant, $T$ the temperature, and $m$ the molecular mass. Use the de Broglie equation to calculate the wavelength $\lambda$ corresponding to $v_{\mathrm{mp}}$.

(b) For a given gas at a given temperature, the relative proportion of molecules with a particular velocity $v$, $r(v)$, may be calculated from the Maxwell distribution law in the simplified form,

$$r(v) = v^2 \exp\left(-\frac{mv^2}{2kT}\right).$$

Use this relationship to work out the relative numbers of neutrons with wavelengths 0·5, 0·7, 1·0, 1·5, 1·78, 2·0, 3·0, and 4·0 Å. Plot a graph (of the form of figure 2.2.1) showing how this relative number varies with wavelength for thermal neutrons of the type in part (a).
(c) A 'monochromatic' neutron beam is obtained by allowing the beam of 'white' neutrons, i.e. neutrons with a Maxwellian distribution of velocities, to fall upon a particular face of a large single crystal of (say) copper, at a Bragg angle set to reflect the desired wavelength.

To secure a maximum neutron intensity, one would naturally choose the wavelength found in part (a). However, as there would then also be in the beam appreciable numbers of neutrons with wavelengths $\lambda/2$, or even $\lambda/3$, these would also be reflected from the copper crystal to yield second-order or even third-order reflections. This would contaminate the monochromatised beam. To avoid this trouble, it is better to choose a rather shorter wavelength, for which contamination by overtones is negligible. Consider this problem in relation to the graph obtained in (b) and suggest a wavelength which will keep harmonic contamination below 2% without wasting too much neutron flux.

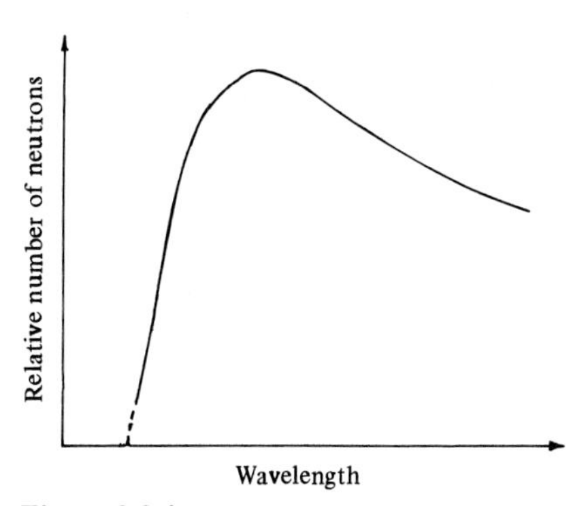


**Figure 2.2.1.**

**Solution**

(a) The rest mass of the neutron, $m_n$, is 1·009 atomic mass units (a.m.u.) i.e. $m_n = 1\cdot009 \times 10^{-3}/N_A$ kg, where $N_A$ is the Avogadro number. At $T = 300$ K, with $k = 1\cdot380 \times 10^{-23}$ J K$^{-1}$ and $N_A = 6\cdot022 \times 10^{23}$ mol$^{-1}$, we obtain from equation (2.2.1a)

$$v_{mp} = \left(\frac{600 \times 1\cdot380 \times 10^{-23} \times 6\cdot022 \times 10^{23}}{1\cdot009 \times 10^{-3}}\right)^{1/2}$$

$$= 2223 \text{ m s}^{-1}.$$

By the de Broglie equation,

$$\lambda = \frac{h}{mv},$$

with $h = 6 \cdot 626 \times 10^{-34}$ J s, we obtain

$$\lambda = \frac{6 \cdot 626 \times 10^{-34} \times 6 \cdot 022 \times 10^{23}}{1 \cdot 009 \times 2223 \times 10^{-3}}$$

$$= 1 \cdot 78 \times 10^{-10} \text{ m}$$

$$= 1 \cdot 78 \text{ Å} .$$

(b) This problem is best worked out in the form of a table, with eight rows and seven columns. In the first column, enter the wavelengths $\lambda$. In the second, enter $v$, which can easily be calculated since, for a given kind of particle (neglecting relativity effects), $v\lambda =$ is fixed, and we already have one pair of values from part (a). In the third, enter $v^2$, which is $15 \cdot 64 \times 10^6$ m$^2$ s$^{-2}$ for $\lambda = 1 \cdot 0$ Å. In the fourth, enter $mv^2/2kT$; this may be obtained by multiplying the value in column three by the constant factor $m/2kT$ which is equal to $2 \cdot 023 \times 10^{-7}$ m J$^{-1}$, giving $3 \cdot 164$ in the row we are following in detail. In the fifth, enter the exponential of $(-mv^2/2kT)$; this has the value $0 \cdot 0423$ for $\lambda = 1$ Å. In the sixth, enter the product of columns three and five ($6 \cdot 609 \times 10^5$ m$^2$ s$^{-2}$). These relative numbers can be normalised by dividing each by the highest value, which, naturally, will be at $1 \cdot 78$ Å. The numbers in the seventh column should then be 0·0001, 0·028, 0·363, 0·937, 1·000, 0·975, 0·672, and 0·441. Plot these, vertically, against the wavelengths, horizontally.
(c) Satisfactory wavelengths may be expected in the range 1·00–1·35 Å.

**2.2.2 Location of hydride ions in sodium hydride**

Sodium hydride, NaH, crystallises in the cubic system with lattice parameter $a = 4 \cdot 89$ Å. It was supposed to have the rock-salt (NaCl) structure. This was certainly true so far as the $Na^+$ ions were concerned, but the positions of the $H^-$ ions could not be verified in early x-ray work because the scattering from hydrogen is very weak.

Verification came in 1948 when (then) recent developments made neutron diffraction practicable. Figure 2.2.2 shows powder-pattern intensities from samples of NaH and NaD, the neutron wavelength being 1·08 Å. Only the first two reflections are covered, and their indexing is given. For the rock-salt type of structure, and when the vibrational or temperature factor [$T_i$ in equation (2.2.0a)] is ignored, the structure-factor expression reduces to (see problem 2.1.4):

$$F(hkl) = 4f_A + 4f_B, \qquad h, k, l \text{ all even;}$$
$$F(hkl) = 4f_A - 4f_B, \qquad h, k, l \text{ all odd;}$$

here $f_A$ and $f_B$ are the atomic scattering factors for the A and B types of ions in the compound AB. Using the information given in table 2.2.0, explain the main features of figure 2.2.2 and hence verify the rock-salt structure for sodium hydride.

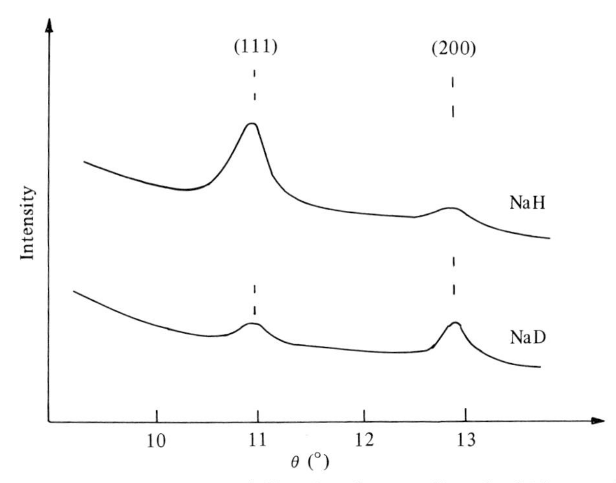


**Figure 2.2.2.** Neutron diffraction from sodium hydride powders, based on a diagram from Shull *et al.* (1948).

**Solution**

Though only two lines are recorded here, their values of $\theta$ and indexing are consistent with a face-centred lattice having $a = 4 \cdot 89$ Å. The calculated structure amplitudes, $|F|$ (fm), are:

| *hkl* | NaH | NaD |
|---|---|---|
| 111 | 28·9 | 12·8 |
| 200 | 0·8 | 40·8 . |

These are in qualitative agreement with each pair of powder-diffraction peaks. Before any quantitative comparison between observed intensities and calculated structure factors can be made, various corrections need to be applied, notably for vibrational ($T$) and geometric factors. These would affect the above numerical values considerably.

**2.2.3 The null matrix**

The hydrogen atoms in a partially deuterated crystal will show a reduced neutron-scattering power, since the scattering amplitudes of H and D have opposite signs. Work out the degree of deuteration needed to produce a 'null matrix', i.e. a crystal from which the hydrogen atoms have, in effect, been eliminated and will have no effect on the neutron diffraction.

**Solution**

$$\frac{[\mathrm{D}]}{[\mathrm{H+D}]} = 35 \cdot 7\% \,.$$

In all but the simplest structures, there will be more than one kind of site for hydrogen atoms. The relative thermodynamic (or kinetic) stabilities of hydrogen and deuterium will, in general, differ for different sites.

When this is so, the deuterium will not be uniformly distributed; it may show a preference for certain positions. Neutron diffraction provides a means of studying this effect.

As we are normally anxious to use neutron measurements to locate hydrogen atoms, the null matrix might seem a singularly futile concept. However the same principle can be applied with other isotopes, when one of these has a negative scattering amplitude. For instance, metallic alloys have been devised which give x-ray-diffraction patterns, but which do not diffract neutrons. They may have useful applications.

**2.2.4 The study of a hydrogen bond by neutrons, and a comparison with the results of x-ray analysis**

Crystals of the acid salt, potassium hydrogen dicrotonate, $KH[C_4H_5O_2]_2$, contain discrete anions of formula

$$[CH_3-CH{=}CH-CO\cdot O\cdots H\cdots O\cdot CO-CH{=}CH-CH_3]^- .$$

The two crotonate residues are connected by an O···H···O bond, and we are to compare the results of x-ray and neutron diffraction studies of this bond. Tables 2.2.4a and 2.2.4b give some results of x-ray work; table 2.2.4c presents results of neutron work (McGregor *et al.*, 1977).

(a) Table 2.2.4a is a section through the electron-density pattern, based on a three-dimensional Fourier synthesis, which used as coefficients $F^x$ (structure factors) for some 1000 x-ray reflections. The section is at $z/c = 25/30$, for it so happens that the three atoms constituting the O···H···O bond all lie close to this level. Plot these densities on a grid of the kind sketched in figure 2.2.4, draw contour lines, and so find positions for the two oxygen atoms and determine the O(2)···O(3) distance. If this distance is much less than twice the van der Waals radius of oxygen (1·40 Å), we may deduce that there is a hydrogen bond between these atoms.

**Table 2.2.4a.** Electron densities. [A section parallel to the (001) face, and at a level $z/c = 25/30$; the grid intervals are $a/50$ along the $x$ direction and $b/30$ along the $y$ direction; the density unit is $e^-$ $Å^{-3}$.]

| $y/b$ | $x/a$ 14 | 15 | 16 | 17 | 18 | 19 | 20 | 21 | 22 | 23 | 24 | 25 | 26 | 27 |
|---|---|---|---|---|---|---|---|---|---|---|---|---|---|---|
| 23 | | | | | | 0·6 | 0·3 | 0·0 | 0·7 | 3·0 | 5·3 | 5·6 | 3·6 | 1·0 |
| 24 | | | | | 0·2 | 0·7 | 0·6 | 0·4 | 1·4 | 4·3 | 7·5 | 8·3 | 5·7 | 1·9 |
| 25 | | | | | 0·0 | 0·6 | 0·8 | 0·7 | 1·5 | 4·3 | 7·9 | 9·2 | 6·8 | 2·7 |
| 26 | 0·8 | 1·0 | 1·1 | 0·6 | 0·1 | 0·3 | 0·8 | 0·7 | 1·1 | 3·1 | 6·3 | 8·0 | 6·4 | 3·0 |
| 27 | 1·7 | 3·0 | 3·6 | 2·5 | 0·8 | 0·2 | 0·6 | 0·7 | 0·5 | 1·4 | 3·6 | 5·2 | 4·6 | 2·6 |
| 28 | 2·3 | 4·9 | 6·3 | 4·9 | 2·1 | 0·4 | 0·5 | 0·7 | 0·2 | 0·1 | 1·1 | 2·2 | 2·2 | 1·5 |
| 29 | 2·5 | 6·0 | 8·1 | 6·8 | 3·4 | 0·9 | 0·5 | 0·9 | 0·6 | 0·0 | 0·0 | 0·2 | 0·2 | |
| 30 | 2·0 | 5·7 | 8·2 | 7·3 | 3·9 | 1·1 | 0·5 | | | | | | | |
| 31 | 1·2 | 4·1 | 6·4 | 6·0 | 3·4 | 0·9 | | | | | | | | |

**Table 2.2.4b.** 'Difference' electron density. Details as in table 2.2.4a.

| $y/b$ | $x/a$ 14 | 15 | 16 | 17 | 18 | 19 | 20 | 21 | 22 | 23 | 24 | 25 | 26 | 27 |
|---|---|---|---|---|---|---|---|---|---|---|---|---|---|---|
| 23 | | | | | | | | | | | | | | |
| 24 | | | | | 0·1 | 0·1 | 0·1 | 0·0 | | | | | | |
| 25 | | | 0·0 | 0·0 | 0·1 | 0·1 | 0·1 | 0·0 | | | | | | |
| 26 | | | 0·0 | 0·1 | 0·2 | 0·2 | 0·2 | 0·0 | | | | | | |
| 27 | | | | 0·0 | 0·3 | 0·3 | 0·3 | 0·1 | 0·0 | | | | | |
| 28 | | | | 0·1 | 0·3 | 0·4 | 0·3 | 0·2 | 0·0 | 0·0 | | | | |
| 29 | | | | 0·1 | 0·3 | 0·4 | 0·3 | 0·2 | 0·0 | 0·0 | | | | |
| 30 | | | | | 0·1 | 0·3 | 0·3 | 0·2 | 0·1 | 0·0 | | | | |
| 31 | | | | | | 0·2 | 0·2 | 0·2 | 0·1 | 0·0 | | | | |

**Table 2.2.4c.** Neutron-scattering densities. The grid intervals are as in table 2.2.4a; the density unit is fm Å$^{-3}$; a bar over the value means a negative density.

| $y/b$ | $x/a$ 14 | 15 | 16 | 17 | 18 | 19 | 20 | 21 | 22 | 23 | 24 | 25 | 26 | 27 |
|---|---|---|---|---|---|---|---|---|---|---|---|---|---|---|
| 23 | | | | | | | | | | 0·1 | 0·8 | 0·9 | 0·1 | |
| 24 | | | | | | | | | 0·0 | 0·5 | 2·0 | 2·5 | 1·1 | |
| 25 | | | | | | | $\overline{0\cdot1}$ | | 0·0 | 0·4 | 2·2 | 3·3 | 1·9 | 0·0 |
| 26 | | | 0·0 | $\overline{0\cdot2}$ | $\overline{0\cdot3}$ | $\overline{0\cdot5}$ | $\overline{0\cdot8}$ | $\overline{0\cdot4}$ | $\overline{0\cdot1}$ | 0·0 | 1·2 | 2·4 | 1·5 | 0·0 |
| 27 | | 0·1 | 0·3 | $\overline{0\cdot1}$ | $\overline{0\cdot5}$ | $\overline{1\cdot0}$ | $\overline{1\cdot5}$ | $\overline{1\cdot0}$ | $\overline{0\cdot3}$ | 0·2 | 0·0 | 0·8 | 0·4 | |
| 28 | 0·0 | 0·8 | 1·4 | 0·6 | $\overline{0\cdot3}$ | $\overline{0\cdot9}$ | $\overline{1\cdot4}$ | $\overline{1\cdot1}$ | $\overline{0\cdot3}$ | $\overline{0\cdot1}$ | | | | |
| 29 | 0·0 | 1·3 | 2·5 | 1·6 | 0·1 | $\overline{0\cdot4}$ | $\overline{0\cdot6}$ | $\overline{0\cdot6}$ | 0·0 | | | | | |
| 30 | 0·0 | 1·0 | 2·4 | 1·9 | 0·4 | 0·0 | 0·0 | | | | | | | |
| 31 | | 0·3 | 1·3 | 1·1 | 0·1 | | | | | | | | | |

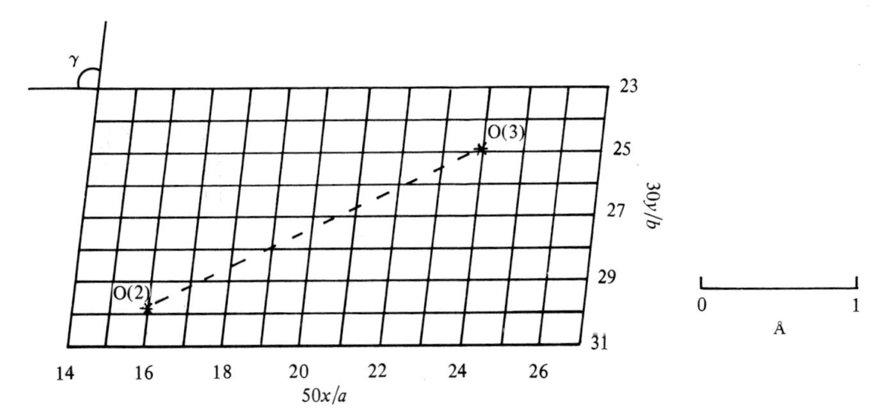


**Figure 2.2.4.** Section of triclinic cell. $a = 12{\cdot}46$ Å, $b = 6{\cdot}05$ Å, $c = 7{\cdot}45$ Å, $\gamma = 97{\cdot}4°$.

(b) For reasons discussed in the introduction, this simple electron-density map (table 2.2.4a) is not good enough to locate the hydrogen atom itself, although a ridge of density can be discerned between atoms O(2) and O(3). A more sensitive device is the 'difference map'. This is produced by a Fourier synthesis with coefficients $(F^{x} - F'_{c})$, where $F^{x}$ is the observed x-ray structure factor as before, and $F'_{c}$ is the structure factor calculated from equation (2.2.0a) by including all the heavier atoms but omitting this hydrogen. The densities presented in table 2.2.4b were derived in this manner—"When the main mountain range is 'taken out', the foothills show up in splendid isolation". The only significant feature of the map, produced when these results are plotted and contoured, is a small not-well-shaped, positive area[1], which shows that there is indeed a hydrogen atom present and gives a rough idea of its position. Measure the O(2)–H distance.

(c) Table 2.2.4c lists the results of a Fourier synthesis with structure factors based on neutron diffraction. Here all three atoms are revealed, the hydrogen nucleus by a negative peak. Plot the densities, draw positive (or, where necessary, negative) contours, and so determine the distances O(2)···O(3), O(2)···H, and O(3)···H.

*Notes on drawing Fourier maps.* The best procedure is to begin by drawing the net carefully on ordinary paper. Any self-consistent scale may be used, but 5 cm (2 in) to 1 Å is convenient. Then, for each density map, place a sheet of tracing paper over the grid, and keep it firmly in position while the appropriate densities are entered (in *small* figures) at the grid intersections. Contour lines may be drawn on the tracing paper. This may be done simply by judicious interpolation; but it may be better to use ordinary graph paper to plot the density against grid number for the lines near the atomic centres (such as, for instance, the row of densities from table 2.2.4a which runs: ... 1·5, 4·3, 7·9, 9·2, 6·8, 2·7), and thus to estimate more accurately where a particular contour crosses the line between two grid intersections. For the map in part (a) contours at 2·0, 4·0, 6·0, and 8·0 $e^{-}$ Å$^{-3}$ are sufficient to reveal the oxygen atoms; for part (b), a more sensitive scale is necessary: 0·2, 0·3, and 0·4 $e^{-}$ Å$^{-3}$, perhaps; for part (c) draw positive contours at 0·5, 1·0, 1·5, 2·0, 2·5, and 3·0 fm Å$^{-3}$ for the oxygen nuclei, and negative contours (using dotted lines) for contours −0·5, −1·0, and −1·5 fm Å$^{-3}$ for the proton. It is wise to start by drawing contours in pencil; the lines can then be erased where necessary and gradually improved until elegant curves are obtained. Remember that the numbers listed in the tables have been rounded off: a value of 0·9 does not mean that this is the exact density at that point, but, merely, that the density lies between 0·85 and 0·94. Further, the density is a physical property which cannot

[1] The imperfections of the hydrogen peak are due to 'noise'. Its location, in this case, is near to, but not beyond, the limits of resolution [see solution 2.2.4(b)].

have discontinuities; the map must represent rounded hills and valleys, and sheer precipices are impossible.

**Solution**

(a) $r[O(2)\cdots O(3)] = 2\cdot52$ Å. This value, and the others deduced from Fourier maps, depend critically on how carefully the contours are drawn.
(b) $r[O(2)-H] \sim 0\cdot88$ Å.
(c) $r[O(2)\cdots O(3)] = 2\cdot50$ Å, $r[O(2)-H] = 1\cdot10$ Å, $r[O(3)\cdots H] = 1\cdot40$ Å.

These results are not the best obtainable from these analyses. There are various minor errors arising from this simple approach based on density sections; one is that the atoms are not all exactly at the level $z/c = 25/30$. We may, however, conclude with confidence that the proton is nearer to O(2) than to O(3). The implication is that the first carboxyl group [carrying O(2)] approximates to $-CO_2H$, whereas the other approximates to $-CO_2^-$.

The O–H distance found in part (b) is unacceptable, since it is shorter than the distance, ~0·96 Å, characteristic of a hydroxyl group not involved in hydrogen bonding, and the effect of hydrogen bonding tends to lengthen O–H, not reduce it. A small error in this sense is normal when x rays are used to locate a hydrogen atom covalently linked to an electronegative neighbour. The reasons for it are broadly understood now, and the reader may care to speculate on them. The results from neutron diffraction in part (c) are more reliable in giving the positions of the atomic nuclei.

Notice that the neutron diffraction peaks from (c) are noticeably smaller and sharper than the corresponding x-ray peaks from (a). This is generally so, and the reason for it is implied in the introduction.

**2.2.5 Discrimination between atoms of nearly equal atomic number, and a quantitative study of disorder in a crystal**

The bases 9-methyladenine and 1-methylthymine form a solid hydrogen-bonded comples whose structure was studied by Hoogsteen in 1963, using x rays. This structure was of special interest because it constituted a possible model for the hydrogen-bonded base pairs, found in DNA and related materials, which play an essential part in the maintenance and replication of the genetic code. Hoogsteen's structure is represented by the left-hand formula drawn in figure 2.2.5. But, as he recognised the right-hand formula differs only in the interchange of carbon and nitrogen atoms at two positions. The x-ray scattering powers of C and N being nearly the same, the 'reversed-Hoogsteen' structure (as it is now called) was difficult to rule out, though the results favoured the Hoogsteen structure.

A precise neutron study of this complex has recently been made (Frey *et al.*, 1973). Owing to the greater difference between the neutron scattering factors for C and N, neutron diffraction is a more powerful means of discrimination. To a good first approximation, it confirmed the Hoogsteen structure. However further refinement revealed a more complicated situation. The crystal is 'disordered' in that a small proportion of molecules having the reversed structure are present, randomly distributed amongst the normal molecules.

The relative proportions of these two sorts of molecule was estimated by the following procedure. In normal refinement, i.e. of the atomic coordinates and the vibrational parameters, the scattering factors of each type of atom are held constant. But, when there is a large excess of observed structure factors, it is also possible to refine the atomic scattering factor, $f^{\mathrm{n}}$, at any site. This was done at the two alternative C/N sites shown in figure 2.2.5:

(a) At the carbon site in the lefthand formula, $f_{\mathrm{C}}^{\mathrm{n}}$ refined to 6·91 fm which lies between the values for $f_{\mathrm{C}}^{\mathrm{n}}$ and $f_{\mathrm{N}}^{\mathrm{n}}$ given in table 2.2.0. Estimate the proportion of the reversed-Hoogsteen structure by linear interpolation.

(b) At the other site, $f_{\mathrm{N}}^{\mathrm{n}}$ refined to 8·84 fm. Use this result to calculate an alternative value for the proportion.

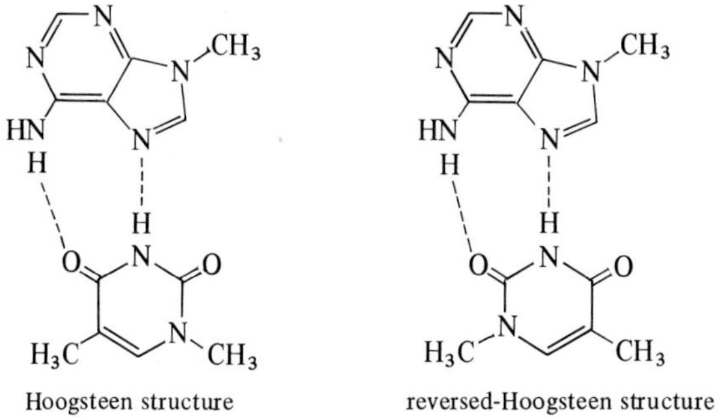

Hoogsteen structure reversed-Hoogsteen structure

**Figure 2.2.5.** Possible structures for hydrogen-bonded 9-methyladenine–1-methylthymine base pairs.

**Solution**

(a) Let $p$ be the proportion of reversed-Hoogsteen structure, then

$$6{\cdot}91 = 6{\cdot}63(1-p)+9{\cdot}40p\ ,$$

therefore $p = 9{\cdot}8\%$.

(b) $p = 20{\cdot}2\%$.

The agreement between these estimates is poor, but Frey *et al.* (1973) found it to be improved when $f_{\mathrm{N}}^{\mathrm{n}}$ for the nondisordered nitrogen nuclei was changed to 9·10 fm (as the reader should check, $p$ then works out to be 10% and 11%). Their procedure is justified as follows.

In structure analysis by normal least-squares refinement, the variables are three positional coordinates for each atom and several parameters to describe its (anisotropic) vibration (see problem 2.1.10). (In addition, there must always be a variable scale factor, since the measured scattering factors $|F_{obs}|$ are not on a true, absolute scale.) A typical analysis may have 150 parameters, but some 2000 observations, $|F_{obs}|$. Provided that such a very large excess of observations is maintained, it is permissible to include extra parameters, as was done when $f_N^n$ was refined to 8·84 fm at the disordered site. But it is equally permissible, if we wish, to refine $f_N^n$ at the other five nondisordered nitrogen sites, rather than to use the 'accepted' value given in table 2.2.0. The reader might well question the ethics of this, and indeed should do so. Are we not 'fudging' $f_N^n$ to improve our agreement? Surely $f_N^n$ = 9·4 fm is a constant of nature, and we ought not to play fast and loose with it?

In fact the situation is less simple than appears at first sight. Unlike x-ray scattering factors, which have a sound theoretical basis, neutron scattering factors are more difficult to evaluate independently. (This is implied in the limited, and varying, numbers of significant figures shown in table 2.2.0.) In particular there is some uncertainty over the best value for $f_N^n$. This consideration justifies including it for refinement along with the other least-squares parameters. To be sure, when this is done, we may, and sometimes do, find different values in different structures. Why are they not all the same? The reason is that our observed data, $|F_{obs}|$, are never quite free from small systematic errors. Such errors may be reflected in differing apparent values for $f_N^n$ in different analyses. But, *for a given analysis*, the actual value found for the other nitrogen nuclei will constitute a sounder basis for comparison with the value at an aberrant site where disorder is suspected.

**References**

Frey, M. N., Koetzle, T. F., Lehmann, M. S., Hamilton, W. C., 1973, *J. Chem. Phys.*, **58**, 2547.

McGregor, D. R., Speakman, J. C., Lehmann, M. S., 1977, *J. Chem. Soc. Perkin Trans. 2*, 1740.

Shull, C. G., Wollan, E. O., Morton, G. A., Davidson, W. L., 1948, *Phys. Rev.*, **73**, 842.

## 2.3 Electron diffraction by gases

**B Beagley** University of Manchester Institute of Science and Technology

**Numerical data required for the problems**

Rest mass of electron, $m_0 = 9\cdot 11 \times 10^{-31}$ kg
Velocity of light, $c = 3\cdot 00 \times 10^{8}$ m s$^{-1}$
Planck constant, $h = 6\cdot 63 \times 10^{-34}$ J s
Avogadro number, $N_A = 6\cdot 02 \times 10^{23}$ mol$^{-1}$
$1$ eV $= 1\cdot 60 \times 10^{-19}$ J

### 2.3.1 Visual method

The fourth minimum of an electron diffraction pattern of gaseous hydrogen chloride occurs at $2\theta = 10°0'$. If the wavelength, $\lambda$, of the electron beam is $0\cdot 0599$ Å, calculate the bond length in the molecule. [Note: $\theta$ is defined so that $s = (4\pi/\lambda)\sin\theta$.]

**Solution**

The way in which the intensity $I$ varies with the diffraction angle, in an electron diffraction experiment on a diatomic molecule, can be expressed in a simplified form as

$$I(s) = I_0(s)\sin sr_{ij}\exp(-\tfrac{1}{2}u_{ij}^2 s^2)\,, \tag{2.3.1a}$$

where $I_0$ is a slowly varying function of $s$ containing the electron scattering factors, $r_{ij}$ is the interatomic distance, and $u_{ij}$ the corresponding amplitude of vibration. It is more convenient to use the variable $s = (4\pi/\lambda)\sin\theta$, than simply the diffraction angle $2\theta$. (In electron diffraction, some workers prefer to use just $\theta$ as the diffraction angle between incident and diffracted beams. The present author prefers $2\theta$ because this nomenclature is consistent with usage in x-ray diffraction, where $\theta$ is then the Bragg angle.)

The form of the intensity curve is effectively that of a sine wave damped by the exponential factor (figure 2.3.1). Minima will occur on the intensity curve whenever $\sin sr_{ij} = -1$; i.e. when $sr_{ij} = \frac{3}{2}\pi, \frac{7}{2}\pi, \frac{11}{2}\pi, \frac{15}{2}\pi, \ldots, \frac{1}{2}(4n+3)\pi$, with $n = 0, 1, 2, \ldots$.

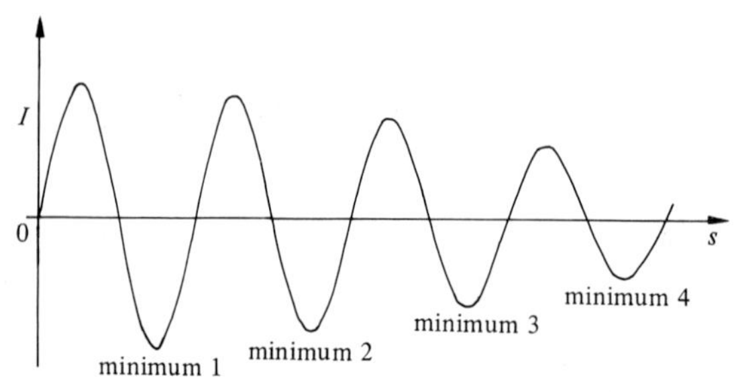


**Figure 2.3.1.** Idealised (molecular) intensity curve.

For the fourth diffraction minimum,

$$s = (4\pi/0{\cdot}0599)\sin 5°0' = 4\pi \times 0{\cdot}08716/0{\cdot}0599\ \text{Å}^{-1}\ .$$

The fourth minimum occurs when $sr_{ij} = \frac{15}{2}\pi$, thus

$$r_{ij} = \frac{15\pi \times 0{\cdot}0599}{2 \times 4\pi \times 0{\cdot}08716} = 1{\cdot}289\ \text{Å}\ .$$

**2.3.2 Wave properties of electrons**

Derive a formula which expresses the wavelength of an electron beam as a function of accelerating voltage. Calculate the wavelength of 40 kV electrons.

**Solution**

An approximate relationship between the wavelength λ of an electron beam and the voltage $V$ through which the electrons have been accelerated can be derived by equating the kinetic energy of an electron, $\frac{1}{2}m_0v^2$, with the energy gained by acceleration, $eV$. Thus the velocity $v$ of the electron is

$$v = \left(\frac{2eV}{m_0}\right)^{1/2}\ . \tag{2.3.2a}$$

Using de Broglie's relationship we obtain

$$\lambda = \frac{h}{m_0v} = \frac{h}{(2m_0eV)^{1/2}} \tag{2.3.2b}$$

For 40 kV electrons, $eV = 40000 \times 1{\cdot}60 \times 10^{-19} = 6{\cdot}40 \times 10^{-15}$ J, and $m_0eV = 9{\cdot}11 \times 6{\cdot}40 \times 10^{-46}$ kg J or kg$^2$ m$^2$ s$^{-2}$ if we use the numerical value of the rest mass. Hence,

$$(2m_0eV)^{1/2} = (116{\cdot}6 \times 10^{-46})^{1/2} = 10{\cdot}8 \times 10^{-23}\ \text{kg m s}^{-1}\ .$$

Equation (2.3.2b) now gives

$$\lambda = \frac{6{\cdot}63 \times 10^{-34}}{10{\cdot}8 \times 10^{-23}}\ \frac{\text{kg m}^2\ \text{s}^{-1}}{\text{kg m s}^{-1}} = 0{\cdot}614 \times 10^{-11}\ \text{m} = 0{\cdot}0614\ \text{Å}\ .$$

From equation (2.3.2a) we find the velocity $v$ of 40 kV electrons:

$$v = \left(\frac{2 \times 6{\cdot}40 \times 10^{-15}}{9{\cdot}11 \times 10^{-31}}\right)^{1/2}\left(\frac{\text{kg m}^2\ \text{s}^{-2}}{\text{kg}}\right)^{1/2} = 1{\cdot}185 \times 10^{8}\ \text{m s}^{-1}\ .$$

This velocity is about one third of the velocity of light; at this velocity the laws of relativity cannot be ignored as they have been in the approximate treatment above. The treatment which follows takes account of the change in mass which occurs when a particle is accelerated to a high velocity. The theory of relativity predicts that the mass $m$ at velocity $v$ is related to the rest mass $m_0$ by

$$m = m_0\left(1-\frac{v^2}{c^2}\right)^{-1/2} . \qquad (2.3.2c)$$

The total energy of an electron accelerated through a potential $V$ is $m_0c^2+eV$, which may be equated with $mc^2$. Thus, using equation (2.3.2c) and rearranging, we obtain

$$\frac{v^2}{c^2} = 1-\frac{m_0^2}{m^2} = 1-\frac{m_0^2c^4}{(m_0c^2+eV)^2} = \frac{2m_0c^2eV+e^2V^2}{(m_0c^2+eV)^2} ,$$

and so

$$v = \frac{c(2m_0c^2eV+e^2V^2)^{1/2}}{(m_0c^2+eV)} . \qquad (2.3.2d)$$

Substituting equation (2.3.2c) into de Broglie's relationship we obtain

$$\lambda = \frac{h}{mv} = \frac{h}{m_0v}\left(1-\frac{v^2}{c^2}\right)^{1/2} = \frac{h}{m_0}\left(\frac{1}{v^2}-\frac{1}{c^2}\right)^{1/2} ,$$

from which $v$ may be eliminated with the aid of equation (2.3.2d) to give

$$\lambda = \frac{h}{m_0}\left[\frac{(m_0c^2+eV)^2}{c^2(2m_0c^2eV+e^2V^2)}-\frac{1}{c^2}\right]^{1/2}$$

$$= \frac{h}{m_0}\left[\frac{m_0^2c^4}{c^2(2m_0c^2eV+e^2V^2)}\right]^{1/2} = \frac{h}{[2m_0eV(1+eV/2m_0c^2)]^{1/2}} . \qquad (2.3.2e)$$

Note that when $eV \ll 2m_0c^2$ (i.e. for small voltages and velocities), equation (2.3.2e) reduces to the approximate form (2.3.2b).

Equation (2.3.2e) may be written as

$$\lambda = \lambda_0\left(1+\frac{eV}{2m_0c^2}\right)^{-1/2} , \qquad (2.3.2f)$$

where $\lambda_0$ is the approximate wavelength given by equation (2.3.2b). Equation (2.3.2f) enables us to obtain $\lambda$ for 40 kV electrons quite simply:

$$2m_0c^2 = 2\times 9{\cdot}11\times 9{\cdot}00\times 10^{-15} = 164{\cdot}0\times 10^{-15}\ \mathrm{J} ;$$

thus, as $eV = 6{\cdot}40\times 10^{-15}$ J,

$$\left(1+\frac{eV}{2m_0c^2}\right)^{1/2} = \left(1+\frac{6{\cdot}40}{164{\cdot}0}\right)^{1/2} = 1{\cdot}019 ,$$

and so

$$\lambda = \frac{0{\cdot}0614}{1{\cdot}019} = 0{\cdot}0603\ \text{Å} .$$

### 2.3.3 Data processing[1]

Table 2.3.3a gives data in the region of a maximum of an electron diffraction pattern of a certain molecule.

The camera distance, $H$, is 1000 mm; the sector-to-plate distance, $\delta$, is 10 mm; the electron wavelength, $\lambda$, is 0·05500 Å; the blackness correction function is $B(D) = 1{\cdot}000 + 0{\cdot}1\,(D-1)$, where $D$ is the optical density, and the sector profile is cut so that the opening is proportional to the cube of its radius. The quantities $f_i$, $f_j$, and $Z_i$, $Z_j$ are the x-ray scattering factors and atomic numbers for atoms $i$ and $j$ respectively.

Process the data to obtain the uphill curve. (Over the range involved it is adequate to assume that $\cos 2\theta$ is constant and that $\sin\theta = \frac{1}{2}\tan 2\theta$.) Derive the molecular intensity curve and correct it to the form whose Fourier transform is the usual radial distribution curve for the molecule. Find graphically the position of the maximum of molecular intensity. If this is the second maximum in the diffraction pattern, find the internuclear distance $r_{ij}$, assuming that most of the molecular scattering arises from this source.

**Table 2.3.3a.**

| Distance from centre of pattern, $R$/mm | Optical density, $D$ | Atomic[a] scattering, $I_{at}$ | $\left(1-\dfrac{f_i}{Z_i}\right)\left(1-\dfrac{f_j}{Z_j}\right)$ |
|---|---|---|---|
| 37·0 | 1·234 | 0·800 | 0·211 |
| 37·2 | 1·253 | 0·805 | 0·213 |
| 37·4 | 1·272 | 0·810 | 0·215 |
| 37·6 | 1·274 | 0·814 | 0·217 |
| 37·8 | 1·265 | 0·819 | 0·219 |
| 38·0 | 1·256 | 0·823 | 0·221 |

[a] $I_{at}$ is scaled to correspond with an uphill curve which is scaled so that $I_{up}$ for $R = 37{\cdot}0$ mm is 1·000.

**Solution**

Before meaningful information can be derived from data such as those presented here, they must be converted to a form which has a simple theoretical equivalent. One such form is the uphill curve $I_{up}$ (Beagley *et al.*, 1967) which for the moment we shall consider as proportional to the sum of atomic scattering $I_{at}$ and molecular scattering $I_{mol}$, i.e.

$$I_{up} \propto I_{at} + I_{mol}\,. \tag{2.3.3a}$$

[1] Data processing is routinely carried out by computer, and the user is often unaware of the steps in the calculations. This problem, in providing an insight into typical procedures which may be incorporated into the computer programs, may also serve as a programmer's guide.

The uphill curve, which is discussed in more detail later, is independent of experimental features particular to a given electron diffraction apparatus, so that the conversion from crude intensity data to the uphill curve eliminates apparatus constants and similar experiment parameters.

The expression we must use in deriving $I_{up}$ from $D$ is

$$I_{up} = \frac{Ds^4 B(D)}{\alpha(s)\cos^3 2\theta} \; . \tag{2.3.3b}$$

Three experimental parameters appear here:
(i) $B(D)$, the blackness correction function, which is necessary to allow for the nonlinear response of photographic emulsions to differing electron currents (i.e. intensity is not exactly proportional to optical density);
(ii) $\cos^3 2\theta$, which appears because the theory is simpler for the case where all diffracted electrons travel the same path length before detection; this condition is not satisfied for *planar* photographic plates, and $\cos^3 2\theta$ is the necessary correction function ('planar plate correction');
(iii) $\alpha(s)$, which is the angular opening of the sector.

We must divide by $\alpha(s)$ to remove from the data the effect of using the rotating sector, and multiply by $s^4$ to bring the data back within a manageable range. The purpose of the sector is to level the intensity right across the plate, so that all regions have intensities which fall within the fairly narrow range which can be measured accurately. Without the sector, the intensities decrease so rapidly that accurate measurement is impossible; the decrease is a function of $s^4$.

Figure 2.3.3a shows the geometry of the diffraction experiment; it is clear that

$$\tan 2\theta = \frac{R}{H} = \frac{x}{H-\delta} = 2\sin\theta \; . \tag{2.3.3c}$$

Equation (2.3.3c) enables us to determine $\sin\theta$ [and hence $s = (4\pi/\lambda)\sin\theta$], and also $x$, which is required in calculating $\alpha(s)$. Note that $\alpha(s) \propto x^3$.

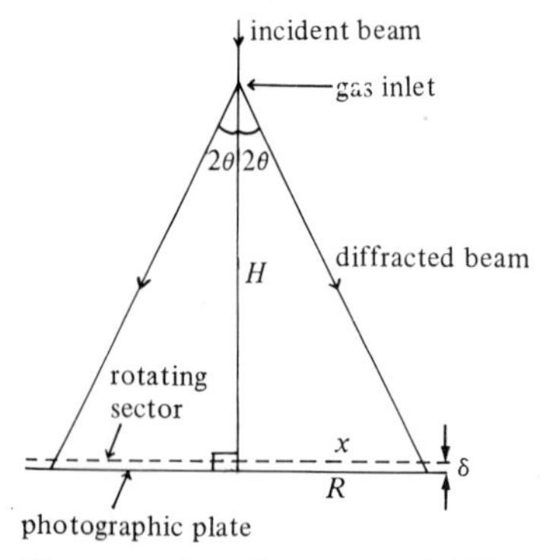


**Figure 2.3.3a.** Geometry of diffraction experiment.

Substituting the given numerical values in equation (2.3.3c), we obtain

$$x = \frac{R(H-\delta)}{H} = 0\cdot990R \, .$$

In table 2.3.3b, $s^4/x^3$ is calculated for each point. In table 2.3.3c, $I_{up}$ is derived and scaled so that the value for the first point is 1·000.

In equation (2.3.3a) which expresses the theoretical form of $I_{up}$,

$$I_{at} = \sum_i [(Z_i - f_i)^2 + S_i] \, , \tag{2.3.3d}$$

and

$$I_{mol} = \sum_{\substack{i \\ i>j}} \sum_j [(Z_i - f_i)(Z_j - f_j)\cos(\eta_i - \eta_j)] \int_0^\infty P_{ij}(r) \frac{\sin sr}{sr} \, dr \, . \tag{2.3.3e}$$

In equation (2.3.3e) $P_{ij}(r)\,dr$ is the probability that the distance between atoms $i$ and $j$ in the molecule is between $r$ and $r + dr$. If the cosine term is omitted, we have the Born approximation for $I_{mol}$, where $Z$ and $f$ are atomic numbers and x-ray scattering factors for the appropriate atoms. The cosine term is discussed further in the solution to problem 2.3.7. In equation (2.3.3d) $S$ is the inelastic part of the atomic scattering.

**Table 2.3.3b.** Calculation of $\dfrac{s^4}{x^3} \propto \dfrac{s^4}{\alpha(s)}$.

| $R$ (mm) | $\tan 2\theta = R/H$ | $\sin\theta = \frac{1}{2}\tan 2\theta$ | $s$ ($Å^{-1}$) | $s^4$ ($Å^{-4}$) | $x = 0\cdot990R$ (mm) | $x^3$ ($mm^3$) | $10^3 \dfrac{s^4}{x^3}$ |
|---|---|---|---|---|---|---|---|
| 37·0 | 0·0370 | 0·0185 | 4·227 | 319·2 | 36·63 | 49149 | 6·495 |
| 37·2 | 0·0372 | 0·0186 | 4·250 | 326·3 | 36·83 | 49958 | 6·531 |
| 37·4 | 0·0374 | 0·0187 | 4·273 | 333·4 | 37·03 | 50776 | 6·566 |
| 37·6 | 0·0376 | 0·0188 | 4·296 | 340·6 | 37·22 | 51562 | 6·606 |
| 37·8 | 0·0378 | 0·0189 | 4·319 | 348·0 | 37·42 | 52398 | 6·641 |
| 38·0 | 0·0380 | 0·0190 | 4·342 | 355·4 | 37·62 | 53242 | 6·675 |

**Table 2.3.3c.** Calculation of $I_{up}$.

| $D$ | $B(D)$ | $I_{up}^{a} \propto \dfrac{DB(D)s^4}{x^3}$ | Scaled $I_{up}$ |
|---|---|---|---|
| 1·234 | 1·023 | 8·199 | 1·000 |
| 1·253 | 1·025 | 8·388 | 1·023 |
| 1·272 | 1·027 | 8·577 | 1·046 |
| 1·274 | 1·027 | 8·643 | 1·054 |
| 1·265 | 1·027 | 8·628 | 1·052 |
| 1·256 | 1·026 | 8·602 | 1·049 |

[a] $\cos^3 2\theta$ is omitted from this column as it is assumed to be constant.

For harmonic vibrations, the integral in equation (2.3.3e) reduces to

$$\exp(-\tfrac{1}{2}u_{ij}^2 s^2)\frac{\sin sr_{ij}}{sr_{ij}} ,$$

where $r_{ij}$ is the interatomic distance between $i$ and $j$ and $u_{ij}$ the root-mean-square amplitude of vibration. Thus, in the Born approximation, and for harmonic vibrations,

$$I_{\mathrm{mol}} = \sum_{\substack{i \; j \\ i>j}}\sum (Z_i - f_i)(Z_j - f_j)\exp(-\tfrac{1}{2}u_{ij}^2 s^2)\frac{\sin sr_{ij}}{sr_{ij}} . \qquad (2.3.3\mathrm{f})$$

It is possible to eliminate to a large extent the scattering due to the electrons in the atoms by multiplying $I_{\mathrm{mol}}$ by the factor $s[1-(f_m/Z_m)]^{-1}[1-(f_n/Z_n)]^{-1}$, which gives

$$I_{\mathrm{m}} = \sum_{\mathrm{all}\, ij} A_{ij}\exp(-\tfrac{1}{2}u_{ij}^2 s^2)\frac{\sin r_{ij}}{r_{ij}} , \qquad (2.3.3\mathrm{g})$$

where

$$A_{ij} = n_{ij}Z_iZ_j\left(1-\frac{f_i}{Z_i}\right)\left(1-\frac{f_j}{Z_j}\right)\left(1-\frac{f_m}{Z_m}\right)^{-1}\left(1-\frac{f_n}{Z_n}\right)^{-1} ,$$

and $n_{ij}$ is the multiplicity of the distance between atoms $i$ and $j$.

Thus $A_{ij} = n_{ij}Z_iZ_j$ when $i$, $j$ are the same atoms as $m$, $n$, and is approximately constant otherwise (i.e. $f$ disappears from the intensity expression). It is the Fourier transform of $I_{\mathrm{m}}$ which gives the radial distribution curve for the molecule (see solutions to problems 2.3.4 and 2.3.5).

The experimental $I_{\mathrm{mol}}$ and $I_{\mathrm{m}}$ intensity curves for the present data are given in table 2.3.3d. Figure 2.3.3b is a plot of $I_{\mathrm{m}}$ versus $s$; the maximum value occurs at $s = 4{\cdot}287$ Å$^{-1}$. Equation (2.3.3g) is to be compared with the simplified expression (2.3.1a). The interatomic distance $r_{ij}$ is given, by analogy with the method of problem 2.3.1, by $sr_{ij} = \frac{5}{2}\pi$; thus

$$r_{ij} = \frac{5\pi}{2\times 4{\cdot}287} = 1{\cdot}832 \text{ Å} .$$

**Table 2.3.3d.** Derivation of $I_{\mathrm{mol}}$ and $I_{\mathrm{m}}$.

| $I_{\mathrm{up}}$ | $I_{\mathrm{at}}$ | $I_{\mathrm{mol}} = I_{\mathrm{up}} - I_{\mathrm{at}}$ | $I_{\mathrm{m}} = sI_{\mathrm{mol}}\left(1-\frac{f_i}{Z_i}\right)^{-1}\left(1-\frac{f_j}{Z_j}\right)^{-1}$ |
|---|---|---|---|
| 1·000 | 0·800 | 0·200 | 4·01 |
| 1·023 | 0·805 | 0·218 | 4·35 |
| 1·046 | 0·810 | 0·236 | 4·69 |
| 1·054 | 0·814 | 0·240 | 4·75 |
| 1·052 | 0·819 | 0·233 | 4·60 |
| 1·049 | 0·823 | 0·226 | 4·44 |

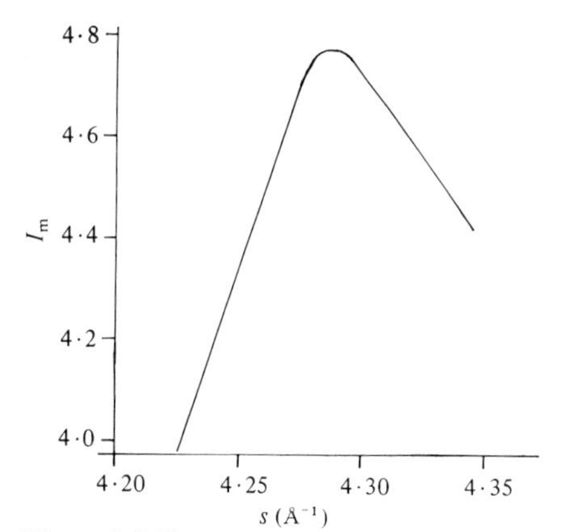


**Figure 2.3.3b.**

### 2.3.4 Structure determination

The molecule $(FS)_3N$, which can be assumed to possess a threefold axis of symmetry, is expected to have one of three possible configurations: (I) completely planar; (II) planar $NS_3$ skeleton with S–F bonds perpendicular to this plane; (III) pyramidal ($\widehat{SNS} \approx 109\frac{1}{2})°$. If S–N = 1·7 Å, S–F = 1·6 Å, $\widehat{NSF}$ = 90°, make rough estimates of the major interatomic distances for each configuration; hence explain how an electron diffraction study of the molecule could distinguish between the possibilities.

### Solution

Figure 2.3.4a shows the three possible configurations. Table 2.3.4 lists the various distances for each configuration and other pertinent information. The experimental radial distribution curve for the molecule provides the evidence which can enable the possible configurations to be distinguished. This curve is obtained by Fourier transformation of the experimental molecular intensity curve $I_m$ described in the solution to

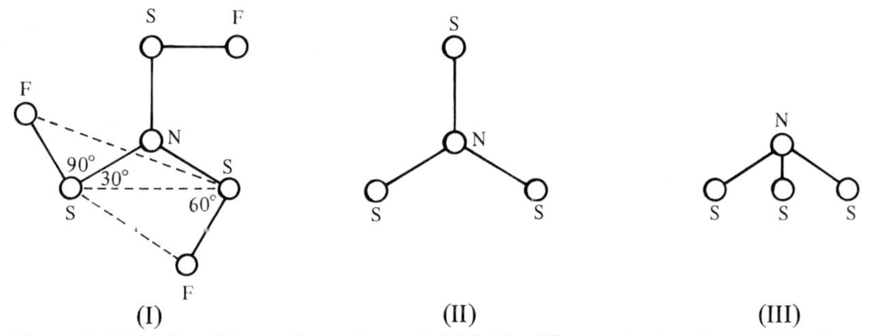


**Figure 2.3.4a.** Possible configurations of $(FS)_3N$: (I) completely planar, (II) planar $NS_3$ skeleton (F atoms above S atoms), (III) pyramidal (positions of F atoms unspecified).

problem 2.3.3. Figure 2.3.4b shows the form which the experimental radial distribution curve would take for each of the configurations. The positions of the peaks on a radial distribution curve give the interatomic distances directly, and the sizes of the peaks are roughly proportional to $n_{ij} Z_i Z_j / r_{ij}$, where $Z_i$, $Z_j$ are the atomic numbers of the two atoms, and $n_{ij}$ is the number of times the distance $r_{ij}$ occurs in the molecule.

The strongest peak will be the composite one for the two bond distances. Configurations (I) and (II) can immediately be distinguished from (III) by the position of the second-strongest peak, at 2·95 Å in the first two cases but at 2·8 Å in the third. Configurations (I) and (II) can readily be distinguished from each other by the positions of the smaller peaks beyond 3 Å; configuration (I) gives rise to a fairly large peak at 4 Å, whilst (II) gives rise to none beyond 3·5 Å.

Clearly the actual configuration of this molecule could be more complex than the simple geometries considered here. Certainly $\widehat{NSF}$ will be larger than 90°, and the orientation of the S—F bonds about S—N, which will depend on bonding and steric factors, could well be intermediate between configurations (I) and (II). There are no grounds, either, to restrict angle $\widehat{SNS}$ to the two values considered here (120° and $109\frac{1}{2}$°); any intermediate value is a possibility. Whether these variations

**Table 2.3.4.** Rough estimates of the interatomic distances for the three configurations.

| Distance | $n_{ij}$ | $r_{ij}$ (Å) | $\frac{n_{ij} Z_i Z_j}{r_{ij}}$ | Remarks |
|---|---|---|---|---|
| **(I) Completely planar configuration** | | | | |
| N—S | 3 | 1·7 | 198 | given |
| S—F | 3 | 1·6 | 270 | given |
| N···F | 3 | 2·3 | 82 | $(1{\cdot}7^2 + 1{\cdot}6^2)^{1/2}$ |
| S···S | 3 | 2·95 | 260 | $1{\cdot}7 \times 3^{1/2}$ |
| short F···S | 3 | 2·6 | 166 | $(2{\cdot}95^2 + 1{\cdot}6^2 - 2 \times 2{\cdot}95 \times 1{\cdot}6 \times \cos 60°)^{1/2}$ |
| long F···S | 3 | 4·0 | 98 | $(2{\cdot}95^2 + 1{\cdot}6^2 - 2 \times 2{\cdot}95 \times 1{\cdot}6 \times \cos 120°)^{1/2}$ |
| F···F | 3 | 4·0 | 61 | $2{\cdot}3 \times 3^{1/2}$ |
| **(II) Planar $NS_3$ skeleton only** | | | | |
| N—S | 3 | 1·7 | 198 | same as (I) |
| S—F | 3 | 1·6 | 270 | same as (I) |
| N···F | 3 | 2·3 | 82 | same as (I) |
| S···S | 3 | 2·95 | 260 | same as (I) |
| F···S | 6 | 3·4 | 254 | $(2{\cdot}95^2 + 1{\cdot}6^2)^{1/2}$ |
| F···F | 3 | 2·95 | 83 | same as S···S |
| **(III) Pyramidal configuration** | | | | |
| N—S | 3 | 1·7 | 198 | same as (I) |
| S—F | 3 | 1·6 | 270 | same as (I) |
| N···F | 3 | 2·3 | 82 | same as (I) |
| S···S | 3 | 2·8 | 274 | $2 \times 1{\cdot}7 \times \sin(109\frac{1}{2}/2)°$ |
| F···S | 6 | - | - | not required |
| F···F | 3 | - | - | not required |

in configuration occur could readily be ascertained by a more careful study of the experimental radial distribution curve than that outlined above.

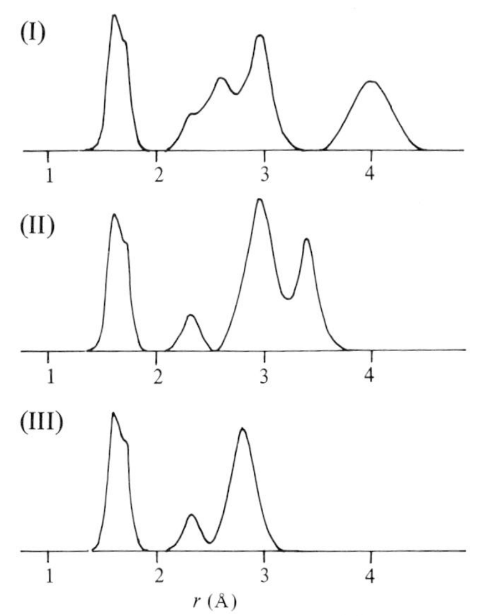


**Figure 2.3.4b.** Idealised radial distribution curves for the possible configurations of $(FS)_3N$: (I) completely planar, (II) planar $NS_3$ skeleton, (III) pyramidal (effect of F ··· S and F ··· F distances not shown).

### 2.3.5 Types of distance

In an electron diffraction study of $(SiH_3)_3P$, the major peaks on the radial distribution curve occur at 1·485, 2·248, and 3·353 Å [$r_g(1)$ values]; the corresponding amplitudes of vibration, $u$, are 0·089, 0·062, and 0·142 Å (Beagley *et al.*, 1968). Interpret the radial distribution curve in terms of the molecular geometry, and calculate the centre-of-gravity distances, and equilibrium bond distances, if the Morse anharmonicity constant, $a$, is 2 Å$^{-1}$ in all cases.

**Solution**

The peaks at 1·485 and 2·248 Å will arise from the Si—H and Si—P bonds in $(SiH_3)_3P$, and that at 3·353 Å will be due to Si···Si. Thus the angle at phosphorus is

$$\widehat{\mathrm{SiPSi}} = 2\sin^{-1}\left(\frac{1}{2}\times\frac{3\cdot353}{2\cdot248}\right) = 2\sin^{-1}0\cdot74577 = 2\times 48°13\tfrac{1}{2}' = 96°27'$$

In an electron diffraction study the experimental radial distribution curve is normally calculated as the damped Fourier transform of the experimental molecular intensity curve $I_m$ (see solutions to problems 2.3.3 and 2.3.4). Analytical Fourier transformation of the theoretical expression for $I_m$ gives the corresponding theoretical form of the radial distribution curve as a probability function

$$f_{tot}(r) = \sum_{\text{all } ij} f_{ij}(r) = \sum_{\text{all } ij} \left[\frac{A_{ij}}{r_{ij} K_{ij}^{1/2}}\right] \exp\left[-\frac{(r_{ij}-r)^2}{4K_{ij}}\right], \tag{2.3.5a}$$

where $A_{ij} \approx nZ_iZ_j$, $K_{ij} = \frac{1}{2}u_{ij}^2 + b$, and the other symbols have the same meanings as in the solutions to problems 2.3.3 and 2.3.4 ($b$ is the damping constant introduced when evaluating the experimental form of the radial distribution curve). Expression (2.3.5a) shows that the peaks on the radial distribution curve are symmetrical about $r_{ij}$, and their widths are related to $K_{ij}$ (and so to $u_{ij}$). Hence these parameters can, of course, be derived from the experimental radial distribution curve. The theoretical treatment shows that the $r_{ij}$ obtained in this way (or from least-squares refinements based on the original molecular intensity curves) are the centres of gravity of the probability functions $f_{ij}(r) = P_{ij}(r)/r^m$, with $m = 1$ (see solution to problem 2.3.3). Bartell (1955) has shown that the centre of gravity $r_g(m)$ of the function $P_{ij}(r)/r^m$ is related to the equilibrium distance $r_e$ by

$$r_g(m) \approx r_e + \tfrac{3}{2}au^2 - \frac{mu^2}{r_e}\ . \tag{2.3.5b}$$

Thus the distances obtained in the electron diffraction experiment are effectively

$$r_g(1) \approx r_e + \tfrac{3}{2}au^2 - \frac{u^2}{r_g(1)}\ . \tag{2.3.5c}$$

[As $u^2/r_e \ll r_e$, $r_g(1)$ can be substituted for $r_e$ in the third term of expression (2.3.5b) without serious error.] Another symbol, $r_a$, is often used instead of $r_g(1)$. The 'centre-of-gravity distance' is normally defined as the centre of gravity of the $P_{ij}(r)$ function (for which $m = 0$) rather than of $P_{ij}(r)/r$. It is thus $r_g(0)$, often abbreviated as $r_g$, which is required in this problem. Equations (2.3.5b) and (2.3.5c) give

$$r_g = r_g(0) \approx r_e + \tfrac{3}{2}au^2 \approx r_a + \frac{u^2}{r_a}\ . \tag{2.3.5d}$$

The numerical results required for $(SiH_3)_3P$ are given in table 2.3.5.

**Table 2.3.5** $r_a$, $r_g$, and $r_e$ values (Å) for trisilylphosphine.

| $r_a$ | $u$ | $\frac{u^2}{r_a}$ | $r_g = r_a + \frac{u^2}{r_a}$ | $\frac{3}{2}au^2$ | $r_e = r_g - \frac{3}{2}au^2$ |
|---|---|---|---|---|---|
| 1·485 | 0·089 | 0·005 | 1·490 | 0·024 | 1·466 |
| 2·248 | 0·062 | 0·002 | 2·250 | 0·012 | 2·238 |
| 3·353 | 0·142 | 0·006 | 3·359 | – | – |

### 2.3.6 Vibration of molecules

Electron diffraction studies of $CH_4$ and $CD_4$ give C—H = $1{\cdot}101_4$ and C—D = $1{\cdot}098_6$ Å [$r_g(1)$ values] with amplitudes of vibration 0·076 and 0·068 Å respectively (Bartell *et al.*, 1961). Show that the difference in amplitudes is almost consistent with the simple harmonic treatment for the vibration of diatomic molecules, and calculate the force constant for the bond. Show that the difference in distances is inconsistent with exactly harmonic vibrations and calculate the Morse anharmonicity constant and the equilibrium internuclear distance.

**Solution**

For a diatomic molecule vibrating with simple harmonic motion, the mean-square amplitude of vibration $\overline{u^2}$ is given by

$$\overline{u^2} = \frac{h}{8\pi^2\nu\mu}\coth\frac{h\nu}{2kT}\,, \tag{2.3.6a}$$

where $h$ is the Planck constant, $k$ is the Boltzmann constant, $T$ is the absolute temperature, $\nu$ is the vibrational frequency, and $\mu$ is the reduced mass of the pair of atoms. At absolute zero the coth term is unity, and, as the theory of simple harmonic motion gives the frequency as $\nu = (1/2\pi)(F/\mu)^{1/2}$, where $F$ is the force constant of the bond, we may write

$$\overline{u^2} = \frac{h}{4\pi(\mu F)^{1/2}}\,. \tag{2.3.6b}$$

Thus if $u_H$, $u_D$, and $\mu_H$, $\mu_D$ are the root-mean-square vibrational amplitudes and reduced masses for C—H and C—D,

$$\frac{u_H^2}{u_D^2} = \left(\frac{\mu_D}{\mu_H}\right)^{1/2}\,. \tag{2.3.6c}$$

($F$ is of course the same for both isotopic species.) Therefore, if the treatment for diatomics at absolute zero can be used for methane,

$$u_H = u_D\left(\frac{\mu_D}{\mu_H}\right)^{1/4}\,. \tag{2.3.6d}$$

Now $\mu_D = 12 \times 2/14N_A$ and $\mu_H = 12 \times 1/13N_A$, where $N_A$ is the Avogadro number, so $\mu_D/\mu_H = 26/14 = 1{\cdot}857$. If we insert the experimental value of $u_D$ in equation (2.3.6d), we obtain a predicted value of $u_H$ of $(1{\cdot}857)^{1/4} \times 0{\cdot}068 = 0{\cdot}079$ Å, which is reasonably close to the experimental value, 0·076 Å. The error is within the limits one might expect for experimental error in the measurement of a difference between two amplitudes. It is reasonable to conclude, therefore, that the observed difference in amplitudes is consistent with the simple harmonic treatment for a diatomic at absolute zero. Using equation (2.3.6b) we can calculate the force constant; a value for $F$ is obtained from each amplitude (table 2.3.6).

This method of determining force constants is clearly not very precise. The fourth-power relationship between $u$ and $F$ magnifies percentage errors in $u$ by a factor of four; thus a 5% error in $u$, which is possible experimentally, gives rise to a 20% error in $F$.

With the values for $r_g(1)$, or $r_a$, of $1 \cdot 101_4$ Å for C—H and $1 \cdot 098_6$ Å for C—D, equation (2.3.5d) enables us to calculate the corresponding $r_g$ distances, which are $1 \cdot 106_6$ and $1 \cdot 102_8$ Å respectively. Equation (2.3.5d) also enables us to write

$$r_g(\text{C—H}) \approx r_e + \tfrac{3}{2} a u_H^2 \tag{2.3.6e}$$

and

$$r_g(\text{C—D}) \approx r_e + \tfrac{3}{2} a u_D^2 \, . \tag{2.3.6f}$$

If we assume that $r_e$ and $a$ are the same for both species, the difference we observe in the $r_g$ distances can be explained as being due to the difference in amplitudes. The anharmonicity constant $a$ cannot be zero, because then the two $r_g$ distances would be the same and equal to $r_e$. Thus we can conclude that the observations are inconsistent with exactly harmonic vibrations.

The values of $r_e$ and $a$ are obtained by solving the simultaneous equations (2.3.6e) and (2.3.6f): by subtraction $0 \cdot 003_8$ Å $= (\tfrac{3}{2}a)(0 \cdot 00115)$; hence $\tfrac{3}{2}a = 3 \cdot 3$ Å$^{-1}$ and $a = 2 \cdot 2$ Å$^{-1}$. Thus $r_e = 1 \cdot 088$ Å.

**Table 2.3.6.** Calculation of force constant.

| Bond | $u$ (m) | $\mu$ (kg) | $\frac{h}{4\pi u^2} = (\mu F)^{1/2}$ (kg s$^{-1}$) | $\mu F$ (kg$^2$ s$^{-2}$) | $F$ (kg s$^{-2}$)[a] |
|---|---|---|---|---|---|
| C—H | $7 \cdot 6 \times 10^{-12}$ | $1 \cdot 53 \times 10^{-27}$ | $0 \cdot 91 \times 10^{-12}$ | $0 \cdot 83 \times 10^{-24}$ | $5 \cdot 4 \times 10^2$ |
| C—D | $6 \cdot 8 \times 10^{-12}$ | $2 \cdot 85 \times 10^{-27}$ | $1 \cdot 14 \times 10^{-12}$ | $1 \cdot 30 \times 10^{-24}$ | $4 \cdot 6 \times 10^2$ |
| | | | | mean | $5 \cdot 0 \times 10^2$ |

[a] The units kg s$^{-2}$ are equivalent to N m$^{-1}$. Note: $10^2$ kg s$^{-2}$ = 1 mdyn Å$^{-1}$.

### 2.3.7 Failure of the Born approximation

If the cosine modulation term present in the theoretical intensity expression to overcome the failure of the Born approximation can be written as $\cos cs$, where $c$ is a constant, show that for a single interatomic distance the corresponding radial distribution curve will be the sum of two peaks separated by a distance equal to $2c$. Calculate the beat-out point in an intensity curve which gives rise to a splitting of $0 \cdot 2$ Å in the radial distribution curve.

**Solution**

If $\cos cs$ replaces the cosine term of equation (2.3.3e), $I_{\text{mol}}$ becomes, for harmonic vibrations

$$I_{\text{mol}} = \sum_{\substack{i \\ i>j}} \sum_{j} (Z_i - f_i)(Z_j - f_j) \exp(-\tfrac{1}{2}u_{ij}^2 s^2) \frac{\sin sr_{ij}}{sr_{ij}} \cos cs\,. \qquad (2.3.7a)$$

The trigonometric functions can be rewritten in terms of sums and differences of angles, that is,

$$\sin sr_{ij} \cos cs = \tfrac{1}{2}\sin s(r_{ij}+c) + \tfrac{1}{2}\sin s(r_{ij}-c)\,. \qquad (2.3.7b)$$

Thus when the Born approximation fails, the appropriate expression analogous to (2.3.3f) is

$$I_{\text{mol}} = \sum_{\substack{i \\ i>j}} \sum_{j} (Z_i - f_i)(Z_j - f_j) \exp(-\tfrac{1}{2}u_{ij}^2 s^2) \frac{\sin s(r_{ij}+c)}{2sr_{ij}} + \sum_{\substack{i \\ i>j}} \sum_{j} (Z_i - f_i)(Z_j - f_j) \exp(-\tfrac{1}{2}u_{ij}^2 s^2) \frac{\sin s(r_{ij}-c)}{2sr_{ij}}\,. \qquad (2.3.7c)$$

It is as though two different interatomic distances $r_{ij}+c$ and $r_{ij}-c$ are present in the molecule, although we know this is not so. Both $r_{ij}+c$ and $r_{ij}-c$ give rise to a peak on the radial distribution curve; the separation is $2c$. Before the modern treatment superseded the Born approximation, it was not realised that two closely spaced peaks may arise from just one distance, and errors of interpretation were made.

For a splitting of 0·2 Å, $c = 0{\cdot}1$ Å. The beat-out point in the molecular intensity curve occurs when $\cos cs = \cos(0{\cdot}1s) = 0$, i.e. when $0{\cdot}1s = \tfrac{1}{2}\pi$. Thus $s = 5\pi = 15{\cdot}7$ Å$^{-1}$. The size of the splitting in the radial distribution curve depends on the difference $|Z_i - Z_j|$; when $Z_i = Z_j$, there is no splitting, but as the difference increases, the single peak broadens, eventually separating into two. A splitting of 0·2 Å corresponds to $|Z_i - Z_j| \approx 45$. The Born approximation is strictly true only when $Z_i = Z_j$.

Figure 2.3.7a shows how the failure of the Born approximation affects the molecular intensity curve; it should be contrasted with figure 2.3.1. Figure 2.3.7b shows the corresponding radial distribution curve.

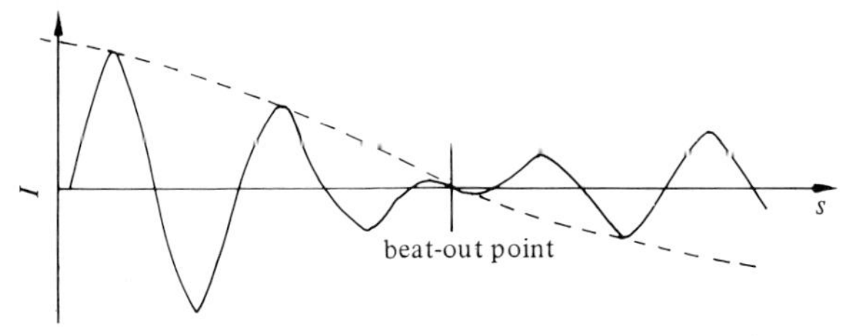


**Figure 2.3.7a.** Section of molecular intensity curve for a single interatomic distance, showing the beat-out point. The broken curve is $\exp(-\tfrac{1}{2}u_{ij}^2 s^2)\cos cs$.

According to the modern theoretical treatment, $Z-f$ in equation (2.3.3e) must be replaced by $s^2|f^e|$ where $|f^e|$ is the amplitude of the electron scattering factor. Angle $\eta$ is the phase associated with this quantity, i.e. the electron scattering factor, $f^e$, is a complex quantity given by

$$f^e = |f^e| \exp(i\eta) .$$

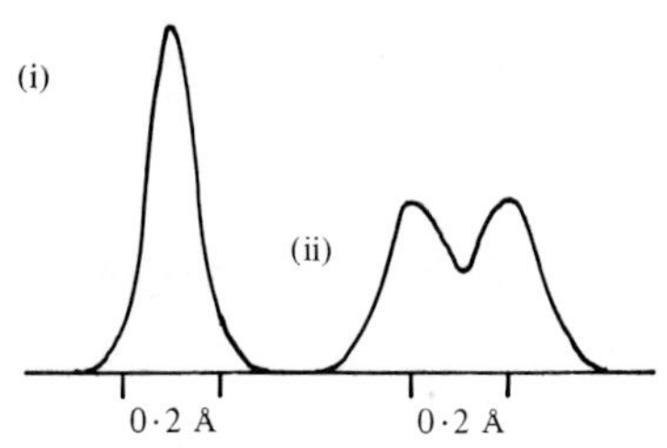


**Figure 2.3.7b.** Radial distribution curves for a single interatomic distance: (i) Born approximation; (ii) failure of Born approximation.

**References**

Bartell, L. S., 1955, *J. Chem. Phys.,* **23**, 1219.

Bartell, L. S., Kuchitsu, K., de Neui, R. J., 1961, *J. Chem. Phys.,* **35**, 1211.

Beagley B., Cruickshank, D. W. J., Hewitt, T. G., Haaland, A., 1967, *Trans. Faraday Soc.,* **63**, 836.

Beagley, B., Robiette, A. G., Sheldrick, G. M., 1968, *J. Chem. Soc. A,* 3002.

## 2.4 Electron diffraction by crystals

**A L Mackay** Birkbeck College, University of London

### 2.4.1 Calibration of the electron microscope for diffraction

Figure 2.4.1 represents an electron diffraction pattern from polycrystalline thallium chloride taken in an electron microscope at approximately 100 kV (wavelength $\lambda = 0\cdot0369$ Å).

(a) Measure up the picture and calculate the camera constant $K$,

$$K = 2L\lambda\,,$$

where $L$ is the effective camera length. The unit-cell dimension, $a$, for TlCl, which is primitive cubic, is $3\cdot842$ Å. Assuming the above value of

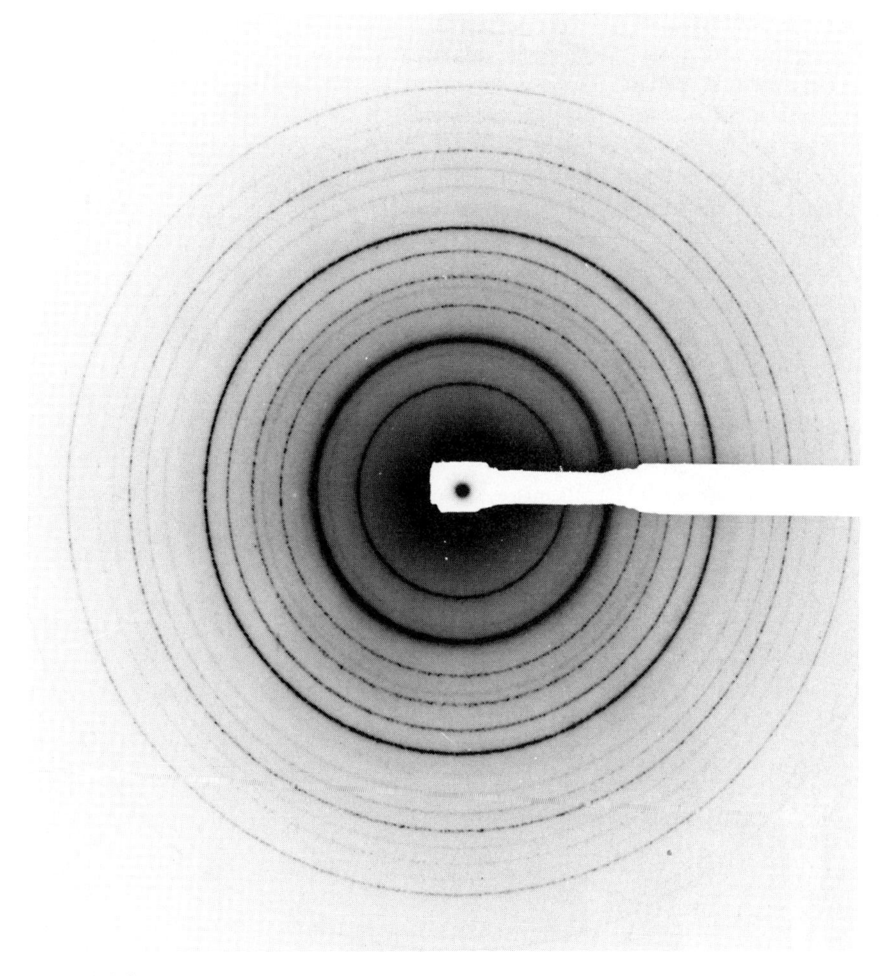

**Figure 2.4.1.**

$\lambda$, estimate $L$. The picture is reproduced from the original plate with a magnification of $1 \cdot 5\times$.
(b) If the focal length of the objective lens is 4 mm calculate the lowest $d$ spacing that would be allowed through a 50 $\mu$m objective aperture.

**Solution**
(a) The data in table 2.4.1 shows that the ring corresponding to planes of index (100), which have the $d$ spacing 3·842 Å, has an average actual diameter of 2·868/1·5 = 1·912 cm; thus

$$K = 2L\lambda = Dd = 1 \cdot 912 \times 3 \cdot 842 = 7 \cdot 35 \text{ cm Å} .$$

If $\lambda = 0 \cdot 0369$ Å, then $L = 99 \cdot 6$ cm.
*Note.*
(i) We confirm, as a check on our indexing, that there is no line with $N = 7$ (which has no possible values of $h$, $k$, $l$).
(ii) There is a slight increase in $K$ with scattering angle, probably due to the neglect of the difference between $\theta$ and $\tan\theta$ in the derivation of $K$. For accurate work it might be better to plot a graph of $K$ against $D$.
(iii) With prolonged irradiation TlCl may decompose giving weak extra lines (see medium-intensity line of diameter ~3·7 cm in figure 2.4.1), perhaps from Tl metal.
(b) Considering the size of the diffraction pattern in the back focal plane of the objective lens, we obtain

$$d_{\min} = 2L\lambda/D ,$$

**Table 2.4.1.** Measurements from figure 2.4.1. Numbers in brackets are for rings that appear too weak for accurate measurement.

| Ring number | Intensity[a] | Diameter, $D$ (cm) | $N$ $(= h^2+k^2+l^2)$ | $D/N^{1/2}$ (cm) |
|---|---|---|---|---|
| 1 | m | 2·86 | 1 | 2·86 |
| 2 | vs | 4·03 | 2 | 2·85 |
| 3 | m | 4·97 | 3 | 2·87 |
| 4 | m | 5·72 | 4 | 2·86 |
| 5 | m | 6·38 | 5 | 2·85 |
| 6 | s | 7·04 | 6 | 2·87 |
| 7 | vw | 8·13 | 8 | 2·87 |
| 8 | vw | 8·65 | 9 | 2·88 |
| 9 | w | 9·10 | 10 | 2·88 |
| 10 | (vw) | (9·6) | (11) | (2·89) |
| 11 | (vw) | (10·0) | (12) | (2·89) |
| 12 | (vw) | (10·4) | (13) | (2·88) |
| 13 | w | 10·81 | 14 | 2·89 |
| | | | Mean | 2·868 |

[a] vw, very weak; w, weak; m, medium; s, strong; vs, very strong.

and therefore,

$$d_{\text{min}} = \frac{2 \times 0\cdot 4 \times 0\cdot 0369}{0\cdot 005} = 5\cdot 9 \text{ Å} ;$$

that is, only planes with spacings greater than 6 Å will contribute to any image formed using a 50 μm aperture and 4 mm focal length.

**2.4.2 Comparison of an object viewed directly with its diffraction pattern**

Figure 2.4.2a shows an electron micrograph of a protein crystal, and figure 2.4.2b represents the electron diffraction pattern obtained from part of this area. The objective aperture was kept in position. Given that the objective aperture, if exactly centred, would represent a spacing of 11·6 Å (that is, would just allow a powder diffraction ring with this *d* value to pass), calculate from the diffraction pattern the periodicities of the crystal.

Check directly from the micrograph by using the 0·1 μm scale bar superimposed.

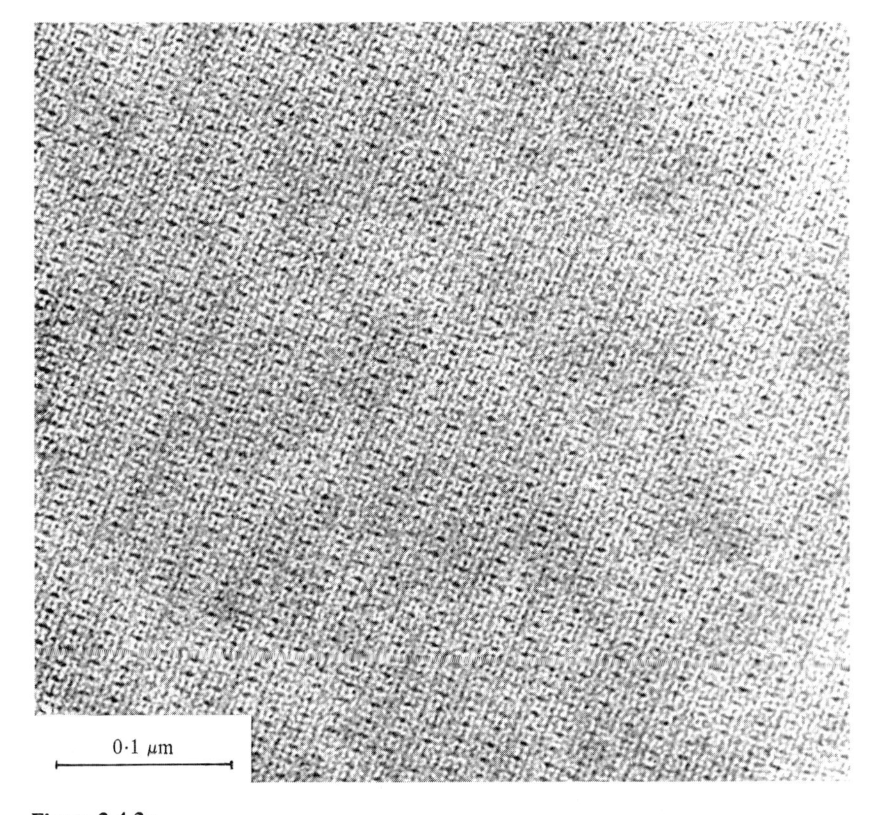


**Figure 2.4.2a.**

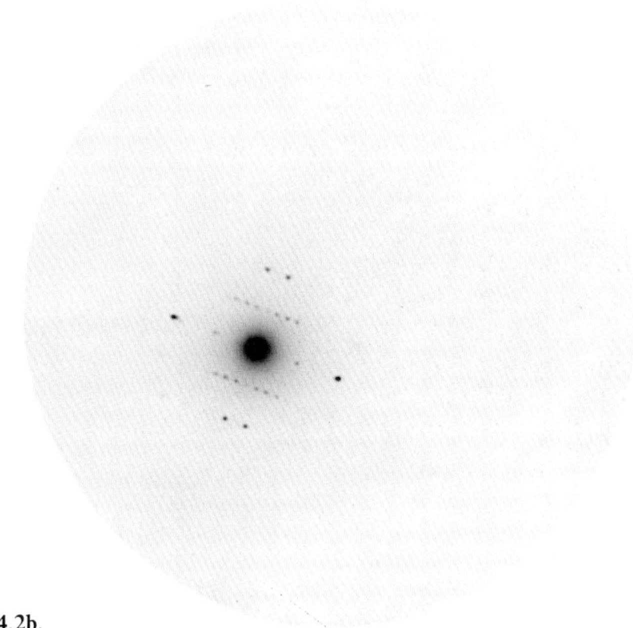

**Figure 2.4.2b.**

**Solution**

The diameter of the aperture shown in figure 2.4.2b is 8·4 cm. The spacings of rows of diffraction spots are as follows

$$D_l = 1\cdot 03 \text{ cm} \qquad \text{(longer spacings)},$$
$$D_s = 0\cdot 30 \text{ cm} \qquad \text{(shorter spacings)}.$$

The aperture represents 11·6 Å; thus

$$2L\lambda = Dd = 8\cdot 4 \times 11\cdot 6 = 97\cdot 4 \text{ cm Å}.$$

Therefore, $D_l$ represents $97\cdot 4/1\cdot 03 = 94\cdot 6$ Å, and $D_s$ represents $97\cdot 4/0\cdot 3 = 325$ Å.

From the observed correspondence in directions in the micrograph and the diffraction pattern, it can be seen that the most prominent layers in the micrograph (figure 2.4.2a) correspond to the eighth order of the longer spacing (and to the prominent diffraction spots). They thus have a spacing of $325/8 = 41$ Å.

On the micrograph (figure 2.4.2a) $80 = 20 \times 4$ of these layers take up 8·5 cm. The 0·1 μm scale bar is 2·50 cm long. Thus the spacing of one layer is $(1000/2\cdot 50) \times (8\cdot 5/80) = 42\cdot 5$ Å.

### 2.4.3 Simple oblique texture

Figure 2.4.3a shows the electron diffraction pattern obtained from a texture of graphite flakes (Aquadag) with the electron beam incident normally on the supporting film carrying the flakes. Show that only

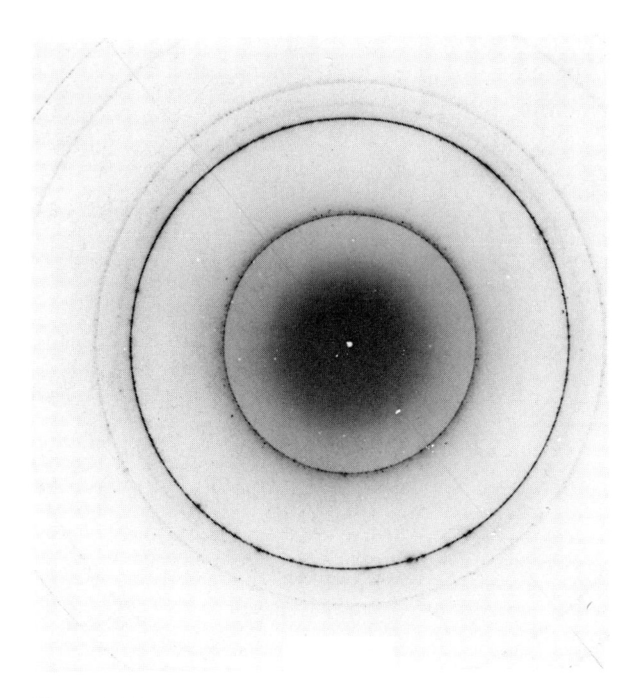

**Figure 2.4.3a.**

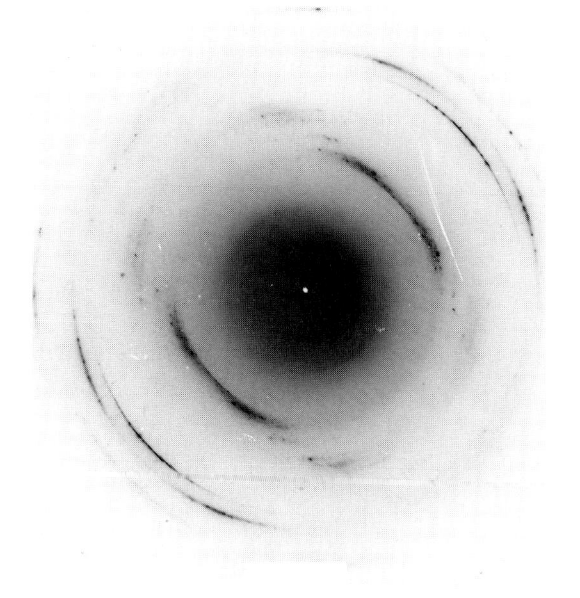

**Figure 2.4.3b.**

$hki0$ reflections should appear. Figure 2.4.3b is the pattern obtained when the same specimen is tilted through 60°. Observe that, through disorder in the texture, a higher-level reflection does appear in figure 2.4.3a and corresponds to a strong arc in figure 2.4.3b. Index the reflections and show that both polymorphs of graphite (two-layer hexagonal and three-layer rhombohedral) are present. [Both forms have $a = 2{\cdot}47$ Å; the hexagonal form has $c = 2 \times 3{\cdot}39$ Å, and the rhombohedral form has $c = 3 \times 3{\cdot}39$ Å. In both cases the flakes are extended parallel to (0001).]

**Solution**

On measuring the diameters of the rings in figure 2.4.3a, we obtain the data given in table 2.4.3. Note the appearance of the $11\bar{2}2$ (hex) ring in figure 2.4.3a. This spreading from $L \neq 0$ layers arises from imperfections in the alignment of the flakes. The value of $a$ is given as $2{\cdot}47$ Å, therefore

$$d_{10\bar{1}0} = 2{\cdot}47 \times (\tfrac{3}{4})^{1/2} = 2{\cdot}14 \text{ Å} ;$$

hence

$$K = 2L\lambda = Dd = 3{\cdot}45 \times 2{\cdot}14 = 7{\cdot}38 \text{ cm Å}.$$

These reflections are seen to occur also along the equator of the plate (figure 2.4.3b) from the tilted specimen. Other reflections fall on layer lines parallel to the equator, although they are considerably drawn out along arcs of circles of constant $\theta$. Row lines project as ellipses.

It is thus difficult to measure the layer-line spacings, but these can be found by triangulation: distances from the centre to the sharp arcs give, for example, $d^*_{11\bar{2}0}$ and $d^*_{11\bar{2}2}$, so that the layer spacing in the reciprocal lattice is obtained from

$$d^*_{0002} = [(d^*_{11\bar{2}2})^2 - (d^*_{11\bar{2}0})^2]^{1/2} .$$

**Table 2.4.3.** Measurements from figure 2.4.3a.

| Ring | Intensity[a] | $D$(cm) | $N$[b] | $D/N^{1/2}$ | Indices |
|---|---|---|---|---|---|
| 1 | m | 3·43 | 1 | 3·43 | $10\bar{1}0$ |
| 2 | vs | 5·96 | 3 | 3·44 | $11\bar{2}0$ |
| 2a[c] | w | 6·36 | – | – | – |
| 3 | w | 6·95 | 4 | 3·475 | $20\bar{2}0$ |
| 4 | w | 9·28 | 7 | 3·508 | $21\bar{3}0$ |
| 5 | w | 10·50 | 9 | 3·50 | $30\bar{3}0$ |

[a] See table 2.4.1; [b] $N = h^2 + k^2 + hk$.
[c] This ring corresponds to $11\bar{2}2$(hex) spreading from a higher layer, as can be seen from figure 2.4.3b.

Flakes of graphite lie on the supporting film with their $c$ axes perpendicular to the substrate. Their angular orientations are random. The reciprocal lattice of the assembly thus consists of circles parallel to the substrate.

Figure 2.4.3c shows a section at 60° to the $c$ axis of this reciprocal lattice for the hexagonal and rhombohedral polytypes.

In figure 2.4.3a, where the beam is normal to the graphite flakes, the diagram obtained is a section perpendicular to $c$ through the origin of reciprocal space. The result is thus concentric circles corresponding to the $hki0$ reflections only (which are the same for both polymorphs). In figure 2.4.3b the texture is tilted by 60°, so that the section cuts through circular cylinders parallel to $c$ and no longer gives circles but ellipses with axes in the ratio 2 : 1. Reflections with the same $hki$ values lie on the same ellipse.

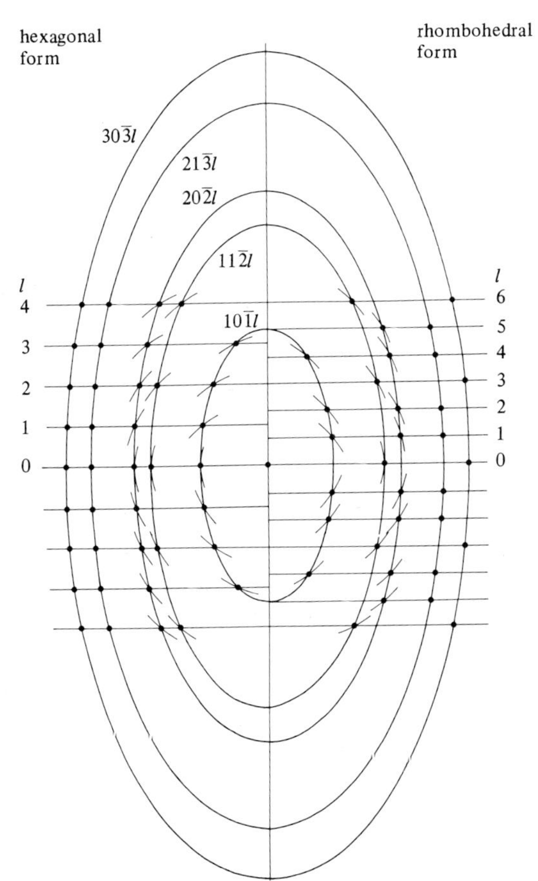


**Figure 2.4.3c.** Section of reciprocal lattice of graphite texture at 60° inclination to $c$ axis. The small arcs represent the reflections visible in figure 2.4.3b.

To measure the $c$ spacings we have, from figure 2.4.3b, with reference to figure 2.4.3c:

$$D_{11\bar{2}0} = 6 \cdot 00 \text{ cm}, \qquad D_{11\bar{2}2}(\text{hex}) = D_{11\bar{2}3}(\text{rhom}) = 6 \cdot 38 \text{ cm} .$$

Thus,

$$\begin{aligned} D_{0002}(\text{hex}) = D_{0003}(\text{rhom}) &= [(6 \cdot 38)^2 - (6 \cdot 00)^2]^{1/2} \\ &= [(0 \cdot 38)(12 \cdot 38)]^{1/2} \\ &= 2 \cdot 17 \text{ cm} . \end{aligned}$$

Since $d_{11\bar{2}0} = 1 \cdot 235$ Å,

$$d_{0002}(\text{hex}) = d_{0003}(\text{rhom}) = \frac{1 \cdot 235 \times 6 \cdot 00}{2 \cdot 17} = 3 \cdot 41 \text{ Å} .$$

For the row lines for which $-h+k = 3n$, the rhombohedral (three-layer) and hexagonal (two-layer) reflections coincide.

### 2.4.4 Complex texture photograph

Figure 2.4.4 represents the electron diffraction pattern obtained when a foil of copper [prepared as a quasi-single crystal by evaporation onto the (100) surface of a heated crystal of NaCl] was oxidised in air. The copper foil has (100) parallel to the substrate and perpendicular to the electron beam. The oxide is $Cu_2O$. Both phases are present and double diffraction effects occur. Find the preferred orientation of the $Cu_2O$ with respect to the copper. The scale of the pattern is 1 cm diameter $\equiv 0 \cdot 1845$ Å$^{-1}$. Copper is face-centered cubic with $a = 3 \cdot 614$ Å, and $Cu_2O$ is also cubic with $a = 4 \cdot 2696$ Å. The wavelength, $\lambda$, is $0 \cdot 0369$ Å.

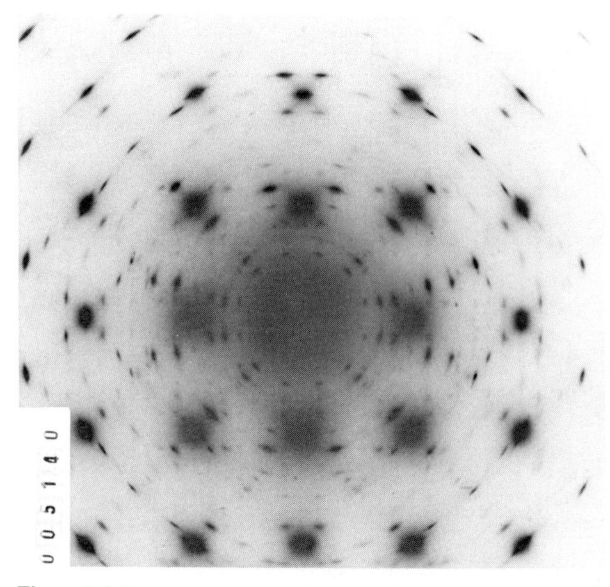

**Figure 2.4.4.**

**Solution**

This diffraction pattern is typical of the complicated patterns obtained from textures of crystallites where more than one phase and more than one orientation occur simultaneously. The pattern consists of two basic components, as follows.

(i) A single-crystal pattern is formed from the copper foil, giving the $hk0$ plane of the reciprocal lattice. Since copper is face-centred cubic, only the reflections for which $h$, $k$, and $l$ are all even or all odd occur. The square grid is indexed by inspection as having the indices 200, 020, 220, 400, ..., and gives a scale to the plate if necessary.

(ii) A powder pattern is obtained from $Cu_2O$ showing strong preferred orientation, there being twelve strong maxima on the first ring. The reference data for cuprite ($Cu_2O$) are $a = 4 \cdot 2696$ Å, space group $Pn3m$, and diffraction lines as shown in table 2.4.4a (JCPDS, 1976).

Double diffraction also takes place. That is, the beams scattered by the copper foil are so strong that they act as new incident beams and repeat the pattern of the $Cu_2O$ with every Cu spot as origin. Stacking faults in the copper foil also give rise to satellite streaks.

The diameter, $D$, corresponding to the 400 spots of copper, which have a $d$ spacing of 0·9035 Å, is 6·00 cm; thus,

$$K = Dd = 6 \cdot 00 \times 0 \cdot 9035 = 5 \cdot 42 \text{ cm Å} .$$

The first rings of the $Cu_2O$ pattern were measured, as shown in table 2.4.4b.

**Table 2.4.4a.** Reference diffraction data for $Cu_2O$ (see text).

| $d$ (Å) | $I$[a] | $hkl$ | $d$ (Å) | $I$[a] | $hkl$ |
|---|---|---|---|---|---|
| 3·020 | 9 | 110 | 1·287 | 17 | 311 |
| 2·465 | 100 | 111 | 1·233 | 4 | 222 |
| 2·135 | 37 | 200 | 1·0674 | 2 | 400 |
| 1·743 | 1 | 211 | 0·9795 | 4 | 331 |
| 1·510 | 27 | 220 | 0·9548 | 3 | 420 |

[a] Relative intensity (arbitrary units).

**Table 2.4.4b.** Measurements of $Cu_2O$ rings from figure 2.4.4. $K = 5 \cdot 42$ cm Å.

| $D$ (cm) | Intensity[a] | $d\,(=K/D)$ (Å) | $hkl$ | $N$ | $a\,(= dN^{1/2})$ (Å) |
|---|---|---|---|---|---|
| 1·80 | s | 3·01 | 110 | 2 | 4·258 |
| 2·20 | w | 2·46 | 111 | 3 | 4·267 |
| 3·10 | w | 1·748 | 211 | 6 | 4·282 |
| 3·57 | vs | 1·518 | 220 | 8 | 4·294 |
| ⋮ | ⋮ | ⋮ | ⋮ | ⋮ | ⋮ |

[a] See table 2.4.1.

The 110 and 220 reflections occur predominantly in twelve equivalent directions; it is necessary to account for this. Since $\lambda = 0{\cdot}0369$ Å, $2\theta$ is only about $0{\cdot}5°$. One [110] direction in $Cu_2O$ is parallel to the [110] direction in Cu but the two cells are not otherwise parallel. By rotating the $Cu_2O$ cell about this [110] axis we see that a set of six (110) planes spaced 60° apart, can be brought near to the reflecting position if a [111] direction is parallel to the electron beam, that is, if a (111) plane of the $Cu_2O$ lies on the (100) plane of Cu which is parallel to the substrate. As the symmetry of the Cu layer is square, there will be two equivalent orientations on it for the $Cu_2O$ layer, which has hexagonal surface symmetry. The twelve 110 reflections are thus explained. No (111) plane lies near the reflecting position, so that, in spite of its high expected strength, this powder ring is weak and not strongly oriented.

If a (110) section of the structure of $Cu_2O$ is drawn out it will be seen that, in the orientation disclosed, it resembes the structure of the (111) plane of Cu. Thus it is likely that, in this case of epitaxy, the (001) plane of $Cu_2O$ grows on the (111) plane of Cu in an azimuth such that the $[\bar{1}\bar{1}1]$ direction of $Cu_2O$ coincides with the [001] direction of Cu.

**2.4.5**

From the three single-crystal diffraction patterns given in figures 2.4.5a, 2.4.5b, and 2.4.5c (all to same scale), find the unit cell and the probable space group of the crystal of guanine shown in figure 2.4.5d. The crystal is a very thin flake and the electron beam is perpendicular to the plate in figure 2.4.5a. In this figure a polycrystalline aluminium film has been used to support the specimen and to provide powder rings for calibration. In figures 2.4.5b and 2.4.5c the crystal has been tilted by +29° and −40°, respectively, about the axis marked in figure 2.4.5a. The material is monoclinic, and the tilting is about the *b* axis (the twofold axis).

**Solution**

The crystal has been tilted about the monoclinic *b* axis so that each of the three diffraction patterns includes this axis and they give different sections through the reciprocal lattice. In figure 2.4.5a (taken with the crystal flake lying perpendicular to the electron beam) it is apparent that the odd-order reflections are extinguished (indicating a $2_1$ screw axis). If the crystal is too large and perfect then double diffraction tends to fill in these absences, as has happened in figures 2.4.5b and 2.4.5c.

From figure 2.4.5a the diameter of the aluminium rings can be measured; results are given in table 2.4.5.

The unit-cell dimension of Al, $a$, is $4{\cdot}05$ Å; thus

$$K = Dd = 2L\lambda = 2{\cdot}32 \times 4{\cdot}05 = 9{\cdot}40 \text{ cm Å} .$$

Hence, in reciprocal space, a distance of 1 cm on this photograph represents $2/9{\cdot}40 = 0{\cdot}2128$ Å$^{-1}$. $d^*_{010}$ is represented on the photograph by $0{\cdot}25$ cm,

Figure 2.4.5a.

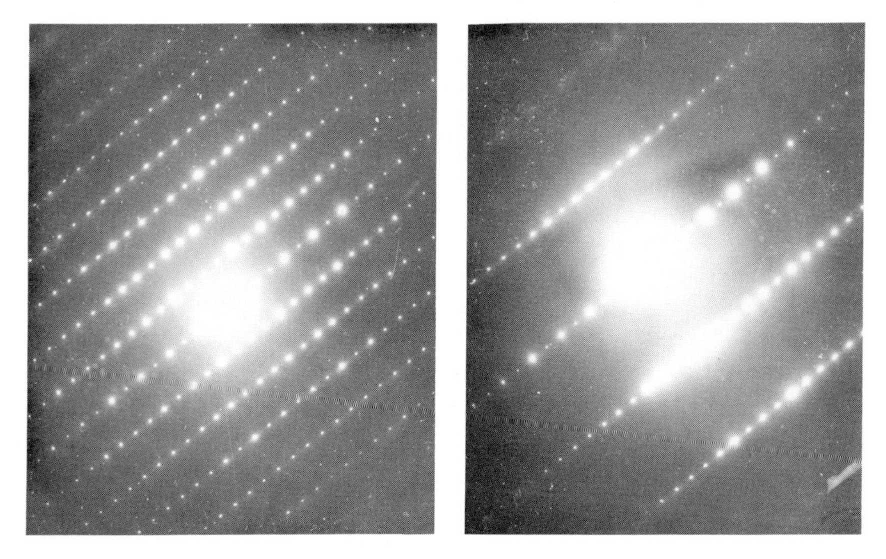

Figure 2.4.5b. Figure 2.4.5c.

so that this is a reciprocal lattice distance of $0{\cdot}25 \times 0{\cdot}2128 = 0{\cdot}0535$ Å$^{-1}$; consequently

$$b = \frac{1}{0{\cdot}0535} = 18{\cdot}7 \text{ Å} .$$

Similarly, the distance to the adjacent row of reflections is $0{\cdot}504$ Å$^{-1}$. The central spot in this row will later be given the index $50\overline{1}$, so that $d^*_{50\overline{1}} = 0{\cdot}504$ Å$^{-1}$.

In figure 2.4.5b, which shows the section through the reciprocal lattice most densely packed with points, $d^*_{010}$ is represented by $0{\cdot}25$ cm, and what turns out to be $d^*_{100}$, the reciprocal lattice repeat in a direction perpendicular to $b^*$, by $0{\cdot}55$ cm. Thus $d^*_{100} = 0{\cdot}12$ Å$^{-1}$.

Similarly in figure 2.4.5c, where the crystal had been tilted about $b$ through 40° in the opposite direction, $d^*_{010}$ is represented by $0{\cdot}27$ cm, and what will be indexed as $d^*_{20\overline{1}}$ by $1{\cdot}30$ cm. Hence $d^*_{20\overline{1}} = 0{\cdot}277$ Å$^{-1}$.

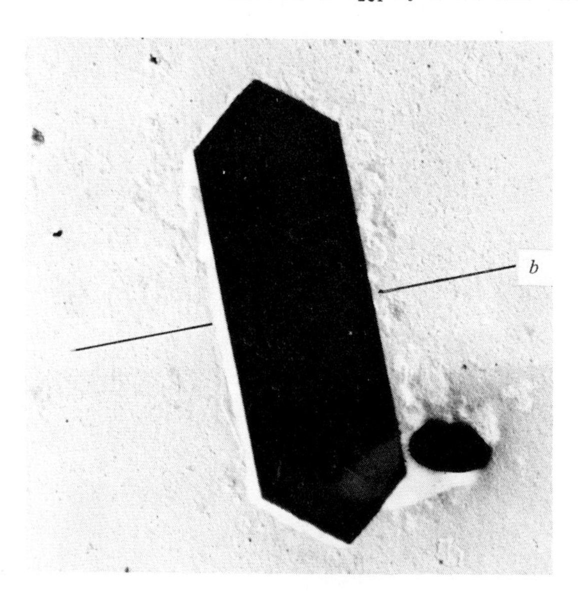


**Figure 2.4.5d.**

**Table 2.4.5.**

| $D$ (cm) | Intensity [a] | $hkl$ | $N$ | $D/N^{1/2}$ |
|---|---|---|---|---|
| 4·05 | m | 111 | 3 | 2·34 |
| 4·61 | w | 200 | 4 | 2·31 |
| 6·53 | vw | 220 | 8 | 2·31 |
| | | | Mean | 2·32 |

[a] See table 2.4.1.

In figure 2.4.5a there are traces of four other rows between the prominent rows of spots, indicating that there is a much more densely packed plane (in fact that revealed in figure 2.4.5b) nearby. The reciprocal-lattice spots are extended in a direction perpendicular to the plane of the crystal because of the thinness of the crystal in that direction.

The reciprocal-lattice section perpendicular to $b$, the axis of tilt, that is the $h0l$ plane, can now be reconstructed by plotting out these measurements, as shown in figure 2.4.5e. From this plot, reciprocal axes $a^*$ and $c^*$, and the angle $\beta^*$, can be allocated and measured. The real axial lengths can then be obtained from $c = d_{001}/\sin\beta$, etc. The unit-cell dimensions are thus found to be

$$a = 9{\cdot}8\ \text{Å}, \qquad b = 18{\cdot}7\ \text{Å}, \qquad c = 4{\cdot}01\ \text{Å}; \qquad \beta = 119^\circ .$$

The only observed extinctions are among the odd $0k0$ reflections, which indicates that the space group may be $P2_1$ or $P2_1/m$.

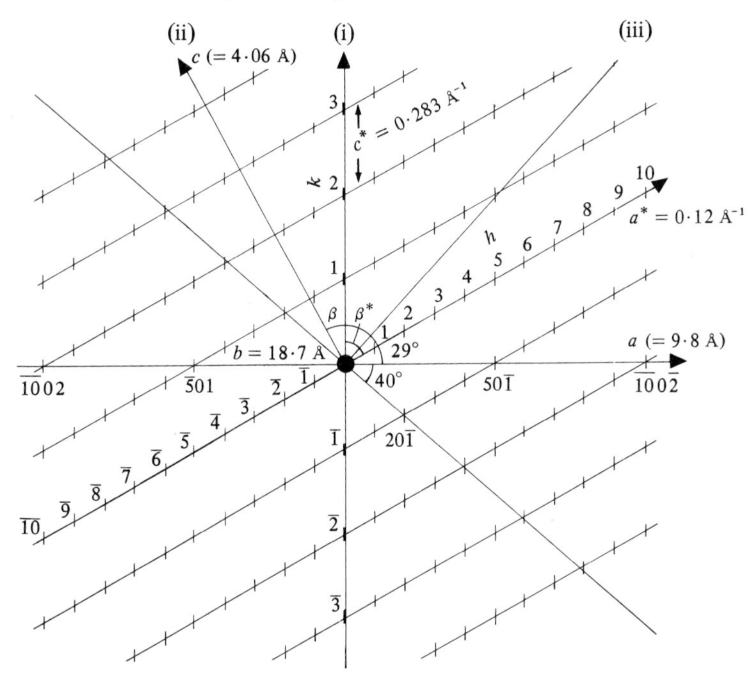


**Figure 2.4.5e.** Plot of $a^*c^*$ plane of the reciprocal lattice for the guanine flake crystal, showing extension of points perpendicular to the plane of the flake. Measurements from figures 2.4.5a–2.4.5c are plotted, with the electron beam direction (i) perpendicular to the plane of the flake, (ii) tilted by 29° about $b$, (iii) tilted 40° in the opposite direction. Scale: 1 cm ~0·23 Å$^{-1}$.

**Reference**

JCPDS, 1976, *Powder Diffraction File Search Manual* (Joint Committee on Powder Diffraction Standards, Swarthmore, Pa) ASTM card number 5-0667.

# 3 Vibration–rotation spectroscopy

## 3.1 Infrared spectroscopy

**N Sheppard, D B Powell** University of East Anglia

### 3.1.1

What do you understand by the normal modes of vibration of a molecule?

Given that such a molecular vibration has neither linear nor angular momentum associated with it, deduce the relative amplitudes of motion of the different types of atoms in the vibrational modes described diagrammatically below.

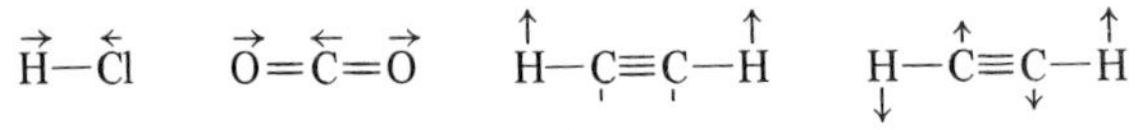

$$\overset{\rightarrow}{\mathrm{H}}\text{—}\overset{\leftarrow}{\mathrm{Cl}} \qquad \overset{\rightarrow}{\mathrm{O}}=\overset{\leftarrow}{\mathrm{C}}=\overset{\rightarrow}{\mathrm{O}} \qquad \overset{\uparrow}{\mathrm{H}}\text{—C}\equiv\text{C—}\overset{\uparrow}{\mathrm{H}} \qquad \underset{\downarrow}{\mathrm{H}}\text{—}\overset{\uparrow}{\mathrm{C}}\equiv\underset{\downarrow}{\mathrm{C}}\text{—}\overset{\uparrow}{\mathrm{H}}$$

(In atomic mass units, a.m.u., $m_{\mathrm{H}} = 1$; $m_{\mathrm{C}} = 12$; $m_{\mathrm{O}} = 16$; $m_{\mathrm{Cl}} = 35$. The C≡C and C—H bond lengths are 0·120 and 0·106 nm respectively.)

*Note*: The amplitudes of motion are linearly proportional to the velocities with which atoms pass through their equilibrium positions.

**Solution**

A normal mode of vibration of a molecule is one in which all atoms in the molecule vibrate with the same frequency and pass simultaneously through their equilibrium positions.

(Note that the second part of the definition holds for all nondegenerate vibrations and for separate components of degenerate vibrations. Degenerate vibrations, such as the two otherwise identical bending modes of vibration of linear XYZ molecules which occur in mutually perpendicular planes, have identical frequencies for reasons of symmetry. However, certain linear combinations of degenerate modes can lead to two-dimensional or to three-dimensional motions of atoms which do not pass through equilibrium positions.)

The requirement that a mode of vibration must involve no angular or linear momentum, i.e. no motion of the molecular centre of gravity or angular rotation of the molecule, imposes limitations on the relative velocities, and the proportional amplitudes of vibration, of different atoms in the molecule. When there is only one mode with a given set of symmetry properties, these considerations define the form of the vibration.

$\overset{\rightarrow}{\mathrm{H}}\text{—}\overset{\leftarrow}{\mathrm{Cl}}$

Let $\dot{x}_{\mathrm{H}}$ and $\dot{x}_{\mathrm{Cl}}$ represent the velocities of vibrating H and Cl atoms as they pass through the equilibrium position. Then the condition of zero net linear momentum requires that $m_{\mathrm{H}}\dot{x}_{\mathrm{H}} + m_{\mathrm{Cl}}\dot{x}_{\mathrm{Cl}} = 0$, or

$$\frac{\dot{x}_{\mathrm{H}}}{\dot{x}_{\mathrm{Cl}}} = -\frac{m_{\mathrm{Cl}}}{m_{\mathrm{H}}} = -\frac{35}{1} \ .$$

As $x_{\mathrm{H}}/x_{\mathrm{Cl}} = \dot{x}_{\mathrm{H}}/\dot{x}_{\mathrm{Cl}}$, where $x_{\mathrm{H}}$ and $x_{\mathrm{Cl}}$ represent the extreme amplitudes of motions during the vibration, we see that, for the relative values of 1 and 35 for $m_H$ and $m_{Cl}$ ($^{35}Cl$ isotope), during the vibration the hydrogen atom moves with an amplitude 35 times that of the Cl atom and in the opposite direction.

$\vec{O}=\overleftarrow{C}=\vec{O}$

By the same reasoning as for HCl, $2m_O\dot{x}_O + m_C\dot{x}_C = 0$, or

$$\frac{\dot{x}_O}{\dot{x}_C} = \frac{x_O}{x_C} = -\frac{m_C}{2m_O} = -\frac{12}{32} = -\frac{3}{8}$$

$\overset{\uparrow}{H}-\underset{\downarrow}{C}\equiv\underset{\downarrow}{C}-\overset{\uparrow}{H}$

Here, $2m_H\dot{y}_H + 2m_C\dot{y}_C = 0$, where $\dot{y}$ is the velocity perpendicular to the molecular axis; hence

$$\frac{\dot{y}_H}{\dot{y}_C} = \frac{y_H}{y_C} = -\frac{m_C}{m_H} = -\frac{12}{1}$$

$\underset{\downarrow}{H}-\overset{\uparrow}{C}\equiv\underset{\downarrow}{C}-\overset{\uparrow}{H}$

In this case it is clear that there is no linear momentum. The angular momentum associated with the motion of an atom X at a distance $r_X$ from the centre of gravity equals $m_X r_X \dot{y}_X$.

For the acetylene molecule the centre of gravity is, by symmetry, at the centre of the C≡C bond, and carbon and hydrogen atoms are at distances $\frac{1}{2}r_{C\equiv C}$ and $(\frac{1}{2}r_{C\equiv C} + r_{C-H})$ from this point, where $r_{C\equiv C}$ and $r_{C-H}$ represent the lengths of the C≡C and C—H bonds respectively. The condition of zero net *angular* momentum then gives

$$2m_C(\tfrac{1}{2}r_{C\equiv C})\dot{y}_C + 2m_H(\tfrac{1}{2}r_{C\equiv C} + r_{C-H})\dot{y}_H = 0 \text{ , or}$$

$$\frac{\dot{y}_H}{\dot{y}_C} = \frac{y_H}{y_C} = -\frac{m_C r_{C\equiv C}}{m_H(r_{C\equiv C} + 2r_{C-H})} .$$

Given that $m_C/m_H = 12/1$, $r_{C\equiv C} = 0{\cdot}120$ nm, and $r_{C-H} = 0{\cdot}106$ nm, then

$$\frac{\dot{y}_H}{\dot{y}_C} = -\frac{12 \times 0{\cdot}120}{(0{\cdot}120 + 0{\cdot}212)} = -4{\cdot}3 .$$

**3.1.2**

Given that, in simple cases, the vibration frequency of a molecule can be calculated from an expression of the type $\nu = (1/2\pi)(k/\mu)^{1/2}$, where $k$ is the appropriate harmonic force constant and $\mu$ is the appropriate 'reduced mass' [for the bond-stretching vibration of HCl, $1/\mu = (1/m_H) + (1/m_{Cl})$] deduce which of the following expressions relates to the symmetrical (I) and antisymmetrical (II) bond-stretching vibrations of the linear and symmetric $CO_2$ molecule.

$\overrightarrow{O}=C=\overleftarrow{O}$ (I) $\qquad$ $\overrightarrow{O}=\overleftarrow{C}=\overrightarrow{O}$ (II)

$$\nu = \frac{1}{2\pi}\left[k\left(\frac{1}{m_O}+\frac{2}{m_C}\right)\right]^{1/2} ; \tag{3.1.2a}$$

$$\nu = \frac{1}{2\pi}\left(\frac{k}{m_O}\right)^{1/2} . \tag{3.1.2b}$$

Calculate values for the bond-stretching force constant, $k$, given that the vibration frequencies for (I) and (II) are approximately 1340 and 2350 $cm^{-1}$ respectively. What reasons can you suggest for the fact that these two values of $k$ are not identical?

($m_C$ = 12; $m_O$ = 16 in a.m.u.; 1 $cm^{-1} \equiv 3{\cdot}00 \times 10^{10}$ Hz; Avogadro constant, $N_A = 6{\cdot}02 \times 10^{23}$ $mol^{-1}$.)

**Solution**

As vibration (I) involves only motions of the oxygen atoms and vibration (II) involves motions of both oxygen and carbon atoms, then equation (3.1.2b) clearly relates to vibration (I), and equation (3.1.2a) to vibration (II).

From equation (3.1.2b)

$$k = 4\pi^2\nu_{(I)}^2 m_O = 39{\cdot}5\nu_{(I)}^2 m_O ,$$

$$\nu_{(I)} = 1340 \times 3 \times 10^{10} \text{ Hz},$$

and

$$m_O = \frac{16 \times 10^{-3}}{6{\cdot}02 \times 10^{23}} = 2{\cdot}66 \times 10^{-26} \text{ kg} .$$

Hence

$$k = 39{\cdot}5 \times (1{\cdot}34)^2 \times 10^6 \times 9 \times 10^{20} \times 2{\cdot}66 \times 10^{-26} = 1{\cdot}70 \times 10^3 \text{ N m}^{-1}.$$

From equation (3.1.2a)

$$k = 4\pi^2\nu_{(II)}^2 m_O m_C/(m_C+2m_O)$$
$$= 39{\cdot}5 \times (2{\cdot}35)^2 \times 10^6 \times 9 \times 10^{20} \times (12/44) \times 2{\cdot}66 \times 10^{-26}$$
$$= 1{\cdot}42 \times 10^3 \text{ N m}^{-1}.$$

Equations (3.1.2a) and (3.1.2b) assume only a single force constant for the bond-stretching vibrations, i.e. that the potential energy, $V$, associated with the motions is given by $V = \frac{1}{2}k(\Delta r_1^2 + \Delta r_2^2)$, where $\Delta r_1$ and $\Delta r_2$ denote the change in the bond lengths of C=O bonds 1 and 2. This is the assumption of a *simple valence force field*.

A *general valence force field* allows for an additional interaction term $k^{-}\Delta r_1 \Delta r_2$ in the expression for $V$, leading to modified equations

$$\nu_{(I)} = \frac{1}{2\pi}\frac{(k+k^{-})^{1/2}}{m_O^{1/2}} , \qquad (3.1.2a')$$

and

$$\nu_{(II)} = \frac{1}{2\pi}\left[(k-k^{-})\left(\frac{1}{m_O}+\frac{2}{m_C}\right)\right]^{1/2} . \qquad (3.1.2b')$$

To this more detailed approximation

$$(k+k^{-}) = 1{\cdot}70 \times 10^3 \text{ N m}^{-1}; \qquad (k-k^{-}) = 1{\cdot}42 \times 10^3 \text{ N m}^{-1}.$$

Thus

$$k = 1{\cdot}56 \times 10^3 \text{ N m}^{-1}; \qquad k^{-} = 0{\cdot}14 \times 10^3 \text{ N m}^{-1}.$$

It is seen that the interaction constant is an order of magnitude smaller than the main force constant.

3.1.3

For simple harmonic motion, classical mechanics gives the following expression for the bond-stretching vibration frequency, $\nu$, of a diatomic molecule AB in terms of the force constant, $k$, and the masses, $m_A$ and $m_B$, of the atoms:

$$\nu = \frac{1}{2\pi}\left[k\left(\frac{1}{m_A}+\frac{1}{m_B}\right)\right]^{1/2} . \qquad (3.1.3a)$$

Given that the stretching frequency of a C—C bond is approximately 1000 cm$^{-1}$ (1 cm$^{-1}$ $\equiv 3 \times 10^{10}$ s$^{-1}$ $= 3 \times 10^{10}$ Hz):

(a) calculate the stretching frequency of C=C and C≡C bonds on the assumption that double and triple bonds have force constants which are respectively two and three times those of a single bond;

(b) calculate the frequencies of C—H and S—S bonds on the assumption that all single-bond force constants are equal.

Compare the frequencies calculated on the above assumptions with the following approximate observed values: C=C, 1650; C≡C, 2200; C—H, 3000; S—S, 500 cm$^{-1}$.

($m_H = 1$; $m_C = 12$; and $m_S = 32$ in a.m.u..)

**Solution**

In this problem we are only concerned with the *relative* masses of the vibrating atoms and the *relative* force constants. In a.m.u. the reciprocal of the reduced mass for the C—C bonds is $(\frac{1}{12}+\frac{1}{12}) = \frac{1}{6}$, and the force constant of the C—C bonds is taken as unity. Thus, from equation (3.1.3a) we obtain the values given in table 3.1.3.

Comparison of the first set of calculated frequencies with the observed values of (1000), 1650, and 2200 $cm^{-1}$ shows that the actual force constants of the double and triple bonds are somewhat greater than two and three times that of the single bond, but that the simpler assumption does give calculated frequencies of the correct general magnitudes.

Comparison of the set of calculated frequencies for single bonds with the observed values of (1000), 3000, and 500 $cm^{-1}$ shows how much of the observed frequency variations from bond to bond can be accounted for in terms of the masses of the vibrating atoms. It demonstrates in particular how the uniquely low mass of the hydrogen atom is primarily responsible for the high frequency of the C–H bond and, more generally, of X–H bonds.

**Table 3.1.3.**

| Bond | Reciprocal relative reduced mass | Relative force constant, $k$ | Frequency | |
|---|---|---|---|---|
| | | | relative | calculated ($cm^{-1}$) |
| C—C | 1 | 1 | 1 | (1000) |
| C=C | 1 | 2 | $2^{1/2}$ | 1414 |
| C≡C | 1 | 3 | $3^{1/2}$ | 1732 |
| C—C | 1 | 1 | 1 | (1000) |
| C—H | $(\frac{1}{12}+1)/\frac{1}{6} = \frac{13}{2}$ | 1 | $(\frac{13}{2})^{1/2}$ | 2550 |
| S—S | $(\frac{2}{32}/\frac{1}{6}) = \frac{3}{8}$ | 1 | $(\frac{3}{8})^{1/2}$ | 612 |

### 3.1.4

Describe the forms of the normal modes of vibration of a linear symmetrical $XY_2$ molecule and deduce, in terms of dipole moment and polarisability changes, which of the modes are infrared and/or Raman active. How are the selection rules modified if the molecule is linear but unsymmetric, of the type XYY or XYZ?

How do infrared vibration–rotation band contours and the degree of polarisation of Raman lines aid in distinguishing bond-stretching from angle-bending frequencies in the vibrational spectra of linear molecules?

**Table 3.1.4.**

| Frequency ($cm^{-1}$) | Infrared | Raman |
|---|---|---|
| 2223·5 | vs, *PR* | s |
| 1285·0 | vs, *PR* | vs |
| 1167·0 | m, *PR* | vw |
| 588·8 | s, *PQR* | - |

vs, very strong; s, strong; m, medium; vw, very weak.
*PR*–*P* and *R* branches in vibration–rotation spectrum; *PQR*–*P*, *Q*, and *R* branches.

Interpret the observed infrared and Raman frequencies for the linear $N_2O$ molecule given in table 3.1.4 (Plyler and Barker, 1931; 1932; Langseth and Nielsen, 1932) in terms of the symmetry of the molecule (i.e. YXY or XYY) and the normal vibration frequencies.

**Solution**

A linear molecule has $3N-5$ vibrational degrees of freedom, i.e. this number of normal modes of vibration, where $N$ is the number of atoms. For $N = 3$ there are therefore 4 vibrational modes, of which 2 are bond-stretching modes and 2 form a doubly degenerate pair of angle-bending modes. For a linear Y—X—Y molecule the bond-stretching modes have in-phase and out-of-phase vibrations of the individual bonds. The forms of the normal modes are therefore:

$\overset{\rightarrow}{Y}-X-\overset{\leftarrow}{Y}$ $(\nu_1)$ $\qquad$ $\overset{\rightarrow}{Y}-\overset{\leftarrow}{X}-\overset{\rightarrow}{Y}$ $(\nu_3)$

$\overset{\uparrow}{Y}-\underset{\downarrow}{X}-\overset{\uparrow}{Y}$ $(\nu_{2a})$ $\qquad$ $\overset{+}{Y}-\overset{-}{X}-\overset{+}{Y}$ $(\nu_{2b})$

where + and – represent vibrational amplitudes perpendicular to the plane of the paper. The numbering of the vibrations is conventional.

*Infrared* activity depends on whether there is a change of *dipole moment* between the extremes of the vibration. The charge asymmetry within the molecule will be of the type $\overset{\delta-}{Y}-\overset{2\delta+}{X}-\overset{\delta-}{Y}$ (or its opposite). Taking into account the displacement arrows associated with the different modes of vibration it is clear that $\nu_1$ does not give rise to a dipole change, $\nu_3$ gives a dipole change parallel ($\parallel$) to the internuclear axis, and $\nu_{2a,\,2b}$ give dipole changes perpendicular ($\perp$) to this axis.

For interaction with electromagnetic radiation (wavelength much greater than molecular dimensions) the electrical polarisabilities of molecules can be represented by ellipsoids. Activity in the *Raman* spectrum depends on a change of polarisability between the extremes of a vibration.

Each mode will now be considered in turn with respect to the *qualitative* polarisability ellipsoids to be expected at various phases of the vibration.

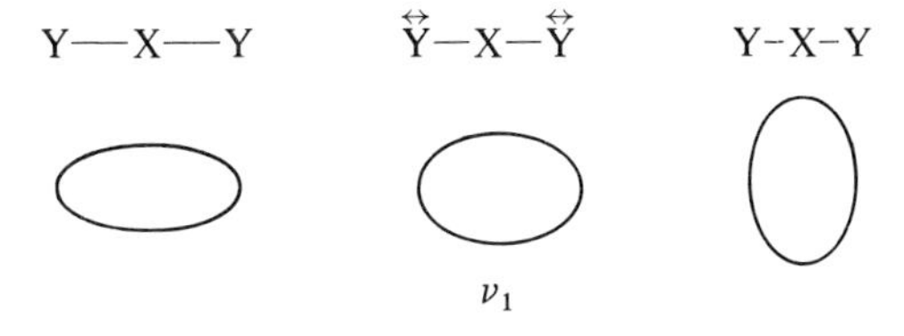


The phases of the vibration and the possible associated polarisability ellipsoids have been represented in an overexaggerated form for clarity; the actual amplitudes and polarisability changes are much smaller than depicted. Clearly for $\nu_1$ the polarisability ellipsoid is different at the two vibrational extremes, and the mode is therefore Raman active.

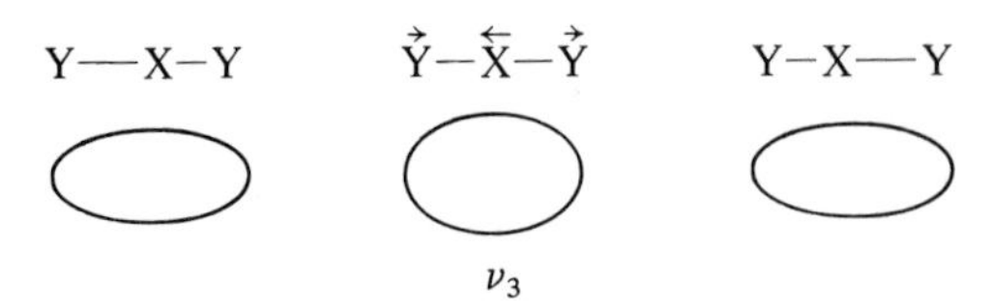


Here there is no change in the polarisability ellipsoid between the two vibrational extremes. These might not have identical polarisability to the molecule in its equilibrium position. However, individual bond-polarisability changes are likely to cancel out. Such residual changes are second order, i.e. they change with twice the frequency of the vibration, and do not contribute intensity to the fundamental. Hence $\nu_3$ is Raman inactive.

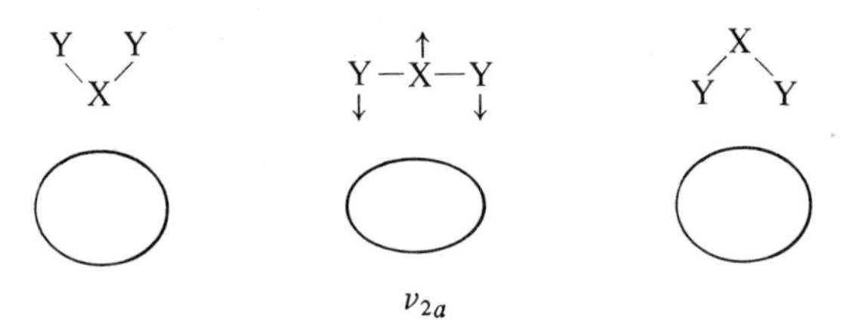


Here again any changes are second order, i.e. $\nu_{2a}$ (and $\nu_{2b}$) are Raman inactive.

In summary, then, $\nu_1$ is infrared inactive and Raman active; $\nu_2$ is infrared active ($\perp$) and Raman inactive; and $\nu_3$ is infrared active ($\|$) and Raman inactive. Such behaviour is also predicted for a centrosymmetric molecule from the 'rule of mutual exclusion'—see problem 3.1.5.

A linear X—Y—Z or X—Y—Y molecule has much less symmetry and there is no reason why either the dipole moment or the polarisability ellipsoid should be the same at the two extremes of any vibration, although the case of the bending mode shows that the extreme ellipsoids differ in orientation rather than in dimensions.

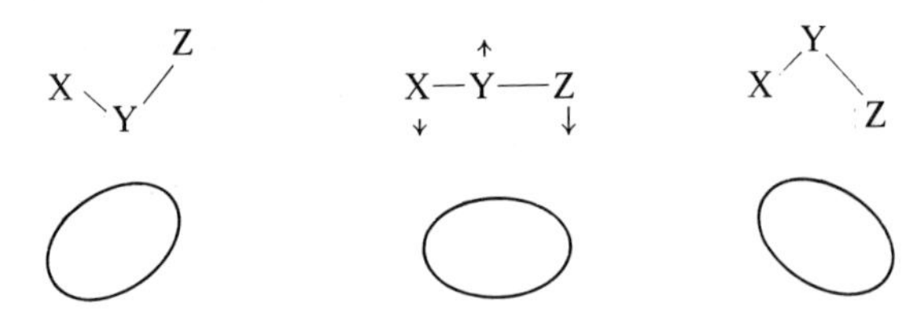


Hence for such molecules every mode of vibration is, in principle, infrared *and* Raman active, although individual ones may be allowed but weak.

*Dipole changes parallel* to the internuclear axis for the vibrations of linear molecules are associated with selection rules $\Delta J = \pm 1$ where $J$ is the total angular momentum quantum number for the rotational energy levels in the gas phase. The corresponding vibration–rotation band has this appearance (schematic):

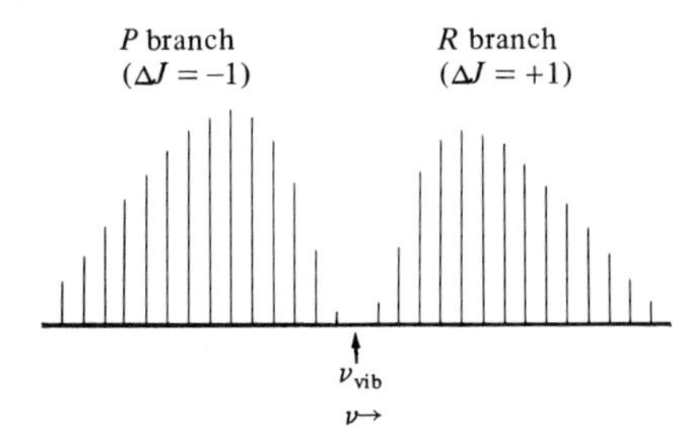


Under poor resolution the doublet form of the overall band contour is still readily recognisable. Bond-stretching modes therefore give rise to this type of vibration–rotation contour.

*Perpendicular dipole changes,* which arise from angle-bending vibrations of linear molecules, have the selection rule $\Delta J = 0, \pm 1$ leading to a strong central $Q$ branch in addition to the $P$ and $R$ branches of parallel modes:

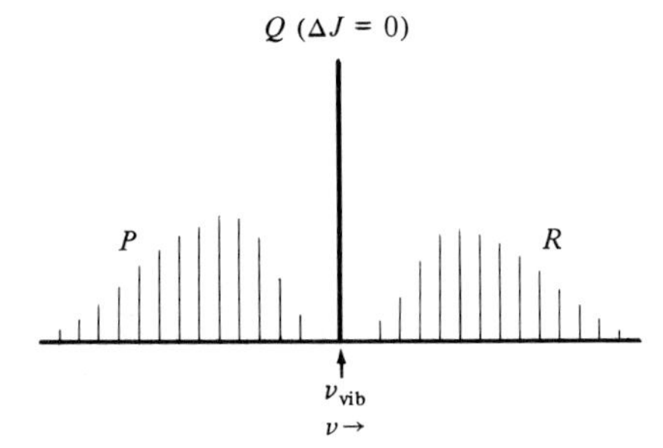


In the Raman effect, completely symmetrical modes of vibration, i.e. those that retain the full symmetry elements of the nonvibrating molecule, give polarised Raman lines; other modes give depolarised lines. For symmetrical Y—X—Y, $\nu_1$ is the only completely symmetrical mode. For X—Y—Z or X—Y—Y the two bond-stretching modes are completely symmetrical and give polarised Raman lines. Angle-bending modes of linear molecules always give depolarised Raman lines.

Another useful consideration in attempting to assign normal modes to particular infrared or Raman frequencies is that, for a given type of bond, angle-bending fundamentals are much lower in frequency than bond-stretching fundamentals.

*The spectrum of $N_2O$ (table 3.1.4)*

Because there are clear-cut coincidences of infrared and Raman frequencies at 2223·5 and 1285·0 $cm^{-1}$, the molecule is clearly not of the symmetry type Y—X—Y (see above discussion of infrared or Raman spectroscopic activity). The formula must therefore be NNO.

The two bond-stretching fundamentals are clearly the two frequencies listed above because they are strong or very strong in both the infrared and Raman spectra, and have the correct *PR* infrared contours for parallel dipole changes.

The 1167·0 $cm^{-1}$ infrared band has a *PR* contour which is not of the *PQR* type expected for the angle-bending mode. The latter should give a perpendicular dipole change. The frequency at 588·8 $cm^{-1}$ does give the correct contour and must therefore be assigned to the angle-bending mode even though it is too weak to appear in the Raman spectrum. The 1167·0 $cm^{-1}$ frequency must therefore correspond to the first overtone of the angle-bending frequency, and the observed frequency is in good agreement with the assignment. Note that the overtone of a vibration does not necessarily retain the same symmetry properties as the fundamental. The overtone of the angle-bending mode can have parallel and perpendicular components. Clearly the former is stronger in the observed infrared spectrum.

**3.1.5**

What do you understand by the 'rule of mutual exclusion' in relation to the infrared and Raman spectra of molecules?

Describe qualitatively the form of the bond-stretching and angle-bending modes of vibration of the acetylene molecule, H—C≡C—H, and deduce which of the normal modes is infrared active. Assuming that the other fundamental vibrations are Raman active, assign the observed infrared (Bell and Nielsen, 1950) and Raman (Feldman *et al.*, 1956) frequencies given in table 3.1.5 to the fundamental vibration frequencies and overtones, combinations, etc.

**Table 3.1.5.**

| Frequency ($cm^{-1}$) | Infrared | Raman[a] |
|---|---|---|
| 3372·5 | - | s, p |
| 3282·5 | s, *PR* | - |
| 2703·6 | m, *PQR* | - |
| 1973·5 | - | vs, p |
| 1328·5 | s, *PR* | - |
| 730·7 | vs, *PQR* | - |
| 613·5 | - | m, dp |

[a] p, polarised line; dp, depolarised line.
See table 3.1.4 for explanation of other abbreviations.

**Solution**

The 'rule of mutual exclusion' concerns the activity of vibrational fundamentals of centrosymmetric molecules in infrared and Raman spectra. It states: "For a centrosymmetric molecule, those vibrations which are infrared active are Raman inactive and vice versa".

As a linear molecule, acetylene has $3N-5$, i.e. 7, degrees of vibrational freedom as $N$, the number of atoms, is 4. Three of these will be bond-stretching modes, as follows:

| | | |
|---|---|---|
| $\nu_1$ | $\overrightarrow{H}-C\equiv C-\overleftarrow{H}$ | C—H stretching, |
| $\nu_3$ | $\overrightarrow{H}-\overleftarrow{C}\equiv\overleftarrow{C}-\overrightarrow{H}$ | C—H stretching, |
| $\nu_2$ | $\overrightarrow{H}-\overrightarrow{C}\equiv\overleftarrow{C}-\overleftarrow{H}$ | C≡C stretching, |

although the first and third of these idealised vibrations may couple together as they have the same symmetry properties.

As is general for linear molecules, the two forms of angle-bending modes as below will each be doubly degenerate:

| | |
|---|---|
| $\overset{\uparrow}{H}-\underset{\downarrow}{C}\equiv\underset{\downarrow}{C}-\overset{\uparrow}{H}$ | $\nu_{4a,4b}$ |
| $\overset{\uparrow}{H}-\underset{\downarrow}{C}\equiv\overset{\uparrow}{C}-\underset{\downarrow}{H}$ | $\nu_{5a,5b}$ |

$\nu_3$ and $\nu_{4a,4b}$ clearly give rise to dipole changes and are infrared active, Raman inactive. $\nu_3$ should be strong with a parallel type *PR* contour, and $\nu_{4a,4b}$ strong with a perpendicular type *PQR* vibration–rotation contour.

$\nu_1$, $\nu_2$, and $\nu_{5a,5b}$ are centrosymmetric and infrared inactive. They are Raman active, with $\nu_1$ and $\nu_2$—as completely symmetrical modes—giving polarised lines, and $\nu_{5a,5b}$ a depolarised one.

From strength and band-contour considerations, $\nu_3 = 3282{\cdot}5\ \text{cm}^{-1}$ and $\nu_{4a,4b} = 730{\cdot}7\ \text{cm}^{-1}$. There is just the right number of Raman lines for the expected fundamentals. Clearly $\nu_1 = 3372{\cdot}5$ (C—H stretching), $\nu_2 = 1973{\cdot}5$ (C≡C stretching) and $\nu_{5a,5b} = 613{\cdot}5\ \text{cm}^{-1}$. The infrared bands at 2703·6 and 1328·5 $\text{cm}^{-1}$ may be interpreted as $\nu_2+\nu_4$ (calculated 2704·2 $\text{cm}^{-1}$) and $\nu_4+\nu_5$ (calculated 1344·2 $\text{cm}^{-1}$) respectively.

**3.1.6**

Describe the form of the normal modes of vibration of a planar trigonal $XY_3$ molecule where $m_X \gg m_Y$. Vibrations will be of the bond-stretching or angle-bending types.

*Note.* In nondegenerate vibrations each Y atom will vibrate in-phase with other Y atoms; in doubly degenerate vibrations of molecules with threefold axes there will be a progressive phase difference of $2\pi/3$ in the amplitudes of adjacent Y atoms.

By assigning partial charges to the X and Y atoms ($\delta-$ to Y and $3\delta+$ to X) show which vibrations are infrared active because of dipole changes. By assigning appropriately shaped polarisability ellipsoids to the extreme forms of each vibration (see problem 3.1.4 for the simple case of a linear molecule), determine the Raman activity of each mode.

The infrared and Raman spectra of $BF_3$ (a mixture of $^{10}BF_3$ and $^{11}BF_3$ in the approximate ratio of 1 : 4) give the frequencies shown in table 3.1.6 (Gage and Barker, 1939; Yost *et al.*, 1938). Assign these to the normal modes of vibration of this molecule, using the selection rules deduced above.

**Table 3.1.6.**

| Frequency ($cm^{-1}$)[a] | Infrared[b] | Raman |
|---|---|---|
| $1497^+$, 1445·9 | vs, ⊥ | w |
| 888 | – | s |
| $719·5^+$, 691·3 | s, ∥ | – |
| $482·0^+$, 480·4 | s, ⊥ | m |

[a] Superscript + denotes weaker component of the doublet.
[b] ⊥, perpendicular band contour; ∥, parallel band contour (with respect to threefold axis). See table 3.1.4 for explanation of other abbreviations.

**Solution**

Assuming that $m_X \gg m_Y$, one can systematically describe the normal modes as follows.

*Nondegenerate modes*

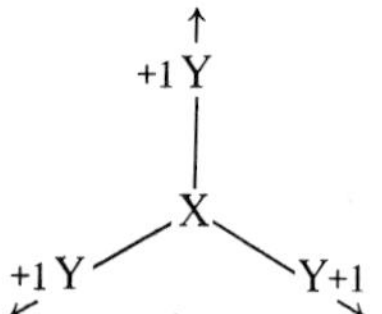

($\nu_1$) (breathing vibration) (a rotation) 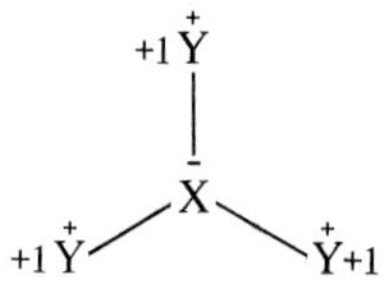 ($\nu_2$) (out-of-plane vibration)

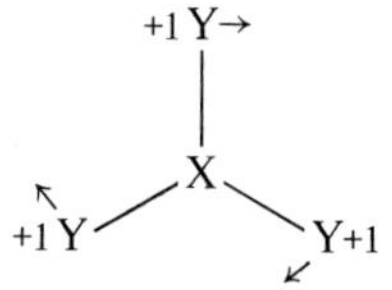

*Doubly degenerate modes*

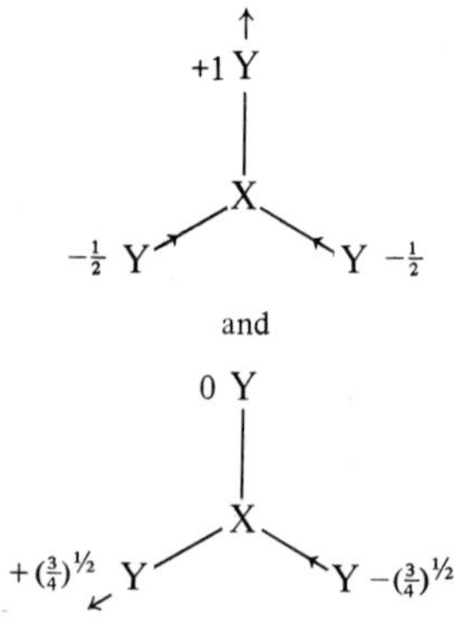

and and and

($\nu_{3a,3b}$) 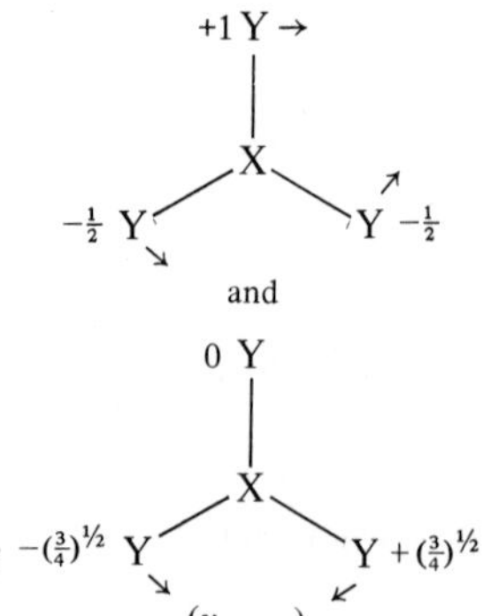 ($\nu_{4a,4b}$) 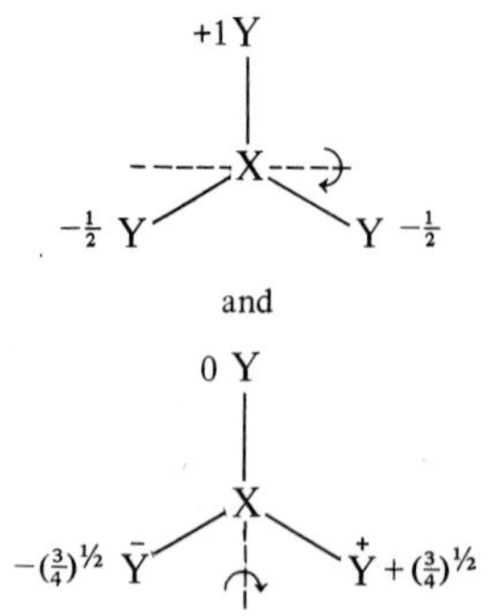 rotations

Here, the numbers positioned by each Y atom give the cosines of the relative phase differences between the amplitudes of these atoms. Note: $\cos 0 = 1$; $\cos\frac{2}{3}\pi = \cos\frac{4}{3}\pi = -\frac{1}{2}$; $\cos\frac{1}{2}\pi = 0$; $\cos\frac{11}{6}\pi = -\cos\frac{7}{6}\pi = \frac{1}{2}(\frac{3}{4})^{1/2}$.

*Infrared activity*

$\nu_1$ has no dipole change.

$\nu_2$ has a dipole change parallel to the threefold axis.

$\nu_{3a,3b}$ and $\nu_{4a,4b}$ have dipole changes perpendicular to the threefold axis.

*Raman activity*

$\nu_1$

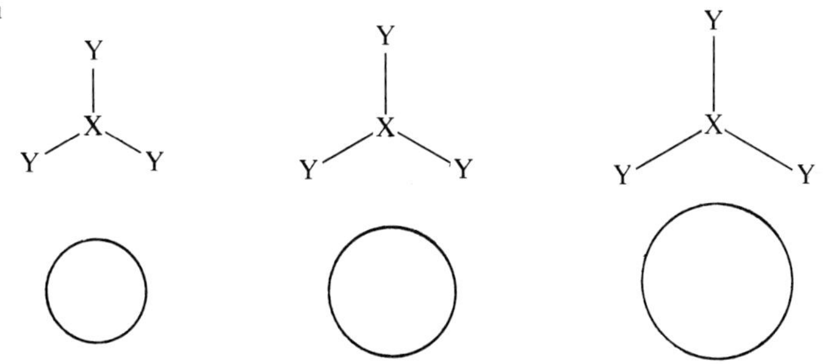


Hence Raman active.

$\nu_2$ Clearly there is no polarisability change between the extremes *in* the plane of the molecule. *Perpendicular* to the plane we have

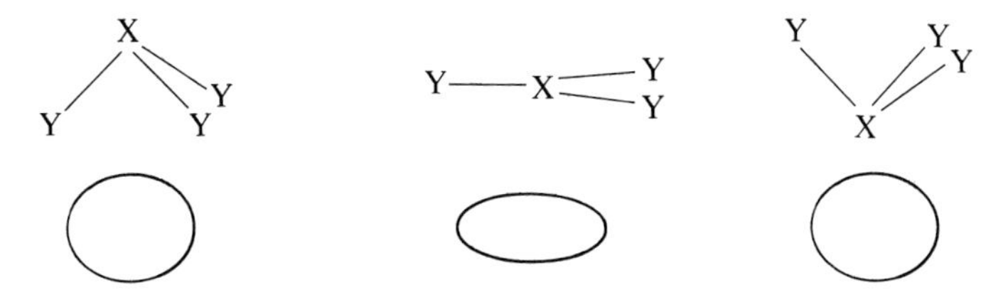


There are only second-order polarisability changes, cf. the angle-bending mode of a linear molecule discussed in problem 3.1.4.

Hence $\nu_2$ is Raman inactive.

$\nu_{3a,3b}$

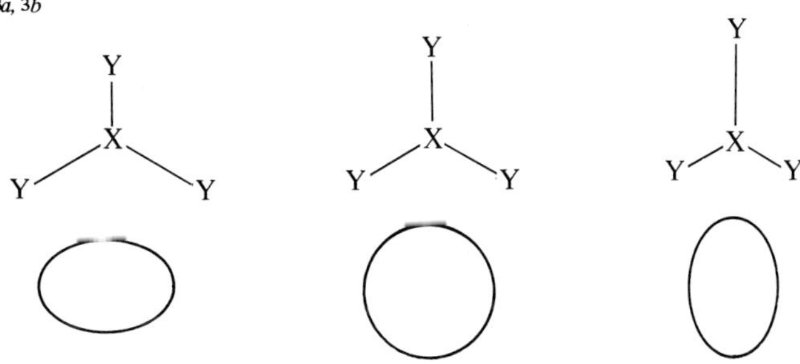

The polarisability ellipsoid rotates between the extremes of the vibration and hence $\nu_{3a,3b}$ is Raman active.

In symmetry terms the same arguments apply to $\nu_{4a,4b}$. In summary we have:

$\nu_1$ infrared inactive: Raman active, polarised.
$\nu_2$ infrared active (parallel); Raman inactive.
$\nu_{3a,3b}$ infrared active (perpendicular); Raman active, depolarised.
$\nu_{4a,4b}$ infrared active (perpendicular); Raman active, depolarised.

From the experimental data, and for the present ignoring the weaker components of the double bands, the polarised Raman line at 888 $cm^{-1}$ is clearly $\nu_1$, and the parallel infrared band at 691·3 $cm^{-1}$ is clearly $\nu_2$. The infrared and Raman active bands at 1445·9 and 480·4 $cm^{-1}$ are clearly the doubly degenerate bond-stretching and in-plane angle-bending frequencies respectively.

For $BF_3$, the central B atom has mass comparable with that of the F atoms, and it will vibrate with considerable amplitudes for all vibrations except $\nu_1$ where, for symmetry reasons, the central atom does not move. Hence $\nu_1$ has a single frequency, but the other fundamentals have strong components at lower frequency from the heavier $^{11}BF_3$ molecules and weak components at higher frequency from $^{10}BF_3$ molecules.

**References**

Bell, E. E., Nielsen, H. H., 1950, *J. Chem. Phys.*, **18**, 1382.
Feldman, T., Shepherd, G. G., Welsh, H. L., 1956, *Can. J. Phys.*, **34**, 1425.
Gage, D. M., Barker, E. F., 1939, *J. Chem. Phys.*, **7**, 455.
Langseth, A., Nielsen, J. R., 1932, *Nature (London)*, **130**, 92.
Plyler, E. K., Barker, E. F., 1931, *Phys. Rev.*, **38**, 1827.
Plyler, E. K., Barker, E. F., 1932, *Phys. Rev.*, **41**, 369.
Yost, D. M., DeVault, D., Anderson, T. F., Lassettre, E. N., 1938, *J. Chem. Phys.*, **6**, 424.

## 3.2 Raman spectroscopy†

**B P Stoicheff** University of Toronto

### 3.2.1

A photograph of the pure rotational Raman spectrum of oxygen is shown in figure 3.2.1a, and the measured frequency shifts are given in table 3.2.1.
(a) Determine the numbering of the rotational lines.
(b) Evaluate the rotational and centrifugal distortion constants for the ground state of oxygen.
(c) Evaluate the internuclear distance $r_0$ for the ground state of oxygen.

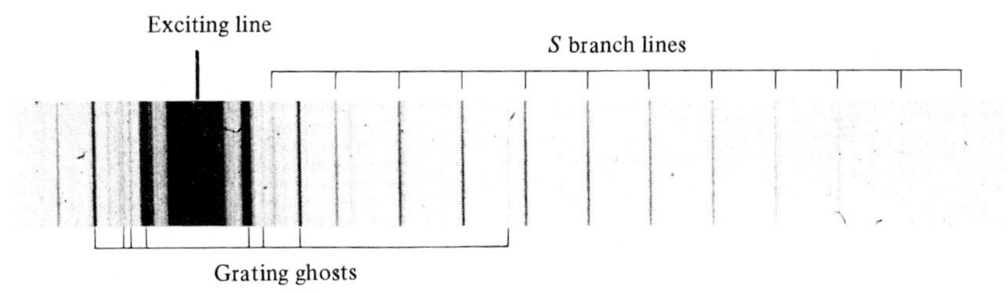


**Figure 3.2.1a.** Pure rotational Raman spectrum of oxygen showing the Stokes side only. (After Stoicheff, 1959.)

**Table 3.2.1.** Observed wavenumber ($\text{cm}^{-1}$) shifts of the rotational lines in the Raman spectrum of oxygen.

| | | | |
|---|---|---|---|
| 14·381 | 48·855 | 83·267 | 117·555 |
| 25·876 | 60·337 | 94·712 | 128·949 |
| 37·369 | 71·809 | 106·143 | |

**Solution**

(a) Oxygen is a homonuclear molecule having a ${}^3\Sigma_g^-$ ground electronic state, and nuclear spin $I = 0$. Thus it can be shown that for the oxygen molecule alternate rotational levels, the even-numbered levels, are missing. For oxygen the resultant angular momentum is designated by $K = \Lambda + N$, where $\Lambda = 0$ (for a $\Sigma$-state) and hence $K \equiv N$. The rotational quantum number $N$, describing rotation about an axis at right angles to the internuclear axis, has the values 1, 3, 5, .... This may be established from the observed frequency shifts as follows:

The rotational energy in the $v = 0$ state is given by

$$F_0(N) = B_0N(N+1) - D_0N^2(N+1)^2 \, .$$

Here, and in other problems in this section, $F_v = E_v/hc$, where $E_v$ is the energy of a level in vibrational state $v$. When the selection rule $\Delta N = 0, \pm 2$

† Author's note: these problems submitted 9 March 1973.

is applied, one obtains for the frequency shifts the expression

$$S(N) = (4B_0 - 6D_0)(N+\tfrac{3}{2}) - 8D_0(N+\tfrac{3}{2})^3 \,. \qquad (3.2.1)$$

Since $D_0 \ll B_0$, $S(N) = 4B_0(N+\frac{3}{2})$ is suitable for our purpose of determining the rotational numbering. It is easily seen that the rotational Raman lines will be separated by $4B_0$ if all values of $N$ are possible; if only odd or only even values of $N$ are possible, adjacent lines are spaced by $8B_0$. Also, the first rotational line (that is, the line closest to the exciting line) will have a frequency shift of $6B_0$, $10B_0$, or $6B_0$, depending on whether $N$ can have all values, only odd, or only even values, respectively. From table 3.2.1 it is seen that the ratio of frequency shift of the first line to the frequency separation of successive lines is ~14·4:11·5 or 1·25:1 which is the predicted ratio (10:8) for $N = 1, 3, 5$ ... . The numbering of the lines given in table 3.2.1 is therefore $N = 1, 3, 5, ..., 21$.

(b) The constants $B_0$ and $D_0$ may be evaluated by fitting the measured frequency shifts to equation (3.2.1).

An alternative solution is to plot the values $S(N)/(N+\frac{3}{2})$ against $(N+\frac{3}{2})^2$ as shown in figure 3.2.1b. The rotational constants are:

$$B_0 = 1{\cdot}43769 \pm (1 \times 10^{-5})\ \mathrm{cm^{-1}}; \qquad D_0 = (4{\cdot}86 \pm 0{\cdot}01) \times 10^{-6}\ \mathrm{cm^{-1}} \,.$$

See Butcher *et al.* (1971).

(c) From the above value of $B_0$ the moment of inertia is

$$I_0 = \frac{h}{8\pi^2 c B_0} = \frac{27{\cdot}9890}{B_0} \times 10^{-40} \text{ or } 19{\cdot}4680 \times 10^{-40}\ \mathrm{g\ cm^2} \,.$$

The internuclear distance $r_0$ is then calculated from $I_0 = \frac{1}{2}mr_0^2$ where $m$ is the oxygen mass, giving $r_0 = 1{\cdot}2106 \times 10^{-8}$ cm.

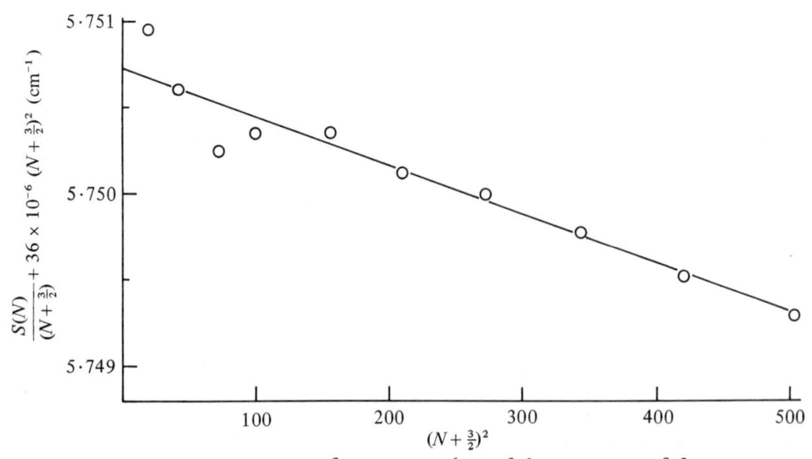


**Figure 3.2.1b.** Plot of $S(N)/(N+\frac{3}{2})+36\times10^{-6}(N+\frac{3}{2})^2$ against $(N+\frac{3}{2})^2$ for oxygen. The term $36\times10^{-6}(N+\frac{3}{2})^2$ was added in order to reduce the slope. (After Butcher *et al.*, 1971.)

3.2.2
The rotational–vibrational Raman spectrum of oxygen is shown in figure 3.2.2a, and the frequency shifts of the observed lines are given in table 3.2.2.
(a) Determine the rotational numbering of the rotation–vibration lines.
(b) Evaluate the rotational coupling constant $\alpha = B_0 - B_1$, and the band centre $\nu_0$.
(c) Evaluate $B_e$ and $r_e$, the equilibrium values of the rotational constant and the internuclear distance, respectively.

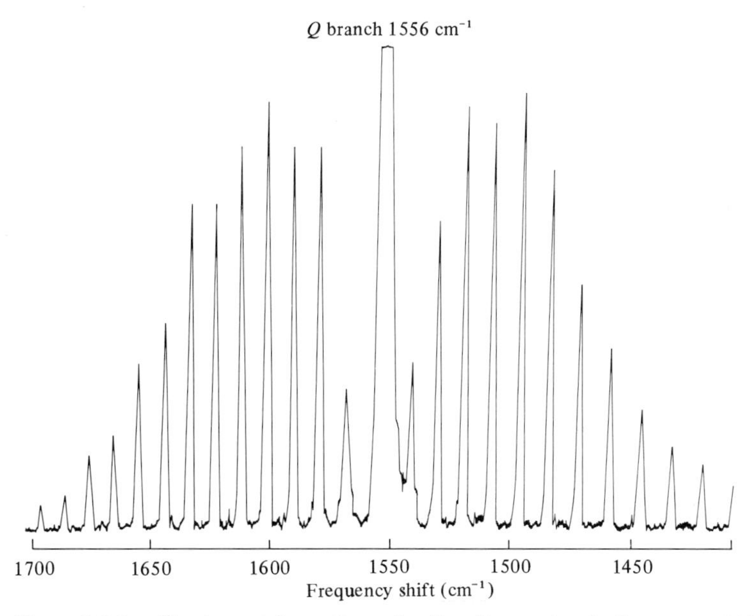


**Figure 3.2.2a.** Fundamental rotation–vibration Raman band of oxygen. (After Barrett and Adams, 1968.)

**Table 3.2.2.** Observed wavenumber shifts ($cm^{-1}$) of the fundamental rotational-vibrational Raman band of oxygen.

| | | | |
|---|---|---|---|
| 1433·17 | 1506·52 | 1581·80 | 1646·15 |
| 1445·58 | 1518·50 | 1592·55 | 1656·26 |
| 1457·71 | 1530·27 | 1603·58 | 1666·50 |
| 1470·10 | 1541·85 | 1614·37 | 1676·21 |
| 1482·27 | 1556·23 ($Q$ branch head) | 1625·20 | |
| 1494·59 | 1570·62 | 1635·78 | |

**Solution**

(a) The sharp $Q$ branch essentially defines the band centre $\nu_0$. By definition, the $S$ branch lines are those for which $\Delta J = +2$ (or $\Delta N = +2$ for oxygen) and therefore they are the lines at higher frequency shifts than the $Q$ branch: the $O$ branch lines have $\Delta N = -2$, and are at lower frequency shifts. Since the accepted rotational numbering, $S(N)$, $O(N)$, is that in which $N$ refers to the $N$ value of the lower level involved in the transition, this necessarily means that the $S$ branch line closest to the $Q$ branch is $S(1)$ at $1570{\cdot}62\ \mathrm{cm}^{-1}$, while the closest $O$ branch line is $O(3)$ at $1541{\cdot}85\ \mathrm{cm}^{-1}$. Thus the rotational–vibrational lines listed in table 3.2.2 are identified as $S(1)$ to $S(21)$ and $O(3)$ to $O(21)$. See Weber and McGinnis (1960).

(b) The energy levels of the $v = 1$ state are given by

$$F_1(N') = \nu_0 + B_1 N'(N'+1) - D_1 N'^2 (N'+1)^2 \,,$$

and those of the $v = 0$ state by

$$F_0(N'') = B_0 N''(N''+1) - D_0 N''^2 (N''+1)^2 \,,$$

where $N'$ and $N''$ refer to the rotational quantum numbers of the upper and lower states, respectively. The frequency shifts of the $S$ and $O$ branch lines are obtained by substituting $N' = N''+2$ ($S$ branch) and $N' = N''-2$ ($O$ branch):

$$S(N) = \nu_0 + 6B_1 + (B_1 - B_0)N^2 + (5B_1 - B_0)N - 8DN^3 - 36DN^2 - 60DN - 36D \,,$$

$$O(N) = \nu_0 + 2B_1 + (B_1 - B_0)N^2 - (3B_1 + B_0)N + 8DN^3 - 12DN^2 + 12DN - 4D \,,$$

where it is assumed that $D_1 = D_0 = D$. The sum $\frac{1}{2}[S(N)+O(N)]$ is $\nu_0 + 4B_1 - 20D_0 + (B_1 - B_0 - 24D_0)N(N+1)$. A plot of $\frac{1}{2}[S(N)+O(N)]$ against $N(N+1)$ then gives a straight line (figure 3.2.2b) with intercept

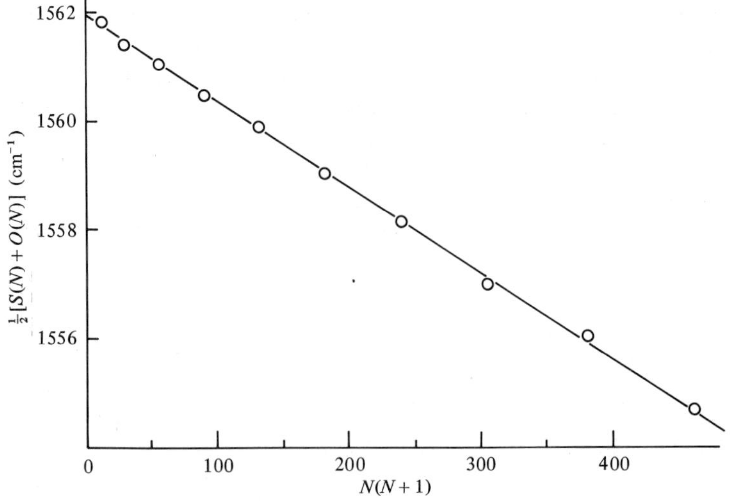


**Figure 3.2.2b.** Plot of $\frac{1}{2}[S(N)+O(N)]$ against $N(N+1)$ for oxygen.

$\nu_0 + 4B_1 - 20D_0 = 1561 \cdot 94$ cm$^{-1}$ and slope $B_1 - B_0 - 24D_0 = -0 \cdot 01593$ cm$^{-1}$. A substitution of $D_0 = 4 \cdot 86 \times 10^{-6}$ cm$^{-1}$ and $B_0 = 1 \cdot 43769$ cm$^{-1}$ gives $\alpha_e = B_0 - B_1 = 0 \cdot 0158$ cm$^{-1}$, $B_1 = 1 \cdot 4219$ cm$^{-1}$ and $\nu_0 = 1556 \cdot 25$ cm$^{-1}$.

(c) Since $B_e = B_v + \alpha_e(v + \frac{1}{2}) - \ldots$ a substitution of $v = 0$, and the above values for $B_0$ and $\alpha_e$ gives $B_e = 1 \cdot 4456$ cm$^{-1}$, $I_e = 19 \cdot 3615 \times 10^{-40}$ g cm$^2$, and therefore $r_e = 1 \cdot 2075 \times 10^{-8}$ cm.

### 3.2.3

The pure rotational Raman spectra of benzene and deuterobenzene have been measured and analyzed (as in problem 3.2.1), yielding for the rotational constants the values

$$0 \cdot 18960 \pm 5 \times 10^{-5} \text{ cm}^{-1} \text{ for } C_6H_6, \quad 0 \cdot 15681 \pm 8 \times 10^{-5} \text{ cm}^{-1} \text{ for } C_6D_6 .$$

Assuming that the benzene molecule has the $D_{6h}$ structure, use this information to determine the internuclear distances $r_0$(C—C) and $r_0$(C—H).

**Solution**

As demonstrated by studies of the vibrational infrared and Raman spectra by Kohlrausch (1936) and by Ingold *et al.* (1936; 1946), the benzene molecule has the highly symmetrical $D_{6h}$ structure. This means that it is planar, that all C–C bond lengths are equal, all C–H bond lengths are equal, and all bond angles are 120°. The molecule is an oblate symmetric top with $I_A = I_B < I_C$, or $A = B > C$.

The pure rotational Raman spectrum arises because of a change in polarizability during the molecular rotation. It is evident that there will be no such change for rotation about an axis perpendicular to the plane of the benzene molecule (the *C*-axis). However, the anisotropy of the polarizability manifests itself for rotation about an axis in the plane of the molecule (say the *B*-axis), and such motion will give rise to the observed spectrum. Thus the given rotational constants refer to this axis, and their reciprocals are proportional to $I_B$, the moment of inertia about an axis in the plane of the molecule (Stoicheff, 1954).

It is easy to derive the expression for $I_B$: for $C_6H_6$ it is

$$I_B = 3\{m_C[r(\text{C—C})]^2 + m_H[r(\text{C—C}) + r(\text{C—H})]^2\} .$$

For $C_6D_6$, $I_B$ is given by a similar expression, but with $m_H$ replaced by $m_D$, and with $r$(C—H) assumed to be equal to $r$(C—D).

From the rotational constants one obtains [see solution 3.2.1(c)]:

$$I_B^0(C_6H_6) = (147 \cdot 59 \pm 0 \cdot 04) \times 10^{-40} \text{ g cm}^2 ,$$

$$I_B^0(C_6D_6) = (178 \cdot 45 \pm 0 \cdot 09) \times 10^{-40} \text{ g cm}^2 .$$

A solution (see figure 3.2.3) for the internuclear distances gives

$$r_0(\text{C—C}) = 1{\cdot}397 \pm 0{\cdot}001\ \text{Å}\,,$$

$$r_0(\text{C—H}) = 1{\cdot}084 \pm 0{\cdot}006\ \text{Å}\,.$$

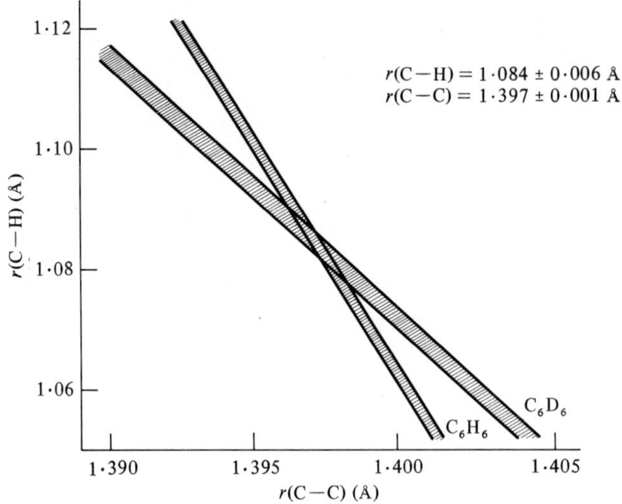


**Figure 3.2.3.** Graphical determination of the internuclear distances in the benzene molecule. The two strips represent the values of the C—H and C—C distances that are consistent with the measured $B_0$ values for $C_6H_6$ and $C_6D_6$. (After Stoicheff, 1954.)

3.2.4[(1)]

The rotational Raman spectrum of an asymmetric top molecule such as ethylene, $C_2H_4$ (point group $D_{2h}$) is characterized by irregularity in line spacing and in intensity distribution (figure 3.2.4a) in contrast to the regularity observed in the spectra of linear and symmetric top molecules. This complexity arises because the $K$ degeneracy is removed for an asymmetric top, and each value of $J$ leads to $2J+1$ sublevels, with energy given by

$$F_0(J_\tau) = \tfrac{1}{2}(A+C)J(J+1) + \tfrac{1}{2}(A-C)E_\tau^J(\kappa) \qquad (3.2.4)$$

in the usual notation (Herzberg, 1945). In spite of this complexity, it has been possible to identify the rotational lines in the spectrum of $C_2H_4$. While most of the 130 observed lines were found to be blends (due to the superposition of several lines arising from different transitions) some lines were identified as single transitions, and these are given in table 3.2.4. The designations for these $S$ branch lines ($\Delta J = 2$; $\Delta K_a = 0$) are $S_{2n+1}(J)$, the transition being $J'_{n,J'-n} \leftarrow J_{n,J-n}$ with $n = 0, 1, 2, \ldots J$, and

(1) Very difficult!

$S_{2n}(J)$, for the transition $J'_{n,J'-n+1} \leftarrow J_{n,J-n+1}$, with $n = 1, 2, 3, \ldots J$. Here the $J$-levels are labelled in the usual $J_{K_a,K_c}$ notation.

(a) Derive the general expression for the frequency shifts $\Delta\nu$ of these $S(J)$ lines.

(b) Use the measured frequency shifts to evaluate the rotational constants $A_0$, $B_0$, $C_0$ for $C_2H_4$.

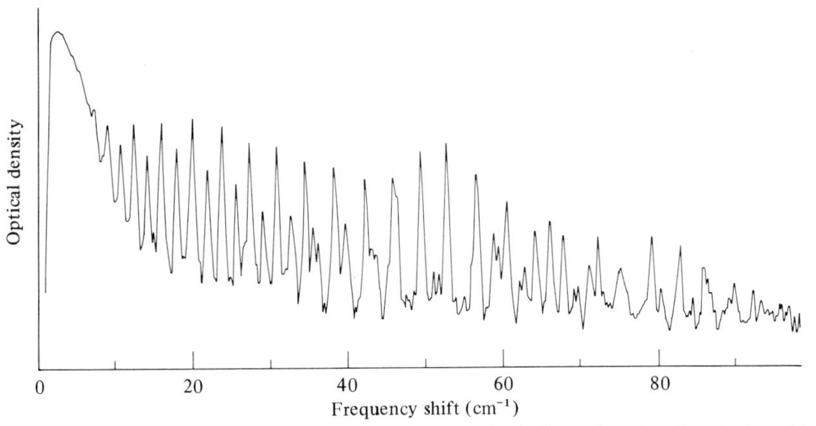


**Figure 3.2.4a.** Pure rotational Raman spectrum of ethylene showing the Stokes side only. (After Dowling and Stoicheff, 1959.)

**Table 3.2.4.** Wavenumber shifts of rotational lines found to arise from single transitions in the Raman spectrum of ethylene.

| Designation | $\Delta\nu$ (cm$^{-1}$) | Designation | $\Delta\nu$ (cm$^{-1}$) |
|---|---|---|---|
| $S_4(10)$ | 41·615 | $S_3(16)$ | 63·521 |
| $S_4(11)$ | 45·139 | $S_7(18)$ | 74·155 |
| $S_4(12)$ | 48·635 | $S_5(19)$ | 77·051 |
| $S_7(13)$ | 54·287 | $S_5(20)$ | 80·474 |
| $S_7(14)$ | 58·234 | $S_5(22)$ | 87·136 |

**Solution**

(a) The expression for the wavenumber displacements of the $S$ branch lines is readily obtained by applying the selection rule $\Delta J = 2$, $\Delta K_a = 0$, to equation (3.2.4):

$$S(J) = (A_0 + C_0)(2J+3) + \tfrac{1}{2}(A_0 - C_0)[E_\tau^{J+2}(\kappa) - E_\tau^{J}(\kappa)]$$

(Dowling and Stoicheff, 1959). Since $E_\tau(\kappa)$ cannot be represented by an explicit formula, the second term is written as shown, to represent the values of $E_\tau^J(\kappa)$ for the upper and lower rotational levels involved in the transition.

(b) For an evaluation of the rotational constants $A_0$, $B_0$, and $C_0$ using the above expression for $S(J)$ and the measured wavenumber shifts, values of $E_\tau^J(\kappa)$ are required. Tabulated values are only available for limited ranges of $J$ and $\kappa$: values for $J \leqslant 12$ with $\kappa$ in increments of 0·01 are given by Townes and Schawlow (1955); Allen and Cross (1963) give values for $J \leqslant 40$ with $\kappa$ in increments of 0·1.

A generally useful method for calculating $E_\tau^J(\kappa)$ has been given by Golden (1948). His formulation may be summarized as follows:

$$E_\tau^J(\kappa) = \left(\frac{\kappa-1}{2}\right)J(J+1) + \left(\frac{3-\kappa}{2}\right)\alpha .$$

The quantity $\alpha$ is given by

$$\alpha = b - [2\theta + C_1\theta' + C_2(\theta')^2] ,$$

where $b$ is the characteristic value of Mathieu's equation corresponding to $4\theta$,

$$\theta(J) = \frac{1}{2}\left|\frac{\kappa+1}{3-\kappa}\right| \left\{J(J+1) - \left[1 + \frac{1}{2J(J+1)}\right]\right\} ,$$

$$\theta'(J) = \frac{1}{2}\left|\frac{\kappa+1}{3-\kappa}\right| \left[1 + \frac{1}{2J(J+1)}\right] ,$$

and $C_1$ and $C_2$ are first- and second-order perturbation corrections. Values of $C_1$ and $C_2$ corresponding to $\theta = 0$ to 10 are tabulated in Golden's paper. This range of values permits the calculation of $E_\tau^J(\kappa)$ up to $J = 27$ for $C_2H_4$.

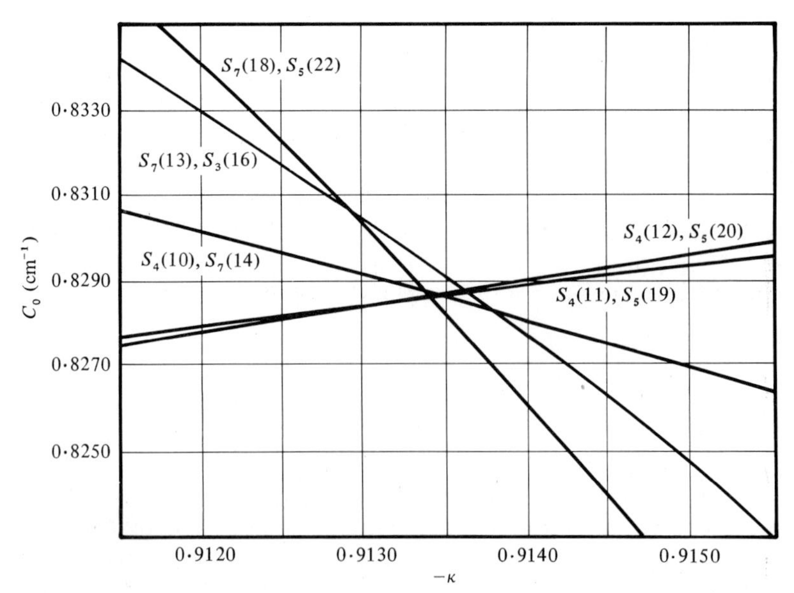


**Figure 3.2.4b.** Graphs used for the determination of $\kappa$ for ethylene. (After Dowling and Stoicheff, 1959.)

The direct solutions of the equations $S(J)$ for the observed lines is not possible since $\kappa$ does not appear explicitly. However, the following procedure, which is customary in the analysis of microwave spectra of asymmetric tops, may be used. For each of the single transitions, the quantities $[E_{\tau'}^{J+2}(\kappa)-E_{\tau}^{J}(\kappa)]$ are calculated for assumed values of $\kappa$. Provisional values of $\kappa$ are obtained from approximate values of $A_0$, $B_0$, $C_0$ and the equation $\kappa = (2B_0-A_0-C_0)/(A_0-C_0)$. The $S(J)$ equations are then taken in pairs (each pair consisting of two different transitions) in order to eliminate $A_0$. The resulting equations then contain $C_0$ alone for various values of $\kappa$. These can be solved for $C_0$ and $\kappa$, or graphs of $C_0$ against $\kappa$ may be plotted as in figure 3.2.4b. Ideally, all the curves should intersect at a single point whose coordinates will be $C_0$ and $\kappa$. However, due to experimental errors, the curves intersect at several points: the average of all the intersections gives a value of $\kappa = -0{\cdot}9135$.

With $\kappa$ evaluated, the values $[E_{\tau'}^{J+2}(\kappa)-E_{\tau}^{J}(\kappa)]$ are recalculated for the single transitions, leading to values for $A_0$ and $C_0$. The constant $B_0$ is then calculated from $\kappa = (2B_0-A_0-C_0)/(A_0-C_0)$. For $C_2H_4$ the values of the rotational constants are:

$$A_0 = 4{\cdot}828 \pm 0{\cdot}009 \text{ cm}^{-1} ,$$

$$B_0 = 1{\cdot}0012 \pm 0{\cdot}0009 \text{ cm}^{-1} ,$$

$$C_0 = 0{\cdot}8282 \pm 0{\cdot}0004 \text{ cm}^{-1} ,$$

$$\kappa = -0{\cdot}9135 \pm 0{\cdot}0003 .$$

**3.2.5**

The infrared and Raman spectra of gaseous diacetylene are shown in figures 3.2.5a and 3.2.5b respectively, and the fundamental frequencies are given in table 3.2.5. The diacetylene molecule has a linear configuration.
(a) Show that the intensity contours of the observed bands labelled with an asterisk are in agreement with the expected shapes of the fundamental vibrational bands.
(b) Make the assignment of the fundamental bands.

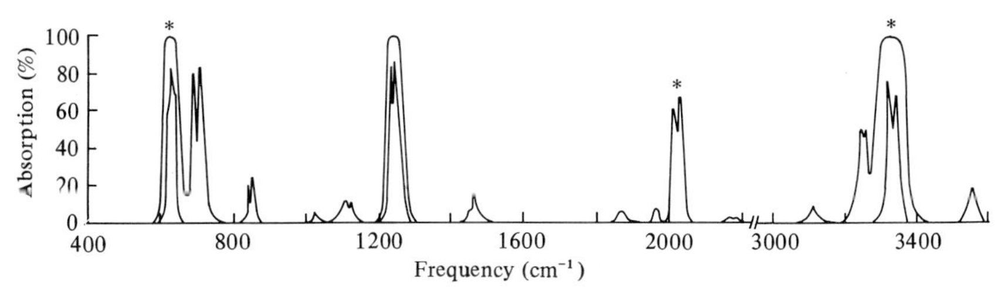


**Figure 3.2.5a.** Infrared spectrum of gaseous diacetylene obtained with absorption path lengths of 10 and 100 cm. The asterisks denote fundamental bands. (After Jones, 1952.)

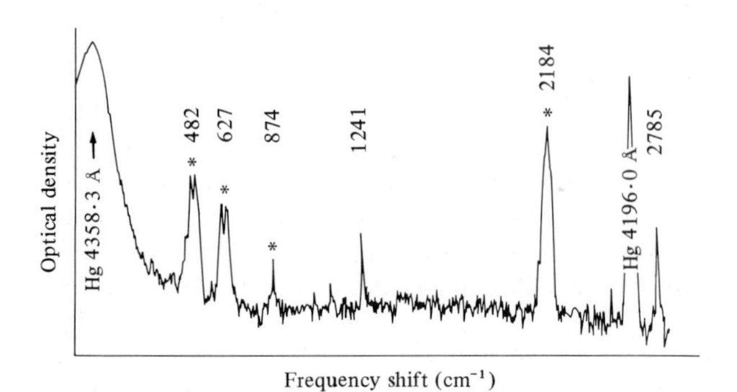


**Figure 3.2.5b.** Raman spectrum of gaseous diacetylene excited by the Hg 4358 Å line. The asterisks denote fundamental bands. (After Jones, 1952.)

**Table 3.2.5.** The fundamental vibrational bands in the spectrum of diacetylene.

| Frequency ($cm^{-1}$) | Band type and remarks |
|---|---|
| 222 | infrared (⊥) |
| 472 max<br>482 min<br>493 max | Raman (gas); [484 $cm^{-1}$, Raman (liq), depolarized] |
| 614 max<br>627 min<br>636 max | Raman (gas); [647 $cm^{-1}$, Raman (liq), depolarized] |
| 620 $P_{max}$<br>630 $Q_{max}$<br>636 $R_{max}$ | infrared (⊥) |
| 874 $Q_{max}$ | Raman (gas); [877 $cm^{-1}$, Raman (liq), polarized] |
| 2010 $P_{max}$<br>2020 min<br>2027 $R_{max}$ | infrared (‖) |
| 2184 $Q_{max}$ | Raman (gas); [2172 $cm^{-1}$, Raman (liq), polarized] |
| (3330) | Raman (gas); [3293 $cm^{-1}$, Raman (liq), polarized] |
| 3320 $P_{max}$<br>3329 min<br>3338 $R_{max}$ | infrared (‖) |

**Solution**

(a) The diacetylene molecule $C_4H_2$ is linear and has a centre of symmetry. Its normal vibrations are shown in figure 3.2.5c. There are five nondegenerate vibrations and four doubly degenerate vibrations. Because of the centre of symmetry the rule of mutual exclusion is expected to hold, that is, Raman active vibrations are infrared inactive, and vice versa.

It is easy to show using table 55 of Herzberg (1945) (cf. section 3.1) that the symmetry types and activities of the normal vibrations are

$3\Sigma_g^+$ and $2\Pi_g$, all Raman active;
$2\Sigma_u^+$ and $2\Pi_u$, all infrared active.

The infrared absorption bands arising from $\Sigma_u^+ \leftarrow \Sigma_g^+$ transitions are those for which there is a change of dipole moment along the internuclear axis during the vibration. For these bands only $\Delta J = \pm 1$ can occur. Therefore these bands have $P$ and $R$ branches but no $Q$ branch (parallel bands). The fundamental bands at 2020 and 3329 $cm^{-1}$ shown in figure 3.2.5a are of this type (Jones, 1952). A change of dipole moment normal to the internuclear axis gives rise to $\Pi_u \leftarrow \Sigma_g^+$ transitions. The selection rules are $\Delta J = 0, \pm 1$; that is, these bands have a $Q$ branch in addition to $P$ and $R$ branches (perpendicular bands). The fundamental band at 630 $cm^{-1}$ is of this type (figure 3.2.5a), and the infrared band at 222 $cm^{-1}$ observed under low resolution is believed to be the second perpendicular fundamental band (Miller *et al.*, 1965).

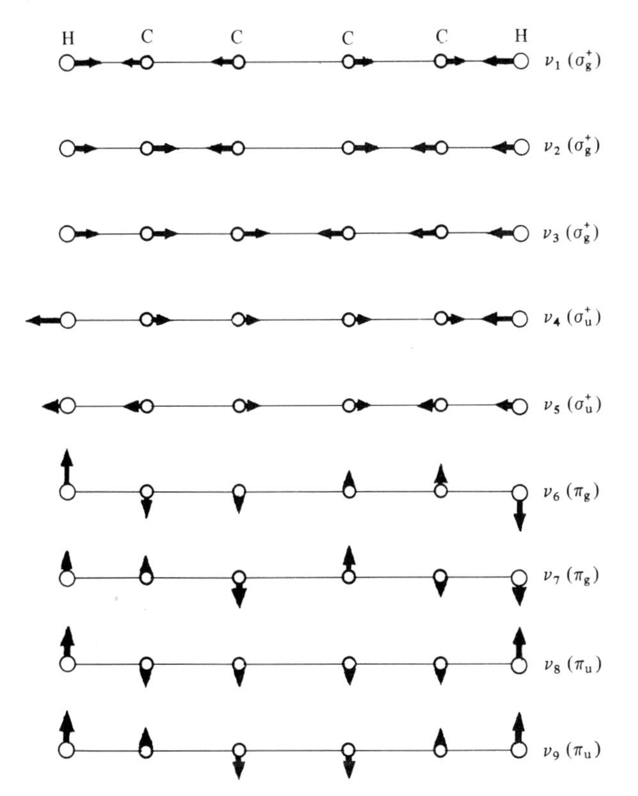


**Figure 3.2.5c.** Normal vibrational modes of the diacetylene molecule. (After Herzberg, 1945, p.324.)

The Raman active, totally symmetric vibrations ($\Sigma_g^+ \leftarrow \Sigma_g^+$ transitions) have selection rules $\Delta J = 0, \pm 2$, and therefore give rise to bands having a sharp $Q$ branch accompanied by much weaker $O$ and $S$ branches. The observed line-like features at 2184 and 874 $cm^{-1}$ are two of these fundamentals, the third being a very weak polarized band observed in the liquid at 3293 $cm^{-1}$. (The other sharp lines at 1241 $cm^{-1}$ and at 2785 $cm^{-1}$ are overtone and combination bands arising from $\Sigma_g^+ \leftarrow \Sigma_g^+$ transitions.) The broad bands with intensity minima at 482 and 627 $cm^{-1}$ correspond to perpendicular type ($\Pi_g \leftarrow \Sigma_g^+$) transitions (Herzberg, 1945, p.399). For such a transition the selection rules $\Delta J = 0, \pm 1, \pm 2$ lead to $O$, $P$, $Q$, $R$, and $S$ branches, but the $Q$ branch is extremely weak, resulting in the band minimum.

(b) A careful study of the band shapes as in (a), and of the force constants of related molecules, by Jones (1952) has led to the following assignment for the fundamental vibrations of $C_4H_2$. The Raman active C—H stretching mode $\nu_1$ is assigned to the weak polarized band observed in the liquid at 3293 $cm^{-1}$; the triple-bond and single-bond stretching modes $\nu_2$ and $\nu_3$ to the sharp lines at 2184 and 874 $cm^{-1}$, respectively. The infrared active, antisymmetric C—H stretching mode $\nu_4$ would be expected to be close to the symmetric mode $\nu_1$ and is ascribed to the parallel band at 3329 $cm^{-1}$. Further, it is inferred that the infrared active triple-bond stretching vibration $\nu_5$ gives rise to the band at 2020 $cm^{-1}$. The perpendicular Raman active bands at 627 and 482 $cm^{-1}$ are correlated with the hydrogen and skeletal bending modes $\nu_6$ and $\nu_7$, respectively. Finally, the infrared active perpendicular bands $\nu_8$ and $\nu_9$ are assigned to the 630 and 222 $cm^{-1}$ bands, respectively.

### 3.2.6

There are two plausible configurations for the symmetric top molecule ethane: the eclipsed model (point group $D_{3h}$) and the staggered model (point group $D_{3d}$).

(a) Use the data of table 3.2.6a, which give the irreducible representation (species) of the components of the dipole moment and of the polarizability, to discuss the possibility of establishing the equilibrium configuration of ethane from a study of infrared and Raman spectra obtained at low resolution.

**Table 3.2.6a.** Irreducible representations (species) of the components of the dipole moment and of the polarizability.

| | $M_x$ | $M_y$ | $M_z$ | $\alpha_{xx}$ | $\alpha_{yy}$ | $\alpha_{zz}$ | $\alpha_{xy}$ | $\alpha_{xz}$ | $\alpha_{yz}$ |
|---|---|---|---|---|---|---|---|---|---|
| $D_{3h}$ | $E'$ | $E'$ | $A_2''$ | $A_1', E'$ | $A_1', E'$ | $A_1'$ | $E'$ | $E''$ | $E''$ |
| $D_{3d}$ | $E_u$ | $E_u$ | $A_{2u}$ | $A_{1g}, E_g$ | $A_{1g}, E_g$ | $A_{1g}$ | $E_g$ | $E_g$ | $E_g$ |

($z$ axis ≡ top axis)

(b) Use the data of table 3.2.6b, giving the selection rules of the quantum number $K$, to distinguish between the $D_{3d}$ and $D_{3h}$ models from spectra taken at high resolution.

**Table 3.2.6b.** Selection rules for (nonzero) components of the dipole moment and polarizability during vibrational transitions.

| | |
|---|---|
| $M_z$ | $\Delta K = 0$ |
| $M_x, M_y$ | $\Delta K = \pm 1$ |
| $\alpha_{zz}, \alpha_{xx} + \alpha_{yy}$ | $\Delta K = 0$ |
| $\alpha_{xz}, \alpha_{yz}$ | $\Delta K = \pm 1$ |
| $\alpha_{xx} - \alpha_{yy}, \alpha_{xy}$ | $\Delta K = \pm 2$ |

**Solution**

(a) At low resolution it should be possible to determine overall band shapes and therefore to identify which infrared and Raman bands are of the parallel or perpendicular type. In addition the state of polarization of Raman bands can be measured. Both types of information are provided in table 3.2.6c, as deduced from the data given in tables 3.2.6a and 3.2.6b. This information, with the measured frequencies, can then be used to make assignments for the twelve fundamental modes of ethane.

Even without the determination and assignment of the normal vibrations of ethane, it is possible to draw some conclusions regarding the ethane configuration. Since the $D_{3d}$ model has a centre of symmetry, transitions active in Raman scattering are not active in infrared absorption, and vice versa. This is shown by the results in table 3.2.6c, namely, that vibrations of species $A_{1g}$ and $E_g$ are only Raman active, and those of species $A_{2u}$ and $E_u$ are only infrared active. For the $D_{3h}$ model, on the other hand, the doubly degenerate vibrations of species $E'$ are active both in Raman scattering and infrared absorption.

**Table 3.2.6c.** Vibrational species, activities, and selection rules for the two models of ethane.

| Configuration | Species | Activity[a] | Selection rule |
|---|---|---|---|
| $D_{3h}$ (eclipsed model) | $A_1'$ | R (polarized) | $\Delta K = 0$ |
| | $A_2'$ | I (∥) | $\Delta K = 0$ |
| | $E'$ | I (⊥) | $\Delta K = \pm 1$ |
| | | R (depolarized) | $\Delta K = \pm 2$ |
| | $E''$ | R (depolarized) | $\Delta K = \pm 1$ |
| $D_{3d}$ (staggered model) | $A_{1g}$ | R (polarized) | $\Delta K = 0$ |
| | $A_{2u}$ | I (∥) | $\Delta K = 0$ |
| | $E_u$ | I (⊥) | $\Delta K = \pm 1$ |
| | $E_g$ | R (depolarized) | $\Delta K = \pm 1, \pm 2$ |

[a] I, infrared; R, Raman.

It should therefore be possible to distinguish between the $D_{3d}$ and $D_{3h}$ models. If there are coincidences in the Raman (depolarized) and infrared (⊥) band frequencies (and there should be three for the three $E'$ normal vibrations in ethane) the structure is clearly $D_{3h}$. However, if no coincidences are observed in the two spectra, and in fact none are (Herzberg, 1945, p.342), this might be because these are weak transitions and not necessarily forbidden because of the symmetry selection rules.
(b) At high resolution one would expect to be able to resolve the $K$ structure (or $Q$ branches) in the Raman and infrared bands. The selection rules for $\Delta K$ are shown in table 3.2.6c. It is seen that the parallel bands will have the same structure for both models ($\Delta K = 0$). Similarly, the infrared perpendicular bands (species $E'$ and $E_u$) will have essentially the same structure ($\Delta K = \pm 1$) for both models. However, the Raman perpendicular bands will have different structures: for $D_{3h}$, the $E''$ bands will have $\Delta K = \pm 1$, and the $E'$ bands $\Delta K = \pm 2$; for $D_{3d}$, the Raman bands of $E_g$ symmetry will have $\Delta K = \pm 1, \pm 2$. Thus if a band with ${}^OQ_k$, ${}^PQ_k$, ${}^RQ_k$, ${}^SQ_k$ branches is observed, this fact would establish the $D_{3d}$ configuration for ethane. In fact, such bands have been found for $C_2H_6$ (Romanko *et al.*, 1955; Lepard *et al.*, 1966) and for $C_2D_6$ (Lepard *et al.*, 1962) and their observation represents the clearest evidence for the $D_{3d}$ structure of ethane.

**References**


Allen, H. C., Jr., Cross, P. C., 1963, *Molecular Vib-Rotors* (John Wiley, New York), p.235.
Barrett, J. J., Adams, N. I., III, 1968, *J. Opt. Soc. Am.*, **58**, 311.
Butcher, R. J., Willetts, D. V., Jones, W. J., 1971, *Proc. R. Soc. London, Ser. A*, **324**, 231.
Dowling, J. M., Stoicheff, B. P., 1959, *Can. J. Phys.*, **37**, 703.
Golden, S., 1948, *J. Chem. Phys.*, **16**, 78.
Herzberg, G., 1945, *Infrared and Raman Spectra of Polyatomic Molecules* (Van Nostrand, New York).
Ingold, C. K., *et al.* (9 papers), 1936, *J. Chem. Soc.*, 912.
Ingold, C. K., *et al.* (13 papers), 1946, *J. Chem. Soc.*, 222.
Jones, A. V., 1952, *Proc. R. Soc. London, Ser. A*, **211**, 285.
Kohlrausch, K. W. F., 1936, *Phys. Z.*, **37**, 58.
Lepard, D. W., Shaw, D. E., Welsh, H. L., 1966, *Can. J. Phys.*, **44**, 2353.
Lepard, D. W., Sweeney, D. M. C., Welsh, H. L., 1962, *Can. J. Phys.*, **40**, 1567.
Miller, F. A., Lemmon, D. H., Witkowski, R. E., 1965, *Spectrochim. Acta*, **21**, 1709.
Romanko, J., Feldman, T., Welsh, H. L., 1955, *Can. J. Phys.*, **33**, 588.
Stoicheff, B. P., 1954, *Can. J. Phys.*, **32**, 339.
Stoicheff, B. P., 1959, "High resolution Raman spectroscopy", in *Advances in Spectroscopy*, volume I, Ed. H. W. Thompson (Interscience, New York), p.91.
Townes, C. H., Schawlow, A. L., 1955, *Microwave Spectroscopy* (McGraw-Hill, New York), p.527.
Weber, A., McGinnis, E. A., 1960, *J. Mol. Spectrosc.*, **4**, 195.

## 3.3 Microwave spectroscopy

**D H Whiffen** University of Newcastle upon Tyne

### 3.3.1 Rotation of carbon monoxide

The pure rotational spectra of diatomic molecules may be interpreted on the basis of the energy-level formula

$$h^{-1}E_{\mathrm{rot}} = BJ(J+1) - DJ^2(J+1)^2 \ ,$$

where $B$ is related to the vibrational quantum number, $v$, by

$$B_v = B_e - \alpha(v+\tfrac{1}{2}) \ .$$

$B_e$ is related to $I_e$, the moment of inertia at the equilibrium separation of the atoms, $r_e$ by

$$B_e = \frac{h}{8\pi^2 I_e} \ .$$

The selection rule for absorption is $\Delta J = 1$, where $J$ is the angular momentum quantum number.

(a) Obtain an expression for the absorption frequencies in terms of $B$, $D$, and $J$ for the lower level.

(b) Three absorption lines of $^{12}C^{16}O$ are at 115·271201, 230·537974 and 345·795900 GHz. Assuming these to be the lowest frequencies and that $D$ is small, suggest the initial and final level for each transition.

(c) Calculate $B$ and $D$ as accurately as possible.

(d) Infrared measurements on the transition to the $v = 1$ level give $B_1 = 57{\cdot}11073$ GHz. Using this value and $B_0$ from (c) calculate $\alpha$ and thence $B_e$ and $I_e$.

(e) Derive from classical dynamics the formula for the moment of inertia of two mass points of masses $m_C$ and $m_O$ separated by a distance $r$.

(f) Thence determine $r_e$ for carbon monoxide.

(g) At what approximate frequency would you expect to find the $J_{1\leftarrow 0}$ absorption of $^{13}C^{16}O$? What information would you require to derive a more accurate value?

(Isotopic masses: $^{12}C = 12$, $^{13}C = 13{\cdot}003354$ and $^{16}O = 15{\cdot}994105$ g mol$^{-1}$.)

See Rosenblum *et al.* (1958); Cowan (1960).

### Solution

(a) The frequency difference is given by

$$h^{-1}(E_{J+1} - E_J) = B[(J+1)(J+2) - J(J+1)] - D[(J+1)^2(J+2)^2 - J^2(J+1)^2]$$

$$= 2B(J+1) - 4D(J+1)^3 \ .$$

(b) The three lines are at approximately 1 x 115, 2 x 115 and 3 x 115 GHz suggesting values of 1, 2, and 3 for $(J+1)$ and $B$ near 57·5 GHz. The transitions may be designated $J_{1\leftarrow 0}$, $J_{2\leftarrow 1}$ and $J_{3\leftarrow 2}$ respectively.
(c) $B$ = 57·635 97 GHz and $D$ = 183·8 kHz.
(d) $\alpha$ = 525·24 MHz and $B_e$ = 57·898 59 GHz. $I_e$ = 1·449 331 x $10^{-46}$ kg m$^2$; this is equivalent to 8·728623 g Å$^2$ mol$^{-1}$.

*Note.* $\alpha$ can in principle be calculated from the molecular force constants, when two major terms arise. Firstly, the reduced value of $r$ (and hence increased $I^{-1}$) for the contraction phase of a harmonic oscillator leads to an increase of $B$ which is not perfectly counterbalanced by the decrease for the stretching phase. This contribution to $\alpha$ is negative. A larger positive contribution arises from the increase of the mean value of $r$ with vibrational amplitude of a real molecular oscillator with an anharmonic potential such as that of the Morse curve.
(e) The centre of mass is at $m_C r/(m_C+m_O)$ from the O atom and $m_O r/(m_C+m_O)$ from the C atom. The moment of inertia is the sum of the products of the masses and the square of their distances from the centre of mass, so that

$$I = \frac{m_O m_C^2 r^2}{(m_C+m_O)^2} + \frac{m_C m_O^2 r^2}{(m_C+m_O)^2} = \frac{m_C m_O r^2}{(m_C+m_O)} .$$

(f) $r_e$ = 1·1283 x $10^{-10}$ m, which is 1·1283 Å.
(g) It is difficult to improve on the value obtained by correcting the $B_0$ value by multiplying by the ratio of the reduced masses, that is by $m(^{12}C)[m(O)+m(^{13}C)]/m(^{13}C)[m(O)+m(^{12}C)]$ = 0·9559150. If the same factor is applied to $D$ it may also be applied to the observed $^{12}C^{16}O$ frequency with the result 110·189 470 GHz. The observed value is 110·201 370 GHz so that the error is 12 MHz or about 0·01%. A complete treatment would require a proper discussion of the mass dependence of $\alpha$ and $D$.

### 3.3.2 Geometry of sulphur dioxide

The values of three rotational constants of $^{32}S^{16}O_2$ are $A_0$ = 60·778 79, $B_0$ = 10·318 10, and $C_0$ = 8·799 96 GHz, and for $^{34}S^{16}O_2$ are 58·991 21, 10·318 40, and 8·761 41 GHz respectively.
(a) Remembering that $A = h(8\pi^2 I_a)^{-1}$ etc., determine the effective moments of inertia $\langle I^{-1}\rangle^{-1}$ for each principal axis of each molecule.
(b) Derive from classical dynamics the formulae for the three moments of inertia in terms of the masses, of $r$ the S—O bond length and of $\theta$ the O—S—O bond angle. What relationship arises in this, and all other planar bodies, between the three moments? To what accuracy is this relationship obeyed by the answers to (a)?
(c) Does the equality of $B_0$ for the two isotopic species imply anything about the molecular symmetry?

(d) Determine $r$ and $\theta$ for each species.
(e) Would you conclude that the equilibrium geometry differs for the two isotopic species?

(Isotopic masses: $^{16}O$ = 15·994015, $^{32}S$ = 31·972074, and $^{34}S$ = 33·970068 g mol$^{-1}$.)
See Kivelson (1954).

**Solution**
(a) Effective values of $I_a$, $I_b$, and $I_c$ for the $^{32}S$ and $^{34}S$ species are 1·380649, 8·132719, 9·535748, 1·422487, 8·132797, and 9·577705 (x $10^{-46}$) kg m$^2$ respectively. These are equivalent to 8·31499, 48·97946, 57·42923, 8·56695, 48·97999, and 57·68192 g Å$^2$ mol$^{-1}$.
(b)

$$I_a = \frac{2m_S m_O r^2 \cos^2 \frac{1}{2}\theta}{m_S + 2m_O}, \qquad I_b = 2m_O r^2 \sin^2 \tfrac{1}{2}\theta ,$$

$$I_c = 2m_O r^2 \left( \sin^2 \tfrac{1}{2}\theta + \frac{m_S \cos^2 \frac{1}{2}\theta}{m_S + 2m_O} \right) .$$

These formulae are obtained by recognizing that the external bisector of angle O–S–O, the internal bisector, and the perpendicular to the plane are parallel to the three principal axes which must pass through the centre of mass.

For all planar bodies the moment of inertia about the perpendicular axis is equal to the sum of the other two: in this case $I_c = I_a + I_b$. Defining $\Delta = I_c - I_a - I_b$, we find $\Delta$ = 2·238 and 2·242 x $10^{-48}$ kg m$^2$ for the $^{32}S$ and $^{34}S$ species respectively. These quantities are called the inertia defects and arise because $\langle I^{-1} \rangle^{-1} \neq I$ when the angular brackets imply averaging over the zero-point vibrational motions.
(c) Yes. If $B$, and hence $I_b$, are independent of the sulphur mass, then the principal axis of inertia must pass through the sulphur atom and the S—O bond lengths must be equal as assumed. This requires $SO_2$ to have the symmetry of the point group $C_{2v}$.
(d) The bond lengths are 1·43220 and 1·43199 x $10^{-10}$ m and the angles 119·536° and 119·566° for the $^{32}S$ and $^{34}S$ species from the numbers given.
(e) No. Because of the zero-point averaging difficulty. The discrepancies are of the magnitude to be expected. Very small errors also arise because of the assumption that the mass of the electrons can be treated as being at the same position as the nuclei. This assumption is more justified than might be supposed because of the slipping motion of the unshared electrons.

### 3.3.3 Dipole moment of isobutane[1]

Symmetric top molecules have a linear Stark effect and the additional energy of any level in the presence of an electric field, $E$, is given by

$$-(pE)KM/J(J+1)\,,$$

where $p$ is the molecular electric dipole moment.

(a) Remembering that $K$ and $M$ may each have all integral values from +J to $-J$, tabulate the additional energies, in multiples of $(pE)$, for all rotational states with $J = 0$, 1, or 2.

(b) If, as is common, the microwave electric field is parallel to the static electric field so that the selection rules are $\Delta J = 1$, $\Delta K = \Delta M = 0$, which transitions between the states tabulated in the answer to (a) are allowed? What multiple of $(pE)$ must be added to the rotational transition energy in the absence of the electric field?

(c) Isobutane, $(CH_3)_3CH$, is an effectively rigid symmetric top for the purpose of Stark effect measurements with $B = 7{\cdot}789\ 45$ GHz. With a Stark electric field parallel to the microwave electric field the $J_{2\leftarrow 1}$ transition at $31{\cdot}157\ 46$ GHz is split into three absorption lines. One is undisplaced and the others are shifted by $\pm\Delta\nu$ in the electric field as shown:

| $E$ (kV m$^{-1}$) | 0 | 35·7 | 48·3 | 72·8 | 85·0 |
|---|---|---|---|---|---|
| $\Delta\nu$ (MHz) | 0 | 7·85 | 10·62 | 16·07 | 18·90 |

By drawing a graph, or otherwise, show that the splitting is indeed linearly proportional to the field.

Evaluate the dipole moment of isobutane from these measurements.

See Lide and Mann (1958).

**Solution**

(a) See table 3.3.3.

**Table 3.3.3.**

| $J$ | $K$ | $M$ | $\frac{KM}{J(J+1)}$ | $J$ | $K$ | $M$ | $\frac{KM}{J(J+1)}$ |
|---|---|---|---|---|---|---|---|
| 0 | 0 | 0 | 0 | 2 | ±2 | ±2 | $+\frac{2}{3}$ |
| | | | | | ±2 | ±1 } | |
| 1 | ±1 | ±1 | $+\frac{1}{2}$ | | ±1 | ±2 } | $+\frac{1}{3}$ |
| | 0, ±1 | 0 } | | | ±1 | ±1 | $+\frac{1}{6}$ |
| | 0 | ±1 } | 0 | | 0, ±1, ±2 | 0 } | |
| | ±1 | ∓1 | $-\frac{1}{2}$ | | 0 | ±1, ±2 } | 0 |
| | | | | | ±1 | ∓1 | $-\frac{1}{6}$ |
| | | | | | ±2 | ∓1 } | |
| | | | | | ±1 | ∓2 } | $-\frac{1}{3}$ |
| | | | | | ±2 | ∓2 | $-\frac{2}{3}$ |

[1] Editors' note: see also problems in section 4.4.

± signs in the $K$ and $M$ columns for the same row are linked. There are $(2J+1)$ values of $K$ and $(2J+1)$ values of $M$ for each value of $J$ and so there are $(2J+1)^2$ states for each $J$, i.e. 1, 9 and 25 for $J$ = 0, 1, and 2 respectively. Check that this statement agrees with table 3.3.3.

(b) The allowed transitions are

| | | | |
|---|---|---|---|
| $J_{1\leftarrow0}$ | $K_{0\leftarrow0}$ | $M_{0\leftarrow0}$ | with Stark energy contribution $0\times(pE)$ |
| $J_{2\leftarrow1}$ | $K_{\pm1\leftarrow\pm1}$ | $M_{\pm1\leftarrow\pm1}$ | with Stark energy contribution $+\frac{1}{3}\times(pE)$ |
| | $K_{0\leftarrow0}$ | Any $\Delta M=0$ | with Stark energy contribution $0\times(pE)$ |
| | Any $\Delta K=0$ | $M_{0\leftarrow0}$ | |
| | $K_{\pm1\leftarrow\pm1}$ | $M_{\mp1\leftarrow\mp1}$ | with Stark energy contribution $-\frac{1}{3}\times(pE)$ |

(c) A graph of $\Delta\nu$ against $E$ is a straight line with slope 221 m s$^{-1}$ V$^{-1}$. The dipole moment $p$ equals $3h\times$ slope of graph. The factor 3 arises from the answer to part (b).

This gives $p = 4{\cdot}39\times10^{-31}$ C m (0·132 D).

It should be noted that no density measurements are required for this method of dipole determination and there is no requirement that the gas be particularly pure.

### 3.3.4 Dipole moment of carbonyl sulphide

Both linear and asymmetric top molecules have a small quadratic Stark effect. For a linear molecule with a dipole moment $p$, the additional energy in an electric field $E$ is given by

$$\frac{p^2E^2[J(J+1)-3M^2]}{2hBJ(J+1)(2J-1)(2J+3)},$$

apart from the special case of $J = 0$ when the additional energy is

$$-\frac{p^2E^2}{6hB}.$$

(a) Tabulate the states and their additional energies as multiples of $p^2E^2/2hB$ for $J$ = 0, 1, 2, and 3.

(b) Use these results to tabulate the contributions to the transition frequencies for the transitions allowed between these states by the selection rules $\Delta J$ = 1, $\Delta M$ = 0, which apply when the static and microwave electric fields are parallel.

(c) For the linear molecule $^{16}O^{12}C^{32}S$, $B_0$ = 6·081 GHz and the displacement of the two lines from their natural field-free positions for the $J_{2\leftarrow1}$ transition near 24·3 GHz are −0·406 and +0·330 MHz with a field of 50 kV m$^{-1}$ and −1·623 and +1·319 MHz for a field of 100 kV m$^{-1}$. Confirm that the effect is second-order and evaluate the dipole moment, $p$.

See Muenter (1968).

**Solution**

(a) See table 3.3.4a.

**Table 3.3.4a.**

| $J$ | $M$ | Energy, $\frac{1}{2}p^2E^2/hB$ | $J$ | $M$ | Energy, $\frac{1}{2}p^2E^2/hB$ |
|---|---|---|---|---|---|
| 0 | 0 | $-\frac{1}{3}$ | 3 | 0 | $+\frac{1}{45}$ |
| 1 | 0 | $+\frac{1}{5}$ | | $\pm1$ | $+\frac{1}{60}$ |
| | $\pm1$ | $-\frac{1}{10}$ | | $\pm2$ | 0 |
| 2 | 0 | $+\frac{1}{21}$ | | $\pm3$ | $-\frac{1}{36}$ |
| | $\pm1$ | $+\frac{1}{42}$ | | | |
| | $\pm2$ | $-\frac{1}{21}$ | | | |

Note how the splittings diminish with increasing $J$ and how the sum of the energies over all $M$ states for each $J$ is zero except for $J = 0$.

(b) See table 3.3.4b.

**Table 3.3.4b.**

| $J$ | $M$ | Frequency, $\frac{1}{2}p^2E^2/h^2B$ | $J$ | $M$ | Frequency, $\frac{1}{2}p^2E^2/h^2B$ |
|---|---|---|---|---|---|
| $1 \leftarrow 0$ | 0 | $+\frac{8}{15}$ | $3 \leftarrow 2$ | 0 | $-\frac{8}{315}$ |
| $2 \leftarrow 1$ | 0 | $-\frac{16}{105}$ | | $\pm1$ | $-\frac{1}{140}$ |
| | $\pm1$ | $+\frac{13}{105}$ | | $\pm2$ | $+\frac{1}{21}$ |

(c) For double the field the frequency shifts are multiplied by four. From the four frequency shifts, the fields, and factors in table 3.3.4b,

$$\frac{p^2}{2h^2B_0} = 1 \cdot 065 \times 10^{-3}\ \mathrm{kg^{-2}\ m^{-2}\ s^5\ A^2}\ ,$$

and thence $p = 2 \cdot 385 \times 10^{-30}$ C m (0·715 D).

Note that, compared to the first-order shifts in isobutane for the same field, 100 kV m$^{-1}$, the frequency separations are much smaller although the dipole moment is larger.

**3.3.5 Chlorine nuclear quadrupole coupling in fluorine chloride**[2]

When a nucleus in a molecule has an electrical quadrupole moment, the energy of interaction between this moment and the electric field gradient at the nucleus must be considered. In the gas phase this requires that the total angular momentum, $F$, be the compounded momentum of the framework, $J$, and the nucleus, $I$. If $I = 0$ or $\frac{1}{2}$ the quadrupole vanishes and the nuclear spin may be disregarded, as with the $^{19}$F nucleus in the problem below.

[2] Editors' note: see also problems in section 5.2.

For given values of $J$ and $I$, $F$ assumes values from $J+I$ to $|J-I|$ in unit steps. Strictly $J$ is no longer a 'good' quantum number, but this difficulty only arises if the quadrupole coupling is so large that its quadratic terms must be retained.

(a) For $^{19}F^{35}Cl$ the spins are $I = \frac{1}{2}$ for $^{19}F$, and $I = \frac{3}{2}$ for $^{35}Cl$. What values of $F$ arise when $J = 0$, 1, or 2? Disregard the spin of the $^{19}F$ nucleus.

(b) The selection rules are $\Delta F = 0, \pm 1$, while $\Delta J = 1$ is still valid, and, of course, $\Delta I = 0$ if no nuclear transformation is involved. Which transitions between the $J = 0$, 1, 2 set of levels are allowed for $^{19}F^{35}Cl$?

(c) The simplest case is that in which the quadrupolar nucleus lies on an axis of effective cylindrical symmetry (as it must for all linear molecules), as the coupling energy can then be expressed as a multiple of $eqQ$, where $e$ is the charge on the proton and $q$ is the second derivative of the electric potential at the nucleus in the axial direction, i.e. $\partial^2 V/\partial z^2$, which is simultaneously the negative of the electric field gradient. $Q$ is the quadrupole moment of the nucleus. The first-order energy contribution is

$$-eqQ\, f(I, J, F)\,,$$

where $f(I, J, F)$ is the Casimir function given by

$$f(I, J, F) = \frac{3C(C+1) - 4I(I+1)J(J+1)}{4(2J-1)(2J+3)2I(2I-1)}\,,$$

in which $C = F(F+1) - I(I+1) - J(J+1)$.

Evaluate $f(I, J, F)$ for the case $I = \frac{3}{2}$, $J = 1$, $F = \frac{5}{2}$. Either evaluate or copy from table 3.3.5a the value of $f(I, J, F)$ for the remaining states listed in (a), as these will be needed subsequently.

(d) When, as here, the first-order case applies, the additional energy may be added to that due to the moment of inertia term, namely $hBJ(J+1)$. Write explicit formulae for the set of $J_{1\leftarrow 0}$ transition frequencies in terms of $B$ and $h^{-1}eqQ$.

(e) For the $J_{1\leftarrow 0}$ transitions of $^{19}F^{35}Cl$ the three observed frequencies are 30·807366, 30·843875, and 30·872963 GHz. Assign these transitions and evaluate $h^{-1}eqQ$ and $B_0$.

(f) For the first vibrationally excited state the transitions are at 30·545994, 30·582614, and 30·611761 GHz. Evaluate $B_1$ and $h^{-1}\, eqQ$ for this excited state.

(g) From the answers to (e) and (f) calculate the rotation–vibration interaction constant, $\alpha$, and thence $B_e$ and the internuclear distance, $r_e$. The atomic masses are $^{19}F = 18{\cdot}998405$ and $^{35}Cl = 34{\cdot}968854$ g mol$^{-1}$.

(h) Given that $Q$ for $^{35}Cl$ is $-7{\cdot}9 \times 10^{-30}$ m$^2$, calculate $q$.

See White (1955).

**Solution**

(a) See columns $F$ and $J$ of table 3.3.5a.

**Table 3.3.5a.**

| $F$ | $J$ | $C$ | $f(I, J, F)$ | $F$ | $J$ | $C$ | $f(I, J, F)$ |
|---|---|---|---|---|---|---|---|
| $\frac{3}{2}$ | 0 | 0 | 0 | $\frac{7}{2}$ | 2 | +6 | $+\frac{1}{14}$ |
| $\frac{5}{2}$ | 1 | +3 | $+\frac{1}{20}$ | $\frac{5}{2}$ | 2 | −1 | $-\frac{5}{28}$ |
| $\frac{3}{2}$ | 1 | −2 | $-\frac{1}{5}$ | $\frac{3}{2}$ | 2 | −6 | 0 |
| $\frac{1}{2}$ | 1 | −5 | $+\frac{1}{4}$ | $\frac{1}{2}$ | 2 | −9 | $+\frac{1}{4}$ |

The net quadrupole coupling summed over all states of a given $J$ must be zero, provided the $(2F+1)$-fold degeneracy is included. Check that the values in table 3.3.5a satisfy this condition.

(b) See columns $\Delta J$ and $\Delta F$ of table 3.3.5b.

**Table 3.3.5b.**

| $\Delta J$ | $\Delta F$ | $c_1$ | $c_2$ | $\Delta J$ | $\Delta F$ | $c_1$ | $c_2$ |
|---|---|---|---|---|---|---|---|
| $1 \leftarrow 0$ | $\frac{5}{2} \leftarrow \frac{3}{2}$ | 2 | $-\frac{1}{20}$ | $2 \leftarrow 1$ | $\frac{5}{2} \leftarrow \frac{3}{2}$ | 4 | $-\frac{3}{140}$ |
| | $\frac{3}{2} \leftarrow \frac{3}{2}$ | 2 | $+\frac{1}{5}$ | | $\frac{3}{2} \leftarrow \frac{3}{2}$ | 4 | $-\frac{1}{5}$ |
| | $\frac{1}{2} \leftarrow \frac{3}{2}$ | 2 | $-\frac{1}{4}$ | | $\frac{1}{2} \leftarrow \frac{3}{2}$ | 4 | $-\frac{9}{20}$ |
| $2 \leftarrow 1$ | $\frac{7}{2} \leftarrow \frac{5}{2}$ | 4 | $-\frac{3}{140}$ | | $\frac{3}{2} \leftarrow \frac{1}{2}$ | 4 | $+\frac{1}{4}$ |
| | $\frac{5}{2} \leftarrow \frac{5}{2}$ | 4 | $+\frac{8}{35}$ | | $\frac{1}{2} \leftarrow \frac{1}{2}$ | 4 | 0 |
| | $\frac{3}{2} \leftarrow \frac{5}{2}$ | 4 | $+\frac{1}{20}$ | | | | |

(c) See columns $C$ and $f(I, J, F)$ of table 3.3.5a.

(d) Transition frequency $= c_1 B + c_2 h^{-1} eqQ$, see columns 3 and 4 of table 3.3.5b for $c_1$ and $c_2$.

(e) The frequencies quoted correspond to $\Delta F$ being $\frac{3}{2} \leftarrow \frac{3}{2}$, $\frac{5}{2} \leftarrow \frac{3}{2}$, and $\frac{1}{2} \leftarrow \frac{3}{2}$ respectively as the last quoted pair correspond to the smallest interval. $h^{-1}eqQ = -145 \cdot 837$ MHz is the exact value obtained by including second-order corrections. Values between $-145 \cdot 4$ and $-146 \cdot 1$ MHz are obtained from the first-order treatment suggested. Note that the sign of $eqQ$ is obtained; this is not found in the pure quadrupole resonance in the solid state.

In the absence of quadrupole coupling the line would be near

$$30 \cdot 843\,875 + h^{-1}eqQ/20 = 30 \cdot 836\,583 \text{ GHz},$$

and hence $B_0 = 15 \cdot 418\,291$ GHz.

(f) $h^{-1}eqQ = -146 \cdot 2$ MHz and $B_1 = 15 \cdot 287\,65$ GHz.

(g) $\alpha = 130 \cdot 64$ MHz and hence $B_e = 15 \cdot 483\,61$ GHz. This gives $I_e = 5 \cdot 419\,550 \times 10^{-46}$ kg m$^2$, the reduced mass $\overline{m} = 12 \cdot 310\,287$ g mol$^{-1}$ and $r_e = 1 \cdot 6283 \times 10^{-10}$ m ($1 \cdot 6283$ Å).

(h) $q = 7{\cdot}6 \times 10^{22}$ V m$^{-2}$ by direct substitution, remembering to multiply the frequency by $h$, as well as to divide by $e$ in coulombs and by $Q$. Although typical for molecules, this value of $q$ is larger than can be produced conveniently from nonuniform macroscopic fields in the laboratory.

**General references**

Useful introduction; Sugden, T. M., Kenney, C. N., 1965, *Microwave Spectroscopy of Gases* (Van Nostrand, London).

Standard text; Townes, C. H., Schawlow, A. L., 1955, *Microwave Spectroscopy* (McGraw-Hill, New York).

Modern text: Gordy, W., Cook, R. L., 1970, *Microwave Molecular Spectra,* being Part II of *Chemical Applications of Spectroscopy,* Ed. W. West, being itself Volume IX of *Techniques of Organic Chemistry,* Ed. A. Weissberger (Interscience, New York).

**References for problems**

**3.3.1**

Cowan, M. J., 1960, University Microfilms Publication No.262; *Diss. Abstr.,* **20,** 4139.

Rosenblum, B., Townes, C. H., Geschwind, S., 1958, *Rev. Mod. Phys.,* **30,** 409.

**3.3.2**

Kivelson, D., 1954, *J. Chem. Phys.,* **22,** 904.

**3.3.3**

Lide, D. R., Mann, D. E., 1958, *J. Chem. Phys.,* **29,** 914.

**3.3.4**

Muenter, J. S., 1968, *J. Chem. Phys.,* 48, 4544.

**3.3.5**

White, R. L., 1955, *Rev. Mod. Phys.,* **27,** 276.

# 4 Electronic properties

## 4.1 Ultraviolet and visible spectroscopy

**G R Eaton, J R Riter Jr** University of Denver
**S S Eaton** University of Colorado at Denver

### 4.1.1 $[Ti(H_2O)_6]^{3+}$

Figure 4.1.1a is the visible spectrum of an aqueous solution of Ti(III).
(a) This figure is exhibited in numerous texts with a statement that the peak is due to a d–d transition. What are the arguments which justify this assignment?
(b) Interpreting the spectrum first as a single peak with a maximum at 20000 $cm^{-1}$, use electrostatic crystal-field theory to assign the ground and excited electronic states involved in the transition.
(c) Considering the selection rules for electronic transitions, demonstrate in what way the transition being discussed is 'allowed'.
(d) Closer examination of the spectrum indicates that there is a 'shoulder' on the low-energy side, i.e. that there is more than one transition contributing to the peak. Assuming that the experiment is correctly reported, i.e. that the species being observed is really $[Ti(H_2O)_6]^{3+}$, explain the existence of more than one transition.

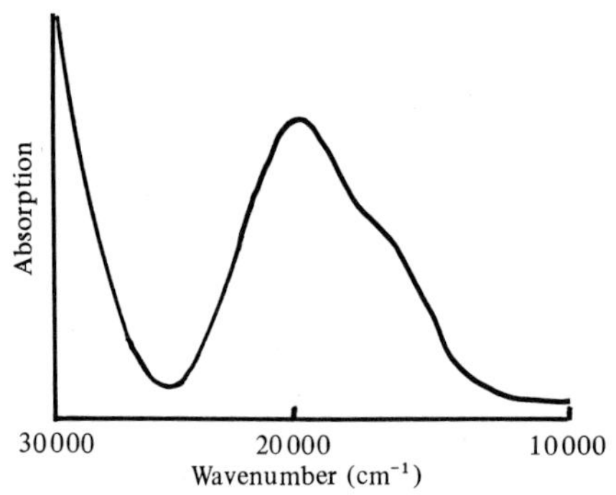


**Figure 4.1.1a.** The visible absorption spectrum of $[Ti(H_2O)_6]^{3+}$ (Nicholls, 1975, p.56).

**Solution**

(a) The problem of being certain that a particular spectral band is properly assigned is a nontrivial one. We will not discuss here the very real problem of knowing what species it is that one has in the flask, but as we discuss the subtle details of spectra it is well to keep in mind that many very sophisticated interpretations have been put forth to explain spectra, assuming an incorrect species. In the present case, the assignment of the 20000 $cm^{-1}$ band in the spectrum of $[Ti(H_2O)_6]^{3+}$ as being due to a d–d transition was made by Ilse and Hartmann (1951) on the basis of agreement between the observed spectrum and a calculation assuming it was a transition between d levels split by the electronic effect of the water molecules. This assignment is of historical importance because it was the

first correct assignment of a d–d band in a transition-metal complex, and it provided the basis for further understanding of the optical spectra of other metal complexes.
(b) The detailed treatment of each step in a thorough logical development even for a 'simple' $d^1$ configuration would require many pages. Most steps are outlined in standard texts (Cotton, 1971; Hall, 1969; Lever, 1968) and will not be reproduced here. The logical development is as follows.
(i) $Ti^{3+}$ is a $d^1$ ion. According to the Russell–Saunders coupling scheme with Hund's rules the ground state of the free ion is $^2D_{3/2}$. It is necessary to appeal to experiment to find that the $J = \frac{3}{2}$ to $J = \frac{5}{2}$ splitting is $384 \cdot 3$ $cm^{-1}$ (see Phillips and Williams, 1965, chapter 24), which justifies considering only the lowest $J$ level. Hence we will drop the $J$ designation.
(ii) Next it is necessary to ascertain the behaviour of a D state in an octahedral field. This is dealt with in problem 1.2.7 where it is shown that in an octahedral field the $^2D$ free-ion state splits into $^2E$ and $^2T_2$ states. Since the $^2T_2$ state arises from the d electron being in a $t_{2g}$ orbital this state is the ground state and has the gerade label ($^2T_{2g}$). Thus the transition is a $^2E_g \leftarrow {}^2T_{2g}$ 'd–d' transition.
(c) The selection rules derive from a consideration of the properties of the transition-moment integral (Lohr, 1972):

$$P = \int \Psi_a^* \mathbf{O} \Psi_b \, d\tau \equiv \langle a | \mathbf{O} | b \rangle ,$$

where $\mathbf{O}$ is the operator appropriate for the type of transition under consideration. $P$ must be invariant under the operations of the group; thus the integral must either be zero or transform as the totally symmetric representation. Assuming separability of the wave function into orbital and spin electronic functions and assuming electronic dipole transitions, we find the selection rules:

$\Delta S = 0$ (no change in total spin);

$$\left.\begin{array}{l} u \not\leftrightarrow u \\ g \not\leftrightarrow g \\ g \leftrightarrow u \end{array}\right\} \quad \text{(the Laporte rule).}$$

Since d-electron states are all g, all d–d transitions are Laporte-forbidden by the analysis so far. When we consider simultaneous electronic and vibrational transitions and examine the symmetry of the vibrational wave functions we find that the transitions become allowed. In $O_h$, the normal modes of vibration of the $ML_6$ array have irreducible representations: $A_{1g} + E_g + T_{2g} + 2T_{1u} + T_{2u}$. Since the electronic wave functions are both g, the dipole moment operator is $T_{1u}$ (in $O_h$), and the ground vibrational state is totally symmetric, the direct product of the irreducible representations for these four must be, or contain, the representation of one of the u

vibrations ($T_{1u}$ or $T_{2u}$) of the molecule. Only in this case can the transition-moment integral be nonzero, i.e. the transition be 'allowed'. It turns out that this occurs for the octahedral $d^1$ system, and we say that the transitions are vibronically allowed.
(d) In the analysis in (b) above, we ended up with a single electron in a multiply degenerate state. There is a theorem due to Jahn and Teller to the effect that when the orbital state of a nonlinear molecule is degenerate the nuclear framework will distort until the symmetry is lowered enough to remove the degeneracy. The distortions usually take the form of one of the normal modes of vibration of the molecule. For example, the $E$ vibration lengthens the two axial bonds and shortens the four equatorial bonds, lowering the symmetry from $O_h$ to $D_{4h}$. In order to remove the degeneracy, the $^2B_{2g}$ level must be lowered relative to the $^2E_g$ level in the ground state. Consideration of electrostatic crystal-field theory predicts that $^2A_{1g}$ will be lower than $^2B_{1g}$ in the excited state, as shown in figure 4.1.1b.

Thus the shoulder in the spectrum is attributed to Jahn–Teller distortion in the excited state. A note of caution should be added here that a Jahn–Teller distortion should be invoked only as a last resort: the decrease in symmetry may result from more mundane effects, such as crystal packing forces or steric effects among polyatomic ligands.

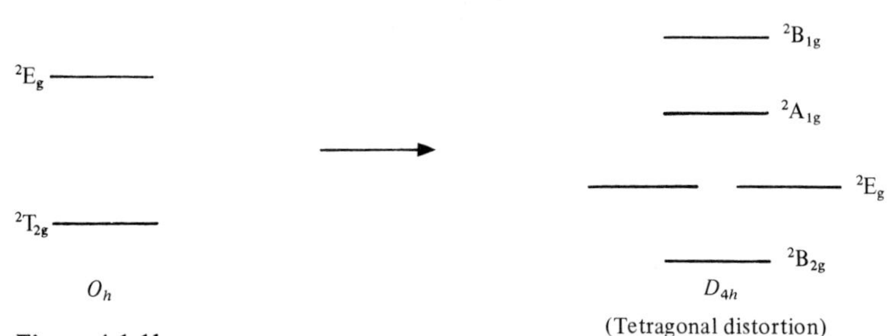


**Figure 4.1.1b.**

### 4.1.2 Effect of ligand-field strength

Figure 4.1.2a displays the electronic spectra of three six-coordinate Ni(II) complexes. Peak maxima for additional compounds are given in table 4.1.2.
(a) Explain how to obtain the ligand-field strength, $10Dq(\Delta_o)$, from the electronic spectra.
(b) Arrange the ligands in order of increasing ligand-field strength.
(c) Why is the middle band split in the aquo-complex but not in the other complexes in figure 4.1.2a?

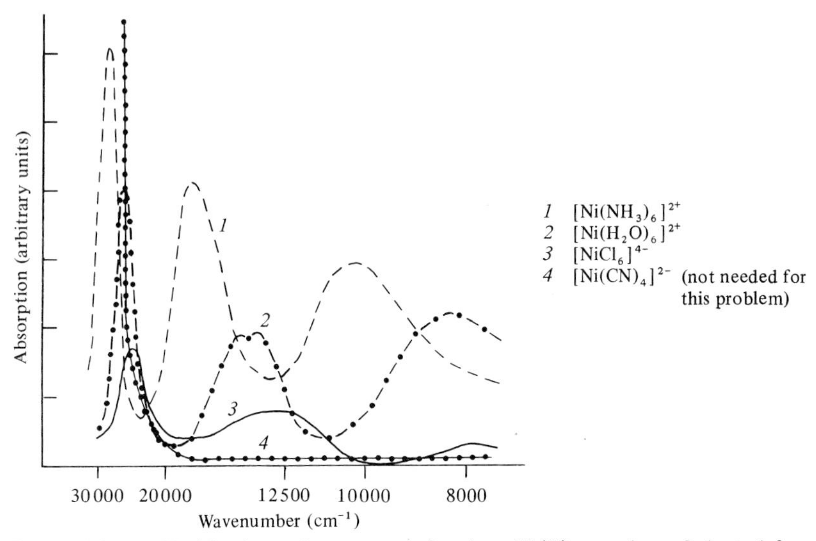


**Figure 4.1.2a.** Visible absorption spectra of various Ni(II) complexes [adapted from Dorain (1965, p.88); Cotton and Wilkinson (1972, p.894)].

**Table 4.1.2.** Electronic spectra of octahedral Ni(II) complexes (peak maxima in $cm^{-1}$). Sources: Lever (1968, p.334); Nicholls (1973, p.94).

| Complex | $^3T_{2g} \leftarrow {^3A_{2g}}$ | $^3T_{1g} \leftarrow {^3A_{2g}}$ | $^3T_{1g}(P) \leftarrow {^3A_{2g}}$ |
|---|---|---|---|
| $NiBr_2$ | 6800 | 11800 | 20600 |
| $NiCl_2$ | 7200 | 11500 | 21900 |
| $KNiF_3$ | 7250 | 12530 | 23810 |
| $[Ni(DMSO)_6]^{2+}$ | 7728 | 12970 | 24038 |
| $[Ni(MeOH)_6]^{2+}$ | 8430 | 14226 | 25000 |
| $[Ni(H_2O)_6]^{2+}$ | 8500 | 13800 | 25300 |
| $[Ni(NCS)_6]^{4-}$ | 9600 | 15950 | 25800 |
| $[Ni(MeCN)_6]^{2+}$ | 10700 | 17400 | 27810 |
| $[Ni(NH_3)_6]^{2+}$ | 10750 | 17500 | 28200 |
| $[Ni(en)_3]^{2+}$ | 11200 | 18350 | 29000 |
| $[Ni(dipy)_3]^{2+}$ | 12650 | 19200 | |
| $[Ni(phen)_3]^{2+}$ | 12700 | 19300 | |

**Solution**

(a) The energy levels for $d^8$ in an octahedral environment in terms of Racah parameters are as shown in figure 4.1.2b (Nicholls, 1975, p.1152). Thus the $^3T_{2g} \leftarrow {^3A_{2g}}$ transition has energy $10Dq$. This feature makes interpretation of octahedral $Ni^{2+}$ spectra quite straightforward; ligand-field strengths can be read directly from the experimental spectra by using the peak maximum of the lowest-energy band.

(b) Table 4.1.2 is arranged in order of increasing ligand-field strength, top to bottom.

(c) According to the Tanabe–Sugano diagrams for octahedral $d^8$ (Cotton, 1971, p.266), when $Dq/B \sim 1$ the $^1E$ and $^3T_1$ (F) levels cross. This occurs for the aquo-complex, but not for other complexes. Lever (1968, p.333) cites the original papers for two possible explanations of the splitting of the middle band in the aquo-complex: "This may be a consequence of the transition to the $^1E_g$ level gaining intensity through configuration interaction with the $^3T_{1g}$ (F) level, although other authors prefer to interpret the structure in terms of spin-orbit coupling".

$E$

$^3P$ —— $^3T_{1g}$ $3Dq + \frac{15}{2}B + \frac{1}{2}[(15B - 6Dq)^2 + 64(Dq)^2]^{1/2}$

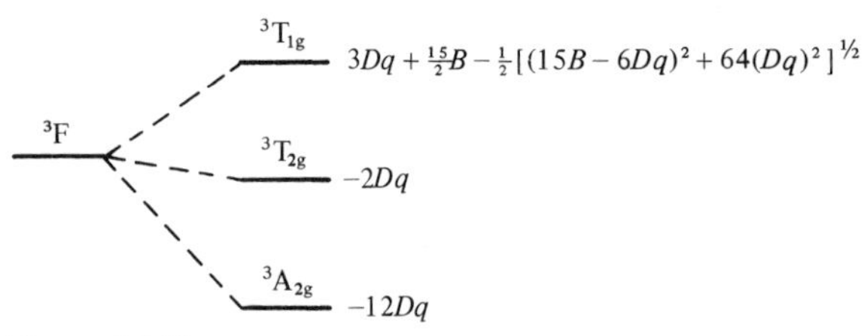


**Figure 4.1.2b.**

**4.1.3 Effect of low symmetry**

Most problems of spectral interpretation encountered in the laboratory involve complexes of symmetry lower than octahedral. This problem addresses spectral interpretation when the symmetry is 'obvious'.

In problem 4.1.2 we assumed that $[Ni(en)_3]^{2+}$ could be analyzed in terms of an octahedral ligand field, and we make the same assumption

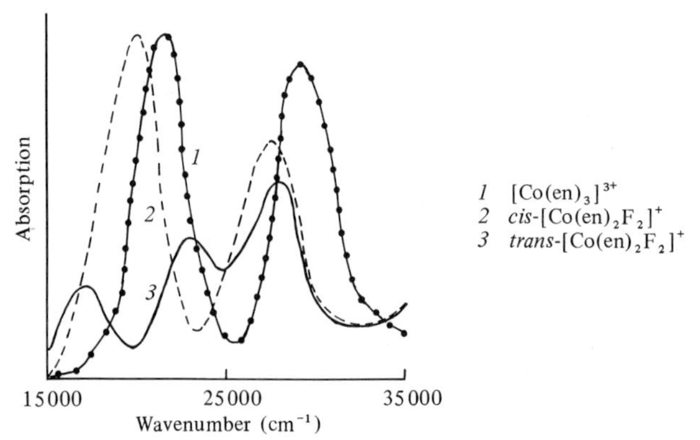


**Figure 4.1.3a.** The visible absorption spectra of three (ethylenediamine)cobalt(III) complexes (Figgis, 1966, p.234).

regarding $[Co(en)_3]^{3+}$ in this problem. Figure 4.1.3a sketches the spectra of $[Co(en)_3]^{3+}$ and the *cis* and *trans* isomers of $[Co(en)_2F_2]^+$. Assign the bands to transitions between states for the appropriate symmetry.

**Solution**

The spectra of complexes with symmetry lower than octahedral can generally be analyzed by considering the lower symmetry as a perturbation of the energy levels in an octahedral (or tetrahedral) field. If only the positions of the donor atoms are considered, the symmetries of the complexes in this problem are:

| | |
|---|---|
| $[Co(en)_3]^{3+}$ | $O_h$ |
| *trans*-$[Co(en)_2F_2]^+$ | $D_{4h}$ |
| *cis*-$[Co(en)_2F_2]^+$ | $C_{2v}$ |

The effect of the two lower symmetries on the energy levels in an octahedral field are shown in figure 4.1.3b, which is not drawn to scale.

The splitting of the $^1T_{1g}$ ($O_h$) level for complexes with ligands which differ considerably in ligand-field strength is larger in $D_{4h}$ than in $C_{2v}$ symmetry; thus two separate bands are observed for $^1E_g \leftarrow {}^1A_{1g}$ and $^1A_{2g} \leftarrow {}^1A_{1g}$ in the *trans* isomer, whereas a single slightly asymmetric band is observed for $^1B_1 \leftarrow {}^1A_1$ and $^1A_2$, $^1B_2 \leftarrow {}^1A_1$ in the *cis* isomer. The splitting of the $^1T_{2g}$ ($O_h$) level is small in both isomers, and the two transitions are not resolved (Cotton and Wilkinson, 1972, p.885).

Evidence for these assignments can be obtained from the polarization of the bands in the visible spectra obtained from a single crystal by use of polarized light. Such an experiment has been done for *trans*-$[Co(en)_2Cl_2]^+$ and the spectra are shown in figure 4.1.3c.

In $D_{4h}$ symmetry the electric dipole operators $x$ and $y$ transform as $E_u$ whereas the electric dipole operator $z$ transforms as $A_{2u}$. To determine which transitions are allowed for which operators we must take the direct products (DP) of the ground and excited electronic wave functions with the dipole moment operators and compare the results with the unsymmetrical

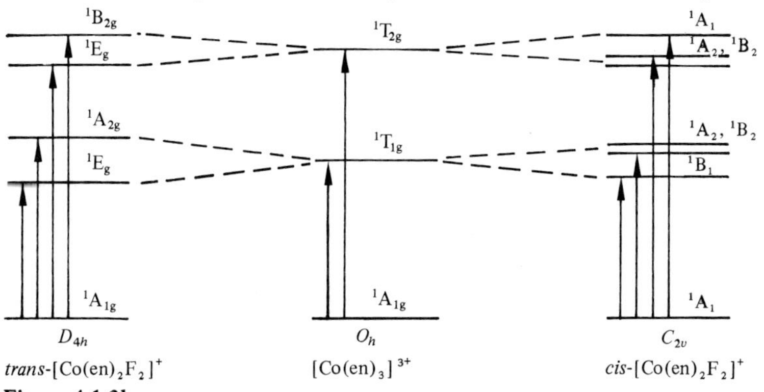


**Figure 4.1.3b.**

vibrational modes of the molecules. These vibrational modes are $2A_{2u}+B_{2u}+3E_u$ for $ML_4X_2$ in $D_{4h}$. The results (Lever, 1968, p.151) are shown in table 4.1.3.

Since the low-energy transition occurs in both the parallel and the perpendicular spectra it is assigned to $^1E_g \leftarrow {}^1A_{1g}$. The second transition occurs only in the perpendicular spectrum so is assigned to $^1A_{2g} \leftarrow {}^1A_{1g}$. The third transition occurs in both spectra so is assigned to the overlapping $^2E_g \leftarrow {}^1A_{1g}$ and $^1B_{2g} \leftarrow {}^1A_{1g}$ transitions.

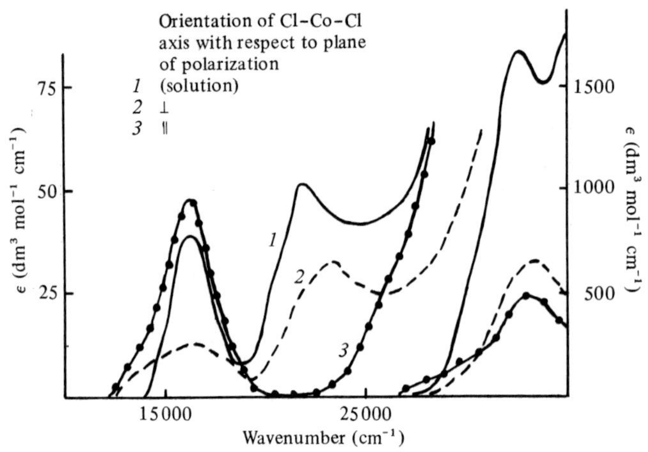


**Figure 4.1.3c.** Electronic spectra of *trans*-$[Co(en)_2Cl_2]^+$ in solution and as a single crystal with specific orientations relative to plane polarized light (Figgis, 1966, p.231).

**Table 4.1.3.**

| Transition | DP of electronic wave functions | DP with dipole moment operators | | Polarization[a] |
|---|---|---|---|---|
| | | $A_{2u}$ | $E_u$ | |
| $^1A_{2g} \leftarrow {}^1A_{1g}$ | $A_{2g}$ | $A_{1u}$ | $E_u$ | ⊥ |
| $^1B_{2g} \leftarrow {}^1A_{1g}$ | $B_{2g}$ | $B_{1u}$ | $E_u$ | ⊥ |
| $^1E_g \leftarrow {}^1A_{1g}$ | $E_g$ | $E_u$ | $A_{1u}+A_{2u}$ $B_{1u}+B_{2u}$ | ⊥, ‖ |

[a] ⊥ refers to $x, y$ polarization, i.e. perpendicular to Cl—Co—Cl axis; ‖ refers to $z$ polarization, i.e. parallel to Cl—Co—Cl axis.

### 4.1.4 Width of ligand-field bands

The spectra of Cr(III) in $K_2Na[CrF_6]$ (figure 4.1.4a), and of $[Mn(H_2O)_6]^{2+}$ and $MnF_2$ (figure 4.1.4b) contain bands that are much sharper than those in the other spectra shown in this section. The metal ion is in an octahedral environment in each case.

(a) Explain the relative bandwidths in the spectra shown in figures 4.1.4a and 4.1.4b.
(b) List the contributions to widths of d–d absorption bands.

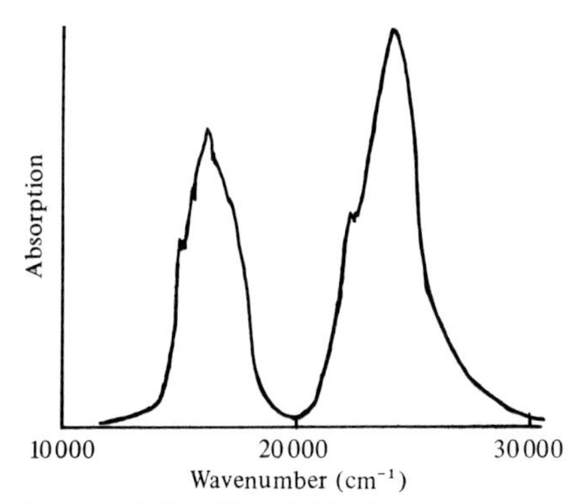


**Figure 4.1.4a.** The visible absorption spectrum of $K_2Na[CrF_6]$ (Ferguson, 1970, p.229).

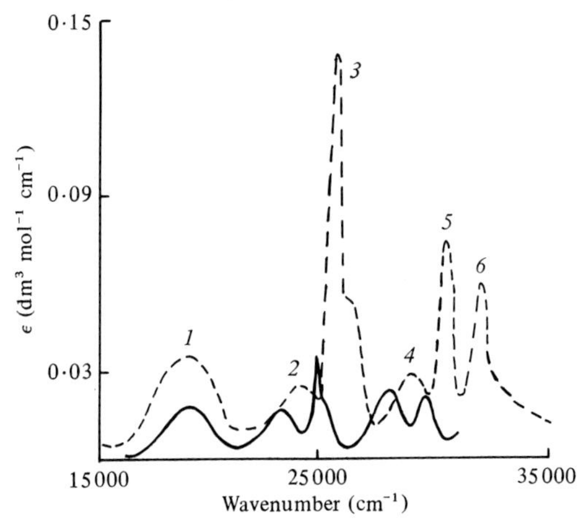


**Figure 4.1.4b.** The visible absorption spectra of $[Mn(H_2O)_6]^{2+}$ (solid line) and $MnF_2$ (dashed line). Assignments for excited states involved in $MnF_2$ transitions are: *1*, $^4T_{1g}$ ($^4G$); *2*, $^4T_{2g}$ ($^4G$); *3*, $^4A_{1g}$, $^4E_g$ ($^4G$); *4*, $^4T_{2g}$ ($^4D$); *5*, $^4E_g$ ($^4D$); *6*, $^4T_{1g}$ ($^4P$) (Phillips and Williams, 1965, p.395).

**Solution**

(a) Molecular vibrations modulate the crystal-field splitting $\Delta$ (= $10Dq$). When the molecule is in the 'compressed' phase of the vibration the effective crystal-field strength of the ligand is increased, and when the molecule is in the 'stretched' phase $10Dq$ is decreased. If the energy separation of the ground state and the excited state depends on ligand-field strength, the transition energy will have a range of values dependent on vibration of the molecule. This is shown in figure 4.1.4c which is a portion of the Tanabe–Sugano diagram for a $d^5$ ion.

The width of the band will thus depend on the relative slopes of the ground state and excited state in a Tanabe–Sugano diagram. If the transition is between levels with the same dependence on $10Dq$, the band will be sharp. Energy levels have the same dependence on $10Dq$ if they have the same occupation of the $e_g$ and $t_{2g}$ orbitals. Thus sharp lines are spin-forbidden (but not all spin-forbidden lines are sharp). For octahedral $Mn^{2+}$ all transitions are spin-forbidden. In the $d^5$ Tanabe–Sugano diagram the $^4A_{1g}$ level runs parallel to the $^6A_{1g}$ ground state, thus $^4A_{1g} \leftarrow {}^6A_{1g}$ is a sharp line in figure 4.1.4b. However, the slope of $^4T_{1g}$ is very different from that of $^6A_{1g}$, and so $^4T_{1g} \leftarrow {}^6A_{1g}$ is a much broader band. The widths of the other bands are explained analogously. Similarly, the sharp lines in the $K_2Na[CrF_6]$ spectrum (and in ruby) are transitions from the $^4A_2$ ground state to doublet excited states.

(b) A discussion of the contributions to bandwidths of d–d transitions can be found in Lever (1968, pp.137–149). The major influences are: (i) the Jahn–Teller effect; (ii) variation of $10Dq$ with vibration; (iii) vibrational structure due to simultaneous excitation of electrons and one or more vibrational modes; (iv) spin–orbit coupling removing the degeneracy of a term (e.g. $^3T_{1g}$); (v) low-symmetry components to the ligand field.

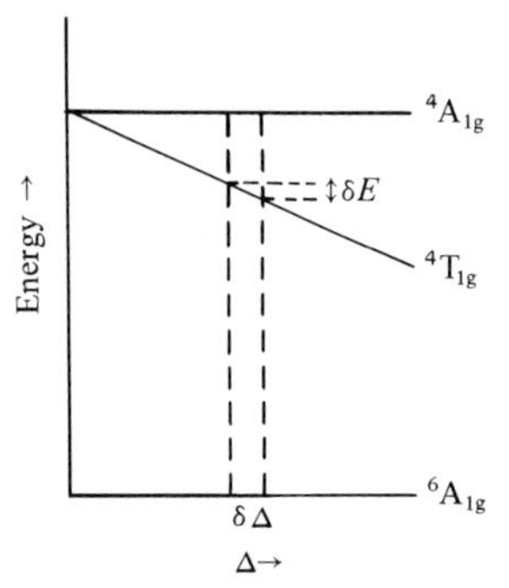


**Figure 4.1.4c.** Diagram of a portion of the Tanabe–Sugano diagram for a $d^5$ ion, illustrating the influence of changes in $\Delta$ on the energy of electronic transitions.

### 4.1.5 Intensity of ligand-field bands

The greater intensity of the spectrum of $[CoCl_4]^{2-}$ than that of $[Co(H_2O)_6]^{2+}$ (figure 4.1.5) is representative of a very general observation regarding intensities of d–d spectra. Another example of this effect can be seen in figure 4.1.3a, where it can be seen that the molar extinction coefficient of the *cis* complex is larger than that of the *trans* complex.

(a) Explain the relative intensities cited above.

(b) List the contributions to the intensities of d–d transitions.

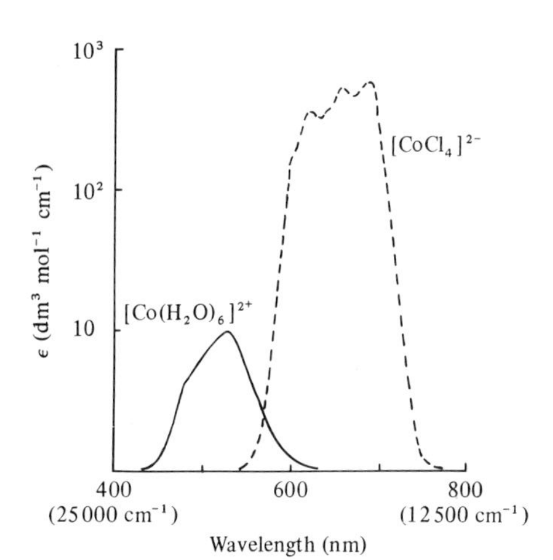


**Figure 4.1.5.** The visible absorption spectra of $[Co(H_2O)_6]^{2+}$ and $[CoCl_4]^{2-}$. Spectra are plotted on a logarithmic scale (Cotton and Wilkinson, 1972, p.881).

**Solution**

(a) As outlined in the solution to problem 4.1.1, transitions between states of the same symmetry (g or u) are forbidden in centrosymmetric molecules. Vibronic coupling is invoked to explain the finite intensities observed for these transitions. Tetrahedral molecules and *cis*-$MX_4Y_2$ molecules do not have a centre of symmetry and thus d–d transitions in these compounds are not parity-forbidden.

Although this argument suffices to explain qualitatively the relative intensities cited, quantitative prediction of the intensities of transitions in low-symmetry compounds has been troublesome. It has been suggested that the discrepancy between theory and experiment can be at least partially removed by "consideration of the perturbation of the ligands by the charge distribution of a pure d-electron transition of the metal ion in a tetrahedral complex" (Gale *et al.*, 1975).

(b) Lever (1968, p.133) summarizes molar extinction coefficients for various types of complexes exhibiting d–d transitions. The extinction coefficients lie in the approximate range 0·01–1·0 dm$^3$ mol$^{-1}$ cm$^{-1}$ for transitions that are both spin-forbidden and Laporte-forbidden, and are about $10^2$–$10^4$ dm$^3$ mol$^{-1}$ cm$^{-1}$ for transitions that are spin-allowed in complexes with ligands having delocalized $\pi$ systems. In addition to spin and parity selection rules, factors such as intensity stealing and degree of covalency contribute to the intensity of the transition

**4.1.6 Electronic spectrum of $[CuCl_4]^{2-}$**

$[CuCl_4]^{2-}$ is a planar centrosymmetric molecule (in certain crystals) which gives a three-line visible spectrum and a Raman spectrum that can be interpreted in terms of a frequency of 276 cm$^{-1}$ for the $a_{1g}$ mode in the ground electronic state. Figure 4.1.6a shows that the electronic spectrum

of $[CuCl_4]^{2-}$ exhibits fine structure (spacing $265 \pm 20\ cm^{-1}$) when the crystal is cooled below ~100 K.
(a) Identify the transitions responsible for the three main peaks.
(b) Explain the occurrence of fine structure on the electronic absorption bands.
(c) State the significance of this observation of fine structure.

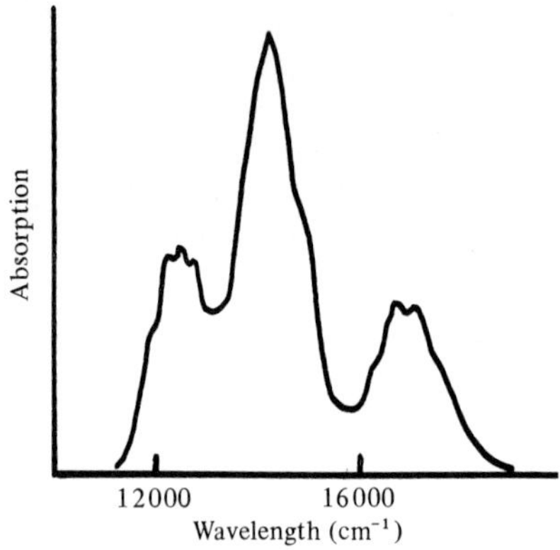


**Figure 4.1.6a.** The electronic absorption spectrum of $[CuCl_4]^{2-}$ at 77 K (Cassidy and Hitchman, 1975).

**Solution**

(a) The energy-level diagram for $d^9$ in a $D_{4h}$ environment does not occur in the standard texts, but the reader should have no trouble justifying the scheme given in figure 4.1.6b (Boudreaux and Mulay, 1976, p.235).

The unpaired electron will be in the $d_{x^2-y^2}$ orbital in the ground state; thus the transitions giving rise to the three bands are given in table 4.1.6, in order of increasing energy. The vibrations required to make these transitions vibronically allowed are given by Hathaway and Billing (1970).

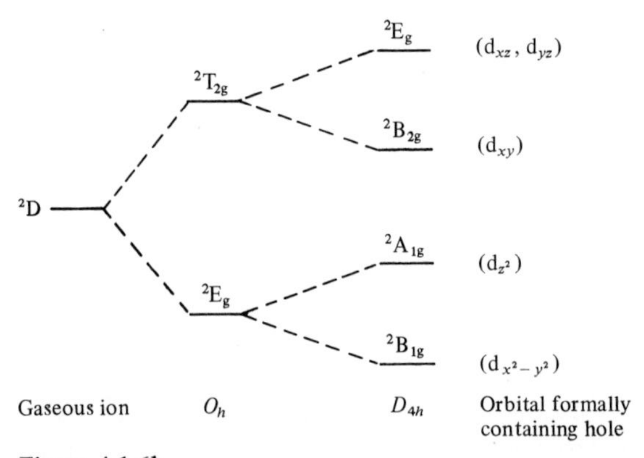


**Figure 4.1.6b.**

(b) The fine structure can be explained with the following assumptions (Cassidy and Hitchman, 1975).
(i) The electronic transition occurs with excitation of one of the 'u' modes of vibration (thus making it vibronically allowed) and with excitation of the $a_{1g}$ vibrational mode of the excited electronic state.
(ii) There is a lengthening of the Cu–Cl bond and lowering of the $a_{1g}$ frequency in the excited state relative to that in the ground state.
(iii) There is low anharmonicity for the $a_{1g}$ mode.
(c) The observation of vibrational fine structure on the electronic transitions is confirmation of the vibronic mechanism that we have invoked to make parity-forbidden transitions allowed. Several other examples of vibrational fine structure on d–d electronic transitions have been reported, e.g. $[PtCl_4]^{2-}$ (Martin *et al.*, 1965), and $[Ni(NH_3)_6]^{2+}$ (Schreiner and Hamm, 1973; 1975).

**Table 4.1.6.**

| Transition | Vibrations needed |
|---|---|
| $^2A_1 \leftarrow {}^2B_1$ | $e_u + b_{2u}$ |
| $^2B_2 \leftarrow {}^2B_1$ | $e_u + a_{1u}$ |
| $^2E \leftarrow {}^2B_1$ | $e_u + a_{1u} + a_{2u} + b_{1u} + b_{2u}$ |

### 4.1.7 Temperature dependence of $[VCl_6]^{4-}$ spectra

Figure 4.1.7 shows that the visible spectrum of $[VCl_6]^{4-}$ is strongly temperature-dependent. Evidence is given in the original paper (McPherson and Freedman, 1976) that the $[VCl_6]^{4-}$ ion is perfectly octahedral in this case.
(a) Assign the bands in the visible spectrum of $[VCl_6]^{4-}$.
(b) Explain the temperature dependence of the bandshape in the spectra of $[VCl_6]^{4-}$.

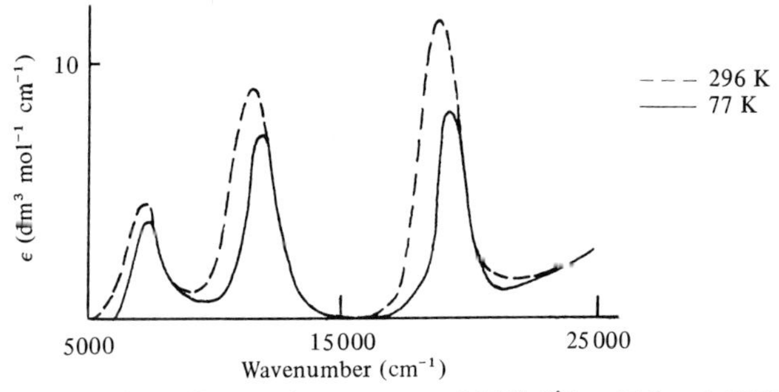


**Figure 4.1.7.** The visible spectrum of $[VCl_6]^{4-}$ at 296 and 77 K. The positions of the peak maxima at 77 K are 7200, 11700, and 19100 $cm^{-1}$ (McPherson and Freedman, 1976).

**Solution**

(a) $[VCl_6]^{4-}$ is a $d^3$ ion. Using the results of crystal-field theory as summarized in the Tanabe–Sugano diagrams (Cotton, 1971; Hall, 1969; Lever, 1968) we find that the ground state is $^4A_2$. The spin selection rule (see problem 4.1.1) predicts that the major bands will be from the $^4A_{2g}$ state to the three quartet excited states, $^4T_{2g}$, $^4T_{1g}$, and $^4T_{1g}$, in order of increasing energy. It can be shown, though not easily, that the separation between $^4A_{2g}$ $(t_{2g}^3)$ and $^4T_{2g}$ $(t_{2g}^2e_g)$ is exactly $10Dq$ (see Lever, 1968, p.182). Hence in this case $10Dq = 7200\ \text{cm}^{-1}$ at 77 K. The relative energy levels, in terms of the Racah $B$ and $C$ parameters for the transitions of interest, are given in table 4.1.7 (Lever, 1968, p.182).

If the interelectronic repulsion parameters $B$ and $C$ are varied until a good fit is achieved between calculated and observed energy levels the values $B = 610\ \text{cm}^{-1}$ and $C = 2500\ \text{cm}^{-1}$ are obtained. By use of these values in the expressions in table 4.1.7 it can be seen that the $^2T_{1g}$ level is very close to the $^4T_{1g}$ level in the Tanabe–Sugano diagram. Presumably the spin-forbidden transition to the $^2T_{1g}$ level 'borrows' intensity from the nearby $^4T_{1g}$ $(t_{2g}^2e_g) \leftarrow {}^4A_{2g}$ transition.

(b) There are three aspects of the temperature dependence of the band shape that must be explained: as the temperature falls (i) the bands become less intense, (ii) the band maxima move toward higher energy, and (iii) the bands become sharper.

$Dq$ usually increases as the temperature decreases. The slopes $dE/d(Dq)$ are positive except for levels approaching crossover. Therefore, the band maxima move toward higher energy. Also the spread of $Dq$ values decreases as the vibrations decrease with decreasing temperature. Therefore the bands become sharper. The band shape of vibronically allowed transitions will change with temperature because of the change in the Boltzmann distribution of vibrational energy levels. As the higher vibrational levels of the ground electronic state become less populated, the lower-energy portion of the band will decrease in intensity; thus the apparent band maximum will shift to higher energy. Since the transition between ground vibrational levels in the two electronic states is forbidden, the oscillator strength (band intensity) decreases as the temperature decreases.

**Table 4.1.7.**

| Configuration | Level | Relative energy |
|---|---|---|
| $t_{2g}^1e_g^2$ | $^4T_{1g}$ | $7{\cdot}5B + 3Dq + \frac{1}{2}(225B^2 + 100Dq^2 - 180DqB)^{1/2}$ |
| $t_{2g}^2e_g^1$ | $^4T_{1g}$ | $7{\cdot}5B + 3Dq - \frac{1}{2}(225B^2 + 100Dq^2 - 180DqB)^{1/2}$ |
| | $^4T_{2g}$ | $-2Dq$ |
| $t_{2g}^3e_g^0$ | $^2T_{1g}$ | $-12Dq + 9B + 3C - 24B^2/10Dq$ |
| | $^4A_{2g}$ | $-12Dq$ |

### 4.1.8 Intensely colored metal complexes

In most of the problems in this section the electronic absorption spectrum has been interpreted in terms of transitions between energy states characterized by a partially filled d-electron manifold. However, most chemists are aware of brightly colored compounds which involve $d^0$ or $d^{10}$ metals, such as $HgI_2$ ($d^{10}$, brick red) and $MnO_4^-$ ($d^0$, intense purple), and of the ferroin indicator system which involves not only a change in color, but also a significant change in intensity of the color:

$$\underset{\text{blood red}}{[Fe(phen)_3]^{2+}} \rightleftharpoons \underset{\text{pale blue}}{[Fe(phen)_3]^{3+}}$$

(a) What is the explanation for the colors of, for example, $HgI_2$ and $MnO_4^-$?
(b) Explain the changes in intensity of color in the ferroin indicator reaction.

**Solution**

(a) The unifying theme of these seemingly disparate examples is a primary contribution of charge-transfer (CT) absorptions to the visible spectra. In principle, all molecules exhibit charge-transfer spectra, but only if the CT absorption extends into the visible region does it affect the color of the compound. An extensive discussion of CT spectra in transition-metal complexes can be found in Lever (1968, chapter 8). The same fundamental principles apply to interpretation of donor–acceptor complexes such as the colored tetracyanoethylene (acceptor) complexes with benzene derivatives (donor). The colors of $HgI_2$ and $MnO_4^-$ result from ligand-to-metal charge transfer. In the case of $MnO_4^-$, the molecular-orbital diagram is given by Wrobleski and Long (1977).
(b) $[Fe(phen)_3]^{2+}$ is low-spin $d^6$ and thus would be expected to have visible spectra similar to Co(III). However, the ease of oxidation of Fe(II) and the availability of low-lying empty orbitals on the phenanthroline combine to give a very intense metal-to-ligand CT transition. Thus many Fe(II) complexes of unsaturated amines are intensely colored.

The ion $[Fe(phen)_3]^{3+}$ is isoelectronic with $d^5$ Mn(II), but the combination of metal charge 3+ and a strong ligand field make $[Fe(phen)_3]^{3+}$ low-spin. The allowed d–d transitions have molar extinction coefficients considerably smaller than for the CT bands in $[Fe(phen)_3]^{2+}$. Since Fe(III) is rather strongly oxidizing, its charge-transfer bands would more likely be ligand-to-metal than metal-to-ligand. However, since the phenanthroline is more suited for metal-to-ligand than ligand-to-metal charge transfer, no charge-transfer bands are observed.

The intensity differences in the colors of the ferroin indicator are thus due to the difference in intensity of charge-transfer and d–d transitions.

### 4.1.9

Figure 4.1.9 and table 4.1.9 are taken from the work of Ohbayashi *et al.* (1977). Their experiments indicate that neon has a very low cross section for electronic quenching of the excited methoxy radical $CH_3O^*$, <1% that of CO, and <0·25% that of the parent compound methyl nitrite.

(a) Why is this so, and why does the addition of neon to the methyl nitrite before photolysis make several bands in the fluorescence spectrum of figure 4.1.9 sharper and better resolved?

(b) From the Deslandres table (table 4.1.9), argue for the assignment of the 293 nm band to the (2, 0) transition. The energy of the (0, 0) band

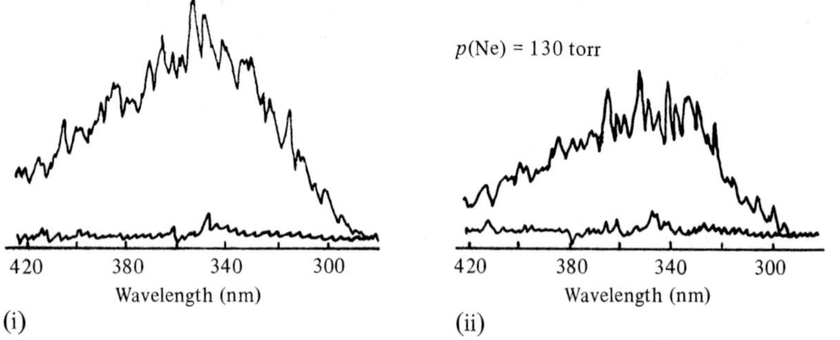


**Figure 4.1.9.** Emission spectra obtained in the photolysis of (i) 0·5 torr of $CH_3ONO$, and (ii) 0·5 torr of $CH_3ONO$ with 130 torr of Ne, by means of an iodine lamp. Background at the bottom. Spectral slit width 1·0 nm. After Ohbayashi *et al.* (1977).

**Table 4.1.9.**

| $v''$ | $v'$ 0 | | | 1 | | | 2 | |
|---|---|---|---|---|---|---|---|---|
| 0 | [32800 | (305) | 800][a] | 33600 | (298) | 500 | 34100 | (293) |
| | 1100 | | | 1300 | | | 1200 | |
| 1 | [31700 | (315) | 600][a] | 32300 | (310) | 600 | 32900 | (304) |
| | 1100 | | | 1000 | | | 1100 | |
| 2 | 30600 | (327) | 700 | 31300 | (320) | 500 | 31800 | (314) |
| | 1100 | | | 1100 | | | 1000 | |
| 3 | 29500 | (339) | 700 | 30200 | (331) | 600 | 30800 | (325) |
| | 1000 | | | 1000 | | | 900 | |
| 4 | 28500 | (351) | 700 | 29200 | (343) | 700 | 29900 | (335) |
| | 1000 | | | 1100 | | | 1100 | |
| 5 | 27500 | (364) | 600 | 28100 | (356) | 700 | 28800 | (347) |
| | 1000 | | | 1000 | | | 1000 | |
| 6 | 26500 | (377) | 600 | 27100 | (365) | 700 | 27800 | (360) |
| | | | | 1000 | | | 1000 | |
| 7 | | | | 26100 | (383) | 600 | 26700 | (374) |

[a] Values estimated from vibrational frequencies.

of the methoxy fluorescence, $E_0(CH_3O^*)$, where

$$E_0(CH_3O^*) \leqslant h\nu_0 - D^0(CH_3O\text{–}NO) \,, \qquad (4.1.9a)$$

turns out to be $\leqslant 4{\cdot}21 \pm 0{\cdot}05$ eV. The first term on the right-hand side of the inequality (4.1.9a) is the threshold energy of incident light necessary to give $CH_3O^*$ fluorescence and is determined in these experiments to be 6·02 eV. The second term is the particular bond dissociation energy of the parent compound and was determined earlier to be 1·81 eV.
*Note.* 1 eV is equivalent to 8065 $cm^{-1}$.
(c) Ab initio calculations of the energies as functions of the geometry for the ground and first excited electronic states of $CH_3O$ (Yarkony *et al.*, 1974) have given the energy difference corresponding to the (0, 0) band as 3·59 eV, in contrast to the experimental value of $4{\cdot}07 \pm 0{\cdot}02$ eV. Does the calculated result violate the variation theorem?
(d) The two vibrational frequencies of table 4.1.9 ($\sim$600 $cm^{-1}$ and $\sim$1000 $cm^{-1}$) are implicitly assumed to be chiefly that of the C–O stretch in the two electronic states concerned. Discuss the implications of these experimental findings in terms of the molecular structure of the methoxy radical in the two electronic states (see Yarkony *et al.*, 1974).

**Solution**

(a) The monatomic gases can absorb energy only by translational or electronic excitation. Selection-rule limitations are not a problem here at all, since for collisions the selection rules can be succinctly summarized: there are none.

Because the resonance electronic transitions are so energetic in neon, it is only transitions between two excited states which can be effective in taking up the $CH_3O^*$ excitation energy by an electronic mechanism. This effect would be rather unimportant in this case, owing to the low population of collisionally excited Ne atoms. (The red color of the ubiquitous neon signs is due to just these excited state transitions in the impact-excited atoms.)

On the other hand, vibrational and rotational quenching of $CH_3O^*$ can take place very readily by interaction with the translational energy of the Ne atoms. The effect of this is to depopulate the higher vibrational levels, so that the lower ones are more highly populated, and hence emission transitions from these levels are enhanced at the expense of those from the upper ones.

(b) If the 293 nm band is assigned as (2, 0), the (0, 0) band would lie at $305 \pm 1$ nm, as shown. If the 293 nm band should be assigned as (2, 1), then the (0, 0) band would be at $294 \pm 1$ nm. This assignment would just satisfy the energy requirement given by inequality (4.1.9a), but leaves no room at all for any excess internal or kinetic energy at the threshold. Slight excitation in any internal or external mode lowers the value of $E_0(CH_3O^*)$, and the (0, 0) band at $294 \pm 1$ nm would then be energetically impossible.

(c) The variation theorem is not violated, since it applies only to the calculated energy as an upper bound to the true energy, both for the ground state and for excited states orthogonal to all lower states. Thus the two differences (calculated minus true energies) could be quite different for the $^2E$ ground state and $^2A_1$ excited state. Depending upon one's own particular *Weltanschauung* of quantum chemistry, the agreement between the calculated and experimental transition energies could be viewed as rather good.
(d) The lower frequency in the upper state implies a weaker, and therefore longer, C–O bond. The ab initio calculations yield lengths of 1·44 and 1·65 Å in the ground and excited states, respectively. Other results of these calculations are interesting. An electron is promoted from the bonding $5a_1$ orbital (C–O bond region) to the slightly antibonding 2e orbital (centered on O atom). The carbon atom undergoes a partial shift from $sp^3$ to $sp^2$ hybridization with the methyl group becoming more planar; O–C–H angles of 109 and 102° optimize the energy of the ground and excited states, respectively.

**4.1.10**
Figure 4.1.10a shows the two rotational configurations for donor–acceptor complexes of tetracyanoethylene (TCNE) with monosubstituted benzene derivatives. From the work of Mobley *et al.* (1977) the double CT bands observed earlier in these $\pi-\pi^*$ electron donor–acceptor complexes are shown to be due to the existence of two rotational isomers in the ground electronic state shown schematically in this figure. The Y and X configurations are thought to correspond to maximum overlap between the lowest vacant acceptor orbital with highest and second-highest donor orbitals, respectively.
(a) The low-energy and high-energy bands are predicted to arise from the Y and X configurations, respectively. Is this prediction consistent with the foregoing?

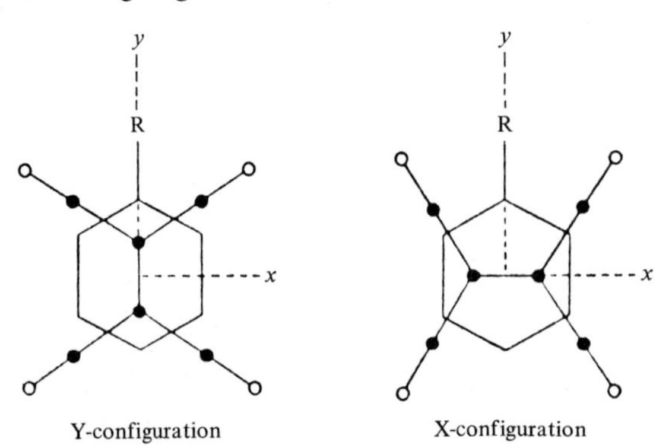


**Figure 4.1.10a.** The two predicted configurations for a TCNE–monosubstituted benzene complex (Mobley *et al.*, 1977).

(b) Assuming Beer's law and a Boltzmann distribution, derive an equation connecting the logarithm of the absorbances of the two bands $A_X$ and $A_Y$ with temperature and the energy difference $\Delta E$, $\Delta E = \pm(E_X - E_Y)$. By assuming that $\ln(\epsilon_X/\epsilon_Y)$ is independent of temperature, $\Delta E$ may be found from the slope ($= -\Delta E/R$) of a plot such as that of figure 4.1.10b.
(c) Verify the value of $\Delta E$ given in figure 4.1.10b for the thioanisole complex in acrylic film; prove that the intercept $\ln(\epsilon_X/\epsilon_Y)$ closely approximates to zero in this case.
(d) To what would one attribute the levelling off of the experimental curve of figure 4.1.10b at lower temperatures? Table 4.1.10 presents the results for other donor molecules.
(e) Does the difference $\Delta E = E_X - E_Y$ correspond to the energy difference between the two charge-transfer bands?

**Table 4.1.10.** Absorbance ratios and configurational energy differences for complexes of TCNE in poly(methyl methacrylate)[a] (after Mobley *et al.*, 1977).

| Donor | CT bands $\bar{\nu}_{max}$ ($10^3$ cm$^{-1}$) | | $A_X/A_Y$ [b] | | $\Delta E$ (kJ mol$^{-1}$) |
|---|---|---|---|---|---|
| | high-energy | low-energy | 296 K | 77 K | |
| Anisole | 26·7 | 20·5 | 0·92 | 0·82 | 0·5 ± 0·1 |
| Thioanisole | 27·1 | 18·4 | 0·41 | 0·28 | 2·2 ± 0·3 |
| *o*-Dimethoxybenzene | 24·1 | 17·6 | 0·62 | 0·52 | |
| *p*-Dimethoxybenzene | 27·0 | 16·1 | 0·68 | 0·57 | 0·8 ± 0·1 |
| 1,2,4-Trimethoxybenzene | 23·5 | 14·7 | 0·47 | 0·34 | 2·0 ± 0·3 |
| *p*-Xylene | 25·6 | 21·6 | 1·08 | 1·17 | −0·5 ± 0·1 |

[a] Bands resolved with a Dupont 310 curve resolver; approximate concentrations 0·5 mol dm$^{-3}$, 1 : 1 donor to acceptor.
[b] Approximate error ±2·5%.

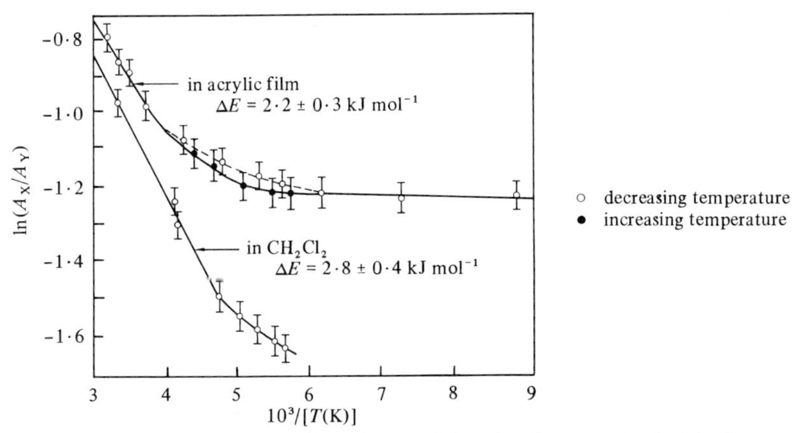


**Figure 4.1.10b.** Temperature dependence of the absorbance ratio $A_X/A_Y$ for thioanisole–TCNE in acrylic film and $CH_2Cl_2$ solution. After Mobley *et al.* (1977).

(f) What does the approximate equality of $\epsilon_X$ and $\epsilon_Y$ imply?
(g) Suggest two factors that might influence the preferred geometry of the complex.
(h) Can you rationalize the reversed sign of $\Delta E$ for *p*-xylene in table 4.1.10?

**Solution**
(a) Yes, as can be seen by drawing an electronic energy-level diagram for the $\pi$, $\pi^*$ levels.
(b) Since the absorbance is proportional to $\epsilon$ and the concentration $c$ for each band (Beer's Law) and the pathlength is the same, then we can write:

$$A_X/A_Y = (N_X/N_Y)(\epsilon_X/\epsilon_Y),$$

with

$$N_X/N_Y = \exp(-\Delta E/RT),$$

where $N_X$, $N_Y$ are the numbers of complexes in each configuration.
(c) For the intercept perhaps the easiest way is to use the point from table 4.1.10 at 296 K and *assume* $A_X/A_Y = 1$ at infinite temperature to compute a new slope. This leads to the result $\Delta E = 2 \cdot 194$ kJ mol$^{-1}$, which is well within the error bounds of the experimental value of $2 \cdot 2 \pm 0 \cdot 3$ kJ mol$^{-1}$.
(d) The 'levelling-off' effect can be attributed to the freezing in of configurations when the energy barrier between them becomes high relative to $kT$, so that true thermodynamic equilibrium is not attained.
(e) No, magnitudes are much different. Recall 1 kJ mol$^{-1}$ is equivalent to $83 \cdot 6$ cm$^{-1}$.
(f) The near equality of $\epsilon_X$ and $\epsilon_Y$ implies that the charge-transfer stabilization is approximately equal in both complex configurations.
(g) Mobley *et al.* (1977) suggest steric effects and exchange repulsion interactions as factors influencing the relative energies of the configurations.
(h) The methyl groups are rather weak electron donors compared with the other substituents listed. Evidence for this can be seen by comparison of the ionization potential of benzene with that of toluene, as quoted by Mobley *et al.* (1977).

**4.1.11**
The electronic absorption spectrum of gaseous pyridine in the vacuum-ultraviolet region is given in figure 4.1.11.
(a) Olsher (1977) discusses the 180 nm $\pi^* \leftarrow \pi$ transition in terms of benzene symmetry. Why? Discuss his assignment of the spectrum of figure 4.1.11 given in table 4.1.11. The fundamental frequency of the $\nu_2$ $(a_1)$ vibrational mode in the ground electronic state of pyridine is 1030 cm$^{-1}$.
(b) Into what state(s) does the degenerate pair $E_{1u}$ transform, when the actual symmetry of pyridine is taken into account?

(c) Earlier workers had predicted various splittings for this (benzene-degenerate) upper state in pyridine. From the spectra in figure 4.1.11, estimate the magnitude of this splitting.
(d) Earlier measurements of the electron impact spectrum of pyridine gave narrow peaks at 7·15 and 7·25 eV. These were interpreted as transitions to the $A_1$ and $B_1$ electronic states. What would you say about them now?

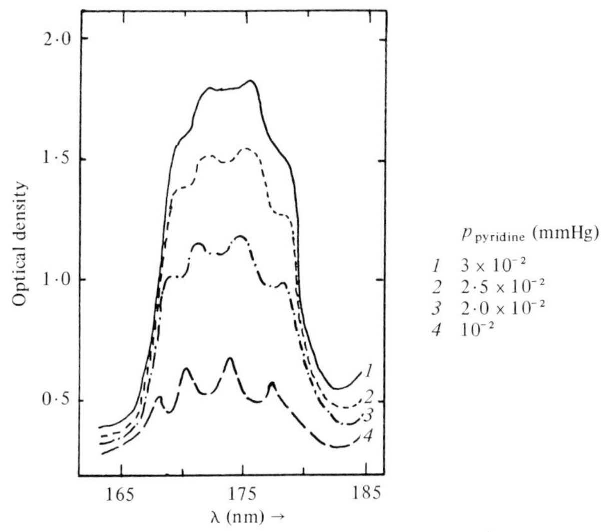


**Figure 4.1.11.** The absorption spectrum of the $\pi^* \leftarrow \pi$ (180 nm) transition of gaseous pyridine at 298 K. After Olsher (1977).

**Table 4.1.11.** Analysis of the absorption spectrum of gaseous pyridine 180 nm $\pi^* \leftarrow \pi$ transition ($^1E_{1u} \leftarrow {}^1A_{1g}$ transition in benzene symmetry). After Olsher (1977).

| Wavelength (nm) | Frequency ($cm^{-1}$) | $\bar{\nu} - \bar{\nu}_{00}$ ($cm^{-1}$) | Vibrational assignment[a] |
|---|---|---|---|
| 176·2 | 56753 | 0 | 0←0 |
| 173·75 | 57560 | 807 | 1←0 |
| 171·2 | 58411 | 1658 | 2←0 |
| 168·75 | 59263 | 2510 | 3←0 |

[a] Numbers for upper state refer to excitation of $\nu_2$ ($a_1$) mode.

**Solution**

(a) One would expect the perturbation of the electronic system, which can be regarded as pushing one proton into a carbon nucleus thereby converting it into a nitrogen nucleus and letting the former C–H $\sigma$ bond now become a nitrogen-atom lone pair, to be rather small. The benzene-state notation is used because of this similarity of electronic states and because the corresponding transition, $^1E_{1u} \leftarrow {}^1A_{1g}$ is symmetry-allowed in benzene.

The assignment of table 4.1.11 seems quite straightforward; one expects the transitions to originate in the ground vibrational state (why? cf. solution to problem 4.1.12(a)] and the 800 cm$^{-1}$ frequency of the $\nu_2$ $(a_1)$ mode in the excited electronic state is somewhat smaller than that of this ring-breathing mode in the ground state, as befits the upper state of the $\pi^* \leftarrow \pi$ transition where a bonding electron has moved to an antibonding orbital. The finite width of the vibronic bands is due to the very closely spaced accompanying rotational transitions. In some cases these rotational lines could not be determined even with infinite resolving power, because of the Doppler width; this is particularly so for molecules with larger moments of inertia [see the remarkable spectrum of azulene and the discussion in the paper of Ramsay (1965)].

(b) The symmetry descent from $D_{6h}$ (benzene) to $C_{2v}$ (pyridine) splits the $E_{1u}$ state:

$$E_{1u}(D_{6h}) \rightarrow A_1 + B_1(C_{2v}) \, .$$

Note that the same result is obtained whether one correlates $C_2(z)$ with $C_2'$ or $C_2''$ of benzene as expected.

(c) The upper state splitting is 300 cm$^{-1}$, the width of a single vibronic band.

(d) The energies of these peaks are identical with those of the middle two vibronic transitions of table 4.1.11. Note that their difference is ~800 cm$^{-1}$.

**4.1.12**

The electronic spectra of binuclear cobalt carbonyl complexes in figure 4.1.12a are taken from the work of Abrahamson *et al.* (1977). The $Co_2(CO)_6L_2$ complexes are viewed as having a two-electron metal–metal

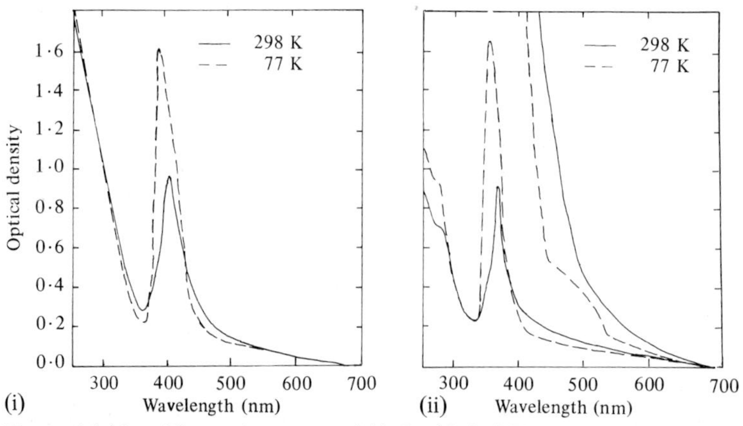


**Figure 4.1.12a.** Electronic spectra of (i) $Co_2(CO)_6(PPh_3)_2$ in 2-methyltetrahydrofuran, and (ii) $Co_2(CO)_6[P(OMe)_3]_2$ in EPA (5:5:2 diethyl ether–isopentane–ethanol, by volume) solution. The high optical density in the $Co_2(CO)_6[P(OMe)_3]_2$ spectrum is for a higher concentration. Changes in the spectra upon cooling are not corrected for solvent contraction. After Abrahamson *et al.* (1977).

$\sigma$ bond, with intense near-u.v. absorptions assigned as $\sigma^* \leftarrow \sigma$ transitions as indicated in figure 4.1.12b. No appreciable quantities of any bridged isomers were found.

(a) One sees from figure 4.1.12a that for both compounds cooling results in an increase in absorbance, a sharpening, and a slight shift of the band to a higher energy. Give a reason for each of these observations.

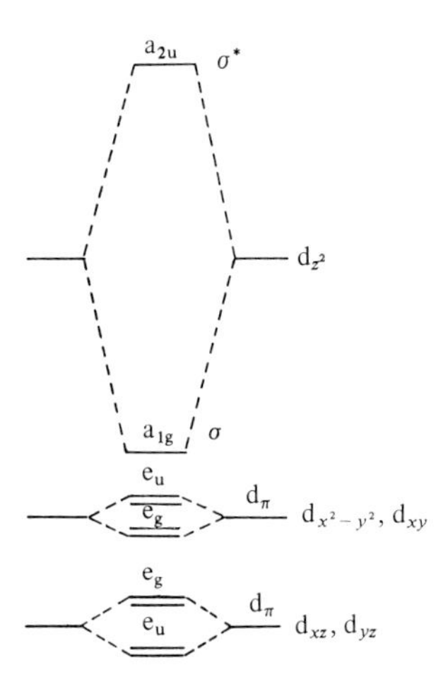


**Figure 4.1.12b.** Relative energies of the molecular orbitals of $D_{3d}$ $Co_2(CO)_6L_2$. The ground state is $^1A_{1g}$ ($e_u^4\, e_g^4\, e_g^4\, e_u^4\, a_{1g}^2$). After Abrahamson *et al.* (1977).

**Table 4.1.12a.** Spectral data recorded at 298 K unless stated otherwise; sh, shoulder.

| Complex | $\bar{\nu}_{max}$ ($cm^{-1}$) | $\epsilon$ ($dm^3\ mol^{-1}\ cm^{-1}$) | Assignment [a] |
|---|---|---|---|
| $Co_2(CO)_8$ [b] | 23470 | 980 | $\sigma^* \leftarrow d_\pi$ } nonbridged |
| | 28570 | 5500 | $\sigma^* \leftarrow \sigma$ } nonbridged |
| | 35460 | 14400 | $\sigma^* \leftarrow \sigma$ } bridged |
| | 26460 [c] | 2200 | $\sigma^* \leftarrow d_\pi$ } bridged |
| | 29000 [c] | 3700 | n.a. |
| | 32100 [c] | 13500 | n.a. |
| | 35460 [c] | 22600 | $\sigma^* \leftarrow \sigma$ bridged |
| $Co_2(CO)_6(PPh_3)$ [d] | 22200 (sh) | 5300 | $\sigma^* \leftarrow d_\pi$ |
| | 25450 | 23800 | $\sigma^* \leftarrow \sigma$ |
| $Co_2(CO)_6[P(OMe)_3]_2$ [e] | 20800 (sh) [f] | – [f] | $\sigma^* \leftarrow d_\pi$ |
| | 27780 | 24000 | $\sigma^* \leftarrow \sigma$ |

[a] Abrahamson *et al.* (1977); n.a., not assigned.
[b] In 2-methylpentane; $\epsilon$ values all reported with respect to total $Co_2$ concentration.
[c] 50 K; $\epsilon$ values uncorrected for solvent contraction.
[d] In dichloromethane.
[e] In 2-methylheptane.
[f] Only observable in low-temperature spectra.

In contrast to the $Co_2(CO)_6L_2$ compounds, $Co_2(CO)_8$ exists at room temperature as roughly a 50 : 50 mixture of bridged and nonbridged isomers. Earlier work demonstrated that the bridged form is the only one existing at lower temperature. Spectral data and assignments for the three complexes are given in table 4.1.12a.

(b) Are the differences in the spectra of $Co_2(CO)_8$ at 298 and 50 K (figure 4.1.12c) consistent with the changes in isomer population with temperature?

(c) What evidence could be used to infer a shorter Co—Co bond length?

(d) What is the assumption behind the calculated $\sigma^* \leftarrow \sigma$ transitional frequencies for the heterodinuclear complexes listed in table 4.1.12b?

Earlier work correlated the position of this $\sigma^* \leftarrow \sigma$ transition with metal-metal bond strength.

(e) What do the calculated results of part (d) and table 4.1.12b imply about the bonding of the heterodinuclear complexes reported in this work?

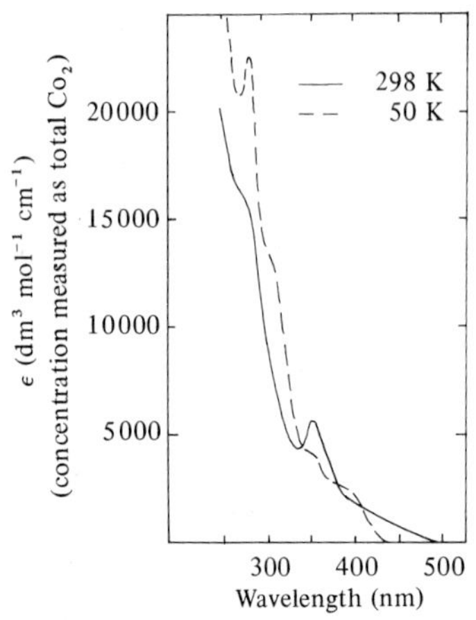


**Figure 4.1.12c.** Electronic spectra of $Co_2(CO)_8$ in 2-methylpentane. The spectral changes are not corrected for solvent contraction. After Abrahamson *et al.* (1977).

**Table 4.1.12b.** Near-u.v. $\sigma^* \leftarrow \sigma$ band maxima for several binuclear metal-metal σ-bonded (nonbridged) complexes. Spectral data recorded at 298 K in aliphatic hydrocarbon solvents. After Abrahamson *et al.* (1977).

| Complex | $\bar{\nu}_{max}$ (cm$^{-1}$) | | Complex | $\bar{\nu}_{max}$ (cm$^{-1}$) |
|---|---|---|---|---|
| | calculated | observed | | observed |
| $(\eta^5\text{-}C_5H_5)Mo(CO)_3Mn(CO)_5$ | 27505 | 26810 | $Mn_2(CO)_{10}$ | 29240 |
| $(\eta^5\text{-}C_5H_5)Mo(CO)_3Re(CO)_5$ | 28860 | 27400 | $Re_2(CO)_{10}$ | 31950 |
| $(\eta^5\text{-}C_5H_5)W(CO)_3Mn(CO)_5$ | 28430 | 27620 | $(\eta^5\text{-}C_5H_5)Mo_2(CO)_6$ | 25770 |
| $(\eta^5\text{-}C_5H_5)W(CO)_3Re(CO)_5$ | 29785 | 30210 | $(\eta^5\text{-}C_5H_5)W_2(CO)_6$ | 27620 |
| $Co(CO)_3(PPh_3)Mn(CO)_5$ | 27345 | 26600 | $Co_2(CO)_6(PPh_3)_2$ | 25450 |
| $(\eta^5\text{-}C_5H_5)Mo(CO)_3Co(CO)_4$ | 27170 | 28170 | $Co_2(CO)_8$ | 28570 |
| $(\eta^5\text{-}C_5H_5)W(CO)_3Co(CO)_4$ | 28095 | 29240 | | |

**Solution**

(a) The same explanation could cover all three phenomena. In problem 4.1.11 we espoused the idea that at room temperature no particular vibration is excited. This of course is really determined by the ratio $h\nu/kT$ and is expressed quantitatively by Boltzmann factors. In the case of these metal–metal complexes we expect a large number of low-frequency vibrations with appreciable excitations at room temperature. Hence cooling to liquid nitrogen temperature depopulates the vibrationally excited state in the electronic ground state, and thus makes almost all electronic transitions originate from the ground vibrational level. This at once explains the increased sharpness and shift to slightly higher energies. To explain the increased absorbance we follow a well-known path and postulate that the Franck–Condon overlap integrals between the single lower vibrational state and the appropriate upper vibrational states are greater in magnitude than the integrals between the set of occupied lower vibrational states and target upper vibrational states. This explanation has two virtues. Firstly, it could be correct, and secondly, it is very difficult to disprove.

(b) From table 4.1.12a, the absorbances at 35460 $cm^{-1}$ (bridged) and 28570 $cm^{-1}$ (nonbridged) are assigned to $\sigma^* \leftarrow \sigma$ transitions of the two isomers. Note that the extinction coefficient for the bridged $\sigma^* \leftarrow \sigma$ transition at 50 K is within 10% of the sum of those for the bridged and nonbridged $\sigma^* \leftarrow \sigma$ transitions at room temperature; notice also that the weak low-temperature band at 29000 $cm^{-1}$ is unassigned, and that Abrahamson *et al.* (1977) are not willing to suggest that this is the residual $\sigma^* \leftarrow \sigma$ nonbridged transition.

It is, of course, possible that other nonbridged isomers are present; their detection would be difficult in this particular set of experiments.

(c) The general correlation is made between increasing transition energies and shorter Co–Co bonds in the bridged form, in comparison to the nonbridged dimer, by Abrahamson *et al.* (1977). The shorter bond length is inferred from the Co–Co stretching frequencies (cf. Onaka and Shriver, 1976).

(d) By inspection, the wavenumber of the $\sigma^* \leftarrow \sigma$ transition for a given heterodinuclear complex, $\bar{\nu}_{\mathrm{M-M'}}$, approximates to the mean of those for the related homodinuclear complexes:

$$\bar{\nu}_{\mathrm{M-M'}} \approx \tfrac{1}{2}(\nu_{\mathrm{M-M}} + \nu_{\mathrm{M'-M'}}) \,.$$

The agreement is rather good, within 5% in each case.

(e) The success of the above prediction indicates there is little ionic bonding in the M–M′ complexes. The bond strengths could then be approximated by the arithmetic mean of the homodinuclear bond strengths when bonding is largely covalent, but not when it is ionic.

Note that nonelectronic spectra may be useful in interpreting the electronic spectrum.

### 4.1.13

The appearance of quasi-lines in optical spectra of several aromatic compounds dissolved in low-temperature n-alkane matrices was apparently first observed by Bolotnikova (1959). These spectra have come to be known as Shpolskii spectra after the originator of the lock and key (or key and hole) principle or rule (Shpolskii, 1962, 1963). The idea is that of matching the long and short axes of the guest and host molecules for linear aromatics such as naphthalene, anthracene, and naphthacene; perhaps for nonlinear condensed aromatic compounds such as phenanthrene and chrysene a match between the long axes of the guest and host molecules is also possible.

Dekkers *et al.* (1977) have put forth very interesting counter arguments to the key and hole rule.

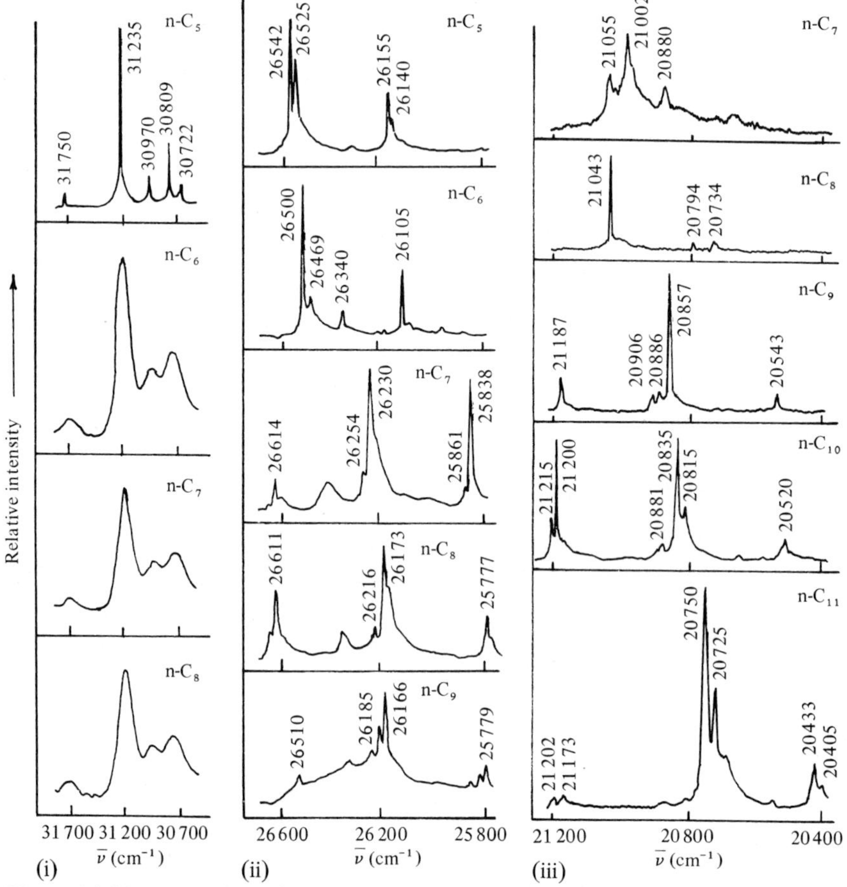


**Figure 4.1.13.** Parts of the fluorescence spectra of (i) naphthalene, (ii) anthracene, and (iii) naphthacene in straight-chain aliphatic hydrocarbon polycrystals at 20 K. After Dekkers *et al.* (1977).

(a) Assuming the key and hole rule for the moment, what explanation can you give for the differences in linewidths of the spectra in figure 4.1.13(i)? Compare the results of figures 4.1.13(ii) and 4.1.13(iii).

Note that Bolotnikova found that the sharpest bands for anthracene guest occurred in n-$C_7$ host, and for naphthacene guest in n-$C_9$ host.

(b) Given the very low solubility of solid naphthalene in the solid n-$C_5$ to n-$C_8$ alkanes, and finite cooling rates of the mixtures, construct an alternate explanation for the variation of guest–host spectral behavior of figure 4.1.13(i). The concentration of naphthalene is about $10^{-4}$ mol $dm^{-3}$, and those of anthracene and naphthacene are similar.

**Solution**

(a) If it is assumed that narrow lines result from a 'good' matching of 'key' and 'hole', it appears that the naphthalene molecule (guest) has about the same dimension as the n-$C_5$ molecule (host) in the lattice, and that the other three 'keys' do not match the 'hole'. What could be simpler? Both groups found that the anthracene bands were sharpest in n-$C_7$ but Dekkers *et al.* (1977) found equally narrow lines for naphthacene in n-$C_7$, n-$C_8$, and n-$C_{10}$.

(b) If the rate of approach to thermodynamic equilibria is faster in naphthalene mixtures with n-$C_6$–n-$C_8$ than in n-$C_5$, then the latter can form a nonequilibrium solid solution, whereas the other three mixtures could attain equilibrium more rapidly with phase separation occurring. Dekkers *et al.* (1977) conclude that small aromatic hydrocarbons seldom display Shpolskii spectra when dissolved in frozen n-alkanes and that the experimental conditions are, in general, less critical for large aromatic molecules than for those intermediate in size. They cite the lack of the Shpolskii effect for benzene and monosubstituted and polysubstituted benzenes, the occurrence of it for larger-sized compounds such as anthracene, naphthacene (as in figure 4.1.13), coronene, perylene, and benz[*a*]perylene, within a larger range of n-alkane crystals.

**4.1.14**

This problem is concerned with photodissociation spectroscopy and the McLafferty rearrangement product in the fragmentation of n-butylbenzene and 2-phenylethanol ions.

Photodissociation spectroscopy has been applied by Dunbar and Klein (1977) to the question of the structure of the ion at $m/e$ = 92 ($C_7H_8$) resulting from the two separate fragmentations:

$$\text{(I)} \xrightarrow{-C_3H_6} \text{(III)} \xrightarrow{?} \text{(IV)}$$

(I) (III) (IV)

(II) $\xrightarrow{-CH_2O}$ (III) $\xrightarrow{?}$ (IV)

by use of an ion cyclotron resonance mass spectrometer.

The product (III) follows from the McLafferty rearrangement—capture of a hydrogen atom by a polar functional group of an odd-electron ion (McLafferty, 1966). The identity of (IV) could then be either (III) itself, (V) or (VI), or possibly still another species.

(V) (VI)

The open circles in the photodissociation spectrum of figure 4.1.14a give a tentative identification of the actual product of the fragmentation process as the methylenecyclohexadiene cation (III). The dotted and dashed curves are earlier results for the cycloheptatriene (V) and toluene (VI) cations, respectively. The spectrum is the same whether either (I) or (II) is the initial reactant.

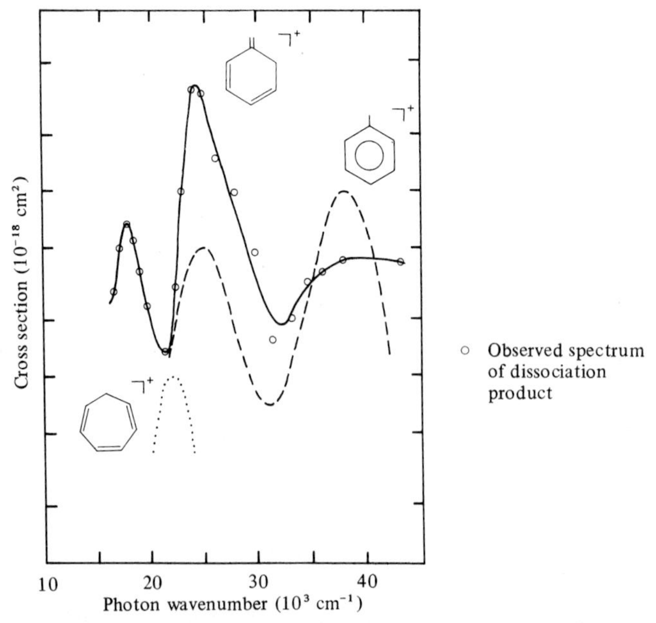


**Figure 4.1.14a.** Photodissociation spectra of isomeric $C_7H_8^+$ radical ions. The indicated methylenecyclohexadiene structure for the product ion (IV) is, of course, uncertain, as discussed in the text. After Dunbar and Klein (1977).

Figure 4.1.14b(i) displays time-resolved curves. The upper pair of curves (two duplicate runs) represent the ion-decay curve obtained for $m/e = 92$ after termination of ion production at time zero. The lower pair of curves are similar data with irradiation at 405 nm commencing at time zero. Figure 4.1.14b(ii) is a logarithmic plot of the results of figure 4.1.14b(i) obtained by point-by-point division of the 'light-on' curve by the 'light-off' curve. The simple exponential decay to more than 75% extent of photodissociation suggests a homogeneous ion population.

(a) From the spectrum of (IV) (figure 4.1.14a), what suggests that (IV) contains no more than 10% of (VI)?

(b) The spectrum of (IV) does not resemble that of (V) either; why is this conclusion more tentative?

As the discussion above indicates, the result of the time-resolved photodissociation experiments is that (IV) consists of at least 75% of a homogeneous ion population.

(c) Given that the model calculations for *trans*-1,3,5-hexatriene radical cation predict an electronic spectrum with an intense peak at 22500 $cm^{-1}$, and smaller peaks at 13000 and 41000 $cm^{-1}$, and that the photodissociation spectrum of this compound gave a strong peak at 27500 $cm^{-1}$, with a smaller peak at 16000 $cm^{-1}$, and another one near 40000 $cm^{-1}$, construct an additional argument to substantiate the conclusion that (IV) is primarily the same as (III).

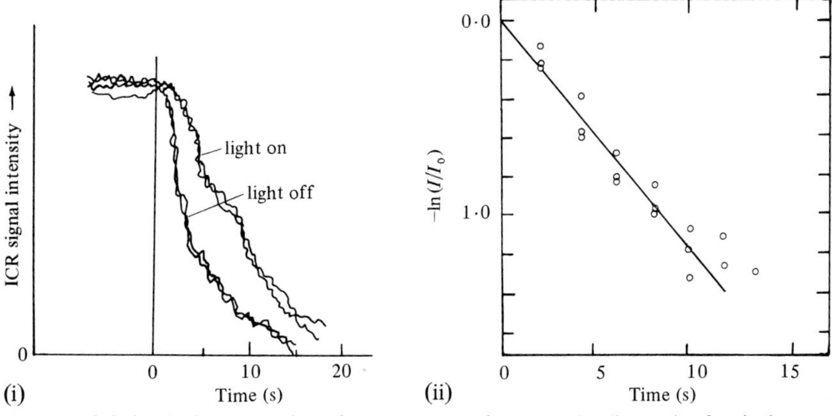


**Figure 4.1.14b.** Exhaustive photodissociation, or 'time-resolved', results for (IV) ions at 405 nm (see text). After Dunbar and Klein (1977).

**Solution**

Figure 4.1.14c, taken from Dymerski *et al.* (1974) demonstrates the difference between photoelectron spectroscopy (p.e.s.) and photodissociation spectroscopy (p.d.s.). These techniques should be viewed as complementary. In addition to the condition that the excited ion produced by these two processes should be structurally similar before comparison can be made, it is apparent that, as in any light-absorption process, symmetry conditions may impose selection rules on the simple Franck–Condon overlap integral. Neutral ionization to electronically excited ion states and direct optical excitation to these same states would be expected to have differing allowed transitions. Thus, while the p.e.s. and p.d.s. processes are expected to be comparable, they are not expected to be identical. These differences are of potential value in interpreting both types of spectra and may aid in unraveling ionic excitation processes and ion structures.

Although the photodissociation spectrum reflects the optical spectrum of the ion, it is worth emphasizing that it is an indirect approach, and the appearance of the peak in the p.d.s. spectrum requires both an optical absorption peak and a subsequent available dissociation mechanism for the excited ion.

In figure 4.1.14c the p.e.s. vertical electronic excitation process involves neutral ionization to an excited ion state, whereas the p.d.s. process involves direct ion excitation. The u.v. neutral excitation process is included for comparison.

(a) The spectrum of (IV) clearly lacks the strong u.v. peak characteristic of (VI), and the 400 nm peak is at least six times more intense for (IV) than for (VI). Note the logarithmic ordinate scale. It is important to observe that both of the peaks in the toluene cation spectrum arose from a single ionic species as was shown earlier.

(b) Because (V) has such a small photodissociation cross section, a large concentration of it would produce a relatively weak spectrum.

(c) Dunbar and Klein (1977) claim that it is evident that qualitatively the spectrum of (IV) fits the expected pattern for a conjugated triene. Quantitative comparison of (IV) with hexatriene cation shows a shift of the red peak of hexatriene toward the blue, and a shift of the near-u.v. peak toward the visible. They note that a similar pattern of peak shifts is seen in conjugated dienes upon going from the straight-chain (predominantly *trans*) to the cyclized (enforced *cis*) radical cation species. They further

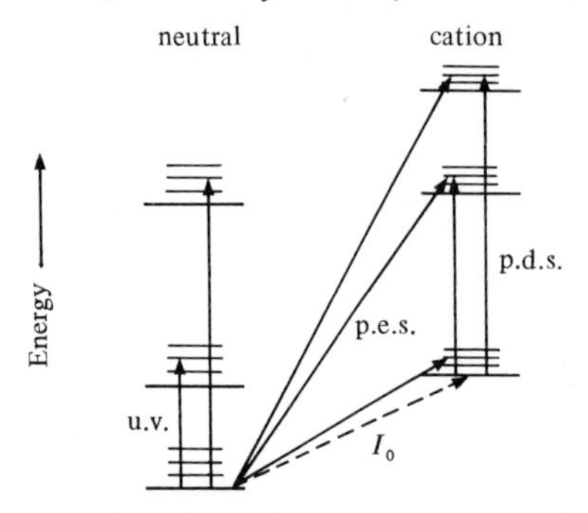


**Figure 4.1.14c.** Schematic representation of transitions observed with p.d.s., p.e.s., and u.v. absorption spectroscopy. $I_0$ represents adiabatic ionization.

conclude that the spectrum of (IV) strongly suggests retention of the methylenecyclohexadiene structure (III), although other structures cannot be ruled out; the very intense peak in the blue is probably inconsistent with structures not having a degree of conjugated unsaturation comparable to (III).

**References**


Abrahamson, H. B., Frazier, C. C., Ginley, D. S., Gray, H. B., Lilienthal, J., Tyler, D. R., Wrighton, M. S., 1977, *Inorg. Chem.*, **16**, 1554.

Bolotnikova, T. N., 1959, *Opt. Spectros. (USSR)*, **8**, 138.

Boudreaux, E. A., Mulay, L. N., 1976, *Theory and Applications of Molecular Paramagnetism* (John Wiley, New York).

Cassidy, P., Hitchman, M. A., 1975, *J. Chem. Soc. Chem. Commun.*, 837.

Cotton, F. A., 1971, *Chemical Applications of Group Theory* (John Wiley, New York), Chapters 9 and 10.

Cotton, F. A., Wilkinson, G., 1972, *Advanced Inorganic Chemistry: A Comprehensive Text*, 3rd edition (John Wiley, New York).

Dekkers, J. J., Hoornweg, G. Ph., Visser, G., MacLean, C., Velthorst, N. H., 1977, *Chem. Phys. Lett.*, **47**, 357.

Dorain, P. B., 1965, *Symmetry in Inorganic Chemistry* (Addison-Wesley, Reading, Mass), p.88.

Dunbar, R. C., Klein, R., 1977, *J. Am. Chem. Soc.*, **99**, 3744.

Dymerski, P. P., Fu, E., Dunbar, R. C., 1974, *J. Am. Chem. Soc.*, **96**, 4109.

Ferguson, J., 1970, *Prog. Inorg. Chem.*, **12**, 159.

Figgis, B. N., 1966 *Introduction to Ligand Fields* (John Wiley, New York).

Gale, R., Godfrey, R. E., Mason, S. F., Peacock, R. D., Stewart, B., 1975, *J. Chem. Soc. Chem. Commun.*, 329.

Hall, L. H., 1969, *Group Theory and Symmetry in Chemistry* (McGraw-Hill, New York), Chapters 10 and 11.

Hathaway, B. J., Billing, D. E., 1970, *Coord. Chem. Rev.*, **5**, 143.

Ilse, F. E., Hartmann, H. Z., 1951, *Z. Phys. Chem. (Leipzig)*, **197**, 239.

Lever, A. B. P., 1968, *Inorganic Electronic Spectroscopy* (Elsevier, Amsterdam).

Lohr, L. L., Jr, 1972, *Coord. Chem. Rev.*, **8**, 241.

McLafferty, F. W., 1966, *Interpretation of Mass Spectra* (Benjamin, New York), p.123.

McPherson, G. L., Freedman, M. R., 1976, *Inorg. Chem.*, **15**, 2299.

Martin, D. S., Jr, Tucker, M. A., Kassman, A. J., 1965, *Inorg. Chem.*, **4**, 1682.

Mobley, M. J., Rieckhoff, K. E., Voigt, E. M., 1977, *J. Phys. Chem.*, **81**, 809.

Nicholls, D., 1973, *Comprehensive Inorganic Chemistry* (Pergamon, New York).

Nicholls, D., 1975 *Complexes and First Row Transition Elements* (Macmillan, London).

Ohbayashi, K., Akimoto, H., Tanaka, I., 1977, *J. Phys. Chem.*, **81**, 798.

Olsher, U., 1977, *J. Chem. Phys.*, **66**, 5242.

Onaka, S., Shriver, D. F., 1976, *Inorg. Chem.*, **15**, 915.

Phillips, C. S. G., Williams, R. J. P., 1965, *Inorganic Chemistry*, Volume 2 (Oxford University Press, London).

Ramsay, D. A., 1965, *J. Chem. Phys.*, **43**, 18 (supplement).

Schreiner, A. F., Hamm, D. J., 1973, *Inorg. Chem.*, **12**, 2037.

Schreiner, A. F., Hamm, D. J., 1975, *Inorg. Chem.*, **14**, 519.

Shpolskii, E. V., 1962, *Sov. Phys. Usp.*, **5**, 522.

Shpolskii, E. V., 1963, *Sov. Phys. Usp.*, **6**, 411.

Wrobleski, J. T., Long, G. J., 1977, *J. Chem. Educ.*, **54**, 75.

Yarkony, D. R., Schaefer (III), H. F., Rothenberg, S., 1974, *J. Am. Chem. Soc.*, **96**, 656.

## 4.2 Electron spin resonance

**D H Whiffen** University of Newcastle upon Tyne

### 4.2.1 Carbon dioxide anion

Simple electron resonance in a free radical such as $CO_2^-$ is described by using a Hamiltonian

$$\mathcal{H} = g\mu_B \boldsymbol{B} \cdot \boldsymbol{S} ,$$

which serves to define the isotropic $g$-factor, $g$.

In many circumstances this operator equation may be replaced[(1)] by

$$E = g\mu_B B M_S$$

where $E$ is the energy, $\mu_B$ is the Bohr magneton, $B$ is the magnetic induction (magnetic flux density) in the $z$ direction, and $M_S$ is the resolved electron spin quantum number.

For the carbon dioxide anion $g = 2{\cdot}000\,65$.

(a) Calculate the energies, $E_{\pm}$, for the states with $M_S = \pm\frac{1}{2}$ for inductions of 0·3 T and of 1·0 T. Calculate the frequency, $\nu$, for resonance at each induction. Calculate also the fractions, $f_{\pm}$, of anions in each spin state at equilibrium at a temperature of 300 K.

(b) Calculate the magnetic inductions required for exact resonance with spectrometers operating at 9·3 GHz and at 36·0 GHz (X-band and Q-band).

For the species $^{13}CO_2^-$ one must also consider the nuclear terms, as $^{13}C$ has a spin $I = \frac{1}{2}$ in contrast to $^{12}C$ and $^{16}O$ which are spin free. This modifies the Hamiltonian to

$$\mathcal{H} = g\mu_B \boldsymbol{B} \cdot \boldsymbol{S} + ha\boldsymbol{I} \cdot \boldsymbol{S} - \frac{h\gamma}{2\pi} \boldsymbol{B} \cdot \boldsymbol{I} ,$$

where $a$ is the hyperfine coupling constant in frequency units, and $\gamma$ is the magnetogyric ratio of the nucleus. This too may be approximated by an energy expression,

$$h^{-1}E = h^{-1}g\mu_B B M_S + aM_I M_S - \frac{\gamma B}{2\pi} M_I .$$

(c) For $^{13}CO_2^-$, $a = 468$ MHz and $\gamma/2\pi = 10{\cdot}8$ MHz T$^{-1}$. List the four permutations of $M_S = \pm\frac{1}{2}$ and $M_I = \pm\frac{1}{2}$ and the related energies—or more usefully the quantities $h^{-1}E$—both with $B = 0{\cdot}3$ T and with $B = 1{\cdot}0$ T.

What are the transition frequencies

(i) for the allowed e.s.r. lines with the selection rules $\Delta M_S = 1$ and $\Delta M_I = 0$;

(ii) for forbidden e.s.r. lines $\Delta M_S = 1$, $\Delta M_I = \pm 1$;

(iii) for the ENDOR lines, $\Delta M_S = 0$, $\Delta M_I = 1$?

[(1)] See Editors' note on treatment of magnetic fields in Appendix 1.

(d) Calculate the magnetic induction required to observe the allowed e.s.r. lines with spectrometers operating at 9·3 GHz and at 36·0 GHz. Does the line separation in a swept magnet experiment depend on the operating frequency? Would a $^{12}CO_2^-$ signal as calculated in (b) appear at the exact midpoint of the $^{13}CO_2^-$ signal if both species were present?

The major approximation made in passing from the Hamiltonian operator to the expression for $E$ is the neglect of the off-diagonal hyperfine contribution, which may be written

$$\Delta\mathcal{H} = ha(\boldsymbol{I}\cdot\boldsymbol{S} - M_I M_S) .$$

(e) Show that the above expression is equivalent to

$$\Delta\mathcal{H} = ha(I_x S_x + I_y S_y) = \tfrac{1}{2}ha(I_+ S_- + I_- S_+) ,$$

where $I_x$ is the $x$ component of the nuclear spin operator, $I_+$ is the raising operator, and so on.

Remembering that, for a spin of $\frac{1}{2}$,

$$\langle M_I = \tfrac{1}{2}|I_+|M_I = -\tfrac{1}{2}\rangle = \langle M_I = -\tfrac{1}{2}|I_-|M_I = \tfrac{1}{2}\rangle = 1 ,$$

and all other elements are zero, and also remembering that $S_+$ and $S_-$ have equivalent properties, where $|M_I = \frac{1}{2}\rangle$ stands for the nuclear spin wave function with a complex conjugate $\langle M_I = \frac{1}{2}|$, use perturbation theory to obtain corrections to the energies and transition frequencies calculated in (c).

(f) Also retaining $\Delta\mathcal{H}$, correct the numerical answers given in answer to (d) and re-answer the final qualitative questions.

(g) Would any of the resonance frequencies or resonance values of the magnetic induction have been altered if $a$ were −468 MHz rather than +468 MHz?

**Solution**

(a) See table 4.2.1a.

**Table 4.2.1a.**

| $B$ (T) | 0·3 | 0·3 | 1·0 | 1·0 |
|---|---|---|---|---|
| $M_S$ | $+\frac{1}{2}$ | $-\frac{1}{2}$ | $+\frac{1}{2}$ | $-\frac{1}{2}$ |
| $E$ ($10^{-24}$ J) | 2·782 86 | −2·782 86 | 9·276 21 | −9·276 21 |
| $E$ (J mol$^{-1}$) | 1·675 99 | −1·675 99 | 5·586 62 | −5·586 62 |
| $f$ | 0·499 664 | 0·500 336 | 0·498 880 | 0·501 120 |
| $\nu$ (GHz) | 8·400 34 | | 28·001 13 | |

The Boltzmann distribution of energy gives

$$f = Q^{-1}\exp - \frac{E}{kT} .$$

$Q$ is the partition function, which at this temperature is 2·000000. The transition frequency is given by

$$\nu = h^{-1}(E_{+\frac{1}{2}} - E_{-\frac{1}{2}}) .$$

(b) At 9·3 GHz, $B$ = 0·33213 T; at 36·0 GHz, $B$ = 1·28566 T.
(c) It is convenient to assign a letter to each of the four states for simplicity of subsequent reference. These are p for ($M_S = +\frac{1}{2}$, $M_I = +\frac{1}{2}$), q for ($+\frac{1}{2}$, $-\frac{1}{2}$), r for ($-\frac{1}{2}$, $+\frac{1}{2}$), and s for ($-\frac{1}{2}$, $-\frac{1}{2}$).

In table 4.2.1b $E_p$ then denotes the energy of state p, and $\nu_{pq}$ the transition frequency from q to p at the given magnetic induction. All values are given in MHz. $B_{pq}$ is the induction for resonance at the given observing frequency.

The three contributions to the first entry in table 4.2.1b arise from the separate terms in the energy expression as

$$+4200·2 + 117·0 - 1·6 = 4315·6 .$$

**Table 4.2.1b.**

| $B$ (T) | 0·3 | 1·0 |
|---|---|---|
| $h^{-1}E_p$ | 4315·6 (4315·6)[a] | 14112·2 (14112·2)[a] |
| $h^{-1}E_q$ | 4084·8 (4091·3) | 13889·0 (13891·0) |
| $h^{-1}E_r$ | −4318·8 (−4325·3) | −14123·0 (−14125·0) |
| $h^{-1}E_s$ | −4081·6 (−4081·6) | −13878·2 (−13878·2) |
| (i) $\nu_{pr}$ | 8634·4 (8640·9) | 28235·2 (28237·2) |
| $\nu_{qs}$ | 8166·4 (8172·9) | 27767·2 (27769·2) |
| (ii) $\nu_{ps}$ | 8397·2 (8397·2) | 27990·4 (27990·4) |
| $\nu_{qr}$ | 8403·6 (8416·6) | 28012·0 (28016·0) |
| (iii) $\nu_{pq}$ | 230·8 (224·3) | 223·2 (221·2) |
| $\nu_{sr}$ | 237·2 (243·7) | 244·8 (246·8) |

[a] Numbers in parentheses denote corrected values, see (e).

*Note.* In this approximation the sum of the two frequencies in (iii), and the difference of the two frequencies in (i) are both exactly the hyperfine coupling, $a$. Also the difference in the frequencies in (iii) and in (ii) are twice the n.m.r. resonance frequency of $^{13}$C in the given magnetic induction.

(d) At 9·3 GHz, $B_{pr}$ = 323·77 mT and $B_{qs}$ = 340·48 mT .
At 36·0 GHz, $B_{pr}$ = 1277·30 mT and $B_{qs}$ = 1294·01 mT .

These figures are most readily obtained by realising that the hyperfine coupling contribution to the transition frequency, $\pm\frac{1}{2}a$, must be compensated by a change of induction given by

$$\Delta B = \mp \frac{ha}{2g\mu_B} .$$

In this approximation the separation of 16·71 mT is independent of the measuring frequency. Indeed the hyperfine coupling is often quoted as if it were an induction, when its units are tesla or gauss (1 G = $10^{-4}$ T). The midpoint is the resonance induction for the nuclear spin free ion, $^{12}CO_2^-$.
(e) $\boldsymbol{I} \cdot \boldsymbol{S}$ is the scalar product of two vector operators and may in the normal way of scalar products be reduced to three products of scalars relating to Cartesian axes, that is

$$\boldsymbol{I} \cdot \boldsymbol{S} = (I_x S_x + I_y S_y + I_z S_z) \, .$$

Only the part $I_z S_z$ is diagonal in the chosen basis states and its contribution is $M_I M_S$. The first two products remain to be included by perturbation theory with the multiplier $ha$. The normal definitions of the raising and lowering operators are

$$I_+ = I_x + \mathrm{i} I_y \quad \text{and} \quad I_- = I_x - \mathrm{i} I_y \, ,$$
$$S_+ = S_x + \mathrm{i} S_y \quad \text{and} \quad S_- = S_x - \mathrm{i} S_y \, ,$$

whence

$$I_x = \tfrac{1}{2}(I_+ + I_-) \quad \text{and} \quad I_y = -\tfrac{1}{2}\mathrm{i}(I_+ - I_-) \, ,$$
$$S_x = \tfrac{1}{2}(S_+ + S_-) \quad \text{and} \quad S_y = -\tfrac{1}{2}\mathrm{i}(S_+ - S_-) \, .$$

Consequently

$$I_x S_x + I_y S_y = \tfrac{1}{4}[(I_+ + I_-)(S_+ + S_-) - (I_+ - I_-)(S_+ - S_-)]$$
$$= \tfrac{1}{2}(I_+ S_- + I_- S_+) \, .$$

*Note.* $\boldsymbol{I}$ and $\boldsymbol{S}$ operate in different spaces and no commutation difficulties arise.

Since $\Delta\mathcal{H}$ is off-diagonal in the basis states, p, q, r, s, it gives rise to no correction in first-order perturbation theory. For second-order corrections we require the off-diagonal matrix elements. As each term of $\Delta\mathcal{H}$ contains both a raising and a lowering operator all elements connecting p and s are zero. All that remain are

$$\langle \mathrm{q} | I_+ S_- | \mathrm{r} \rangle = \langle \mathrm{r} | I_- S_+ | \mathrm{q} \rangle = 1 \, ,$$

by using the formula of the question twice, once for the nuclear and once for the electron part of the product wave function. Therefore

$$\langle \mathrm{q} | \Delta\mathcal{H} | \mathrm{r} \rangle = \langle \mathrm{r} | \Delta\mathcal{H} | \mathrm{q} \rangle = \tfrac{1}{2} ha$$

are the only nonzero elements. Only states q and r are modified, being raised and lowered by $(\frac{1}{2}ha)^2/h\nu_{\mathrm{qr}} = \frac{1}{4}ha^2/\nu_{\mathrm{qr}}$ respectively by this second-order correction.

For the present case this correction is ±6·52 MHz at $B = 0{\cdot}3$ T, and ±1·96 MHz when $B = 1{\cdot}0$ T. These corrections are incorporated in table 4.2.1b. The hyperfine coupling is still given by the difference in the frequencies in (i) or the sum in (iii). The relationship with the n.m.r. frequency is, however, lost.

If the first-order correction to the wave function is retained, it can be shown that the transition q ← r is now weakly allowed with the microwave magnetic field *parallel* to that of the magnet, which is not the normal arrangement. Transition p ← s remains forbidden.
(f) Transitions p ← r and q ← s are both increased in frequency at fixed induction, and reduced in induction at fixed measuring frequency. Converting the correction in (e) to induction gives $h^2a^2/4g^2\mu_B^2B$ if one neglects the small nuclear term in the denominator. The answers in (d) give sufficiently accurate $B$ values for use in the corrections, which are now different for each resonance line because of the change of magnetic induction.

The corrections to $B_{pr}$ and $B_{qs}$ are 0·216 mT and 0·205 mT at 9·3 GHz, and 0·055 mT and 0·054 mT at 36·0 GHz. The net positions are therefore:

at 9·3 GHz, $B_{pr} = 323{\cdot}55$ mT and $B_{qs} = 340{\cdot}28$ mT;
at 36·0 GHz, $B_{pr} = 1277{\cdot}25$ mT and $B_{qs} = 1293{\cdot}96$ mT.

The line separations are essentially unchanged from the coupling of 16·71 mT, being 16·73 mT and 16·71 mT at the two frequencies. The residual difference arises because the two observations are made at different magnetic inductions at the two frequencies.

However, the midpoint is significantly below that induction required for $^{12}CO_2^-$ resonance, by 0·21 mT and 0·05 mT, at the two frequencies. If the second-order corrections are not applied to the observations when calculating $g$-factors, the apparent values would be too high by 0·00126 at 9·3 GHz, and 0·00008 at 36·0 GHz, giving $g$ (apparent) values of 2·00191 and 2·00073. Note that the errors are proportional to the inverse square of the spectrometer frequency.
(g) No. This is not obvious, as many of the labels are permuted. The essential point is that, if two resonances are observed, one does not know which to assign to $\nu_{pr}$ and which to $\nu_{qs}$. The positive sign of $a$ in this instance is required on structural grounds associated with anisotropic measurements in the solid state.

**4.2.2 Triplet state of naphthalene**

For a triplet state with a net electron spin $S = 1$, there is a zero-field contribution to the spin Hamiltonian which some authors write

$$hDS_Z^2 + hE(S_X^2 - S_Y^2),$$

and others

$$-hXS_X^2 - hYS_Y^2 - hZS_Z^2, \qquad \text{with } X+Y+Z = 0.$$

Here, $D$ and $E$, or $X$, $Y$ and $Z$, are the zero-field parameters expressed as frequencies. Capital suffixes have been used for $S$ to indicate molecule-fixed axes, in contrast to small suffixes which imply space-fixed axes with a field, if any is present, parallel to $z$.

(a) Show that these two forms have the same operator dependence if

$$X = \tfrac{1}{3}D - E\ , \quad Y = \tfrac{1}{3}D + E\ , \quad Z = -\tfrac{2}{3}D\ ,$$

and the energy origin is shifted by $\tfrac{2}{3}hD$.

A good way to treat problems in this subject, especially numerical problems, is to choose a basis set of eigenfunctions, usually those for which $M_S = +1$, 0, and $-1$, and express the Hamiltonian in the form of a 3 x 3 matrix.
(b) With the relative axis arrangements $x \parallel X, y \parallel Y$, and $z \parallel Z$, express the zero-field Hamiltonian,

$$h^{-1}\mathcal{H} = -XS_X^2 - YS_Y^2 - ZS_Z^2\ ,$$

in this matrix form.
Remember that for $S = 1$,

$$\langle +1 | S_+ | 0 \rangle = \langle 0 | S_+ | -1 \rangle = \langle 0 | S_- | +1 \rangle = \langle -1 | S_- | 0 \rangle = 2^{1/2}\ ,$$

where $|0\rangle$ is the eigenfunction for $M_S = 0$, etc.
Solve this matrix Hamiltonian for the zero-field eigenvalues and eigenvectors.

For naphthalene in its lowest electronically excited triplet state, $X = +1413$ MHz, $Y = +592$ MHz, and $Z = -2005$ MHz, where the $Z$ axis is perpendicular to the ring and the $X$ axis is parallel to the long axis of the molecule.
(c) Evaluate the energies, divided by $h$, and the transition frequencies in the radio-frequency region for zero magnetic field.

In the presence of a magnet there is an additional term,

$$g\mu_B \boldsymbol{B} \cdot \boldsymbol{S}\ ,$$

in the Hamiltonian. Normally $g$ would be a tensor expressed in molecular axes, but experiment shows that for this state of naphthalene $g$ is isotropic with the value 2·0030 so the above form is sufficient.
The spectrum depends on the orientation of the magnetic induction, $\boldsymbol{B}$, with respect to the molecular axes. Naphthalene is commonly examined as aligned molecules in solid solution in a single crystal of durene, so that the induction may be made parallel to any desired direction in the naphthalene by rotation of the crystal.
(d) Calculate the eigenvalues, divided by $h$, and the transition frequencies for inductions of 0·3 T and 1·0 T aligned parallel to the $X$, $Y$, and $Z$ molecular axes in turn.
(e) Which of the observable frequencies in (c) and (d) would be different if the parameters $X$, $Y$, and $Z$ were all reversed in sign?
The inverse problem, fitting observed resonance inductions to zero-field parameters and $g$-factors, is more difficult. It is usually performed by

making initial estimates, calculating the frequencies for the known resonance magnetic inductions, and then correcting by trial and error or proper iteration until an overall fit is obtained.

**Solution**

(a) The quantity

$$S^2 = S_X^2 + S_Y^2 + S_Z^2 = S(S+1) = 2$$

is a constant for the triplet state. If a constant multiple of zero in the form

$$C(S_X^2 + S_Y^2 + S_Z^2 - 2)$$

is added to the first expression for the Hamiltonian, one obtains

$$h^{-1}\mathcal{H} = (C+E)S_X^2 + (C-E)S_Y^2 + (C+D)S_Z^2 - 2C\,.$$

The last term is the shift of energy origin which cannot affect any observations and, if this Hamiltonian is to match the second form with $X+Y+Z = 0$, one requires $3C+D = 0$ or $C = -\frac{1}{3}D$. Matching the coefficients now gives

$$X = \tfrac{1}{3}D - E\,, \qquad Y = \tfrac{1}{3}D + E\,, \qquad Z = -\tfrac{2}{3}D\,,$$

and an origin shifted downwards by $\frac{2}{3}D$.

(b) It is useful first to tabulate in matrix form three simple operators:

$$S_z = \begin{bmatrix} 1 & 0 & 0 \\ 0 & 0 & 0 \\ 0 & 0 & -1 \end{bmatrix}, \qquad (S_+ + S_-) = \begin{bmatrix} 0 & 2^{1/2} & 0 \\ 2^{1/2} & 0 & 2^{1/2} \\ 0 & 2^{1/2} & 0 \end{bmatrix},$$

$$(S_+ - S_-) = \begin{bmatrix} 0 & 2^{1/2} & 0 \\ -2^{1/2} & 0 & 2^{1/2} \\ 0 & -2^{1/2} & 0 \end{bmatrix}.$$

It is the squares of these matrices that are required to match the operators in the Hamiltonian, namely

$$S_X^2 = S_x^2 = \tfrac{1}{4}(S_+ + S_-)^2 = \begin{bmatrix} \frac{1}{2} & 0 & \frac{1}{2} \\ 0 & 1 & 0 \\ \frac{1}{2} & 0 & \frac{1}{2} \end{bmatrix},$$

$$S_Y^2 = S_y^2 = -\tfrac{1}{4}(S_+ - S_-)^2 = \begin{bmatrix} \frac{1}{2} & 0 & -\frac{1}{2} \\ 0 & 1 & 0 \\ -\frac{1}{2} & 0 & \frac{1}{2} \end{bmatrix},$$

$$S_Z^2 = S_z^2 = \begin{bmatrix} 1 & 0 & 0 \\ 0 & 0 & 0 \\ 0 & 0 & 1 \end{bmatrix}.$$

Hence the matrix of the Hamiltonian as $h^{-1}\mathcal{H}$ can be obtained by inserting the coefficients and adding to give

$$\begin{bmatrix} -\tfrac{1}{2}X-\tfrac{1}{2}Y-Z & 0 & -\tfrac{1}{2}X+\tfrac{1}{2}Y \\ 0 & -X-Y & 0 \\ -\tfrac{1}{2}X+\tfrac{1}{2}Y & 0 & -\tfrac{1}{2}X-\tfrac{1}{2}Y-Z \end{bmatrix}.$$

This can be simplified by using $X+Y+Z = 0$ to give

$$\begin{bmatrix} -\tfrac{1}{2}Z & 0 & -\tfrac{1}{2}X+\tfrac{1}{2}Y \\ 0 & Z & 0 \\ -\tfrac{1}{2}X+\tfrac{1}{2}Y & 0 & -\tfrac{1}{2}Z \end{bmatrix}.$$

By inspection the state with $M_S = 0$ is an eigenstate since all its off-diagonal elements are zero; its eigenenergy is clearly $hZ$ from the central element of the matrix.

The remaining diagonal elements are equal, so the two basis states must be equally admixed to give the eigenstates as

$$2^{-\frac{1}{2}}(|+1\rangle+|-1\rangle) \qquad \text{and} \qquad 2^{-\frac{1}{2}}(|+1\rangle-|-1\rangle).$$

By multiplication into the Hamiltonian the eigenvalues are seen to be $hY$ and $hX$ respectively. These last two values could have been predicted from the first result since there is nothing to distinguish the $Z$ direction especially in zero field: thus one might have chosen $z$ parallel to $X$.
(c) For naphthalene the $h^{-1}E$ values are +1413, +592, and −2005 MHz. The three transitions which could be observed are therefore at 3418, 2597, and 821 MHz in the high radio-frequency or low microwave region.
(d) If the magnetic induction is parallel to $X$ one must permute $X$, $Y$, and $Z$ from the last given matrix and add the coupling to the induction, whence $h^{-1}\mathcal{H}$ becomes

$$\begin{bmatrix} h^{-1}g\mu_B B-\tfrac{1}{2}X & 0 & -\tfrac{1}{2}Y+\tfrac{1}{2}Z \\ 0 & X & 0 \\ -\tfrac{1}{2}Y+\tfrac{1}{2}Z & 0 & -h^{-1}g\mu_B B-\tfrac{1}{2}X \end{bmatrix}.$$

Numerically at 0·3 T this is, in units of MHz,

$$\begin{bmatrix} 7703\cdot7 & 0 & -1298\cdot5 \\ 0 & 1413\cdot0 & 0 \\ -1298\cdot5 & 0 & -9116\cdot7 \end{bmatrix},$$

which has solutions 7803·4, 1413·0, and −9216·4 MHz.

The full set of solutions requested is given in table 4.2.2.

Check that in each case the sum of the three energies is zero on this energy origin. Note also that for a principal direction, as here, one energy is independent of $B$. The two low frequencies in each case are around the value expected for $g = 2\cdot00$, namely 8·4 and 28·0 GHz for the two values of $B$. The exact values vary appreciably with direction so that a very

broad line would result from a powdered sample. In contrast the highest frequency correlates approximately with a transition with $\Delta M_S = 2$, and is fairly constant in position giving a fairly narrow line in a powder. It appears at twice the frequency for $g = 2{\cdot}00$ lines and consequently at half the magnetic induction for a fixed frequency experiment.
(e) None. The absolute signs are obtained by cooling to a temperature of 4 K or lower, when some transitions weaken because the upper levels are depopulated. Which transitions weaken depends on the absolute signs of the zero-field parameters.

**Table 4.2.2.** Solutions to part (b). All values in $MH_z$.

| $B$ (T) | 0·3 | | 1·0 | |
|---|---|---|---|---|
| | $h^{-1}E$ | $\nu$ | $h^{-1}E$ | $\nu$ |
| $B \parallel X$ | 7803·4 | 17019·8 | 27357·6 | 56128·2 |
| | 1413·0 | 10629·4 | 1413·0 | 30183·6 |
| | −9216·4 | 6390·4 | −28770·6 | 25944·6 |
| $B \parallel Y$ | 8286·1 | 17164·2 | 27790·1 | 56172·1 |
| | 592·0 | 9470·1 | 592·0 | 28974·1 |
| | −8878·1 | 7694·1 | −28382·1 | 27198·1 |
| $B \parallel Z$ | 9422·7 | 16840·4 | 29039·5 | 56074·0 |
| | −2005·0 | 11427·7 | −2005·0 | 31044·5 |
| | −7417·7 | 5412·7 | −27034·5 | 25029·5 |

**4.2.3 *trans*-15,16-Dimethyldihydropyrene semiquinone**

Figure 4.2.3 shows the first derivative of the electron resonance absorption spectrum of *trans*-15,16-dimethyldihydropyrene semiquinone in alkaline solution. One resonance form of the structure is indicated in the inset.

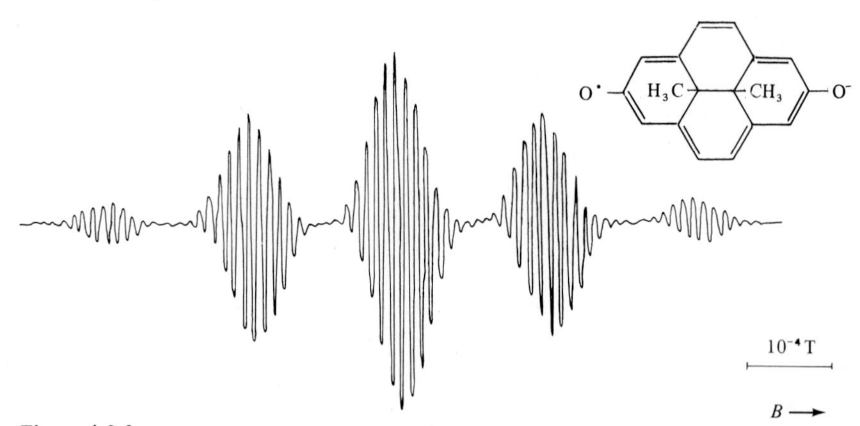


**Figure 4.2.3.**

(a) The centre of the pattern is at a magnetic induction of 331·24 mT with a spectrometer operating at exactly 9·3 GHz. What is the $g$-factor for the radical?
(b) The spectrum clearly separates into five equally spaced groups of narrowly spaced lines. Comment on this grouping, the relative intensities of the groups, and evaluate any coupling constants invoked given that the separation of the groups is 166$\mu$T.
(c) Close inspection of the figure shows that each of these groups consists of fifteen lines spaced by 11$\mu$T. Interpret these facts and deduce further hyperfine couplings and suggest their assignment.
(d) Calculate the relative intensities of all 75 lines in the spectrum and compare with the figure. Comment on possible reasons for any discrepancies.

**Solution**
(a) The regular formula applies here, namely

$$h\nu = g\mu_B B \ ,$$

and insertion of the values gives $g = 2{\cdot}0060$.

The value is slightly above the free spin value 2·0023 which is typical for aromatic radicals in solution. The presence of the quinone oxygen, which has a higher spin-orbit coupling than carbon, leads to an even higher value than in hydrocarbon anions which have a value typically nearer 2·004.
(b) Only hydrogen, $I = \frac{1}{2}$, amongst the nuclei present, has a nonzero spin, and the groups must be due to a large coupling to hydrogen. Since the groups are equally spaced the hydrogens are probably equivalent, and five groups implies four hydrogens. The groups should have intensities in proportion to the binomial coefficients $\binom{4}{n}$, that is, as $1:4:6:4:1$. This matches the figure. The intensities correspond to the ways in which a nuclear spin resolved angular momentum, $M_I$, can have the values −2, −1, 0, +1, +2, when compounded from four nuclei of spin $\frac{1}{2}$. The coupling will be 166$\mu$T, or 4·65 MHz if we take $h^{-1}g\mu_B = 28{\cdot}0$ GHz T$^{-1}$.

One cannot easily say whether this coupling belongs to the four hydrogens on the central rings, or to the four adjacent to the oxygen atoms. A Hückel type molecular-orbital treatment suggests that the greatest spin population is at the former position, and the 4·65 MHz coupling is assigned to these central hydrogens.
(c) Fifteen lines equally spaced might suggest that there are fourteen equivalent hydrogen atoms, but there are only ten further hydrogen atoms in the radical, the six methyl hydrogens and the four adjacent to the oxygens. If the coupling to the latter is double that to the former set, the five groups of seven lines intermesh and overlap with some exactness to give fifteen lines of roughly the observed intensities. The couplings are

then 11 $\mu$T to the six methyl hydrogens and 22 $\mu$T to the ring hydrogens (310 and 620 kHz respectively).

(d) Each set of seven lines due to the methyl hydrogens has intensities in the proportions $\binom{6}{n}$, namely $1:6:15:20:15:6:1$. They overlap as indicated in (c) to give the intensity pattern of the fifteen lines arising thus

$$\begin{array}{ccccccccccccccc}
1: & 6: & 15: & 20: & 15: & 6: & 1 & & & & & & & & \\
 & & 4: & 24: & 60: & 80: & 60: & 24: & 4 & & & & & & \\
 & & & & 6: & 36: & 90: & 120: & 90: & 36: & 6 & & & & \\
 & & & & & & 4: & 24: & 60: & 80: & 60: & 24: & 4 & & \\
 & & & & & & & & 1: & 6: & 15: & 20: & 15: & 6: & 1 \\
1: & 6: & 19: & 44: & 81: & 122: & 155: & 168: & 155: & 122: & 81: & 44: & 19: & 6: & 1\ .
\end{array}$$

The total intensity is 1024 ($= 2^{10}$), as it should be for ten nuclei of spin $\frac{1}{2}$ with the weakest line taken to be of unit intensity.

These lines do not overlap the next group of fifteen and the groups have the intensity pattern $1:4:6:4:1$ as discussed under (b).

The net intensity scheme is then

$1:6:19:44:81:122:155:168:155:122:81:44:19:6:1:4:24:76:176:$
$324:488:620:672:620:488:324:176:76:24:4:6:36:114:264:486:$
$732:930:1008:$etc.

The last line listed is the strongest at the centre of the symmetrical pattern, being the 38th line of the 75.

Inspection of the spectrum suggests that this pattern is not faithfully obeyed. The following comments illustrate the possible difficulties, rather than give an exact interpretation, which would require spectra at other frequencies and at other temperatures.

(i) Purely instrumental errors leading to noise and nonlinearity.

(ii) Differing line shapes. Strictly the intensity is the area of the absorption curve. This is proportional to the height of the derivative times the square of the line width. One obvious possibility for differing line widths arises if the smaller couplings are not precisely in the ratio $2:1$ so that the overlap is not perfect.

(iii) The natural abundance of $^{13}$C, with spin $\frac{1}{2}$, is $1\cdot1\%$, so that for a radical with eighteen carbon atoms 18% of the radicals have at least one $^{13}$C atom and hence a different spectrum. There are several sets of four equivalent sites for $^{13}$C, and this means that the strongest individual line will have intensity $4 \times 0\cdot011 \times (\frac{1}{2}) \times 1008 = 24$ on the intensity scale used. The factor of $\frac{1}{2}$ arises as there will be a new splitting due to the $^{13}$C spin, giving twice as many lines of half the intensity. The stronger lines will be towards the middle of the spectrum and none are clearly resolved, but they confuse the intensity pattern a little.

(iv) No account has been taken of the cation. This might associate preferentially with one oxygen; if it associated permanently with one oxygen it would make the two ends of the molecule nonequivalent, which is not in agreement with the observed spectrum. The line widths are very roughly 5$\mu$T or 140 kHz, and the reciprocal of the line width in radians per second is $10^{-6}$ s. If the cation residence were for about this length of time a complicated line-width problem arises.
(v) It may be that such a large ion is not tumbling rapidly enough to average all anisotropic components. This too gives a complicated line-width problem, as the anisotropy arises both in the coupling and in the $g$-factor, and a cross term arises between these two sources of broadening. This term spoils the symmetrical nature of the spectrum about its own centre. There is some evidence of this in the spectrum; for instance the 31st line, intensity 6, is much clearer than its partner, the 45th line, which is not really visible. These lines are the wing lines of the central group.
(vi) Even if there are line-width features masked by the instrumental line width, differences may nevertheless appear for strong microwave measuring fields, which cause power saturation, which is not necessarily a proportional weakening for all lines.

**4.2.4 Electron resonance of type 1b diamond**

Type 1b diamonds give fairly strong electron resonance spectra. Figure 4.2.4a shows schematically the strong lines observed with a spectrometer operating at 9·3 GHz. The magnetic induction, $B$, of the central line resonance is 331·74 mT and the relative line intensities are 1:3:4:3:1.

The diamond crystal belongs to the cubic system and is essentially a single molecule of tetrahedrally linked carbon atoms. Figure 4.2.4b shows a five atom section of the structure arranged in a cube, one atom being at the centre and the other four at nonadjacent corners and linked to the central atom with normal covalent bonds. Three important directions are indicated. The spectrum of figure 4.2.4a is observed when $\boldsymbol{B}$ is parallel to $\langle 111 \rangle$, that is, one of the body diagonals. (Remember $\langle 111 \rangle$ refers to any of the symmetry-related directions, that is, it includes $[1\bar{1}\bar{1}]$, where the square brackets imply the specific direction out of the set. The observed spectrum must of course be the same when $\boldsymbol{B}$ is parallel to any one of a set of symmetry-related directions.)

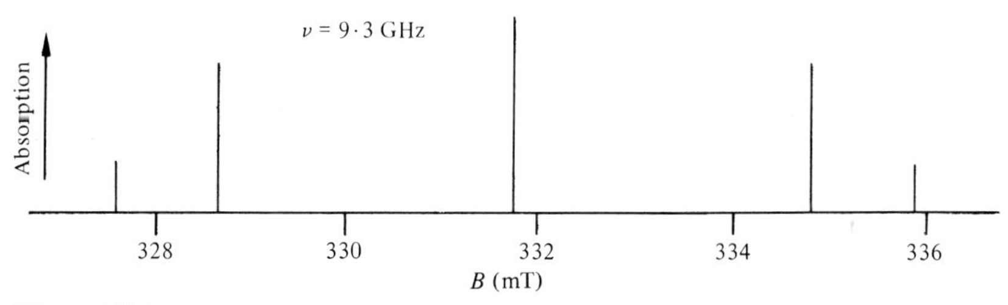


**Figure 4.2.4a.**

(a) Does the spectrum appear to belong to doublet free radicals? If so what is their mean $g$-factor?
(b) Can the entire spectrum belong to one hyperfine pattern? If so, how many nuclei are involved and with what spins?
(c) How may the spectrum be reasonably assigned to overlapping patterns? What spins and abundances would be required?
(d) Diamond contains mostly $^{12}C$ nuclei of spin 0 with 1·1% abundance of $^{13}C$ of spin $\frac{1}{2}$. The major impurity is nitrogen with its dominant isotope $^{14}N$ having spin 1. Does this additional information confirm any of the suggestions under (c) as being the most likely interpretation?
(e) If the direction of $\boldsymbol{B}$ with respect to the crystal is changed slightly, the lines of intensity 3 can be split into three lines of intensity 1. Comment on the reason for the special overlapping in the figure and the number of distinct chemical species implied.
(f) When $\boldsymbol{B}$ is parallel to ⟨100⟩ there are three equally spaced lines of intensity 4:4:4, and when $\boldsymbol{B}$ is parallel to ⟨1 1 0⟩ there are five lines of intensities 2:2:4:2:2 and the general spacing is qualitatively similar to that of the figure. Suggest the symmetry and orientation of the paramagnetic site. Suggest a likely electronic structure for this paramagnetic centre.
(g) For $\boldsymbol{B}$ parallel to ⟨111⟩ as in figure 4.2.4a, the wing line separations from the centre are ±4·17 and ±3·07 mT. For the single spectrum with $\boldsymbol{B}$ parallel to ⟨1 00⟩ the separations are ±3·38 mT, and for $\boldsymbol{B}$ parallel to ⟨1 1 0⟩ they are ±3·79 and ±2·90 mT. Calculate the parallel and perpendicular hyperfine coupling tensor elements, $a_{\parallel}$ and $a_{\perp}$, and check that all the numerical data are self-consistent.

*Hint.* For the accuracy quoted here one may assume the form of the separation to be suitable for high $B$ and isotropic $g$, namely

$$a_{\text{eff}} = (a_{\parallel}^2 \cos^2\theta + a_{\perp}^2 \sin^2\theta)^{1/2} ,$$

where $\theta$ is the angle between $\boldsymbol{B}$ and the axis of the cylindrically symmetrical tensor. $a_{\text{eff}}$ is the effective or observed hyperfine coupling for the given direction.

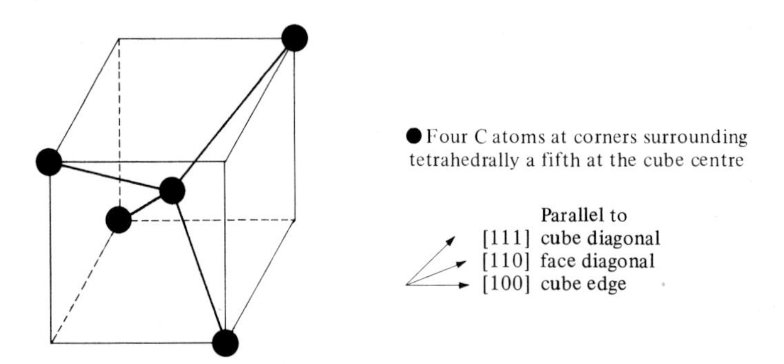


**Figure 4.2.4b.** The diamond lattice.

**Solution**

(a) Yes. The pattern seems simple and the $g$-factor, 2·0030, is very close to the free spin value 2·0023. Triplet states or other high spin species would be expected to cover a much wider range of magnetic induction.

(b) No. Any attempt to assign the outermost separation between the lines of intensity 1 and 3 to a hyperfine coupling leads to the prediction of further lines inside those of intensity 3 with the *same* separation from them as the outer lines.

(c) There are various possibilities including assigning all lines to distinct species. A simpler possibility is to assign the central line to a species with zero spin, and the outer lines as doublets associated with coupling to an isotopic version with spin $\frac{1}{2}$.

(d) The information does not agree with the last answer of (c). If the lines of spin $\frac{1}{2}$ are to be assigned to $^{13}C$, their *total* intensity relative to the central line should be only 0·011, which is far less than the observed value of 2.

If, however, nuclei with $I = 1$ are involved with their separate hyperfine patterns of three equally spaced, equally intense lines associated with $M_I = +1, 0, -1$, the spectrum is explicable in terms of two patterns of intensities 1 : 1 : 1 and 3 : 3 : 3, with the central line of the two patterns superposed. This suggests a nitrogen centre.

Since nitrogen atoms carry seven electrons, an odd number, the presence of an isolated nitrogen atom as an impurity must give a paramagnetic centre if it is electrically neutral.

(e) If the lines change position with the orientation of the crystal, the hyperfine coupling must be anisotropic as is commonly found in rigid crystals. There will be many equivalent individual impurity centres in the macroscopic crystal and these must be related to each other by the symmetry operations of the crystal class, including translation into other unit cells.

If translation only is involved, the signals from the different sites superpose since translation does not affect the coupling tensor. However, if additionally rotation or reflection symmetry elements are required to relate the sites, they may transform the orientation of the principal axes of the coupling tensor and lead to a different spectrum for the given direction of $\boldsymbol{B}$.

How many such different overlapping spectra occur depends on the symmetry of the impurity centre and the orientation of its tensor in the crystal. For the most general case in a cubic crystal there could be twenty-four patterns overlapping but there is no difficulty in envisaging special arrangements with a smaller number.

(f) The unique spectrum from all sites when $\boldsymbol{B}$ is parallel to $\langle 100 \rangle$ shows that the principal axes of all tensors must make equal angles with these directions. This is only true if the unique tensor direction is one of the set $\langle 111 \rangle$. There are four directions in this set, namely $[111]$, $[1\bar{1}\bar{1}]$,

$[\bar{1}1\bar{1}]$ and $[\bar{1}\bar{1}1]$, which are the four directions of the C—C bonds. There will therefore be four overlapping patterns if $\boldsymbol{B}$ is in a general direction. Such a unique direction is parallel to a three-fold axis of symmetry, which must also be a symmetry element of the coupling tensor. But since the tensor is only second order, a three-fold symmetry axis implies cylindrical symmetry to the tensor and two of its three principal values are equal.

Such impurity sites agree with the other observations. When $\boldsymbol{B}$ is parallel to $[1\,1\,1]$, it is parallel to the unique axis of one quarter of the impurity sites and makes the same angle with $[1\bar{1}\bar{1}]$, $[\bar{1}1\bar{1}]$, and $[\bar{1}\bar{1}1]$, the unique directions of the remaining sites, which will therefore have the same pattern for this special direction of $\boldsymbol{B}$. When $\boldsymbol{B}$ is parallel to $[1\,1\,0]$ it is perpendicular to $[1\bar{1}\bar{1}]$ and $[\bar{1}1\bar{1}]$ and makes equal angles with the other two unique axes, $[1\,1\,1]$ and $[\bar{1}\bar{1}1]$, so that two equal intensity patterns overlap.

There is not really room for an interstitial nitrogen in a diamond lattice and substitution for a carbon seems much more probable. There is one extra electron to be accommodated compared to the perfect lattice and this can only reasonably be placed in an antibonding orbital. This would tend to lengthen one N—C bond compared to the other three; which one would be lengthened would be a matter of chance, giving equal numbers of centres with the long bond, or unique tensor direction, parallel to each of the four C—C bond directions of the perfect diamond, as observed.

There is a more sophisticated description of the same idea which says that if the N atom occupied the exact position vacated by the C atom, the four N—C bond orbitals could be reduced under the local tetrahedral symmetry as $t_2 + a_1$. The energy order is expected to be $a_1 < t_2 < t_2^* < a_1^*$, where the asterisk denotes antibonding. The first eight valence electrons will occupy completely $a_1 + t_2$, and the ninth electron would occupy the triply degenerate $t_2^*$. The Jahn–Teller theorem therefore applies and the system will distort spontaneously so that one N—C bond becomes longer than the other three.

(g) Table 4.2.4 outlines the angles and calculations. In fact $a_\parallel$ and $a_\perp$ are the largest and smallest effective couplings observed, and

$$a_\parallel = 4{\cdot}17 \text{ mT or } 114 \text{ MHz},$$
$$a_\perp = 2{\cdot}90 \text{ mT or } 81 \text{ MHz}.$$

**Table 4.2.4.**

| $\boldsymbol{B}$ Parallel to | Intensity | $\cos^2\theta$ | $\theta$ | $a_{\text{eff}}$ (mT) | $a_{\text{eff}}$ (calculated) |
|---|---|---|---|---|---|
| ⟨111⟩ | 1 | 1 | 0 | 4·17 | $a_\parallel$ |
| | 3 | $\frac{1}{9}$ | 109·5 | 3·07 | $(\frac{1}{9}a_\parallel^2 + \frac{8}{9}a_\perp^2)^{1/2}$ |
| ⟨100⟩ | 4 | $\frac{1}{3}$ | 54·7 | 3·38 | $(\frac{1}{3}a_\parallel^2 + \frac{2}{3}a_\perp^2)^{1/2}$ |
| ⟨110⟩ | 2 | $\frac{2}{3}$ | 35·3 | 3·79 | $(\frac{2}{3}a_\parallel^2 + \frac{1}{3}a_\perp^2)^{1/2}$ |
| | 2 | 0 | 90 | 2·90 | $a_\perp$ |

More accurate values are obtainable from ENDOR measurements, which give 113·984 and 81·344 MHz respectively as well as the quadrupole coupling constant. This method shows that $a_{\parallel}$ and $a_{\perp}$ are the same sign, which cannot be demonstrated from the e.s.r. spectrum at 9·3 GHz. The absolute signs are taken to be positive to agree with the chemical structure proposed, but this is not confirmed experimentally.

### 4.2.5 Myoglobin

With metals having incompletely filled d shells, quite complex electron resonance behaviour can arise. One of the simpler cases is that of the iron resonance in myoglobin. Free $Fe^{3+}$ has five unpaired electrons, $S = \frac{5}{2}$, and a $^6S_{\frac{5}{2}}$ ground state. But in this crystal the ion is by no means free, being coordinated by four nitrogens of the porphyrin plane. It is also attached to the globin moiety and some other group, such as water in the acid metmyoglobin studied, which fills the sixth place so that the iron is octahedrally surrounded.

Because of the dominance of the porphyrin plane there is essential cylindrical symmetry and a large zero-field splitting. If the $z$ axis is taken to be perpendicular to this plane then, in the zero-field Hamiltonian,

$$-hXS_x^2\,, \quad -hYS_y^2\,, \quad -hZS_z^2\,,$$

$Z$ is about $-50$ GHz.

(a) Calculate the zero-field energy level pattern.

(b) What is the effect on the lowest levels of applying a magnetic induction, $\boldsymbol{B}$: (i) parallel to $z$; (ii) perpendicular to $z$; and (iii) at an angle $\theta$ to $z$? Show that the splitting is the same as that for an 'effective' Hamiltonian for a simple case with $S = \frac{1}{2}$, $g_{\parallel} = g_0$, and $g_{\perp} = 3g_0$, where $g_0$ is the free spin $g$-factor of the electron.

With this value of $Z$ and magnetic inductions, $B$, supplied by normal laboratory magnets, say about 0·3 T, the higher levels may be neglected in calculating the resonance.

(c) When the magnetic induction lies in the $ab$ crystallographic plane, $g$ varies from 6·00 to a minimum of 2·64 as the magnet is rotated. Calculate the angle between $z$ and the normal to this crystallographic plane.

**Solution**

(a) Given that there is cylindrical symmetry, $X = Y$, and taking the mean energy as zero, that is $X+Y+Z = 0$, then the Hamiltonian is

$$hZ(\tfrac{1}{2}S_x^2 + \tfrac{1}{2}S_y^2 - S_z^2) = \tfrac{1}{2}hZ(S^2 - 3S_z^2)\,.$$

This is diagonal in a representation with the eigenfunctions of $S_z$ as basis functions, with the energies given by

$$\begin{aligned} h^{-1}E = \tfrac{1}{2}Z[S(S+1) - 3S_z^2] &= -5Z && \text{if} \quad S_z = \pm\tfrac{5}{2}\,, \\ &= Z && \text{if} \quad S_z = \pm\tfrac{3}{2}\,, \\ &= 4Z && \text{if} \quad S_z = \pm\tfrac{1}{2}\,. \end{aligned}$$

In agreement with Kramers' theorem the electric fields of the ligands are unable to remove completely the degeneracy, and the six states form three degenerate pairs with the $S_z = \pm\frac{1}{2}$ lying the lowest. If $Z = -50$ GHz, this level lies 150 GHz below the $S_z = \pm\frac{3}{2}$ states.
(b) The additional Hamiltonian term is

$$g_0 \mu_B \boldsymbol{B} \cdot \boldsymbol{S}$$

where $g_0$ is the $g$-factor of the free electron spin and there is no orbital contribution for the $^6$S state.
(i) If $\boldsymbol{B}$ is parallel to the $z$ axis, then the Hamiltonian is

$$g_0 \mu_B B S_z$$

and is diagonal. The $\pm\frac{1}{2}$ states have additional energies of $\pm\frac{1}{2} g_0 \mu_B B$ and a transition frequency $h^{-1} g_0 \mu_B B$, all of which corresponds to a spin $\frac{1}{2}$ case with $g_\parallel = g_0$ $(= 2{\cdot}00)$.
(ii) If $\boldsymbol{B}$ is perpendicular to $z$, say parallel to $x$, then one requires the matrix elements of $S_x$ in a representation in which $S_z$ is diagonal and only the $\pm\frac{1}{2}$ states are retained, as the others have a much higher diagonal energy from the zero-field term. The relevant matrix elements are

$$\langle \tfrac{1}{2} | S_x | \tfrac{1}{2} \rangle = \langle -\tfrac{1}{2} | S_x | -\tfrac{1}{2} \rangle = 0 \, ,$$
$$\langle \tfrac{1}{2} | S_x | -\tfrac{1}{2} \rangle = \langle -\tfrac{1}{2} | S_x | \tfrac{1}{2} \rangle = \tfrac{1}{2}[S(S+1) - (\tfrac{1}{2})(-\tfrac{1}{2})]^{1/2} = \tfrac{3}{2} \, .$$

The Hamiltonian in matrix form is then

$$\tfrac{1}{2} g_0 \mu_B \begin{bmatrix} 0 & 3 \\ 3 & 0 \end{bmatrix} ,$$

which has eigenvectors that are equal admixtures of the $\pm\frac{1}{2}$ states, and eigenenergies $\pm\frac{3}{2} g_0 \mu_B B$. The transition frequency would be $h^{-1} 3 g_0 \mu_B B$. Just the same consequence would apply for a case with spin $\frac{1}{2}$ with $g_\perp = 3 g_0$ $(= 6{\cdot}00)$.
(iii) In the general case one can resolve the inductions parallel and perpendicular to $z$ and add the matrix contributions to give the Hamiltonian

$$\tfrac{1}{2} g_0 \mu_B \begin{bmatrix} \cos\theta & 3\sin\theta \\ 3\sin\theta & -\cos\theta \end{bmatrix} ,$$

with its eigenenergies $\pm\frac{1}{2} g_0 \mu_B B (\cos^2\theta + 9\sin^2\theta)^{1/2}$.

For $S = \frac{1}{2}$, rather than $\frac{5}{2}$, and $g_\parallel = g_0$, $g_\perp = 3 g_0$, one would write the effective Hamiltonian

$$g_0 \mu_B B (S_z \cos\theta + 3 S_x \sin\theta) \, .$$

But whereas $S_z$ is still represented by the matrix

$$\begin{bmatrix} \frac{1}{2} & 0 \\ 0 & -\frac{1}{2} \end{bmatrix} ,$$

for $S = \frac{1}{2}$, $\frac{1}{2}[S(S+1)-(\frac{1}{2})(-\frac{1}{2})]^{1/2} = \frac{1}{2}$, and now $S_x$ is represented by the matrix

$$\begin{bmatrix} 0 & \frac{1}{2} \\ \frac{1}{2} & 0 \end{bmatrix} .$$

The total matrix for $B$ at an angle $\theta$ to $z$ is then

$$\tfrac{1}{2}g_0\mu_B B\begin{bmatrix} \cos\theta & 3\sin\theta \\ 3\sin\theta & -\cos\theta \end{bmatrix} ,$$

which is precisely that obtained above. This shows that the consequences of using the 'effective' Hamiltonian are the same as those obtained by the fuller treatment. In more complex cases the effective form is simpler when an unknown spectrum is being analysed.
(c) When $g = 2\cdot64 = 1\cdot32g_0$, then $\cos^2\theta + 9\sin^2\theta = 1\cdot32^2$, and

$$\theta = \arcsin\{\tfrac{1}{8}[(1\cdot32)^2 - 1]\}^{1/2} = 17\cdot7^\circ .$$

Consequently the $z$ axis lies $17\cdot7^\circ$ from the closest direction in the *ab* plane and hence $72\cdot3^\circ$ from the perpendicular to *ab*.

Historically the determination of the porphyrin plane direction by such means preceded the successful complete X-ray crystallographic analysis of such crystals.

**General references**

Useful introduction: Carrington, A., McLachlan, A. D., 1967, *Introduction to Magnetic Resonance* (Harper and Row, New York).
Most appropriate textbook: Wertz, J. E., Bolton, J. P., 1972, *Electron Spin Resonance* (McGraw-Hill, New York).

**References for problems**

**4.2.1**

Ovenall, D. W., Whiffen, D. H., 1961, *Mol. Phys.,* **4**, 135.

**4.2.2**

Hutchison, C. A., Mangum, B. W., 1961, *J. Chem. Phys.,* **34**, 908.

**4.2.4**

Cook, R. J., Whiffen, D. H., 1966, *Proc. R. Soc. London Ser. A,* **295**, 99.

**4.2.5**

Bennett, J. E., Gibson, J. F., Ingram, D. J. E., 1957, *Proc. R. Soc. London Ser. A,* **240**, 67.

# 4.3 Magnetic moments

**R W Jotham** University of Nottingham

### 4.3.0 Some useful data

As the problems in this section all use CGS units, the following values will be needed: Boltzmann's constant, $k$, $1\cdot3807 \times 10^{-16}$ erg K$^{-1}$ mol$^{-1}$; Bohr magneton, $\mu_B$, $9\cdot274 \times 10^{-21}$ erg G$^{-1}$ = $4\cdot668 \times 10^{-5}$ cm$^{-1}$ G$^{-1}$. The following quantities also appear frequently:

$$N_A\mu_B^2 = 0\cdot2607 \text{ cm}^{-1} \text{ erg G}^{-2} \text{ mol}^{-1} ;$$

$$\frac{3k}{N_A\mu_B^2} = 7\cdot997 \text{ mol G}^2 \text{ erg}^{-1} \text{ K}^{-1} .$$

The CGS unit of molar susceptibility is erg G$^{-2}$ mol$^{-1}$; the SI unit of molar susceptibility is m$^3$ mol$^{-1}$. To convert erg G$^{-2}$ mol$^{-1}$ to m$^3$ mol$^{-1}$, multiply by $4\pi \times 10^{-6}$ (1 erg G$^{-1}$ = $10^{-3}$ A m$^2$). See also Appendix 1.

The symbol $\chi'_M$ is frequently used in these problems to signify a *molar* magnetic susceptibility which has been already *corrected for the underlying diamagnetic contribution.* $\chi'_M$ normally includes the temperature-independent paramagnetic (TIP) distribution, $N_A\alpha$. Magnetic moments, $\mu$, are normally calculated from the molar susceptibility by use of the Langevin equation given below (note the relationship to the Curie Law), but occasionally, as will be explained in the problem concerned, it is advantageous to subtract the TIP contribution before calculating $\mu$.

The Langevin equation is

$$\chi'_M = \frac{N_A\mu_B^2\mu^2}{3kT} , \qquad (4.3.0a)$$

*when $\mu$ is expressed here, and throughout this section, in Bohr magnetons.* From this we obtain

$$\mu = (7\cdot997\chi'_M T)^{1/2} , \qquad (4.3.0b)$$

with all other quantitites expressed in the standard c.g.s units given above.

### 4.3.1

The magnetic moment of a sample of tris(2,2,6,6-tetramethyl-3,5-hepta-dionato)europium dimer, $Eu_2(thd)_6$ (m.w.1402), was determined by the Gouy method; the experimental results and other relevant data are listed in table 4.3.1.

(a) Calculate the magnetic moment of the dimer and the magnetic moment per Eu(III) ion.

(b) Calculate also the internal shift of tetramethylsilane (tms) protons in p.p.m. relative to an external standard of tms in a $CCl_4$ solution containing $0\cdot02$ M $Eu_2(thd)_6$, given that the mass susceptibility of $CCl_4$ is $-0\cdot433 \times 10^{-6}$ erg G$^{-2}$ g$^{-1}$.

**Table 4.3.1.** Experimental data (author's unpublished results) for Gouy determination of the magnetic moment of the $Eu_2(thd)_6$ dimer. All balance readings taken at ambient temperature of 297·3 K.

| Sample | Balance reading (g) at applied magnetic field | | |
|---|---|---|---|
| | 0 G | 5000 G | 10000 G |
| Empty tube | 8·74963 | 8·74871 | 8·74595 |
| Tube + water | 8·94743 | - | - |
| Tube + $Ni(en)_3(S_2O_3)$ | 8·87011 | 8·87264 | - |
| Tube + $Eu_2(thd)_6$ | 8·91095 | 8·91280 | 8·91836 |
| **Pascal's constants**[a] ($10^{-6}$ erg $G^{-2}$ $mol^{-1}$) | | | |
| $Eu^{3+}$ | −35 | H | −2·93 |
| acac ($C_5H_7O_2^-$) | −52 | C | −6·0 |

[a] Source: Earnshaw (1968).

**Solution**

(a) In a field of 5000 G the tube itself is repelled by a force equivalent to 0·92 mg as a consequence of its diamagnetism. Thus the attractive forces experienced by the tube when it is packed with 0·1205 g of the calibrant, nickel(ethylenediamine) thiosulphate $[Ni(en)_3(S_2O_3)]$, and 0·1613 g of the europium dimer are equivalent to $2 \cdot 53 + 0 \cdot 92 = 3 \cdot 34$ mg, and $1 \cdot 85 + 0 \cdot 92 = 2 \cdot 77$ mg, respectively.

The volume of the sample tube is determined from its weight when filled with water and, from this, the paramagnetic susceptibility of the air displaced by the sample is $0 \cdot 029 \times 0 \cdot 198 \times 10^{-6} = 0 \cdot 006 \times 10^{-6}$ erg $G^{-2}$ per gram (see Earnshaw, 1968). Now, the mass susceptibility of a specimen in this tube is given by

$$\chi = \frac{(0 \cdot 006 \times 10^{-6}) + w\beta'}{m} \text{ erg } G^{-2} \text{ g}^{-1},$$

where $m$ is the mass of the sample, $w$ is the increase in weight when the magnet is switched on (after correction for the diamagnetism of the tube itself) and $\beta'$ is a constant characteristic of the tube, which may be calculated approximately from the dimensions of the sample in the magnetic field and from the applied magnetic field.

The value of $\beta'$ is best determined by calibration with a standard such as $Ni(en)_3(S_2O_3)$, which has a susceptibility of $10 \cdot 87 \times 10^{-6}$ erg $G^{-2}$ $g^{-1}$ at 297 K (Earnshaw, 1968). Substitution of the experimental data in the above formula immediately yields $\beta' = 0 \cdot 378 \times 10^{-6}$ erg $G^{-2}$ $mg^{-1}$.

We now proceed to calculate the unknown susceptibility of $Eu_2(thd)_6$:

$$10^6\chi = \frac{0 \cdot 006 + (0 \cdot 378 \times 2 \cdot 77)}{0 \cdot 1613}$$

$$= 6 \cdot 53 \text{ erg } G^{-2} \text{ g}^{-1}.$$

The molecular weight of $Eu_2(thd)_6$ is 1402, whence the molar susceptibility, $\chi_M$, is $6 \cdot 53 \times 10^{-6} \times 1402 = 9155 \times 10^{-6}$ erg $G^{-2}$ $mol^{-1}$, or $4578 \times 10^{-6}$ erg $G^{-2}$ per mole of Eu atoms.

However, this molar susceptibility includes a diamagnetic contribution from the paired electrons. Ideally, this contribution is best determined by measurement on a closely related diamagnetic complex such as $La_2(thd)_6$. Alternatively, the diamagnetic contribution may be estimated by calculation from Pascal's constants (Earnshaw, 1968). In this case, it seems most appropriate to use the tabulated constants for the europium ion, the acetylacetonate ligand, and the additional methylene groups to convert acac to thd:

| | |
|---|---|
| $Eu^{3+}$ | $-35 \times 10^{-6}$ |
| $3 \times$ (acac) | $-156 \times 10^{-6}$ |
| $18 \times (CH_2)$ | $-214 \times 10^{-6}$ |
| total | $-405 \times 10^{-6}$ erg $G^{-2}$ $mol^{-1}$. |

Thus, the paramagnetic contribution to the total molar susceptibility, $\chi'_M$, is $(4578+405) \times 10^{-6} = 4983 \times 10^{-6}$ erg $G^{-2}$ per mole of Eu atoms.

The magnetic moments may now be calculated from the Langevin equation. Substitutition of the above value for $\chi'_M$ in equation (4.3.0b) yields the magnetic moment per Eu atom,

$$\mu_{Eu} = (7 \cdot 997 \times 4983 \times 10^{-6} \times 297 \cdot 3)^{1/2}$$
$$= 3 \cdot 44 \mu_B \; ;$$

thus the magnetic moment of the dimer is $3 \cdot 44 \times 2^{1/2} = 4 \cdot 86 \mu_B$.

We may check that the magnetic susceptibility is independent of the magnetic field by comparing the data at the two different field values. If the magnetic field at the edges of the sample is nearly zero, then the calibration constant, $\beta'$, depends inversely on the square of the magnetic field—i.e., in our case, $\beta' = 0 \cdot 0945$ at 10000 G. It is readily seen that the balance readings alter four times as much in the 10000 G field as they do in the 5000 G field and so, indeed, the susceptibility of $Eu_2(thd)_6$ is independent of the magnetic field.

(b) The determination of magnetic susceptibility by the n.m.r. frequency difference between lines of the same proton in external and internal references was first advocated by Evans (1959). He gave the following expression for the mass susceptibility of the dissolved substance, $\chi$:

$$\chi = \frac{3\Delta f}{2\pi f c_m} + \chi_0 + \frac{\chi_0(d_0 - d_s)}{c_m} ,$$

where $\Delta f$ is the frequency difference between the two n.m.r. lines; $f$ is the frequency at which the protons are being studied; $c_m$ is the concentration of the solution expressed as mass of substance per unit volume of solution; $\chi_0$ is the mass susceptibility of the solvent; $d_0$ and $d_s$ are the densities of the solvent and of the solution, respectively. The third term may be safely neglected for dilute solutions.

For a 0·02 M solution of $Eu_2(thd)_6$,

$$c_m = \frac{0{\cdot}02 \times 1402}{1000} = 0{\cdot}0280 \text{ g cm}^{-3} .$$

The measured value of $\chi$ was $6{\cdot}53 \times 10^{-6}$ erg $G^{-2}$ $g^{-1}$,

$$6{\cdot}53 \times 10^{-6} = \frac{3\Delta f}{2\pi f \times 0{\cdot}0280} - 0{\cdot}43 \times 10^{-6} ,$$

$$\Delta f/f = 6{\cdot}96 \times 10^{-6} \times 0{\cdot}0280 \times \tfrac{2}{3}\pi$$

$$= 0{\cdot}408 \times 10^{-6} = 0{\cdot}408 \text{ p.p.m.},$$

i.e. the frequency shift is 24·5 Hz on a 60 MHz instrument, which is easily measurable. Although this calculation is indicative of the value of the Evans method, it should be pointed out that in this particular case it might be necessary to add a base, which would facilitate the solution of the rather insoluble $Eu_2(thd)_6$, but the measured susceptibility would now be that of the (base) . $Eu(thd)_3$ complex.

**4.3.2**

The data in table 4.3.2 are measured susceptibility values (corrected for diamagnetism) for two complexes of Mn(II) and Ni(II) which are expected to be six-coordinate, with symmetry close to octahedral. Use the data to determine graphically the constants $C$ and $N_A \alpha$ in the equation

$$\chi'_M = \frac{C}{T} + N_A \alpha . \qquad (4.3.2a)$$

Interpret the values you obtain for the constants on the assumption that the values of the free-ion spin–orbit coupling constant, $\zeta$, for $Mn^{2+}$ and $Ni^{2+}$ are 300 and 630 $cm^{-1}$, respectively, and that the value of the ligand-field splitting parameter $10Dq$, obtained from a spectroscopic analysis of the Ni(II) complex, is 9000 $cm^{-1}$.

**Table 4.3.2.**

| Mn(II) complex | | Ni(II) complex | |
|---|---|---|---|
| $T$ (K) | $\chi'_M$ ($10^{-6}$ erg $G^{-2}$ $mol^{-1}$) | $T$ (K) | $\chi'_M$ ($10^{-6}$ erg $G^{-2}$ $mol^{-1}$) |
| 300·3 | 14620 | 297·5 | 4634 |
| 273·1 | 16170 | 273·1 | 5044 |
| 242·5 | 18145 | 258·6 | 5314 |
| 211·6 | 20613 | 231·7 | 5812 |
| 181·8 | 24303 | 202·6 | 6657 |
| 140·3 | 31401 | 183·4 | 7416 |
| 121·1 | 36294 | 151·7 | 8853 |
| 90·3 | 48427 | 130·5 | 10381 |
| 77·1 | 57050 | 94·0 | 14047 |
| | | 77·1 | 17154 |

**Solution**

The form of equation (4.3.2a) suggests that the constants $C$ and $N_A\alpha$ may be determined graphically from plots of $\chi'_M$ against $1/T$, as depicted in figure 4.3.2a. However, it is rarely possible to make susceptibility measurements at a high enough temperature to determine the small value of the constant $N_A\alpha$ in this way. The values of $C$ for the Mn(II) and Ni(II) complexes are found to be 4·40 and $1\cdot30_5$ erg $G^{-2}$ $mol^{-1}$ K, respectively, from these plots.

Nevertheless, the presence of a nonzero $N_A\alpha$ term may be detected by plotting $1/\chi'_M$, against $T$, since this is only linear when $N_A\alpha = 0$, as in the case of the Mn(II) complex (see figure 4.3.2b). The nonlinearity observed for the Ni(II) complex is a clear indication of a significant temperature-independent paramagnetism. The dashed line in figure 4.3.2b corresponds to the equation $1/\chi'_M = T/1\cdot305$ (usual units), which has been formulated by use

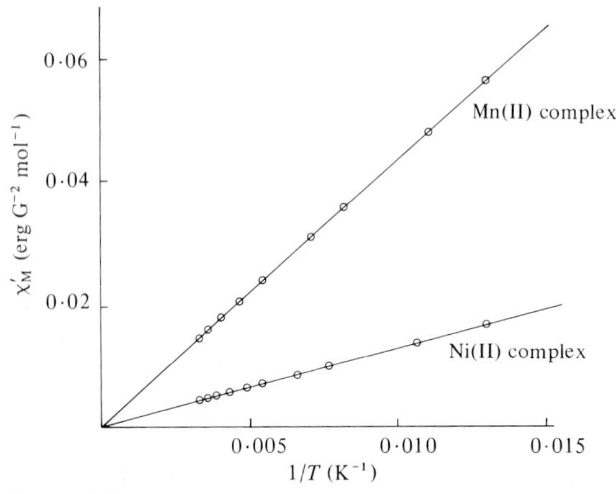


**Figure 4.3.2a.**

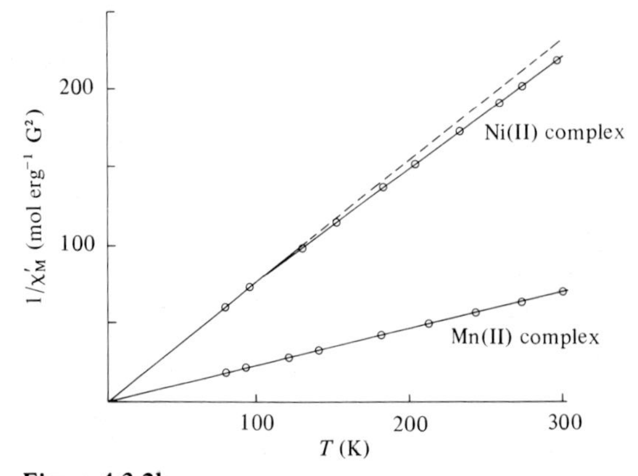


**Figure 4.3.2b.**

of the value of $C$ found from figure 4.3.2a; the values of $\chi'_M(T)$ that are calculated from this equation indicate that $N_A\alpha$ has a value of ~225 ± 25 erg $G^{-2}$ $mol^{-1}$ for the Ni(II) complex. More exact values of $C$ and $N_A\alpha$ may be found by standard numerical techniques—a procedure that is now popularly carried out by computer.

Any temperature-independent paramagnetism will normally be incorporated into an experimental determination of a magnetic moment. However, this procedure of itself leads to a somewhat temperature-dependent moment. If the susceptibility is corrected for the temperature-independent term before the value of $\mu$ is calculated, this value may be simply interpreted in terms of the two equations:

$$\mu^2 = \frac{3kT\chi'_M}{N\mu_B^2} = \frac{3kC}{N\mu_B^2}$$

$$= \frac{C}{7\cdot997} \text{ [from equation (4.3.0b)] ;} \tag{4.3.2b}$$

$$\mu = \mu_s\left(\frac{1-\alpha\lambda}{10Dq}\right) = 2[S(S+1)]^{1/2}\left(\frac{1-\alpha\lambda}{10Dq}\right), \tag{4.3.2c}$$

where $\mu_s$ is the spin-only contribution to $\mu$, $\lambda$ is the spin–orbit coupling constant, and $\alpha$ is a numerical constant which depends upon the symmetry species of the ground state.

This simple analysis only applies when the complex has a nondegenerate or doubly degenerate ground state. Cubic complexes with triply degenerate ground states are dealt with in some of the later problems in this section. However, the ground state of an octahedral Ni(II) complex is invariably $^3A_{2g}$, whereas that of a Mn(II) complex is $^6A_{1g}$ in a weak ligand field and $^2T_{2g}$ in the strong-field case (which clearly does not apply here since the Mn(II) complex is evidently strongly paramagnetic).

The values of $\mu$ calculated from equation (4.3.2b) are 5·93 $\mu_B$ and 3·23 $\mu_B$ for the Mn(II) and Ni(II) complexes, respectively; from equation (4.3.2c), the spin-only values for spin-triplets and sextets are 2·83 $\mu_B$ and 5·92 $\mu_B$, respectively. The value of $\mu$ for the Mn(II) complex is thus effectively equal to the spin-only value, whereas that of the Ni(II) complex lies between the spin-only value and the value of $[L(L+1)+4S(S+1)]^{1/2}$ [i.e. 4·47 $\mu_B$, since the $^3A_{2g}$ ground state arises from a $^3F$ free-ion term ($L = 3$)].

Only triply degenerate states of an octahedral complex may make an orbital contribution to the magnetic moment, since the angular momentum operators $L_x$, $L_y$, $L_z$ transform as rotations about molecular axes, i.e. as $T_{1g}$, which is only contained in the symmetric direct products of $T_{1g}$, $T_{2g}$, $T_{1u}$, and $T_{2u}$ with themselves. However, excited states of suitable symmetry may be mixed into the ground state by spin–orbit coupling. There is no suitable excited state which may be mixed in with the $^6A_{1g}$ ground state of a Mn(II) complex in a weak field and, consequently, the observed moment agrees with that predicted by the spin-only formula. The zero value of $\alpha$ is also pertinent to the temperature-independent paramagnetism, since this is given by the formula $2\alpha N_A \mu_B^2/10Dq$ for complexes with A and E ground states, i.e. a weak-field complex of Mn(II) should indeed have no temperature-independent contribution to the magnetic susceptibility. In contrast, the $^3A_{2g}$ ground state of a Ni(II) complex may be mixed with the nearby $^3T_{2g}$ state by spin–orbit coupling; this leads to a value $\alpha = 4$ in the above equations (Mabbs and Machin, 1973). The spin–orbit coupling constant, $\lambda$, is given by the value of $-\zeta/2S = -315$ cm$^{-1}$ for a Ni(II) complex. Substituting the values of $\alpha$, $\lambda$, and $10Dq$ for Ni(II) in the above equations, we obtain $\mu = 3{\cdot}22\ \mu_B$ and $N_A\alpha = 231$ erg G$^{-2}$ mol$^{-1}$, in excellent agreement with the experimental data.

**4.3.3**

The ground states of Cr(II) and Cu(II) complexes with a weak octahedral ligand field are $^5E_g$ and $^2E_g$, respectively. If the possibility of a Jahn–Teller distortion is neglected, account for the fact that the room-temperature magnetic moments of the majority of complexes of these types cluster around the regions $4{\cdot}85\ \mu_B$–$4{\cdot}90\ \mu_B$ and $1{\cdot}90\ \mu_B$–$1{\cdot}95\ \mu_B$, respectively. (The values of the single-electron spin–orbit coupling constant, $\zeta$, for the $Cr^{2+}$ and $Cu^{2+}$ ions are 230 and 830 cm$^{-1}$, respectively. The values of $10Dq$ for typical weak-field octahedral complexes are of the order of 14000 cm$^{-1}$ in both cases.)

**Solution**

The spin-only values of the magnetic moment for quintet and doublet states are found from the formula $\mu_s = [4S(S+1)]^{1/2}\mu_B$ to be $4{\cdot}90\ \mu_B$ and $1{\cdot}73\ \mu_B$, respectively. It is therefore necessary to account only for small observed departures from these values which are, nevertheless, opposite in sense and strikingly greater in magnitude for a Cu(II) complex than for a Cr(II) complex. The origin of these deviations from the spin-only values of $\mu$ is the incomplete quenching of the orbital contribution to the magnetic moment. Excited states are mixed in with the $^2E_g$ and $^5E_g$ ground states; this gives the following expressions for the magnetic susceptibility and moment (Mabbs and Machin, 1973; Earnshaw, 1968):

$$\chi'_M = \frac{1}{7{\cdot}997T}\left[\mu_s\left(\frac{1-2\lambda}{10Dq}\right)\right]^2 + \frac{4N_A\mu_B^2}{10Dq}\,, \qquad (4.3.3a)$$

$$\mu = (7{\cdot}997\chi'_M T)^{1/2} = \left\{\left[\mu_s\left(\frac{1-2\lambda}{10Dq}\right)\right]^2 + \frac{4N_A\mu_B^2(7{\cdot}997T)}{10Dq}\right\}^{1/2}, \qquad (4.3.3b)$$

where $\lambda$ is the spin–orbit coupling constant, and the second term in each expression arises from the temperature-independent paramagnetism. From the derivation of equation (4.3.0a), it can be seen that the constant $7{\cdot}997$ is the value of $3k/N_A\mu_B^2$ in the CGS system, so that we can simplify equation (4.3.3b) to:

$$\mu = \left\{\left[\mu_s\left(\frac{1-2\lambda}{10Dq}\right)\right]^2 + \frac{12kT}{10Dq}\right\}^{1/2} . \tag{4.3.3c}$$

In the cases of Cr(II) and Cu(II) complexes at room temperature, the value of the last term in braces in equation (4.3.3c) will be of the order of

$$\frac{12 \times 0\cdot 695 \times 300}{14000} = 0\cdot 18\mu_B^2 ;$$

this is equivalent to a contribution of <5% of the typical experimental value of $\mu$ in either case, but large enough to be noticeable.

Clearly, however, the difference between the two cases must hinge upon the value of $\lambda$, since this is given by the relation:

$$\lambda = \pm\frac{\zeta}{2S} ,$$

where $\lambda$ is positive for shells that are less than half full and negative for shells which are more than half full—as reflected, for example, in the order of the $J$ levels that arise from a Russell–Saunders term (Golding, 1969). Thus we obtain for $Cr^{2+}$,

$$\lambda = 230/4 = 58 \text{ cm}^{-1} ,$$

and for $Cu^{2+}$,

$$\lambda = -830/1 = -830 \text{ cm}^{-1} .$$

The differences in sign and magnitude are immediately apparent, and we may now attempt to calculate idealised values of $\mu$ from these values of $\lambda$: for Cr(II),

$$\begin{aligned}\mu &= \{[4\cdot 90(1-116/14000)]^2 + 0\cdot 18\}^{1/2} \\ &= [(4\cdot 86)^2 + 0\cdot 18]^{1/2} \\ &= (23\cdot 62 + 0\cdot 18)^{1/2} \\ &= 4\cdot 88\mu_B ;\end{aligned}$$

for Cu(II),

$$\begin{aligned}\mu &= \{[1\cdot 73(1+1660/14000)]^2 + 0\cdot 18\}^{1/2} \\ &= [(1\cdot 94)^2 + 0\cdot 18]^{1/2} \\ &= (3\cdot 76 + 0\cdot 18)^{1/2} \\ &= 1\cdot 98_5\mu_B .\end{aligned}$$

These idealised values agree well with the experimental data, rather better in the case of Cr(II) than that of Cu(II). We can, however, go a little further than this by making some allowance for covalency in the complexes.

This is now commonly done by the introduction of another parameter, the orbital reduction factor $\kappa$, into the magnetic field[(1)] and spin–orbit coupling operators; that is, the operators $\mu_B(\boldsymbol{L}+2\boldsymbol{S})\cdot\boldsymbol{B}$ and $\lambda\boldsymbol{L}\cdot\boldsymbol{S}$ are replaced by $\mu_B(\kappa\boldsymbol{L}+2\boldsymbol{S})\cdot\boldsymbol{B}$ and $\lambda\kappa\boldsymbol{L}\cdot\boldsymbol{S}$, respectively (Gerloch and Miller, 1968). Introduction of $\kappa$ leads to a revised formula for the magnetic moment, namely

$$\mu = \left\{\left[\mu_s\left(1-\frac{2\kappa\lambda}{10Dq}\right)\right]^2+\frac{12\kappa^2 kT}{10Dq}\right\}^{1/2} .$$

The introduction of such a parameter facilitates improved agreement between experiment and theory. A typical value of $\kappa$ is $0\cdot8$ and, if this value is used to recalculate idealised moments for the Cr(II) and Cu(II) complexes, we obtain, for Cr(II),

$$\begin{aligned}\mu &= \{[4\cdot90(1-93/14000)]^2+0\cdot12\}^{1/2}\\ &= [(4\cdot87)^2+0\cdot12]^{1/2}\\ &= 4\cdot88\mu_B ,\end{aligned}$$

and for Cu(II),

$$\begin{aligned}\mu &= \{[1\cdot73(1+1330/14000)]^2+0\cdot12\}^{1/2}\\ &= [(1\cdot89)^2+0\cdot12]^{1/2}\\ &= 1\cdot92\mu_B .\end{aligned}$$

The agreement between experimental and theoretical moments is now excellent. It is immediately apparent that, because the terms within the square root cooperate in the case of a more than half-filled shell, $\kappa$ is a much more significant parameter for Cu(II) complexes than it is for Cr(II) complexes or, for example, Ti(III) complexes.

### 4.3.4

The $^2$D term of a $d^1$ ion is split into a lower $^2T_{2g}$ and an upper $^2E_g$ state in an octahedral ligand field. Calculate formulae for the molar susceptibility and the magnetic moment of such a compound in terms of the spin–orbit coupling parameter, $\lambda$. What differences will arise in the similar case of a $d^9$ ion in a tetrahedral field?

#### Solution

The procedure is to consider first the influence of spin–orbit coupling and then the effect of a magnetic field as successive perturbations on the energies of the six $^2T_{2g}$ functions, since these effects give rise to energy changes of the order of 100 cm$^{-1}$ and 1 cm$^{-1}$, respectively. The spin–orbit coupling operator, $\lambda\boldsymbol{L}\cdot\boldsymbol{S}$, splits $^2T_{2g}$ into a doubly degenerate level of symmetry $E_{5/2g}$ (double-group representation) and a quadruply degenerate

[(1)] See Editors' note on treatment of magnetic fields in Appendix I.

level of symmetry $U_{3/2g}$ with energies $\lambda$ and $-\frac{1}{2}\lambda$, respectively. This result is obtained by use of the operator $\lambda \boldsymbol{L} \cdot \boldsymbol{S}$ and the $^2T_{2g}$ functions expressed in terms of orbital and spin angular momenta. The procedure is very simple but lengthy to describe (see Golding, 1969; Mabbs and Machin, 1973) and may be applied to any system with a triply degenerate ground state (i.e. a $T_1$ or $T_2$ ground state in a cubic field). The magnetic field operator, $\mu_B(\boldsymbol{L}+2\boldsymbol{S})\cdot\boldsymbol{B}$, lifts the degeneracy of the two $E_{5/2g}$ states by a splitting of $2\mu_B B$. Interestingly, none of the $U_{3/2g}$ states are modified in this way, although two of them interact with the $E_{5/2g}$ states in the presence of the magnetic field. The magnitude of this interaction is $2^{1/2}\mu_B B$ and, since the splitting of the $E_{5/2g}$ and $U_{3/2g}$ levels is $\frac{3}{2}\lambda$, the second-order energy changes due to the magnetic field are given by $\pm(2^{1/2}\mu_B B)^2/(\frac{3}{2}\lambda) = \pm\frac{4}{3}\mu_B^2 B^2/\lambda$. These energy changes are summarised in figure 4.3.4a.

The expression for the molar magnetic susceptibility is obtained by considering the thermal population of the energy levels in the magnetic field. The energy changes involving the first power in the magnetic field give rise to a temperature-dependent contribution to the susceptibility, whereas the second-power terms give rise to a temperature-independent contribution. The standard formula into which the energy levels in figure 4.3.4a may be substituted was given by Van Vleck (1932):

$$\chi'_M = \frac{N_A \sum_i [E_i^2(1)/kT - 2E_i(2)] \exp[-E_i(0)/kT]}{\sum_i \exp[-E_i(0)/kT]},$$

where the $E_i(0)$ are the energies in the absence of the magnetic field, and the $E_i(1)$ and $E_i(2)$ are the multipliers of the energy changes in $B$ and $B^2$, respectively.

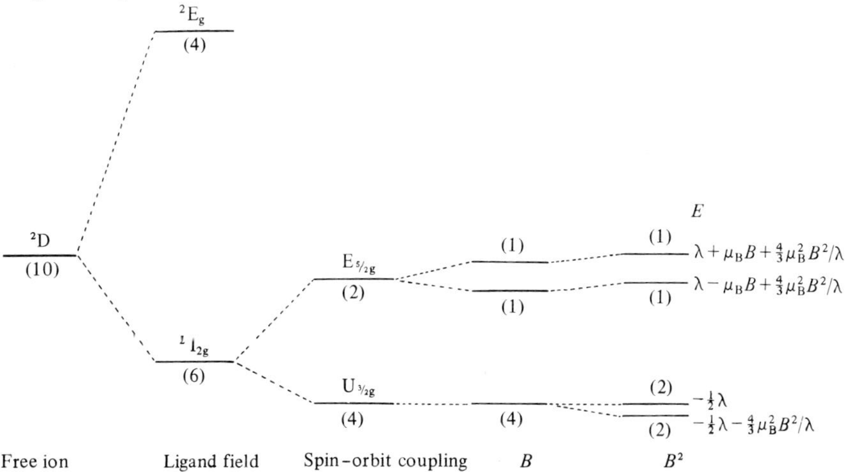


**Figure 4.3.4a.** Splitting of the triply degenerate electronic ground state of a $^2D$ term in an octahedral ligand field by an external magnetic field, to first-order ($B$) and second-order ($B^2$) perturbations. The degeneracy of each level is given in parentheses.

Substituting from the expressions for $E$ given in figure 4.3.4a, we obtain

$$\chi'_M = \frac{N_A\{2[(\mu_B^2/kT) - (8\mu_B^2/3\lambda)]\exp(-\lambda/kT) + (16\mu_B^2/3\lambda)\exp(\lambda/2kT)\}}{2[\exp(-\lambda/kT) + 2\exp(\lambda/2kT)]} .$$

Setting $x = \lambda/kT$ and multiplying through by $\frac{1}{2}\exp(-x/2)$, we obtain

$$\chi'_M = \frac{N_A\mu_B^2}{3kT}\left[\frac{(3-8/x)\exp(-3x/2) + 8/x}{\exp(-3x/2)+2}\right] .$$

The magnetic moment, $\mu_e$, is immediately obtained by substitution of the above expression into the Langevin formula (4.3.0a):

$$\mu_e = \left\{\frac{8 + (3x-8)\exp(-3x/2)}{x[2+\exp(-3x/2)]}\right\}^{1/2} .$$

The spin-orbit coupling constant, $\lambda$, is positive for a less than half-filled shell like $d^1$, but negative for a more than half-filled shell like $d^9$. The profound influence of the inversion of the relative energies within the

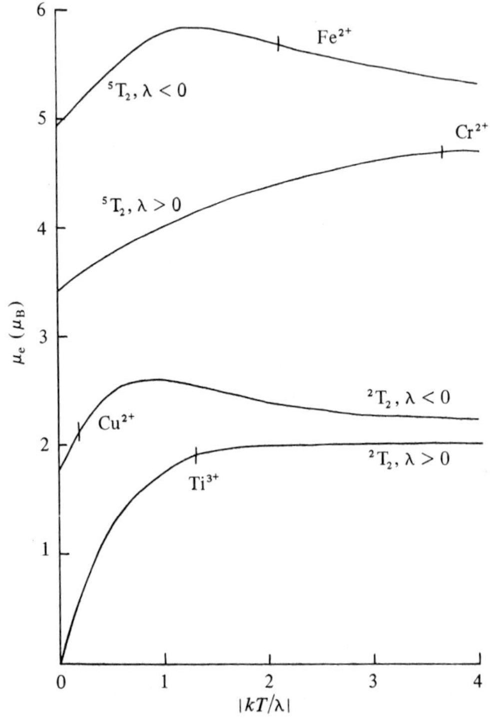


**Figure 4.3.4b.** The variation of magnetic moment for complexes with a $^2T_2$ ground state, for positive and negative values of the spin-orbit coupling constant, $\lambda$. The moments of $Ti^{3+}$ and $Cu^{2+}$ ions at 300 K are marked using the free-ion values of $\lambda$ (155 and $-830$ $cm^{-1}$, respectively). The corresponding curves for complexes with a $^5T_2$ ground state are shown for comparison.

manifold is illustrated in figure 4.3.4b. At infinite temperature or in the absence of spin–orbit coupling both cases converge to $\mu_e = 2 \cdot 24\ \mu_B$, the value given by the formula $\mu_e = [4S(S+1)+L(L+1)]^{1/2}$. Interestingly, the best-known case of the $Ti^{3+}$ ion has a value of $kT/\lambda$ near unity at room temperature, which corresponds to a magnetic moment near the spin-only value $1 \cdot 73\ \mu_B$.

The formula obtained for the molar susceptibility in the $^5T_2$ cases is much more complicated than that for $^2T_2$ cases (see Earnshaw, 1968), but certain similarities in the dependence of the magnetic moment upon $|kT/\lambda|$ may be immediately perceived from figure 4.3.4b.

4.3.5

The magnetic moments of complexes of Pr(III), Eu(III) and Dy(III) at room temperature are near $3 \cdot 5\ \mu_B$, $3 \cdot 5\ \mu_B$, and $10 \cdot 5\ \mu_B$, respectively. Account for these values in terms of the ground states of the $M^{3+}$ ions, which are $^3H_4$, $^7F_0$ and $^6H_{15/2}$ for M = Pr, Eu and Dy, respectively. The $^3H_5$, $^7F_1$ and $^6H_{13/2}$ levels lie 2200, 255, and 3500 $cm^{-1}$, respectively, above the ground state in the free ion, and spectroscopic data suggest similar values for the complexes of interest.

**Solution**

Van Vleck (1932) gave the following formula for the magnetic susceptibility of an ion in cases where the intervals between the levels defined by the total angular momentum quantum number, $J$, are comparable to $kT$:

$$\chi'_M = \frac{N_A \sum_{J=|L-S|}^{L+S} \{[g_J^2 \mu_B^2 J(J+1)/3kT] + \alpha_J\}(2J+1)\exp(-E_J^0/kT)}{\sum_{J=|L-S|}^{L+S} (2J+1)\exp(-E_J^0/kT)} \quad (4.3.5a)$$

The $E_J^0$ are the energy levels in the absence of a magnetic field; the constants $g_J$ and the temperature-independent factors $N_A\alpha_J$ are given by

$$g_J = \tfrac{3}{2} + \frac{S(S+1)-L(L+1)}{2J(J+1)} ;$$

$$N_A\alpha_J = \frac{N_A\mu_B^2}{6(2J+1)}\left[\frac{F(J+1)}{E_{J+1}^0 - E_J^0} - \frac{F(J)}{E_J^0 - E_{J-1}^0}\right],$$

where

$$F(J) = \frac{[(S+L+1)^2 - J^2][J^2 - (S-L)^2]}{J} .$$

This is a complicated formula, but Van Vleck (1932) also showed that, if the multiplet intervals are large compared with $kT$, the formula simplifies to:

$$\chi'_M = \frac{N_A g_J^2 \mu_B^2 J(J+1)}{3kT} + N_A\alpha ; \quad (4.3.5b)$$

and, indeed, in this case the temperature-independent term is small and may be safely neglected. Clearly this condition applies to the $Pr^{3+}$ and $Dy^{3+}$ ions and indeed to all rare-earth ions except $Eu^{3+}$ and, to a lesser degree, $Sm^{3+}$. If this susceptibility expression, neglecting $N_A\alpha$, is substituted in the Langevin equation (4.3.0a), we obtain the very simple expression

$$\mu = g_J[J(J+1)]^{1/2} .$$

We may tabulate the key quantities in the calculation of the magnetic moments, $\mu_{calc}$, of the $Pr^{3+}$, $Eu^{3+}$, and $Dy^{3+}$ ions as in table 4.3.5a.

The results are clearly satisfactory for the Pr(III) and Dy(III) complexes, but quite inappropriate for the Eu(III) case. Interestingly, if the thermal population of the low-lying $^7F_1$ and $^7F_2$ levels are taken into account, the magnetic moment of a Eu(III) complex is still calculated to be only $1{\cdot}7\ \mu_B$ (the $^7F_1$ and $^7F_2$ levels would lead to magnetic moments of $2{\cdot}12\ \mu_B$ and $3{\cdot}36\ \mu_B$, respectively, if each were the only level populated). In fact, in the case of Eu(III) complexes, a very large contribution to the room-temperature moment arises from the temperature-independent terms, $N_A\alpha_J$. This is evident when we substitute the relevant quantities in the unsimplified susceptibility equation (4.3.5a). These quantities are given in table 4.3.5b, where $Q$ denotes the quantity

$$Q = \frac{1}{255}\left[\frac{F(J+1)}{E^0_{J+1} - E^0_J} - \frac{F(J)}{E^0_J - E^0_{J-1}}\right] ,$$

required for the calculation of $N_A\alpha_J$.

**Table 4.3.5a.**

| Ion | Ground state | $J$ | $S$ | $L$ | $g_J(\mu_B)$ | $\mu_{calc}(\mu_B)$ | $\mu_{expt}(\mu_B)$ ($T = 300$ K) |
|---|---|---|---|---|---|---|---|
| $Pr^{3+}$ | $^3H_4$ | 4 | 1 | 5 | $\frac{4}{5}$ | 3·58 | 3·5 |
| $Eu^{3+}$ | $^7F_0$ | 0 | 3 | 3 | – | 0·00 | 3·5 |
| $Dy^{3+}$ | $^6H_{15/2}$ | $\frac{15}{2}$ | $\frac{5}{2}$ | 5 | $\frac{4}{3}$ | 10·63 | 10·5 |

**Table 4.3.5b.**

| Level | $E^0_J$ (cm$^{-1}$) | $J$ | $S$ | $L$ | $g_J(\mu_B)$ | $F(J)$ | $Q$ (cm$^2$) | $g_J^2J(J+1)(\mu_B^2)$ |
|---|---|---|---|---|---|---|---|---|
| $^7F_0$ | 0 | 0 | 3 | 3 | – | – | 48 | 0 |
| $^7F_1$ | 255 | 1 | 3 | 3 | $\frac{3}{2}$ | 48 | −3 | 4·5 |
| $^7F_2$ | 765 | 2 | 3 | 3 | $\frac{3}{2}$ | 90 | −5 | 13·5 |
| $^7F_3$ | 1530 | 3 | 3 | 3 | $\frac{3}{2}$ | 120 | −7 | 27 |
| $^7F_4$ | 2550 | 4 | 3 | 3 | $\frac{3}{2}$ | 132 | $E^0_J$ values too high for significant population at room temperature | |
| $^7F_5$ | 3775 | 5 | 3 | 3 | $\frac{3}{2}$ | 120 | | |
| $^7F_6$ | 5355 | 6 | 3 | 3 | $\frac{3}{2}$ | 78 | | |

Substitution of the data from table 4.3.5b into equation (4.3.5a) yields

$$\chi'_M = \frac{24+(13\cdot5x-1\cdot5)\exp(-x)+(67\cdot5x-2\cdot5)\exp(-3x)+(189x-3\cdot5)\exp(-6x)+...}{7\cdot997xT[1+3\exp(-x)+5\exp(-3x)+7\exp(-6x)+...]},$$

where $x = 255\ cm^{-1}/kT = 365\ K/T$. Substituting in the Langevin equation, we obtain

$$\mu = \left\{\frac{24+(13\cdot5x-\mathit{1\cdot5})\exp(-x)+(67\cdot5x-\mathit{2\cdot5})\exp(-3x)+(189x-\mathit{3\cdot5})\exp(-6x)+...}{x[1+3\exp(-x)+5\exp(-3x)+7\exp(-6x)+...]}\right\}^{1/2},$$

where the temperature-independent contributions have been italicised. At $T = 300$ K, $x = 1\cdot22$; thus we obtain

$$\begin{aligned}\mu &= \left\{\frac{24+15\cdot0\exp(-1\cdot22)+79\cdot9\exp(-3\cdot66)+227\cdot1\exp(-7\cdot32)+...}{1\cdot22[1+3\exp(-1\cdot22)+5\exp(-3\cdot66)+7\exp(-7\cdot32)+...]}\right\}^{1/2}\\ &= \left\{\frac{24+(15\cdot0\times0\cdot295)+(79\cdot9\times0\cdot026)+(227\cdot1\times0\cdot001)+...}{1\cdot22[1+(3\times0\cdot295)+(5\times0\cdot026)+(7\times0\cdot001)+...]}\right\}^{1/2}\\ &= 3\cdot51\mu_B\,.\end{aligned}$$

This result is very satisfactory. In the last few lines of the calculation the diminishing contribution of the higher levels is easily recognised. The importance of the term $N_A\alpha_0$ is immediately apparent from a comparison of the first and second expressions given for $\mu$ above. In fact, this term contributes about 80% of the magnetic susceptibility at 300 K.

### 4.3.6

Three four-coordinate complexes with a distorted tetrahedral structure, namely calcium copper tetra-acetate hexahydrate [$CaCu(OAc)_4.6H_2O$], bis(*N*-isopropylsalicyladiminato)copper(II) [$Cu(IPSA)_2$], and *N*,*N*′-ethylene-bis(acetylacetoneiminato)copper(II) [Cu(EAA)], were included in an investigation of magnetic anisotropy by Figgis *et al.* (1968). The compounds crystallise in tetragonal, orthorhombic, and monoclinic space groups, respectively, but in each case the molecular symmetry is approximately or exactly axial ($D_{2d}$).

The anisotropies of the principal crystal magnetic susceptibilities with respect to orthogonal axes at 300 K, as determined by the critical torque method, and the average powder susceptibilities of all three complexes are given in table 4.3.6a.

The magnetic susceptibilities along the three molecular axes chosen to optimise the axial symmetry are defined as $K_1$, $K_2$, and $K_3$ which is along the principal axis. The $K_3$ axis of $CaCu(OAc)_4.6H_2O$ corresponds to the crystallographic $\overline{4}$ axis. The direction cosines of the optimum molecular axes with respect to the most convenient orthogonal set of crystal axes are given in table 4.3.6b for the other two compounds.

Assuming an effective axial magnetic symmetry in each case (i.e. $K_3 = K_\parallel$, $K_1 = K_2 = K_\perp$), calculate the principal molecular susceptibilities at 300 K and the corresponding magnetic moments, $\mu_\parallel$ and $\mu_\perp$, for each compound.

**Table 4.3.6a.** Anisotropies in crystal magnetic susceptibility ($10^{-6}$ erg $G^{-2}$ $mol^{-1}$) and average (powder) values, $\bar{\chi}$ ($10^{-6}$ erg $G^{-2}$ $mol^{-1}$), for the three copper complexes. The subscripts $A$, $B$, and $C$ represent the crystallographic $a$, $b$, and $c$ axes in the tetragonal and orthorhombic cases; in the monoclinic case the orthogonal axes $a'$, $b$ (unique axis), and $c$ are represented.

| | $(\chi_A - \chi_B)$ | $(\chi_B - \chi_C)$ | $(\chi_C - \chi_A)$ | $\bar{\chi}$ |
|---|---|---|---|---|
| $CaCu(OAc)_4.6H_2O$ | 0 | −450 | −450 | 1550 |
| $Cu(IPSA)_2$ | 226 | −205 | −21 | 1395 |
| Cu(EAA) | −126 | −52 | 127 | 1625 |

**Table 4.3.6b.** Direction cosines of the $K_i$ axes with respect to the crystallographic axes of two copper complexes. In the monoclinic case the orthogonal axes $A$, $C$, and $B$ are chosen to correspond to the three components of the magnetic ellipsoid, $\chi_1$, $\chi_2$, and $\chi_3$. By definition and symmetry, $B$ and $\chi_3$ correspond to the unique crystal axis, $b$.

| | $Cu(IPSA)_2$ | | | Cu(EAA) | | |
|---|---|---|---|---|---|---|
| | $a$ | $b$ | $c$ | $A$ | $B (= b)$ | $C$ |
| $K_1$ | −0·6817 | 0·0589 | 0·7292 | −0·9754 | 0·1908 | −0·1104 |
| $K_2$ | 0·0750 | −0·9859 | 0·1497 | 0·2100 | 0·6521 | −0·7285 |
| $K_3$ | −0·7277 | −0·1568 | −0·6677 | 0·0670 | 0·7337 | 0·6761 |

**Solution**

The first step in the calculation is to convert the powder susceptibility, $\bar{\chi}$, and the crystal anisotropies into the principal susceptibilities, $\chi_1$, $\chi_2$ and $\chi_3$, of the triaxial magnetic ellipsoid. For a crystal of orthorhombic or higher symmetry the $\chi_i$ coincide with the crystal axes, $a$, $b$, and $c$. By solving the simultaneous equations obtained from the equality $\bar{\chi} = \frac{1}{3}(\chi_1 + \chi_2 + \chi_3)$ and from the anisotropies, we obtain: for $CaCu(OAc)_4.6H_2O$,

$$\chi_3 = 1850 \times 10^{-6} \text{ erg G}^{-2} \text{ mol}^{-1}, \quad \chi_1 = \chi_2 = 1400 \times 10^{-6} \text{ erg G}^{-2} \text{ mol}^{-1};$$

and for $Cu(IPSA)_2$,

$$\chi_3 = \chi_c = 1456 \times 10^{-6} \text{ erg G}^{-2} \text{ mol}^{-1};$$
$$\chi_2 = \chi_b = 1251 \times 10^{-6} \text{ erg G}^{-2} \text{ mol}^{-1};$$
$$\chi_1 = \chi_a = 1477 \times 10^{-6} \text{ erg G}^{-2} \text{ mol}^{-1}.$$

In a monoclinic system one axis, $\chi_3$, of the triaxial magnetic ellipsoid of the crystal is coincident with the unique crystallographic axis $b$; the other two lie in the (010) plane with $\chi_2$ at some angle $\theta$ to $c$, and, by definition, with $\chi_2 > \chi_1$. The following relationships have been derived for the case of a monoclinic crystal (see, for example, Figgis and Lewis, 1964).

$$\bar{\chi} = \tfrac{1}{3}(\chi_1 + \chi_2 + \chi_3) \, ; \tag{4.3.6a}$$

$$\chi_c - \chi_{a'} = \chi_2 - \chi_1 \, ; \tag{4.3.6b}$$

$$\chi_b - \chi_c = \pm|\chi_1 \sin^2\theta + \chi_2 \cos^2\theta - \chi_3| \, ; \tag{4.3.6c}$$

$$\chi_{a'} - \chi_b = \pm|\chi_1 \cos^2\theta + \chi_2 \sin^2\theta - \chi_3| \, . \tag{4.3.6d}$$

An immediate problem lies in the ambiguity of sign in two of the above equations. This may be clarified by attempting a calculation of $\theta$ from these equations. Subtracting equation (4.3.6d) from equation (4.3.6c) we obtain:

$$(\chi_1 \sin^2\theta + \chi_2 \cos^2\theta) - (\chi_1 \cos^2\theta + \chi_2 \sin^2\theta) = \pm(\chi_b - \chi_c) \mp (\chi_{a'} - \chi_b) \, ;$$

$$(\chi_2 - \chi_1) \cos 2\theta = \pm(\chi_b - \chi_c) \mp (\chi_{a'} - \chi_b) \, ;$$

$$\cos 2\theta = \frac{\pm(\chi_b - \chi_c) \mp (\chi_{a'} - \chi_b)}{(\chi_c - \chi_{a'})} \, . \tag{4.3.6e}$$

From inspection of the values in table 4.3.6a, it is immediately clear that $(\chi_b - \chi_c)$ and $(\chi_{a'} - \chi_b)$ must have the same sign, and since $\chi_c > \chi_{a'}$ it follows that $\chi_c > \chi_b > \chi_{a'}$, and so $(\chi_b - \chi_c)$ and $(\chi_{a'} - \chi_b)$ are both negative. Now, from equation (4.3.6e), we can write

$$\cos 2\theta = \pm(-52 + 126)/127$$
$$= \pm 0 \cdot 5826 \, ,$$

whence $\theta = 27° \; 11'$ or $152° \; 49'$.

In the cases where $\chi_b$ is intermediate between $\chi_{a'}$ and $\chi_c$, $\theta$ may be eliminated from equations (4.3.6a)–(4.3.6d) to give:

$$\chi_1 = \bar{\chi} + \tfrac{1}{6}[(\chi_b - \chi_c) - 3(\chi_c - \chi_{a'}) + (\chi_{a'} - \chi_b)] \, ;$$

$$\chi_2 = \bar{\chi} + \tfrac{1}{6}[(\chi_b - \chi_c) + 3(\chi_c - \chi_{a'}) + (\chi_{a'} - \chi_b)] \, ;$$

$$\chi_3 = \bar{\chi} - \tfrac{1}{3}[(\chi_b - \chi_c) + (\chi_{a'} - \chi_b)] \, .$$

Substituting the values from table 4.3.6a, we obtain $\chi_1 = 1532 \times 10^{-6}$, $\chi_2 = 1659 \times 10^{-6}$, and $\chi_3 = 1684 \times 10^{-6}$ erg $G^{-2}$ $mol^{-1}$ for the complex Cu(EAA).

The principal crystal magnetic susceptibilities, $\chi_1$, $\chi_2$, and $\chi_3$, are converted into molecular susceptibilities, $K_1$, $K_2$, and $K_3$, by considering the relationship between the molecular and crystal axes. In the case of $CaCu(OAc)_4.6H_2O$ the $K_3$ axis coincides with the crystallographic $\bar{4}$ axis, and we may immediately write for this complex:

$$K_3 = \chi_3 = 1850 \times 10^{-6} \text{ erg } G^{-2} \text{ mol}^{-1};$$
$$K_1 = K_2 = \chi_1 = 1400 \times 10^{-6} \text{ erg } G^{-2} \text{ mol}^{-1}.$$

For an orthorhombic complex we have the general relationships:

$$\chi_1 = \sum_{i=1}^{3} \alpha_i^2 K_i \ ; \qquad \chi_2 = \sum_{i=1}^{3} \beta_i^2 K_i \ ; \qquad \chi_3 = \sum_{i=1}^{3} \gamma_i^2 K_i \ ;$$

where the $\alpha_i$, $\beta_i$, and $\gamma_i$ represent the direction cosines of the $i$th molecular axis with reference to the crystal axes, $a$, $b$, and $c$. From the data for $Cu(IPSA)_2$ in table 4.3.6b, we therefore obtain

$$\chi_1 = (-0{\cdot}6817)^2 K_1 + (0{\cdot}0750)^2 K_2 + (-0{\cdot}7277)^2 K_3 \ ,$$

$$\chi_2 = (0{\cdot}0589)^2 K_1 + (-0{\cdot}9859)^2 K_2 + (0{\cdot}1568)^2 K_3 \ ,$$

$$\chi_3 = (-0{\cdot}7292)^2 K_1 + (0{\cdot}1497)^2 K_2 + (-0{\cdot}6677)^2 K_3 \ ;$$

i.e.

$$0{\cdot}4648 K_1 + 0{\cdot}0056 K_2 + 0{\cdot}5295 K_3 = 1477 \times 10^{-6} \text{ erg G}^{-2} \text{ mol}^{-1} \ ,$$

$$0{\cdot}0035 K_1 + 0{\cdot}9720 K_2 + 0{\cdot}0246 K_3 = 1251 \times 10^{-6} \text{ erg G}^{-2} \text{ mol}^{-1} \ ,$$

$$0{\cdot}5317 K_1 + 0{\cdot}0224 K_2 + 0{\cdot}4458 K_3 = 1456 \times 10^{-6} \text{ erg G}^{-2} \text{ mol}^{-1} \ .$$

The solutions of these equations are:

$$K_1 = 1351 \times 10^{-6} \text{ erg G}^{-2} \text{ mol}^{-1}; \qquad K_2 = 1242 \times 10^{-6} \text{ erg G}^{-2} \text{ mol}^{-1};$$
$$K_3 = 1592 \times 10^{-6} \text{ erg G}^{-2} \text{ mol}^{-1}.$$

To establish axial magnetic symmetry for this case, as is found in the tetragonal case considered above, we assume that

$$K_{\parallel} = K_3 = 1592 \times 10^{-6} \text{ erg G}^{-2} \text{ mol}^{-1},$$

and that

$$K_{\perp} = \tfrac{1}{2}(K_1 + K_2) = 1297 \times 10^{-6} \text{ erg G}^{-2} \text{ mol}^{-1} \ .$$

The monoclinic case is more complicated. However, if we assume in the extreme that the molecule has axial magnetic symmetry with $K_{\parallel} > K_{\perp}$ (see Figgis and Lewis, 1964), we may write, for the complex Cu(EAA):

$$K_{\parallel} = \chi_2 + \chi_3 - \chi_1 = 1811 \times 10^{-6} \text{ erg G}^{-2} \text{ mol}^{-1};$$

$$K_{\perp} = \chi_1 = 1532 \times 10^{-6} \text{ erg G}^{-2} \text{ mol}^{-1}.$$

We may also attempt to calculate the values of $K_1$, $K_2$, and $K_3$ from the direction cosines given in table 4.3.6b for Cu(EAA). In this table the directions of the optimised molecular axes are expressed with respect to those of the principal components of the magnetic ellipsoid, $\chi_1$, $\chi_2$, and $\chi_3$ which, by symmetry, corresponds to the unique axis, $b$. We may therefore write:

$$\chi_1 = (-0{\cdot}9754)^2 K_1 + (0{\cdot}2100)^2 K_2 + (0{\cdot}0670)^2 K_3 \ ,$$

$$\chi_2 = (-0{\cdot}1104)^2 K_1 + (-0{\cdot}7285)^2 K_2 + (0{\cdot}6761)^2 K_3 \ ,$$

$$\chi_3 = (0{\cdot}1908)^2 K_1 + (0{\cdot}6521)^2 K_2 + (0{\cdot}7337)^2 K_3 \ ;$$

i.e.

$$0{\cdot}9514K_1+0{\cdot}0441K_2+0{\cdot}0045K_3 = 1532 \times 10^{-6} \text{ erg G}^{-2} \text{ mol}^{-1} ,$$

$$0{\cdot}0122K_1+0{\cdot}5307K_2+0{\cdot}4571K_3 = 1659 \times 10^{-6} \text{ erg G}^{-2} \text{ mol}^{-1} ,$$

$$0{\cdot}0364K_1+0{\cdot}4252K_2+0{\cdot}5383K_3 = 1684 \times 10^{-6} \text{ erg G}^{-2} \text{ mol}^{-1} .$$

The solutions of these simultaneous equations are:

$$K_1 = 1531 \times 10^{-6} \text{ erg G}^{-2} \text{ mol}^{-1}; \quad K_2 = 1523 \times 10^{-6} \text{ erg G}^{-2} \text{ mol}^{-1};$$
$$K_3 = 1821 \times 10^{-6} \text{ erg G}^{-2} \text{ mol}^{-1}.$$

Thus $K_\parallel = K_3 = 1821 \times 10^{-6}$ erg G$^{-2}$ mol$^{-1}$, and

$$K_\perp = \tfrac{1}{2}(K_1+K_2) = 1527 \times 10^{-6} \text{ erg G}^{-2} \text{ mol}^{-1}.$$

The results of the two calculations agree well [the values of $K_1$, $K_2$, and $K_3$ obtained by Figgis *et al.* (1968) differ slightly from these, because of the sensitivity of the last calculation to the exact values of the data], and we therefore select the mean values, $K_\parallel = 1816 \times 10^{-6}$ erg G$^{-2}$ mol$^{-1}$ and $K_\perp = 1529 \times 10^{-6}$ erg G$^{-2}$ mol$^{-1}$, for Cu(EAA).

We finally calculate the magnetic moments from equation (4.3.0b), and obtain

$$\mu = (7{\cdot}997KT)^{1/2} = (2399K)^{1/2} ,$$

since all data are at 300 K. The final results are tabulated in table 4.3.6c.

**Table 4.3.6c.** Molecular and bulk susceptibilities ($10^{-6}$ erg G$^{-2}$ mol$^{-1}$), and magnetic moments ($\mu_B$), for copper complexes.

| | $K_\parallel$ | $K_\perp$ | $\bar{\chi}$ | $\mu_\parallel$ | $\mu_\perp$ | $\bar{\mu}$ |
|---|---|---|---|---|---|---|
| $CaCu(OAc)_4.6H_2O$ | 1850 | 1400 | 1550 | 2·10 | 1·83 | 1·93 |
| $Cu(IPSA)_2$ | 1592 | 1297 | 1395 | 1·95 | 1·76 | 1·83 |
| Cu(EAA) | 1816 | 1529 | 1625 | 2·09 | 1·92 | 1·97 |

**4.3.7**

Copper(II) acetate monohydrate has a structure consisting of discrete dimeric units. The magnetic susceptibility–temperature data obtained by Figgis and Martin (1956) are given in table 4.3.7a. Given that the susceptibility of a binuclear Cu(II) complex (per mole of copper atoms) can be expressed by the formula

$$\chi'_M = \frac{N_A g^2 \mu_B^2}{kT}\left[\frac{1}{3+\exp(-J/kT)}\right] + N_A\alpha , \qquad (4.3.7a)$$

and that the average value of $g$ obtained from the e.s.r. spectrum is 2·17, calculate an optimum value for the exchange parameter, $J$, which represents the energy difference between the singlet and triplet states which arise from the coupling of two spin doublets (a negative value of $J$ indicates that the ground state is a singlet). The value which is most generally accepted for the temperature-independent contribution, $N_A\alpha$, is $60 \times 10^{-6}$ erg $G^{-2}$ $mol^{-1}$.

**Table 4.3.7a.** Magnetic susceptibility of copper(II) acetate monohydrate (expressed per mole of copper atoms).

| $T$ (K) | $10^6\chi'_M$ (erg $G^{-2}$ $mol^{-1}$) | $T$ (K) | $10^6\chi'_M$ (erg $G^{-2}$ $mol^{-1}$) |
|---|---|---|---|
| 93·5 | 266 | 239·7 | 893 |
| 100·1 | 302 | 258·6 | 898 |
| 120·2 | 451 | 277·9 | 895 |
| 143·5 | 612 | 294·2 | 889 |
| 160·5 | 712 | 300·3 | 878 |
| 181·0 | 787 | 321·3 | 876 |
| 200·9 | 837 | 349·5 | 854 |
| 220·0 | 873 | 396·5 | 797 |

**Solution**

The simplest method of determining an approximate value for the exchange parameter, $J$, is to substitute each pair of values for $\chi'_M$ and $T$ into equation (4.3.7a), together with the values of the other relevant parameters. This has been done as indicated in table 4.3.7b, in which two rows have been left blank so that the reader can follow the calculation through if

**Table 4.3.7b.** Determination of the exchange parameter, $J$, for copper(II) acetate monohydrate ($N_A g^2 \mu_B^2/k = 1{\cdot}753$ erg $G^{-2}$ $mol^{-1}$ K).

| $T$ (K) | $10^6\chi'_M$ (erg $G^{-2}$ $mol^{-1}$) | $(\chi'_M - N_A\alpha)T$ $(= \mu^2/7{\cdot}997)$ (erg $G^{-2}$ $mol^{-1}$ K) | $\dfrac{N_A^2 g^2 \mu_B^2}{kT(\chi'_M - N_A\alpha)}$ $[= 3 + \exp(-J/kT)]$ | $-J/kT$ | $J$ ($cm^{-1}$) |
|---|---|---|---|---|---|
| 93·5 | 266 | 0·0193 | 91·01 | 4·478 | −291·0 |
| 100·1 | 302 | 0·0242 | 72·39 | 4·239 | −294·8 |
| 120·2 | 451 | 0·0470 | 37·30 | 3·534 | −295·2 |
| 143·5 | 612 | | | | −294·4 |
| 160·5 | 712 | 0·1046 | 16·76 | 2·622 | −292·5 |
| 181·0 | 787 | 0·1316 | 13·32 | 2·334 | −293·6 |
| 200·9 | 837 | 0·1561 | 11·22 | 2·107 | −294·1 |
| 220·0 | 873 | 0·1788 | 9·80 | 1·917 | −293·1 |
| 239·7 | 893 | 0·1996 | 8·78 | 1·755 | −292·3 |
| 258·6 | 898 | 0·2167 | 8·09 | 1·627 | −292·4 |
| 277·9 | 895 | | | | −292·6 |
| 294·2 | 889 | 0·2439 | 7·19 | 1·433 | −292·9 |
| 300·3 | 878 | 0·2456 | 7·14 | 1·421 | −296·4 |
| 321·3 | 876 | 0·2622 | 6·68 | 1·303 | −290·9 |
| 349·5 | 854 | 0·2776 | 6·32 | 1·200 | −291·4 |
| 396·5 | 797 | 0·2922 | 5·95 | 1·076 | −296·6 |
| | | | | Mean | −293·4 ± 1·7 |

desired[2]. In practice, a computer fitting procedure would probably be used to find the optimum value of $J$. Some workers minimise the sum of the values of $[\chi_{expt} - \chi_{calc}]^2$; others prefer to minimise the sum of the values of $[(\chi_{expt} - \chi_{calc})T]^2$, which tends to give a smaller weighting to the low-temperature data. These calculations can be done by hand, but they are rather tedious. They give optimum values of $J$ in the region of $-295 \pm 10\ \mathrm{cm}^{-1}$, depending upon the values chosen for the other parameters and the function which is statistically minimised.

### 4.3.8

Calculate an expression for the magnetic susceptibility of a polynuclear complex with three Cr(III) ions in an equilateral triangular arrangement, assuming a dipolar coupling between pairs of ions of the form:

$$\mathcal{H} = -J \sum_{i>j} \boldsymbol{S}_i \cdot \boldsymbol{S}_j ,$$

where $J$ is the exchange parameter, and $S_i$ is the electronic spin operator for ion $i$.

**Solution**

Each Cr(III) ion in the cluster has a formal spin quantum number, $S$, of $\frac{3}{2}$. Thus the possible spin states for the trinuclear complex are those for which the total spin, $S'$, is $\frac{9}{2}$, $\frac{7}{2}$, $\frac{5}{2}$, $\frac{3}{2}$, or $\frac{1}{2}$. Since an individual Cr(III) ion would be in a quartet state, the various spin states of the trinuclear complex must span a total of $4^3 = 64$ components. To find the number of times a given value of $S'$ occurs, $q(S')$, we use the expression

$$q(S') = w(S') - w(S'+1) ,$$

where $w(S')$ is the coefficient of $x^{S'}$ in the expansion:

$$(x^S + x^{S-1} + \ldots + x^{-S+1} + x^{-S})^n ,$$

where $n$ is the number of interacting ions. Thus, for $S = \frac{3}{2}$, $n = 3$ and $S' = \frac{5}{2}$,

$$(x^{3/2} + x^{1/2} + x^{-1/2} + x^{-3/2})^3 = x^{9/2} + 3x^{7/2} + 6x^{5/2} + 10x^{3/2} + 12x^{1/2} + \ldots + x^{-9/2} ,$$

and

$$q(S') = 6 - 3 = 3 .$$

The energy of each of the spin states specified by a spin $S'$ is given by the expression

$$E(S') = \frac{zJ}{2(n-1)}[nS(S+1) - S'(S'+1)] ,$$

where $z$ is the number of nearest neighbours, i.e. $z = 2$ in this case.

[2] Remember that $\log_e x = 2{\cdot}303 \log_{10} x$.

Thus, for the system discussed here,

$$E(S') = \tfrac{1}{2}J[\tfrac{45}{4} - S'(S'+1)] .$$

The readily calculated values of $S'$, $q(S')$, and $E(S')$ are given in table 4.3.8. A check reveals that the total number of components, $\Sigma(2S'+1)q(S')$, is 64 and that the centre of gravity of the energy levels is maintained—that is, $\Sigma(2S'+1)E(S') = 0$. However, since only the relative values of $E(S')$ are significant, it is convenient to displace the energy zero by a factor $27J/4$, as shown in the last column of this table.

Apart from the temperature-independent contribution, $N_A\alpha$, an expression for the magnetic susceptibility per metal ion is now found by substitution of the relevant quantities in the Van Vleck equation:

$$\chi'_M = \frac{1}{n}\left\{\frac{N_A g^2\mu_B^2}{3kT}\left[\frac{\sum_{S'} S'(S'+1)(2S'+1)q(S')\exp(-E'(S')/kT)}{\sum_{S'}(2S'+1)q(S')\exp(-E'(S')/kT)}\right] + N_A\alpha\right\} .$$

i.e., for the present case,

$$\chi'_M = \frac{N_A g^2\mu_B^2}{18kT}\left[\frac{495+504\exp(-\frac{9}{2}x)+315\exp(-8x)+120\exp(-\frac{21}{2}x)+6\exp(-12x)}{10+16\exp(-\frac{9}{2}x)+18\exp(-8x)+16\exp(-\frac{21}{2}x)+4\exp(-12x)}\right] + N_A\alpha ,$$

where $x = J/kT$. Simplifying slightly, we obtain

$$\chi'_M = \frac{N_A g^2\mu_B^2}{3kT}\left[\frac{165+168\exp(-\frac{9}{2}x)+105\exp(-8x)+40\exp(-\frac{21}{2}x)+2\exp(-12x)}{20+32\exp(-\frac{9}{2}x)+36\exp(-8x)+32\exp(-\frac{21}{2}x)+8\exp(-12x)}\right] + N_A\alpha$$

as the required expression for the magnetic susceptibility per Cr(III) ion of a trinuclear complex such as basic chromic acetate,

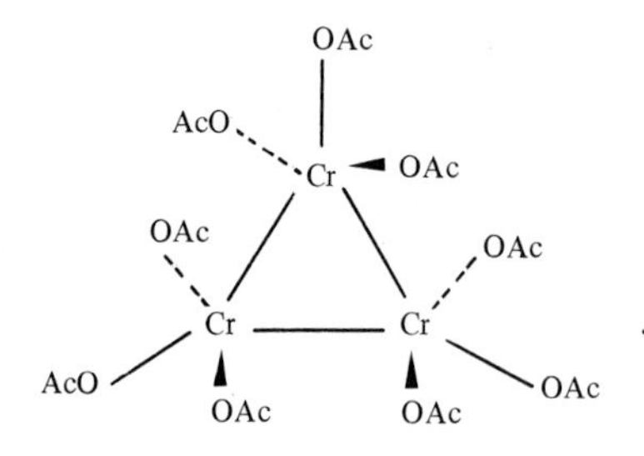


**Table 4.3.8.**

| $S'$ | $q(S')$ | $E(S')$ | $E'(S')$ [$= E(S') + 27J/4$] |
|---|---|---|---|
| $\frac{1}{2}$ | 2 | $21J/4$ | $12J$ |
| $\frac{3}{2}$ | 4 | $15J/4$ | $21J/2$ |
| $\frac{5}{2}$ | 3 | $5J/4$ | $8J$ |
| $\frac{7}{2}$ | 2 | $-9J/4$ | $9J/2$ |
| $\frac{9}{2}$ | 1 | $-27J/4$ | 0 |

**References**

Earnshaw, A., 1968, *Introduction to Magnetochemistry* (Academic Press, London).
Evans, D. F., 1959, *J. Chem. Soc.,* 2003.
Figgis, B. N., Gerloch, M., Lewis, J., Slade, R. C., 1968, *J. Chem. Soc. A.,* 2028.
Figgis, B. N., Lewis, J., 1964, *Prog. Inorg. Chem.,* **6**, 37.
Figgis, B. N., Martin, R. L., 1956, *J. Chem. Soc.,* 3837.
Gerloch, M., Miller, J. R., 1968, *Prog. Inorg. Chem.,* **10**, 1.
Golding, R. M., 1969, *Applied Wave Mechanics* (Van Nostrand, London).
Mabbs, F. E., Machin, D. J., 1973, *Magnetism and Transition Metal Complexes* (Chapman and Hall, London).
Van Vleck, J. H., 1932, *Electric and Magnetic Susceptibilities* (Oxford University Press, London).

## 4.4 Molecular electric dipole and quadrupole moments

**A D Buckingham** University of Cambridge

### 4.4.0 Units

The original literature on electric and magnetic properties of molecules mainly employs unrationalized CGS units (i.e. either e.s.u. or e.m.u.) However, most of the modern texts (e.g. Bleaney and Bleaney, 1976) have adopted the internationally accepted SI units, and these have the great advantage of coherence. The following problems and answers are written in SI units, but to assist those who are not familiar with them a conversion table is provided at the end of this book. The symbols used conform to those currently recommended by the Symbols Committee of the Royal Society. Dipole moments are often quoted in debye units (D); $1\ \mathrm{D} = 10^{-18}$ e.s.u. cm $= 3{\cdot}33564 \times 10^{-30}$ C m.

### 4.4.1

(a) Calculate the root-mean-square dipole moment of the molecule $XH_2C{\cdot}CH_2X$ if the $CH_2X$ group has a dipole moment $\mu_0$ inclined at an angle $\theta$ to the C–C bond and if:
(i) the molecule is in the staggered ($C_{2h}$) conformation;
(ii) the molecule is in the eclipsed ($C_{2v}$) conformation;
(iii) there is free rotation about the C–C bond.
(b) A mole of gaseous $(CH_2X)_2$ occupies $0{\cdot}1\ \mathrm{m}^3$ at 300 K and has a relative permittivity $\epsilon_r$ and optical refractive index $n$ given by

$$\epsilon_r - 1 = 2 \times 10^{-3}\ ,$$
$$n - 1 = 2 \times 10^{-4}\ .$$

Calculate the root-mean-square dipole moment of $(CH_2X)_2$ at 300 K. Assume $\mu_0 = 6 \times 10^{-30}$ C m and $\theta = 70°$. Comment on the result.

**Solution**

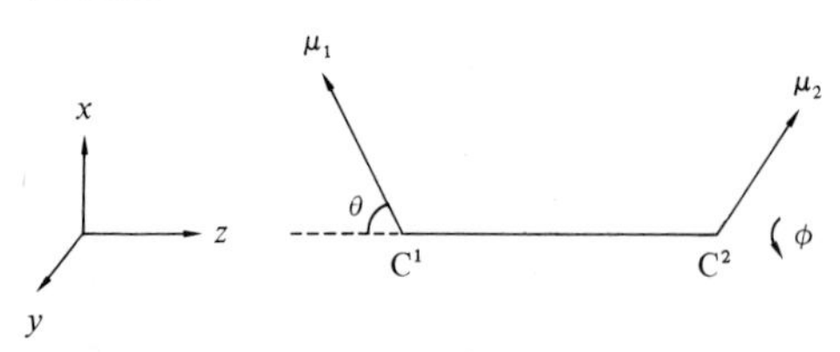


**Figure 4.4.1.** Coordinate system for the molecule $(CH_2X)_2$.

(a) If the $C^1$–$C^2$ bond is in the $z$ direction, the dipole $\mu_1$ in the $xz$ plane and $\mu_2$ rotated through an angle $\phi$ from the $xz$ plane, then

$$\mu_z = \mu_{1z} + \mu_{2z} = 0\ ,$$
$$\mu_x = \mu_{1x} + \mu_{2x} = \mu_0 \sin\theta(1 + \cos\phi)\ ,$$
$$\mu_y = \mu_{1y} + \mu_{2y} = \mu_0 \sin\theta\, \sin\phi\ .$$

Hence

$$\mu^2 = \mu_x^2 + \mu_y^2 + \mu_z^2 = 2\mu_0^2 \sin^2\theta(1+\cos\phi) . \qquad (4.4.1a)$$

(i) In the staggered conformation, $\phi = 180°$ and $\mu = 0$.
(ii) In the eclipsed conformation, $\phi = 0°$ and $\mu = 2\mu_0 \sin\theta$ .
(iii) If there is free rotation, $\cos\phi$ averages to zero and $\mu = 2^{1/2}\mu_0 \sin\theta$.
(b) For a gas at a low density,

$$\epsilon_r - n^2 = \frac{N\overline{\mu^2}}{3\epsilon_0 kT} ,$$

where $N$ is the number of molecules in unit volume and $\epsilon_0 = 8{\cdot}8542 \times 10^{-12}$ J$^{-1}$ C$^2$ m$^{-1}$ is the permittivity of a vacuum. Hence for $(CH_2X)_2$ at 300 K

$$(\overline{\mu^2})^{1/2} = 5{\cdot}406 \times 10^{-30} \text{ C m} = 1{\cdot}621 \text{ D} .$$

If $\mu_0 = 6 \times 10^{-30}$ C m and $\theta = 70°$, equation (4.4.1a) gives $\overline{\cos\phi} = -0{\cdot}5402$.

The negative value for $\cos\phi$ indicates that the staggered form is favoured over the eclipsed. If the potential barrier to free rotation about the C–C bond has the simple form

$$U = \tfrac{1}{2}U_0(1+\cos\phi) ,$$

$$\overline{\cos\phi} = \frac{\displaystyle\int_0^{2\pi} \cos\phi \exp(-U/kT)\, d\phi}{\displaystyle\int_0^{2\pi} \exp(-U/kT)\, d\phi} = -\frac{I_1(U_0/2kT)}{I_0(U_0/2kT)} ,$$

where $I_n$ is the modified Bessel function of order $n$ (see Smith, 1955, p.287; Abramowitz and Stegun, 1964, chapter 9). From tables of $I_x(x)$, and for $\overline{\cos\phi} = -0{\cdot}5402$, one finds that $U_0/2kT = 1{\cdot}29$, and hence $U_0 = 1{\cdot}07 \times 10^{-20}$ J $= 6{\cdot}4$ kJ mol$^{-1}$.

Some of the deficiencies of the assumed form for $U$, and of the classical averaging, are discussed by Smyth *et al.* (1931).

4.4.2
Discuss the following facts:
(a) The apparent dipole moment of 1,2-dichloroethane increases from $3{\cdot}7 \times 10^{-30}$ C m (1·1 D) at 30°C to $5{\cdot}1 \times 10^{-30}$ C m (1·5 D) at 270°C.
(b) The orientation polarization of gaseous ammonia indicates that its molecule possesses a permanent dipole moment of $4{\cdot}90 \times 10^{-30}$ C m (1·47 D), but the microwave spectrum of this symmetric rotor does not show a first-order Stark splitting.
(c) Some of the molecules in a calcium difluoride beam in a nonuniform electric field are deflected in a direction of *decreasing* electric field strength, whereas all the molecules in a zinc difluoride beam move in a direction of *increasing* field strength.

**Solution**

(a) Molecules of 1,2-dichloroethane exhibit hindered internal rotation about the C–C bond. The ground state has the non-dipolar staggered $C_{2h}$ conformation, but there are thermally accessible excited states which are polar. The increasing importance of these excited states with rising temperature yields an apparent dipole moment which increases with $T$; the apparent dipole moment $\mu$ is defined by the Debye equation

$$\frac{\epsilon_r - 1}{\epsilon_r + 2} - \frac{n^2 - 1}{n^2 + 2} = \frac{N\mu^2}{9\epsilon_0 kT} .$$

Using a bond dipole model, as in problem 4.4.1, we obtain the apparent dipole:

$$\mu = 2^{1/2}\mu_0 \sin\theta(1 + \overline{\cos\phi})^{1/2} .$$

At low temperatures, the non-dipolar $C_{2h}$ form ($\cos\phi = -1$) is favoured, but as the temperature is raised there is an increasing tendency towards free rotation ($\overline{\cos\phi} = 0$).

(b) The function giving the potential energy of the ammonia molecule for various positions of the nitrogen atom above and below the plane of the three hydrogens at the corners of an equilateral triangle has two minima, as shown in figure 4.4.2. When the vibrational energy is insufficient to permit the nitrogen atom to mount the peak between the two equilibrium positions, the energy levels are split by the quantum mechanical 'tunnelling' effect. The vibrational frequency is slowed down by the potential hump and occurs at microwave frequencies. The stationary states are combinations of $\psi_r$ and $\psi_\ell$, the 'right' and 'left' classical structures in which the nitrogen atom is on the right or the left of the plane of the hydrogens; since these right and left forms are degenerate, the appropriate states are the symmetric and antisymmetric combinations

$$\psi_0 = 2^{-1/2}(\psi_r + \psi_\ell)$$
$$\psi_1 = 2^{-1/2}(\psi_r - \psi_\ell)$$

shown in figure 4.4.2.

The molecule in state $\psi_1$ has an energy of $\sim 0{\cdot}8\ \mathrm{cm}^{-1} \equiv 24000$ MHz above the ground state $\psi_0$, and this corresponds to an $NH_3$ inversion frequency of 24000 MHz. In both the states $\psi_0$ and $\psi_1$ the molecule has no permanent dipole moment, for

$$\langle\psi_0|z|\psi_0\rangle = 0 = \langle\psi_1|z|\psi_1\rangle .$$

Thus there is no first-order Stark effect (that is, no change in energy proportional to the first power of the electric field strength $E_z$). However, there is a transition dipole between the states $\psi_0$ and $\psi_1$, that is

$$\langle\psi_0|z|\psi_1\rangle = \tfrac{1}{2}(\langle\psi_r|z|\psi_r\rangle - \langle\psi_\ell|z|\psi_\ell\rangle) = \langle\psi_r|z|\psi_r\rangle \neq 0 ,$$

and an electric field therefore mixes $\psi_0$ and $\psi_1$, yielding a large second-order Stark splitting (that is, an energy change proportional to $E_z^2$) and hence an orientation polarization (see problem 4.4.10). The apparent dipole moment of the $NH_3$ molecule is just the dipole $|\mu_r| = |\mu_\ell|$ of a classical form, and has the value $4\cdot90 \times 10^{-30}$ C m = $1\cdot47$ D.

Very strong electric fields (such that $|\mu_r E_z|$ is large compared to the energy difference between the levels $\psi_1$ and $\psi_0$) convert the molecules into the dipolar classical forms $\psi_r$ and $\psi_\ell$ and the Stark splitting then tends to become proportional to the first power of $E_z$.

(c) If a molecule is dipolar, the Stark splitting of the rotational states lowers the energy of some states and raises those in which the dipole tends to be opposed to the direction of the field. Those molecules whose energy is lowered by the field are deflected from the beam in the direction of increasing field strength, whereas those whose energy is raised are deflected in a direction of decreasing field strength. If the molecule is non-dipolar, the moment induced by the field due to distortion of the ground electronic and vibrational states is always in the direction of reducing the energy and all non-dipolar molecules are deflected in the direction of increasing field strength.

It has been shown by Büchler *et al.* (1964) that $CaF_2$ molecules are dipolar whereas $ZnF_2$ molecules are non-dipolar, and therefore have a linear symmetric structure.

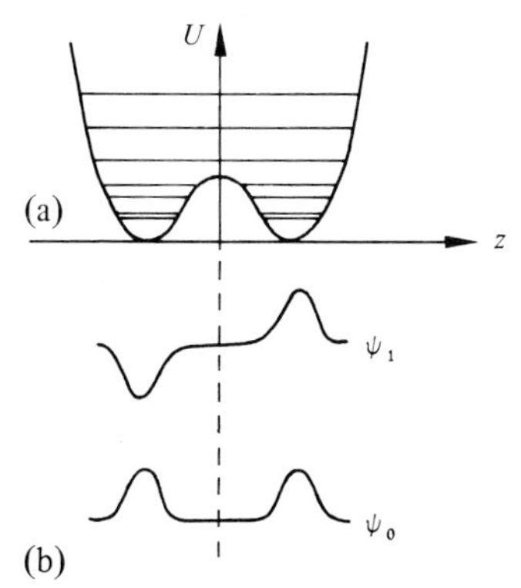


**Figure 4.4.2.** (a) The potential curve for $NH_3$. The plane of the hydrogen atoms is at $z = 0$. Some of the eigenvalues are shown to illustrate the splittings of the levels below the peak. (b) The form of the wave functions for the lowest ($\psi_0$) and the first excited vibrational state associated with the 'tunnelling' motion.

**4.4.3**

Prove that the electric dipole moment of a molecule in a state of definite parity is zero. Hence show that, if a free molecule possesses a first-order Stark effect, it must be in a state which is degenerate. Determine the dipole moments of a hydrogen atom in the stationary states of principal quantum number $n = 2$.

**Solution**

The parity operator $\mathcal{P}$ inverts the coordinates of all particles in the molecule through its centre of mass, and $\mathcal{P}$ commutes with the Hamiltonian operator $\mathcal{H}_0$ for the molecule in the absence of an external field. Since $\mathcal{P}^2 = 1$, the eigenvalues of $\mathcal{P}$ are $+1$ or $-1$ and the corresponding eigenstates have even or odd parity. The electric dipole moment operator $\boldsymbol{\mu} = \sum_i e_i \boldsymbol{r}_i$ is odd under parity $\mathcal{P}$. Hence the product $\Psi^{(i)}\mu\Psi^{(i)}$ is antisymmetric with respect to inversion for both even $\Psi^{(i)}$ and odd $\Psi^{(i)}$, where $\Psi^{(i)}$ is the wave function describing the internal state of the molecule. The expectation value of $\mu$ is zero, i.e.

$$\langle\Psi^{(i)}|\mu|\Psi^{(i)}\rangle = 0 .$$

In the presence of an external field $E_z$ the Hamiltonian is $\mathcal{H} = \mathcal{H}_0 - \mu_z E_z$, and the perturbation $-\mu_z E_z$ does *not* commute with $\mathcal{P}$ (provided that the charges giving rise to $E_z$ are not inverted), and parity is no longer conserved. If there is a first-order Stark effect, then

$$\langle\Psi^{(i)}|\mu_z|\Psi^{(i)}\rangle \neq 0$$

and the stationary state $\Psi^{(i)}$ must be of mixed parity. For this to be possible in the limit $E_z \to 0$, the energy level must be degenerate (orientational degeneracy corresponding to different values of $M$, where $-J \leqslant M \leqslant J$, is *not* sufficient, for states differing only in their values of $M$ have the same parity).

In the presence of the electric field the stationary states of the hydrogen atom with $n = 2$ are:

$$\begin{aligned}
\Psi_+ &= 2^{-1/2}(\psi_{2s} + \psi_{2p_z}) , \\
\Psi_- &= 2^{-1/2}(\psi_{2s} - \psi_{2p_z}) , \\
\Psi_{\pi_+} &= 2^{-1/2}(\psi_{2p_x} + i\psi_{2p_y}) , \\
\Psi_{\pi_-} &= 2^{-1/2}(\psi_{2p_x} - i\psi_{2p_y}) .
\end{aligned}$$

Since $\Psi_{2,s}$ and $\Psi_{2,p}$ are of even and odd parity, respectively, $\Psi_\pm$ are of mixed parity and their dipole moments are

$$\begin{aligned}
\mu_\pm &= \langle\Psi_\pm|\mu_z|\Psi_\pm\rangle = \pm\langle\Psi_{2s}|\mu_z|\Psi_{2p_z}\rangle \\
&= \pm 3ea_0 = \pm 25{\cdot}43 \times 10^{-30} \text{ C m} .
\end{aligned}$$

The two $\pi$ states are of odd parity and therefore have zero permanent dipole moment.

**4.4.4**

Why do we normally *measure* molecular moments relative to a laboratory-fixed coordinate frame, but *quote* the values relative to a molecule-fixed frame?

If the electric dipole and quadrupole moments of a molecule are defined, respectively, as

$$\boldsymbol{\mu} = \sum_i e_i \boldsymbol{r}_i \,,$$

$$\boldsymbol{\Theta} = \tfrac{1}{2}\sum_i e_i (3\boldsymbol{r}_i\boldsymbol{r}_i - r_i^2 \boldsymbol{1})$$

where the point charge $e_i$ is at $\boldsymbol{r}_i$ relative to the origin O, find the effect on $\boldsymbol{\mu}$ and $\boldsymbol{\Theta}$ of a movement of the origin of coordinates to O′ at $\boldsymbol{r}'$ from O.

What information would be obtainable from a measurement of the difference in the quadrupole moments of DF and HF relative to their centres of mass?

**Solution**

We measure a molecular dipole moment by observing the effects of a laboratory electric field on the spectrum, motion or dielectric polarization of a gas. Thus, normally we measure the space-fixed components $\mu_x$, $\mu_y$, $\mu_z$ of the molecular dipole moment vector $\boldsymbol{\mu}$. However, since our interest is generally in the structure of the molecule itself, we wish to obtain the components $\mu_1$, $\mu_2$, $\mu_3$ in some convenient Cartesian frame fixed in the molecule. Thus, even though the components $\mu_x$, $\mu_y$, $\mu_z$ may vanish in the absence of a field, as in the case of gaseous HCl because of the rotation of the molecule (or its isotropic behaviour in the nonrotating state $J = 0$), the internal components may be nonzero. In HCl the dipole component $\mu_3$ along the internuclear axis is of interest, the perpendicular components $\mu_1$ and $\mu_2$ being zero because of the symmetry of the molecule. If the molecule is rigid, its dipole moment is independent of the rotational quantum numbers; also it varies to only a small extent ($\sim$1%) if the vibrational quantum number is changed. However, if the molecule is nonrigid, the dipole moment may change considerably with state, as in 1,2-dichloroethane.

If vectors relating to the origin O′ are given primes, then

$$\boldsymbol{r}_i' = \boldsymbol{r}_i - \boldsymbol{r}' \,,$$

$$\boldsymbol{\mu}' = \sum_i e_i \boldsymbol{r}_i' = \boldsymbol{\mu} - \boldsymbol{r}'\sum_i e_i = \boldsymbol{\mu} - Q\boldsymbol{r}' \,,$$

and

$$\boldsymbol{\Theta}' = \boldsymbol{\Theta} - (\tfrac{3}{2}\boldsymbol{\mu}\boldsymbol{r}' + \tfrac{3}{2}\boldsymbol{r}'\boldsymbol{\mu} - \boldsymbol{\mu}\cdot\boldsymbol{r}'\boldsymbol{1}) + \tfrac{1}{2}Q(3\boldsymbol{r}'\boldsymbol{r}' - r'^2\boldsymbol{1}) \,.$$

In the neutral linear molecule DF, $\Theta_{33}' = \Theta_{33} - 2\mu_3 r_3'$, and the centre of mass is $\frac{19}{420}R$ further from the F nucleus than in HF, where $R$ is the internuclear distance in both molecules. Since $|\mu_3| = 6{\cdot}07 \times 10^{-30}$ C m and $R = 0{\cdot}917 \times 10^{-10}$ m,

$$\Theta_{\mathrm{DF}} - \Theta_{\mathrm{HF}} = -0{\cdot}50 \times 10^{-40}\ \mathrm{C\ m^2} \,,$$

if the sense of the dipole moment is $F^-H^+$ and is of opposite sign in the unlikely event that it is $F^+H^-$. Thus a measurement of $\Theta_{DF} - \Theta_{HF}$ would yield the sense of the dipole moment of HF.

**4.4.5**

Derive an expression for the character of the set of atomic orbitals of angular momentum $l$ under a rotation through an angle $\alpha$. Hence determine the number of independent components of the electric charge, dipole, quadrupole, and octopole moments of a charge distribution of symmetry $D_3$.

**Solution**

The irreducible representation of the three-dimensional rotation group (the point group of an atom) associated with integral angular momentum $l$ has dimension $2l+1$. The matrix representing a rotation through an angle $\alpha$ about the axis of quantization is diagonal with elements $\exp(im_l\alpha)$ and trace

$$\sum_{m_l=-l}^{l} \exp(im_l\alpha) = \frac{\sin(l+\frac{1}{2})\alpha}{\sin\frac{1}{2}\alpha} ,$$

and this is the required character.

A charge transforms as an s atomic orbital ($l = 0$), a dipole as the set of atomic orbitals having $l = 1$, a quadrupole as $l = 2$, and an octopole as $l = 3$. Hence the following character table may be constructed for $Q$, $\mu$, $\Theta$, $\Omega$ in $D_3$:

| $D_3$ | $E$ | $2C_3$ | $3C_2$ |
|---|---|---|---|
| $Q\ (l = 0)$ | 1 | 1 | 1 |
| $\mu\ (l = 1)$ | 3 | 0 | −1 |
| $\Theta\ (l = 2)$ | 5 | −1 | 1 |
| $\Omega\ (l = 3)$ | 7 | 1 | −1 |

From this table and the character table for $D_3$ in Appendix 2, the irreducible representations are obtained as follows:

$$Q = A_1 , \qquad \Theta = A_1 \oplus 2E ,$$
$$\mu = A_2 \oplus E , \qquad \Omega = A_1 \oplus 2A_2 \oplus 2E .$$

A nonvanishing moment is unaffected by a symmetry operation and must therefore transform as the totally symmetric representation $A_1$. Hence a molecule of $D_3$ symmetry may have one charge, no dipole, one independent quadrupole, and one octopole.

**4.4.6**

A rigid symmetric rotor molecule (e.g. $CH_3F$) is in the stationary state $\Psi^{(iJKM)}$, where i describes the internal vibronic state of the molecule, and $JKM$ its rotation in space with total angular momentum $J$ and components $K$ about its symmetry axis and $M$ along the axis of quantization. If the molecule has 'internal' dipole and quadrupole moments

$$\mu^{(i)} = \langle \Psi^{(i)} | \mu_a | \Psi^{(i)} \rangle ,$$
$$\Theta^{(i)} = \langle \Psi^{(i)} | \Theta_{aa} | \Psi^{(i)} \rangle ,$$

along its axis (defined by the unit vector $\boldsymbol{a}$) relative to its centre of mass, use the following direction-cosine matrix elements to determine the general expectation values of $\boldsymbol{\mu}$ and $\Theta$.

$$\langle J = 0, K = 0, M = 0 | a_\alpha | J = 0, K = 0, M = 0 \rangle = 0, \qquad (\alpha = x, y, z) ,$$

$$\langle J, K, M | a_z | J, K, M \rangle = \frac{KM}{J(J+1)} , \qquad (J \geqslant \tfrac{1}{2})$$

$$\langle J, K, M | a_z | J+1, K, M \rangle = \frac{[(J+1)^2 - K^2]^{1/2} [(J+1)^2 - M^2]^{1/2}}{(J+1)[(2J+1)(2J+3)]^{1/2}} ,$$

$$\langle J, K, M | a_z | J-1, K, M \rangle = \frac{[J^2 - K^2]^{1/2} [J^2 - M^2]^{1/2}}{J[(2J-1)(2J+1)]^{1/2}} , \qquad (J \geqslant 1)$$

$$\langle J, K, M | a_x \mp i a_y | J, K, M \pm 1 \rangle = \frac{K[(J \mp M)(J + 1 \pm M)]^{1/2}}{J(J+1)} , \qquad (J \geqslant \tfrac{1}{2})$$

$$\langle J, K, M | a_x \mp i a_y | J+1, K, M \pm 1 \rangle$$
$$= \mp \frac{[(J+1)^2 - K^2]^{1/2} [(J + 1 \pm M)(J + 2 \pm M)]^{1/2}}{(J+1)[(2J+1)(2J+3)]^{1/2}} ,$$

$$\langle J, K, M | a_x \mp i a_y | J-1, K, M \pm 1 \rangle$$
$$= \pm \frac{[J^2 - K^2]^{1/2} [(J \mp M)(J - 1 \mp M)]^{1/2}}{J[(2J-1)(2J+1)]^{1/2}} , \qquad (J \geqslant 1) .$$

**Solution**

The $\alpha$ component ($\alpha = x, y, z$) in space of the expectation value of the dipole moment operator is

$$\mu_\alpha = \langle \Psi^{(iJKM)} | \mu_\alpha | \Psi^{(iJKM)} \rangle = \mu^{(i)} \langle J, K, M | a_\alpha | J, K, M \rangle ,$$

where $\boldsymbol{a}$ is the unit vector along the axis of symmetry of the molecule. The only component of $\boldsymbol{a}$ having a nonzero diagonal matrix element is $a_z$, so, if $\boldsymbol{k}$ is the unit vector in the $z$ direction (the direction of the axis of quantization),

$$\mu_\alpha = \mu^{(i)} \langle J, K, M | a_z | J, K, M \rangle k_\alpha = \mu^{(i)} \left[ \frac{KM}{J(J+1)} \right] k_\alpha , \qquad (J \geqslant \tfrac{1}{2})$$

and

$$\mu_\alpha = 0 , \qquad (J = 0)$$

$$\Theta_{\alpha\beta} = \langle \Psi^{(iJKM)} | \Theta_{\alpha\beta} | \Psi^{(iJKM)} \rangle = \Theta^{(i)} \langle J, K, M | \tfrac{3}{2} a_\alpha a_\beta - \tfrac{1}{2} \delta_{\alpha\beta} | J, K, M \rangle ,$$

where $\delta_{\alpha\beta}$, the Kronecker symbol, is zero for $\alpha \neq \beta$ and unity for $\alpha = \beta$.

Thus

$$\Theta_{\alpha\beta} = \Theta^{(i)}\left(\tfrac{3}{2}\sum_{J',M'}\langle J,K,M|a_\alpha|J',K,M'\rangle\langle J',K,M'|a_\beta|J,K,M\rangle - \tfrac{1}{2}\delta_{\alpha\beta}\right)$$

$$= \Theta^{(i)}\left\{\frac{J^2+J-3M^2}{(2J-1)(2J+3)}\left[1-\frac{3K^2}{J^2+J}\right]\right\}\left(\tfrac{3}{2}k_\alpha k_\beta - \tfrac{1}{2}\delta_{\alpha\beta}\right), \qquad (J \geqslant 1)$$

and

$$\Theta_{\alpha\beta} = 0, \qquad (J = 0 \text{ or } \tfrac{1}{2}).$$

Hence the only nonzero components are $\mu_z$ and $\Theta_{zz} = -2\Theta_{xx} = -2\Theta_{yy}$. It is of interest to note that for a system to possess a quadrupole moment it must be in a state with $J \geqslant 1$. Thus an atom in a $^2P_{1/2}$ state has no quadrupole moment, unlike an atom in a $^2P_{3/2}$ state. This is consistent with the well-known fact that nuclei of spin 0 or $\frac{1}{2}$ have zero quadrupole moment.

4.4.7

Show that the quadrupole moment in the direction of the axis of quantization of a spinless hydrogenic atom with nuclear charge $Ze$ in the state with principal quantum number $n$, orbital quantum number $l$, and magnetic quantum number $m$ is

$$\Theta_{nlm} = -\frac{n^2[5n^2+1-3l(l+1)][l(l+1)-3m^2]}{2(2l-1)(2l+3)}ea_0^2 Z^{-2}.$$

Hence, show that

$$\Theta_{2p_z} = -2\Theta_{2p_\pi} = \tfrac{1}{3}\Theta_{3d_{z^2}} = \tfrac{2}{3}\Theta_{3d_\pi} = -\tfrac{1}{3}\Theta_{3d_\delta} = -53\cdot 83 \times 10^{-40} Z^{-2}\ \mathrm{C\,m^2}.$$

**Solution**

The required component of the quadrupole moment is $\Theta_{zz}$ and

$$\Theta_{nlm} = -\frac{\frac{1}{2}e\langle n,l,m|r^2(3\cos^2\theta-1)|n,l,m\rangle}{\langle n,l,m|n,l,m\rangle}.$$

The required result follows from the well-known properties of Legendre functions and from the radial integrals given by Condon and Shortley (1935, p.117).

For $p_z$ states $l = 1$, $m = 0$,
for $p_\pi$ states $l = 1$, $m = \pm 1$,
for $d_{z^2}$ states $l = 2$, $m = 0$,
for $d_\pi$ states $l = 2$, $m = \pm 1$,
for $d_\delta$ states $l = 2$, $m = \pm 2$.

### 4.4.8

Show that the force and torque acting on a rigid molecule of charge $Q$, dipole $\boldsymbol{\mu}$, and quadrupole $\boldsymbol{\Theta}$ in an external electrostatic field are

$$F_\alpha = QE_\alpha + \mu_\beta E'_{\alpha\beta} + \tfrac{1}{3}\Theta_{\beta\gamma} E''_{\alpha\beta\gamma} + \ldots ,$$
$$T_\alpha = \epsilon_{\alpha\beta\gamma}(\mu_\beta E_\gamma + \tfrac{2}{3}\Theta_{\beta\delta} E'_{\gamma\delta} + \ldots) ,$$

where $\boldsymbol{\mu}$ and $\boldsymbol{\Theta}$ are referred to an origin at which the external field strength is $\boldsymbol{E}$ and the first and second derivatives of $\boldsymbol{E}$ are $\boldsymbol{E}'$ and $\boldsymbol{E}''$; $\epsilon_{\alpha\beta\gamma}$ is the unit skew-symmetric tensor which is +1 (−1) if $\alpha$, $\beta$, $\gamma$ is an even (odd) permutation of $x$, $y$, $z$, and is zero if any two of $\alpha$, $\beta$, $\gamma$ are the same. Hence $\epsilon_{\alpha\beta\gamma}\mu_\beta E_\gamma \equiv [\boldsymbol{\mu} \times \boldsymbol{E}]_\alpha$.

Find $\boldsymbol{F}$ and $\boldsymbol{T}$ if the molecule is axially symmetric.

**Solution**

The energy of a rigid charge system in an external field is

$$W = \sum_i e_i \phi_i = \sum_i e_i [\phi_o + (\nabla_\alpha \phi)_o r_{i\alpha} + \tfrac{1}{2}(\nabla_\alpha \nabla_\beta \phi)_o r_{i\alpha} r_{i\beta} + \ldots] ,$$

where $e_i$ is the element of charge at $\boldsymbol{r}_i$ relative to an origin O; the subscript ‘o’ indicates a value at O. The external potential $\phi$ satisfies Laplace’s equation

$$\nabla^2 \phi = \nabla_\alpha \nabla_\alpha \phi = 0 ,$$

so

$$W = Q\phi_o - \mu_\alpha E_\alpha - \tfrac{1}{3}\Theta_{\alpha\beta} E'_{\alpha\beta} - \ldots ,$$

where $E_\alpha = -(\nabla_\alpha \phi)_o$ and $E'_{\alpha\beta} = -(\nabla_\alpha \nabla_\beta \phi)_o$ are the field and field gradient at O.

The force acting on the system is

$$F_\alpha = -\nabla_\alpha W = \sum_i e_i E_{i\alpha} = QE_\alpha + \mu_\beta E'_{\alpha\beta} + \tfrac{1}{3}\Theta_{\beta\gamma} E''_{\alpha\beta\gamma} + \ldots ,$$

and the torque is

$$T_\alpha = \epsilon_{\alpha\beta\gamma} \sum_i e_i r_{i\beta} E_{i\gamma} = \epsilon_{\alpha\beta\gamma}[\mu_\beta E_\gamma + \tfrac{2}{3}\Theta_{\beta\delta} E'_{\gamma\delta} + \ldots] .$$

If the molecule is symmetric about its $\boldsymbol{a}$ axis,

$$\mu_\alpha = \mu a_\alpha , \quad \Theta_{\alpha\beta} = \tfrac{1}{2}\Theta(3a_\alpha a_\beta - \delta_{\alpha\beta}) ,$$

and

$$W = Q\phi_o - \mu E_a - \tfrac{1}{2}\Theta E'_{aa} ,$$
$$F_\alpha = QE_\alpha + \mu E'_{\alpha a} + \tfrac{1}{2}\Theta E'_{\alpha aa} + \ldots ,$$
$$T_\alpha = \epsilon_{\alpha a\beta}[\mu E_\beta + \Theta E'_{\beta a} + \ldots] .$$

**4.4.9**

Find the potential and electric field strength at a point $\boldsymbol{r}$ from a molecule at a point O relative to which its charge is $Q$, dipole $\mu$, and quadrupole $\Theta$.

**Solution**

The potential is the sum of contributions from the elements of charge $e_i$ at $\boldsymbol{r}_i$ from O:

$$\Phi = \sum_i \frac{e_i}{|\boldsymbol{r}-\boldsymbol{r}_i|} = \sum_i e_i(r^2 - 2r_{i\alpha}r_\alpha + r_i^2)^{-1/2}$$

$$= \sum_i e_i r^{-1}[1 + r_{i\alpha}r_\alpha r^{-2} + \tfrac{3}{2}r_{i\alpha}r_\alpha r_{i\beta}r_\beta r^{-4} - \tfrac{1}{2}r_i^2 r^{-2} + ...]$$

$$= \sum_i e_i r^{-1}\left[1 + \sum_{n=1}^{\infty} r_i^n r^{-n} P_n(\cos\theta_i)\right]$$

$$= Qr^{-1} + \mu_\alpha r_\alpha r^{-3} + \Theta_{\alpha\beta}r_\alpha r_\beta r^{-5} + \ldots,$$

where $P_n(\cos\theta_i)$ is the $n$th Legendre polynomial of the cosine of the angle between $\boldsymbol{r}_i$ and $\boldsymbol{r}$.

$$E_\alpha = -\nabla_\alpha\Phi = Qr_\alpha r^{-3} + \mu_\beta(3r_\alpha r_\beta - r^2\delta_{\alpha\beta})r^{-5}$$

$$+ \Theta_{\beta\gamma}(5r_\alpha r_\beta r_\gamma - r^2 r_\gamma\delta_{\alpha\beta} - r^2 r_\beta\delta_{\alpha\gamma} - r^2 r_\alpha\delta_{\beta\gamma})r^{-7} + \ldots$$

$$= -Q\nabla_\alpha(r^{-1}) + \mu_\beta\nabla_\alpha\nabla_\beta(r^{-1}) - \tfrac{1}{3}\Theta_{\beta\gamma}\nabla_\alpha\nabla_\beta\nabla_\gamma(r^{-1}) + \ldots.$$

The term $-\Theta_{\beta\gamma}r_\alpha\delta_{\beta\gamma}r^{-5}$, which is zero because $\Theta$ is traceless, has been added to the quadrupolar contribution to $\boldsymbol{E}$ to make it expressible in terms of $\nabla_\alpha\nabla_\beta\nabla_\gamma(r^{-1})$.

If the molecule is axially symmetric

$$\Phi = Qr^{-1} + \mu r^{-2}\cos\theta + \Theta r^{-3}(\tfrac{3}{2}\cos^2\theta - \tfrac{1}{2}) + \ldots,$$

where $\theta$ is the angle between the axis of the molecule and $\boldsymbol{r}$.

**4.4.10**

(a) Prove that the total dipole moment of a molecule is minus the derivative of its energy with respect to the field strength, i.e.

$$\mu_z = -\frac{\partial W}{\partial E_z}.$$

(b) Show that a linear molecule in a $^1\Sigma$ state with a permanent dipole moment $\mu$ and polarizability components $\alpha_\parallel$ and $\alpha_\perp$ has no first-order Stark energy, and to second order the energy is

$$W^{(JM)} = (J^2+J)B - \tfrac{1}{6}(\alpha_\parallel + 2\alpha_\perp)E_z^2$$

$$+ \frac{J^2+J-3M^2}{(2J-1)(2J+3)}\left[\frac{\mu^2}{2J(J+1)B} - \tfrac{1}{3}(\alpha_\parallel - \alpha_\perp)\right]E_z^2 \qquad (\text{for } J \neq 0),$$

and

$$W^{(00)} = -\tfrac{1}{6}(\alpha_\parallel + 2\alpha_\perp)E_z^2 - \tfrac{1}{6}\mu^2 E_z^2 B^{-1} \ .$$

(c) Use the results in parts (a) and (b) to show that the orientation dielectric polarization of a gas of linear molecules in a $^1\Sigma$ state arises entirely from the molecules in the nonrotating state.

**Solution**

(a) The energy $W^{(i)}$ of a molecule in the quantum state $\Psi^{(i)}$ is, for a normalized wave function $\Psi^{(i)}$,

$$W^{(i)} = \langle \Psi^{(i)} | \mathcal{H} | \Psi^{(i)} \rangle \ ,$$

where $\mathcal{H}$ is the Hamiltonian operator for the molecule in the external field. Hence

$$\begin{aligned} \frac{\partial W^{(i)}}{\partial E_z} &= \left\langle \Psi^{(i)} \middle| \frac{\partial \mathcal{H}}{\partial E_z} \middle| \Psi^{(i)} \right\rangle + \left\langle \frac{\partial \Psi^{(i)}}{\partial E_z} \middle| \mathcal{H} \middle| \Psi^{(i)} \right\rangle + \left\langle \Psi^{(i)} \middle| \mathcal{H} \middle| \frac{\partial \Psi^{(i)}}{\partial E_z} \right\rangle \\ &= -\langle \Psi^{(i)} | \mu_z | \Psi^{(i)} \rangle + W^{(i)} \left[ \left\langle \frac{\partial \Psi^{(i)}}{\partial E_z} \middle| \Psi^{(i)} \right\rangle + \left\langle \Psi^{(i)} \middle| \frac{\partial \Psi^{(i)}}{\partial E_z} \right\rangle \right] \\ &= -\mu_z^{(i)} \ . \end{aligned}$$

The expression in square brackets is zero, as it is equal to the derivative with respect to $E_z$ of $\langle \Psi^{(i)} | \Psi^{(i)} \rangle \equiv 1$.

(b) The effective Hamiltonian for a linear molecule in a $^1\Sigma$ state in an electric field $E_z$ is

$$\mathcal{H} = \mathcal{H}_{\text{rot}} - \mu a_z E_z - \tfrac{1}{3}(\alpha_\parallel - \alpha_\perp)E_z^2(\tfrac{3}{2}a_z^2 - \tfrac{1}{2}) - \tfrac{1}{6}(\alpha_\parallel + 2\alpha_\perp)E_z^2 \ ,$$

where $\boldsymbol{a}$ is the unit vector along the axis of the molecule and $\mathcal{H}_{\text{rot}}$ is the Hamiltonian of the free rotator.

Using the matrix elements given in problem 4.4.6 for $K = 0$, we obtain the energy to first order:

$$\langle J, M | \mathcal{H} | J, M \rangle = (J^2 + J)B - \tfrac{1}{3}(\alpha_\parallel - \alpha_\perp)E_z^2 \frac{J^2 + J - 3M^2}{(2J-1)(2J+3)} - \tfrac{1}{6}(\alpha_\parallel + 2\alpha_\perp)E_z^2 \qquad (\text{for } J \neq 0).$$

and

$$\langle 0, 0 | \mathcal{H} | 0, 0 \rangle = -\tfrac{1}{6}(\alpha_\parallel + 2\alpha_\perp)E_z^2 \qquad (\text{for } J = 0) \ ,$$

where $B$ is the rotational constant.

The second-order perturbation energy in $\mu^2 E_z^2$ is

$$\mu^2 E_z^2 \sum_{J' \neq J} \frac{\langle J, M | a_z | J', M \rangle^2}{(J^2 + J - J'^2 - J')B} = \begin{cases} \dfrac{J^2 + J - 3M^2}{(2J-1)(2J+3)} \dfrac{\mu^2 E_z^2}{2B(J^2+J)} & (\text{for } J \neq 0) \ , \\[2ex] -\dfrac{6\mu^2 E_z^2}{B} & (\text{for } J = 0) \ . \end{cases}$$

Addition of the first-order energy and the second-order energy yields the required results.

(c) The dielectric polarization of a gas is proportional to the mean dipole moment of a molecule:

$$\epsilon - 1 = \frac{N}{\epsilon_0 E_z Z} \sum_i \mu_z^{(i)} \exp\left[-\frac{W^{(i)}}{kT}\right],$$

where $\epsilon_0$ is the permittivity of free space, $N$ is the number of molecules in unit volume, and $Z$ is the partition function:

$$Z = \sum_i \exp\left[-\frac{W^{(i)}}{kT}\right].$$

Writing the energy as a power series in $E_z$, we obtain

$$W^{(i)} = W_0^{(i)} + W_1^{(i)} E_z + W_2^{(i)} E_z^2 + \ldots,$$

$$\mu_z^{(i)} = -\frac{\partial W^{(i)}}{\partial E_z} = -W_1^{(i)} - 2W_2^{(i)} E_z,$$

and in weak fields

$$\epsilon - 1 = -\frac{N}{\epsilon_0 Z} \sum_i \left\{2W_2^{(i)} - \frac{[W_1^{(i)}]^2}{kT}\right\} \exp\left[-\frac{W_0^{(i)}}{kT}\right].$$

For molecules in $^1\Sigma$ states, $W_1^{(i)} = 0$ and the orientation contribution to $W_2^{(i)}$ is proportional to $(J^2 + J - 3M^2)\mu^2$ for $J \neq 0$, and this reduces to zero on summing over $M$, since

$$\sum_{M=-J}^{J} M^2 = \tfrac{1}{3}(J^2 + J)(2J + 1).$$

Thus, only those molecules with $J = 0$ contribute to the orientation polarization. This is attributed to the fact that for $J \neq 0$ there are some $M$ states which are lowered in energy by the field (e.g. $M = \pm J$) and others are raised (e.g. $M = 0$)—the average energy change is zero. However, the lowest rotational state ($J = 0$) is distorted from its isotropy in the absence of the field and it is thereby lowered in energy by $6\mu^2 E_z^2/B$.

**4.4.11**

A gas comprised of linear molecules of quadrupole moment $\Theta$ experiences an external electric field gradient of the form

$$\boldsymbol{E}' = \begin{pmatrix} E'_{xx} & 0 & 0 \\ 0 & -E'_{xx} & 0 \\ 0 & 0 & 0 \end{pmatrix},$$

as at a point midway between two charged parallel wires. If the molecules have polarizability components $\alpha_\parallel$ and $\alpha_\perp$ along and at right angles to their axes, show that their partial alignment by the field gradient leads to an anisotropy $n_{xx} - n_{yy}$ in the refractive index of the gas:

$$n_{xx} - n_{yy} = \frac{NE'_{xx}(\alpha_{\parallel} - \alpha_{\perp})\Theta}{15\epsilon_0 kT} ,$$

where $N$ is the number of molecules per unit volume. Contrast this result with the anisotropy induced in a dipolar gas by a uniform electric field $E_x$.

**Solution**

The refractive index of a gas having $N$ molecules in unit volume is related to the mean molecular polarizability $\bar{\alpha}$ by

$$n - 1 = \frac{N}{2\epsilon_0}\bar{\alpha} .$$

For a linear molecule

$$\alpha_{\alpha\beta} = \alpha_{\perp}\delta_{\alpha\beta} + (\alpha_{\parallel} - \alpha_{\perp})a_{\alpha}a_{\beta} ;$$

hence

$$n_{xx} - n_{yy} = \frac{N}{2\epsilon_0}(\alpha_{\parallel} - \alpha_{\perp})(\overline{a_x^2} - \overline{a_y^2}) ,$$

where $\boldsymbol{a}$ is the unit vector along the axis of the molecule. The quadrupolar potential energy in the field gradient $\boldsymbol{E}'$ is (see problem 4.4.8)

$$W = W_0 - \tfrac{1}{2}\Theta E'_{aa} = W_0 - \tfrac{1}{2}\Theta E'_{xx}(a_x^2 - a_y^2) .$$

For a slightly perturbed classical rotator we can therefore write:

$$\overline{a_x^2} - \overline{a_y^2} = \frac{\int (a_x^2 - a_y^2)\exp(-W/kT)\,\mathrm{d}\boldsymbol{a}}{\int \exp(-W/kT)\,\mathrm{d}\boldsymbol{a}} = \frac{2\Theta E'_{xx}}{15kT} .$$

Hence

$$n_{xx} - n_{yy} = \frac{N(\alpha_{\parallel} - \alpha_{\perp})\Theta E'_{xx}}{\epsilon_0 \quad 15kT} ,$$

and this equation is the basis of the experimental determination of molecular quadrupole moments by the induced-birefringence method.

In a uniform electric field, the potential energy of a dipole moment is

$$W = W_0 - \mu E_a = W_0 - \mu E_x a_x ,$$

and

$$\overline{a_x^2} - \overline{a_y^2} = \frac{\mu^2 E_x^2}{15k^2T^2} .$$

Thus, the orienting influence of an electric field gradient is proportional to the first power of the product of the applied voltage and the molecular quadrupole moment, while that of a uniform field is proportional to the square of the product of the voltage and the dipole moment.

**4.4.12**

Suppose that a rigid molecule may be represented by the sum of its constituent 'bonds' each of which has a bond dipole moment $\mu_i$ along the bond. Write down formulae for the total dipole, quadrupole, and octopole moments of the molecule. Hence show that the quadrupole moment $\Theta_{zz}$ of two equal but opposite bond dipoles $\mu$ separated by a distance $2R$ is $4\mu R$ and that the octopole moment of four tetrahedrally disposed bonds, each bearing a dipole moment $\mu$ at a distance $R$ from the centre, is $\Omega = \Omega_{xyz} = (10/3^{1/2})\mu R^2$, where the $x$, $y$, $z$ axes are the sides of a cube enclosing the tetrahedron whose bonds point along the diagonals through (1, 1, 1), (1, −1, −1), (−1, 1, −1), and (−1, −1, 1).

**Solution**

If the multipole moments of a molecule are represented as a sum of the contributions of the bond dipole moments $\mu_i$ at the position $r_i$ relative to the origin O, then

$$\mu_\alpha = \sum_i \mu_{i\alpha} ,$$

$$\Theta_{\alpha\beta} = \sum_i [\tfrac{3}{2}\mu_{i\alpha} r_{i\beta} + \tfrac{3}{2}\mu_{i\beta} r_{i\alpha} - \mu_{i\gamma} r_{i\gamma} \delta_{\alpha\beta}] ,$$

$$\begin{aligned}\Omega_{\alpha\beta\gamma} = \tfrac{5}{2} \sum_i (\mu_{i\alpha} r_{i\beta} r_{i\gamma} + \mu_{i\beta} r_{i\alpha} r_{i\gamma} + \mu_{i\gamma} r_{i\alpha} r_{i\beta}) \\ - \sum_i \mu_{i\delta} r_{i\delta} (r_{i\alpha}\delta_{\beta\gamma} + r_{i\beta}\delta_{\gamma\alpha} + r_{i\gamma}\delta_{\alpha\beta}) \\ -\tfrac{1}{2} \sum_i r_i^2 (\mu_{i\alpha}\delta_{\beta\gamma} + \mu_{i\beta}\delta_{\gamma\alpha} + \mu_{i\gamma}\delta_{\alpha\beta}) ,\end{aligned}$$

where the octopole moment is defined as a sum of the contributions of all the point charges $e_j$ in the molecule:

$$\Omega_{\alpha\beta\gamma} = \tfrac{1}{2} \sum_j e_j [5 r_{j\alpha} r_{j\beta} r_{j\gamma} - r_j^2 (r_{j\alpha}\delta_{\beta\gamma} + r_{j\beta}\delta_{\alpha\gamma} + r_{j\gamma}\delta_{\alpha\beta})] .$$

Hence the quadrupole moment of two equal but opposite bond dipoles $\mu$ at $\pm R$ on the $z$ axis is

$$\Theta_{zz} = 4\mu R .$$

There is only one independent octopole moment of a tetrahedral molecule and that is $\Omega_{xyz} = \Omega$. Each bond contains a dipole $\mu$ at a distance $R$ from the centre and directed outwards along the bond. The dipoles are located at the four points $(3^{-1/2}R, 3^{-1/2}R, 3^{-1/2}R)$, $(3^{-1/2}R, -3^{-1/2}R, -3^{-1/2}R)$, $(-3^{-1/2}R, 3^{-1/2}R, -3^{-1/2}R)$, and $(-3^{-1/2}R, -3^{-1/2}R, 3^{-1/2}R)$ and each contributes equally to $\Omega_{xyz}$. Hence

$$\Omega_{xyz} = (10/3^{1/2})\mu R^2 .$$

The bond dipole model works reasonably well, since a bond normally carries only a small charge or no charge at all. If the bond were charged, its dipole moment would depend on its position (see problem 4.4.4) and the moments of the molecule as a whole would not be additive in the constituent bonds. One might expect bonds to have significant quadrupole moments in addition to their dipoles; in that case the molecular quadrupole moment would become

$$\Theta_{\alpha\beta} = \sum_i \Theta_{i\alpha\beta} + \sum_i \left[\tfrac{3}{2}\mu_{i\alpha} r_{i\beta} + \tfrac{3}{2}\mu_{i\beta} r_{i\alpha} - \mu_{i\gamma} r_{i\gamma} \delta_{\alpha\beta}\right] .$$

This more general result has its difficulties, for few if any bond quadrupoles are known and there is the problem of finding the origin for $\mu_i$ and $\Theta_i$ in the bond.

**References**

Abramowitz, M., Stegun, I. A., 1964, *Handbook of Mathematical Functions* (US National Bureau of Standards, Washington, DC).
Büchler, A., Stauffer, J. L., Klemperer, W., 1964, *J. Am. Chem. Soc.,* **86**, 4544.
Condon, E. U., Shortley, G. H., 1935, *The Theory of Atomic Spectra* (Cambridge University Press, London).
Smith, J. W., 1955, *Electric Dipole Moments* (Butterworths, Sevenoaks, Kent).
Smyth, C. P., Dornte, R. W., Wilson, E. B., 1931, *J. Am. Chem. Soc.,* **53**, 4242.

**Bibliography**

The following are recommended for use in conjunction with this section:
Bleaney, B. I., Bleaney, B., 1976, *Electricity and Magnetism*, 3rd edition (Oxford University Press, Oxford).
Buckingham, A. D., 1959, *Q. Rev. Chem. Soc.,* **13**, 183.
Buckingham, A. D., 1970, "Electric moments of molecules", in *Physical Chemistry: An Advanced Treatise,* volume 4, Eds H. Eyring, D. Henderson, W. Jost (Academic Press, New York), pp.349-386.
Debye, P., 1929, *Polar Molecules* (Chemical Catalog Company, New York).
Le Fèvre, R. J. W., 1953, *Dipole Moments,* 3rd edition (Methuen, London).
Smith, J. W., 1955, *Electric Dipole Moments* (Butterworths, Sevenoaks, Kent).
Smyth, C. P., 1955, *Dielectric Behaviour and Structure* (McGraw-Hill, New York).
Townes, C. H., Schawlow, A. L., 1955, *Microwave Spectroscopy* (McGraw-Hill, New York).
Van Vleck, J. H., 1932, *Theory of Electric and Magnetic Susceptibilities* (Oxford University Press, London).

## 4.5 Optical rotatory dispersion and circular dichroism

**S F Mason** King's College, University of London

### 4.5.0 Introduction: optical activity

An optically active compound has different refractive indices, and different extinction coefficients, for left- and right-circularly polarised light. The plane of linearly polarised light, which is composed of left- and right-circular components with equal amplitudes, undergoes a rotation on traversing a transparent optically active medium owing to the circular birefringence. The circular birefringence, and thus the molecular rotation $[M]$ of an optical isomer, is related to the circular dichroism and the frequencies of all of the absorption bands in the spectrum of the molecule. For light of frequency $\nu$ we have the approximate relationship:

$$[M]_\nu = \sum_i [M_i]_\nu = \frac{96\pi N}{hc} \sum_i \frac{R_i \nu^2}{\nu_i^2 - \nu^2}, \tag{4.5.0a}$$

where $R_i$ is the rotational strength and $\nu_i$ the frequency of a particular electronic transition of the optical isomer. The rotational strength $R_i$ is given by the area of the circular dichroism band due to the transition with the frequency $\nu_i$.

The circular dichroism (CD) spectrum of a chiral molecule, presenting the separate contribution of each individual electronic transition to the optical activity of the isomer, is generally more informative than the optical rotatory dispersion (ORD) curve, which generally represents the summed contributions of all of the transitions of the molecule. At a radiation frequency $\nu$ close to a transition frequency $\nu_i$ the corresponding rotational strength $R_i$ provides the principal contribution to the rotation $[M]_\nu$, although the latter does not become infinite at $\nu = \nu_i$, as the simple relation (4.5.0a) suggests. In practice a factor neglected in equation (4.5.0a), the finite width of the CD band representing $R_i$, results in a change of $[M]$ from a large positive to a large negative value as $\nu$ progresses to higher frequencies through $\nu_i$, if $R_i$ is positive. The sigmoid ORD curve in the region of $\nu_i$ is then 'anomalous' in a positive sense, and either the CD or the ORD curve represents a positive Cotton effect (figure 4.5.0a). Conversely, a negative CD band is associated with a negative sigmoid ORD curve (figure 4.5.0b).

The main chemical applications of ORD and CD measurements are stereochemical and spectroscopic, and both are based upon the theory of the molecular origin of optical activity. Here we are concerned with problems of molecular structure, notably of absolute stereochemical configuration, for which only the general aspects of the theory of electronic spectroscopy and optical activity are required. The absorption of light by a chiral molecule gives rise to a helical charge displacement in the promotion of an electron from the ground state to an excited state with a different charge distribution. A right-handed helical displacement of charge results in a positive Cotton effect, whereas a left-handed displacement produces a negative CD absorption or ORD anomaly. The displacement of an electron through a helical path involves the linear motion of charge along the helix axis, generating a transient electric dipole moment; and a rotation of charge about that axis, producing a transient magnetic dipole moment. For a finite Cotton effect the electric and magnetic moment of an electronic transition, or components of those moments, are collinear, the rotational strength $R_{0n}$ of the transition $\psi_n \leftarrow \psi_0$ representing the scalar product of the electric, $\mu_{0n}$, and magnetic, $m_{n0}$, transition moments:

$$R_{0n} = -i\langle\psi_0|\boldsymbol{\mu}|\psi_n\rangle\langle\psi_n|\boldsymbol{m}|\psi_0\rangle \tag{4.5.0b}$$

$$\approx \mu_{0n} m_{n0} \cos\theta \,, \tag{4.5.0c}$$

where $\theta$ is the angle between the directions of the two moments. The two moments are parallel for a positive and antiparallel for a negative rotational strength.

The particular left- or right-handed chirality of a charge displacement during an electronic transition depends upon the stereochemical dissymmetry of the molecule and upon the orientation in the molecular

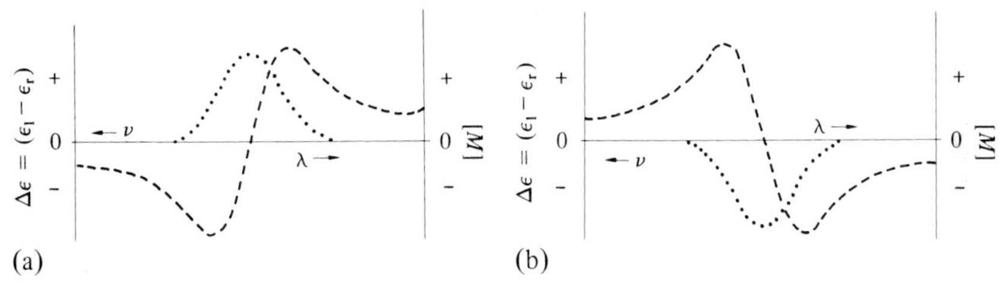


**Figures 4.5.0a and 4.5.0b.** The relation between the wavelength ($\lambda$) or frequency ($\nu$) dispersion of the molecular rotation $[M]$ (dashed line) and the circular dichroism $(\epsilon_l - \epsilon_r)$ absorption (dotted line) for (4.5.0a) a positive Cotton effect and (4.5.0b) a negative Cotton effect.

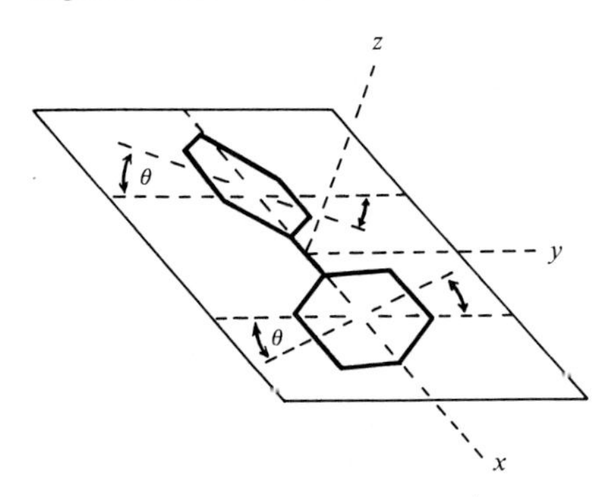


**Figure 4.5.0c.** Dissymmetric biphenyl with a dihedral angle $2\theta$ between the molecular planes of the benzene rings.

coordinate frame of the axis of the helical electronic motion, that is, of the direction of the electric and magnetic transition moments. A given optical isomer presents complementary stereochemical chiralities along different axes of its coordinate frame and the rotational strengths of electronic transitions directed along these axes are then of opposite sign.

In the case of the biphenyl isomer depicted (figure 4.5.0c) the molecule has the stereochemical form of a right-handed two-bladed propeller viewed along the $z$ axis, but along the $y$ axis the molecule has a left-handed propeller form. An electronic transition of biphenyl with the component electric transition moments (I) in the two benzene rings produces a right-handed helical charge displacement along and around the $z$ axis, giving a positive rotational strength, whereas the transition with the component moments (II) has a negative rotational strength due to a left-handed helical motion of charge along and about the $y$ axis. The rotational strengths of the transitions of biphenyl (I) and (II), derived from a particular benzene excitation, are equal in magnitude and opposite in sign, illustrating the general sum rule which specifies that the rotational strengths of an optical isomer sum to zero over all of the electronic transitions of the molecule. The motions of the electrons in a molecule are tied to the molecular framework, and the sum rule for rotational strengths reflects the complementary stereochemical chiralities presented by an optical isomer along orthogonal reference axes.

*Nomenclature*

In order to define the absolute stereochemical configuration of an optical isomer by means of the basic helix description, a reference axis is required to avoid the ambiguity inherent in the dual chirality of a dissymmetric molecule. For molecules with the stereochemical form of a regular helix, such as the helicene series of aromatic hydrocarbons or polypeptides in the α-helix conformation, the screw axis serves as a reference direction and, in the nomenclature of Cahn *et al.* (1966), the P configuration describes an optical isomer with a plus or right-handed helical form and the M configuration the enantiomer with a minus or left-handed helical structure.

The extension of this nomenclature and equivalent descriptions to molecules with a propeller form, such as the dissymmetric biaryls or chelated metal-ion complexes, requires the specification of a reference axis or the introduction of a convention to describe the chiral elements of an optical isomer. Thus the biphenyl illustrated in figure 4.5.0c has the P($z$) or M($y$) configuration. Similarly, the $(+)_D$-tris(ethylenediamine)cobalt(III) ion (III) has the M($C_3$) or P($C_2$) configuration, where $C_3$ refers to a threefold and $C_2$ to a twofold rotational symmetry axis of the complex.

Dissymmetric molecules devoid of elements of symmetry providing reference axes for a helical description of configuration may possess an element or elements of chirality which serve to define the configuration. In an octahedral metal complex containing two or more chelate rings, each pair of rings has a mutual chiral relationship if the two rings do not share a vertex or possess a common mean plane. For such coordination compounds a convention has been adopted (IUPAC, 1970) whereby the octahedral edge spanned by one chelate ring serves as the helix axis used to specify the relative chirality of another ring. A left-handed helical relationship of the two chelate rings defines the Λ configuration and a right-handed relationship the Δ configuration.

In a metal complex containing more than one pair of chelate rings the overall configuration is defined as Λ if the number of Λ ring-pairs exceeds the number of pairs with a Δ chirality, or as Δ if the converse inequality holds (IUPAC, 1970). Each of the three pairs of chelate rings in the complex ion $(+)_D$-$[Co(en)_3]^{3+}$, (III), exhibits the Λ chiral relationship, and the complex (III) has the Λ configuration overall. An advantage of this convention is that a Λ or Δ configuration may be specified for dissymmetric chelated complexes lacking an element of symmetry. In an extension of the IUPAC (1970) convention the conformation of a chelate ring is specified as λ or δ, depending upon the left- or right-handed helical relationship of the line joining the coordinated atoms of the ligand to the line connecting the atoms bonded to those ligators, e.g. the N···N line and the C—C bond of ethylenediamine.

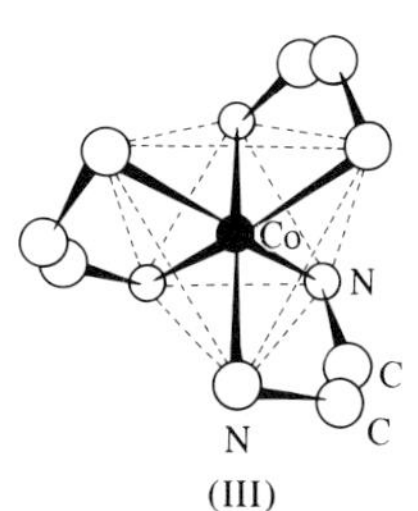

(III)

A helical stereochemical form is not obvious in molecules owing their dissymmetry to an asymmetric centre, although a helix model has been proposed (Brewster, 1967) for compounds of the type X*abcd* in terms of the sequence of the polarisabilities and of the relative dispositions of the groups *a*, *b*, *c*, and *d*, around the asymmetric centre X. The configuration of such a molecule is conveniently specified by the convention of Cahn *et al.* (1966). In this convention the atoms in *a*, *b*, *c*, and *d* directly bonded to the asymmetric centre X are arranged in order of decreasing mass number. The atom with the largest mass number has the highest priority in the sequence, and the atom with the smallest mass number, or a lone-pair, which is taken as a pseudosubstituent with zero mass number, has the lowest precedence.

If the priority order in the molecule X*abcd* is $a > b > c > d$, and the sequence $a \rightarrow b \rightarrow c$ follows a clockwise direction around the X—*d* bond, pointing away from the observer (IV), the sequence corresponds to a right-handed helix with the X—*d* bond as an axis, and the molecule has the (*R*) configuration (*rectus*, right). Conversely if the sequence $a \rightarrow b \rightarrow c$ follows an anticlockwise direction around the X—*d* bond (V), corresponding to a left-handed helix, the configuration of the molecule is specified as (*S*) (*sinister*, left).

If two similar atoms are bonded to the asymmetric centre the next-nearest neighbouring atoms are considered in assigning precedence. If the second atom out from the centre affords no choice, the third and then subsequent atoms are taken into account, following the bonds between atoms with the highest mass number. Further subsidiary sequence rules have been proposed (Cahn *et al.*, 1966) to cover multiply-bonded or stereoisomeric groups attached to an asymmetric centre, and to include molecules with axial chirality, such as the dissymmetric biaryls and allenes, or with planar chirality, e.g. the optically active paracyclophanes.

(IV) (V)

*Types of optical activity*

Detailed theories of optical activity range between the two limiting cases of the inherently dissymmetric chromophore and the symmetric chromophore in a chiral molecular environment. The first case is exemplified by the helicene series of overcrowded aromatic hydrocarbons which contain a helical $\sigma$-bond framework and a $\pi$-electron system. Each $\pi^* \leftarrow \pi$ transition of a helicene isomer involves a helical charge displacement either along the twofold rotational symmetry axis of the molecule, or the helix axis perpendicular to the mean molecular plane, or the third Cartesian molecular axis. The sign of the rotational strength of a given $\pi^* \leftarrow \pi$ transition depends upon the absolute configuration of the helicene isomer and the polarisation direction of that transition. If the latter is identified, the sign of the corresponding CD band or ORD anomaly gives the configuration of the isomer.

In the second limiting case the chromophore is symmetric in the sense that it possesses secondary elements of symmetry, an inversion centre, reflection plane, or rotation-reflection axes generally, and thus is not intrinsically optically active. If the secondary elements of symmetry are removed by substituents as in the replacement of the ammonia ligands of $[Co(NH_3)_6]^{3+}$ by ethylenediamine to form $[Co(en)_3]^{3+}$, (III), the molecule becomes optically active, owing to the mixing of the electric dipole with the magnetic dipole electronic transitions of the chromophore under the perturbing field of the substituents. The field is regarded as a static Coulombic potential if the electronic absorption spectrum is not appreciably changed by the substitution. The spatial form of the potential provides a regional rule relating the position of a substituent conferring dissymmetry on the molecule with the sign of the rotational strength induced in a given electronic transition of the chromophore.

The spatial form of the dissymmetric field produced by the substituents in a chiral molecule containing a symmetric chromophore may be obtained by two methods. In the first of these, adopted in the derivation of the octant rule (Moffitt *et al.*, 1961) for the optical activity of the $\pi^* \leftarrow n$ transition of dissymmetric carbonyl compounds, the nodal surfaces of the molecular orbitals connected by the electronic transition in the unperturbed chromophore provide the boundary planes of the dissymmetric field. The $\pi^* \leftarrow n$ transition of the carbonyl group involves the rotatory charge displacement of an electron from the $2p_y$ lone-pair orbital of the oxygen atom, with a node in the $xz$ plane, to the $\pi_x^*$ orbital of the carbonyl group, with one node in the $xy$ plane bisecting the carbonyl bond and another in the $yz$ molecular plane.

A substituent placed in a nodal plane does not perturb the $\pi^* \leftarrow n$ transition of the carbonyl group. Disposed in one of the octants defined by the three nodal planes of the orbitals considered, a substituent perturbs and mixes the electronic transitions of the chromophore, adding to the intrinsic magnetic dipole of the $\pi^* \leftarrow n$ transition a contribution from an allowed electric dipole transition with a collinear moment. The direction of the electric dipole moment induced by a substituent reverses relative to the intrinsic magnetic moment if the substituent is transferred from one side of a nodal surface to the other, so that the Cotton effect of the carbonyl $\pi^* \leftarrow n$ transition changes sign if the substituent is transferred to an adjacent octant with a common face. The sign of the Cotton effect induced by a substituent in a particular octant is obtained either empirically, by reference to the ORD or CD curve of a chiral ketone with a known absolute configuration, or theoretically on the assumption that a particular electric dipole transition, e.g. the $3d_{yz} \leftarrow 2p_y$ transition of the oxygen atom, is mixed with the magnetic dipole $\pi^* \leftarrow n$ transition.

The second method of deriving regional rules relating the stereochemistry to the optical activity of a chiral molecule, due to Schellman (1966, 1968), is based upon the symmetry properties of the perturbation due to a substituent bonded to a symmetric chromophore. The potential required to mix the electric with the magnetic dipole transitions of a symmetric chromophore, giving a helical charge displacement overall, has the symmetry properties of a helix in the molecular point group of the symmetry elements which the chromophore possesses. These properties are spanned by the pseudoscalar representation of the point group. A pseudoscalar property is antisymmetric, changing sign, under the secondary symmetry operations of the group—inversion, reflection, and rotation-reflection generally—but retains its sign and is symmetric with respect to primary symmetry operations, namely pure rotations.

The carbonyl chromophore, belonging to the point group $C_{2v}$, has a primary element of symmetry in the twofold rotation axis $C_2(z)$ along the carbonyl bond, and two secondary elements, the molecular $yz$ plane and the $xz$ plane, bisecting the chromophore. The simplest pseudoscalar potential for the carbonyl chromophore is thus a function, $\mathrm{f}(XY)$, of the coordinates of the substituent in the Cartesian frame of the chromophore, giving a quadrant rule for chiral ketones. A more complex pseudoscalar potential for the group $C_{2v}$ is the substituent-coordinate function $\mathrm{f}(XYZ)$ giving an octant rule.

A distinction between a quadrant, an octant, or a more complex regional rule for optical activity based upon the pseudoscalar functions of a point group cannot be made on symmetry grounds alone, and this question is resolved either empirically or by detailed theoretical treatment. However, any regional rule for the optical activity of a symmetric chromophore in a dissymmetric molecular environment must correspond to a substituent-coordinate function which transforms under the pseudoscalar representation of the point group to which the chromophore belongs.

Between the limiting cases of the inherently dissymmetric chromophore and the symmetric chromophore in a chiral molecular environment lie a range of coupling cases in which the substituent assumes the role of a subsidiary chromophore, or even of an equivalent chromophore. An alternative to the static-coupling treatment (Moffitt *et al.*, 1961; Schellman, 1966, 1968) of chiral alkyl-substituted ketones is the dynamic-coupling approach of Höhn and Weigang (1968), in which the charge distribution of the carbonyl group $\pi^* \leftarrow \mathrm{n}$ transition induces a transient electric dipole in the alkyl substituent. This approach gives the correct sign and form of the octant rule for chiral ketones.

In chiral $\beta\gamma$-unsaturated ketones the carbonyl chromophore is heavily perturbed, since the isotropic absorption intensity of the $\pi^* \leftarrow \mathrm{n}$ band is greatly enhanced, and the band shifts to the red. The olefinic substituent here has become a subsidiary chromophore, and the principal source of the optical activity connected with the $\pi^* \leftarrow \mathrm{n}$ band of chiral $\beta\gamma$-unsaturated ketones is the coupling of a contribution from the electric dipole $\pi^*_{\mathrm{CC}} \leftarrow \pi_{\mathrm{CC}}$ transition of the olefin group to the magnetic dipole $\pi^*_{\mathrm{CO}} \leftarrow \mathrm{n}$ transition of the carbonyl group through the Coulombic interaction between the respective transition charge distributions (Moscowitz *et al.*, 1964). Reciprocally the $\pi^*_{\mathrm{CC}} \leftarrow \pi_{\mathrm{CC}}$ olefinic transition of a chiral $\beta\gamma$-unsaturated ketone attains an equal and oppositely-signed rotational strength by the Coulombic coupling, which mixes in a contribution from the carbonyl $\pi^*_{\mathrm{CO}} \leftarrow \mathrm{n}$ transition.

If the carbonyl group of a chiral $\beta\gamma$-unsaturated ketone is replaced by an olefin group, the two chromophores in the resultant molecule become equivalent. The $\pi^*_{\mathrm{CC}} \leftarrow \pi_{\mathrm{CC}}$ transitions of the two ethylene chromophores are degenerate in the zero order, but the instantaneous Coulombic interaction between the transition charge distributions lifts the degeneracy, raising the energy of one coupling mode and lowering that of the other. In each coupling mode both of the chromophores are equally excited, contributing components with the same magnitude to the total transition moment of the mode. For a chiral two-chromophore system the overall charge displacement in each coupling mode is helical, right-handed for one mode and left-handed for the other. The determination of the relative energies of the two coupling modes thus provides, from an inspection of the ORD or CD curve, an assignment of the absolute configuration of a chiral two- or multi-chromophore system.

The relative energies of the two coupling modes are readily found in the point-dipole approximation. For a given coupling mode the component transition moments may be represented approximately by a point dipole at the centre of each chromophore, e.g. (I) and (II) for the two coupling modes of a benzene excitation in biphenyl. In the coupling mode with the lower energy the classical Coulombic energy of the point dipoles in the two chromophores is a minimum, e.g. (I). The coupling mode (I) has a positive rotational strength for the P($z$) configuration of biphenyl (figure 4.5.0c) whilst the higher-energy mode (II) has a negative rotational strength. Thus a chiral biphenyl with the P($z$) configuration is expected to give a positive CD band followed by a negative CD band at higher frequency in the region of an isotropic absorption band due to a short-axis polarised transition of biphenyl, e.g. (I) and (II).

Strictly speaking a biaryl contains an inherently dissymmetric chromophore covering the $\pi$-system of the molecule as a whole. However, in chiral systems where the conjugation is small between otherwise distinct chromophoric groups the coupled-chromophore or exciton treatment, in which it is assumed that there is no electron exchange between the interacting groups, suffices generally for the determination of absolute configuration.

### 4.5.1 Regional rules for the carbonyl chromophore

(a) Obtain the absolute signs of the octant and the quadrant rule for chiral ketones, discussed in the previous section, from the observation that (*R*)-(+)-3-methylcyclohexanone exhibits a positive Cotton effect at 280 nm in the wavelength region of the carbonyl $\pi^* \leftarrow$ n absorption.
(b) Is a distinction between the quadrant and the octant rule feasible from the report (Djerassi and Klyne, 1963) that 5$\alpha$-cholestan-1-one exhibits a small negative Cotton effect, whereas the configurational analogue, (9*R*)-9-methyl-*trans*-1-decalone, gives a positive Cotton effect in the carbonyl $\pi^* \leftarrow$ n region?

**Solution**

(a) In the coordinate frame of the carbonyl chromophore (VI) the substituent coordinate function of the quadrant rule ($XY$) is positive, and that of the octant rule ($XYZ$) is negative, for the methyl group of ($R$)-(+)-3-methylcyclohexanone (VII). The sign of the Cotton effect induced by a substituent in the $\pi^* \leftarrow$ n carbonyl absorption thus follows the sign of the substituent coordinate function ($+XY$), according to the quadrant rule, or ($-XYZ$) according to the octant rule.

(b) To an observer viewing the molecule along the carbonyl bond axis from the $+z$ direction (VI), (9$R$)-9-methyl-*trans*-1-decalone (VIII) presents projection (IX) with all of the substituents of the carbonyl chromophore in the $-z$ coordinate region of the rear octants. The 9-methyl group, lying in a positive octant, is closer to the carbonyl chromophore than the 5- and 6-methylene groups of the B ring, which lie in a negative octant (IX), so that the dissymmetric perturbation of the chromophore due to the 9-methyl group is the larger, accounting for the positive Cotton effect observed.

From the corresponding projection (X) for 5α-cholestan-1-one (XI), a larger positive Cotton effect would be expected if the quadrant rule were valid. The octant rule, on the other hand, indicates that (XI) has a smaller positive Cotton effect than (VIII), owing to the disposition of atoms of the C and D rings, and the groups substituted in those rings, in a negative front octant (X). The optical data for (VIII) and (XI), and for related systems (Djerassi and Klyne, 1963) are thus the more consistent with an octant than a quadrant rule for chiral ketones.

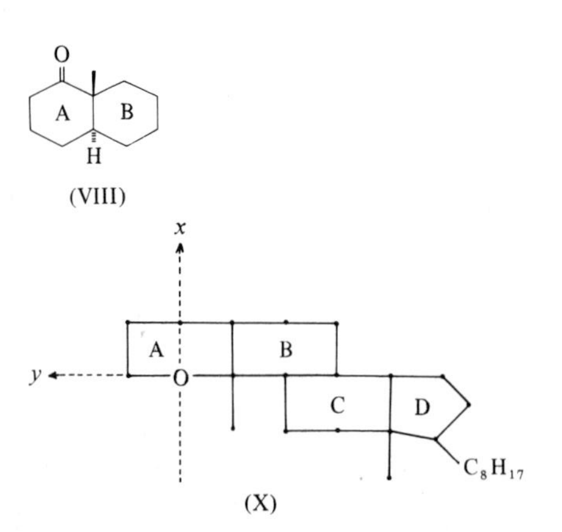


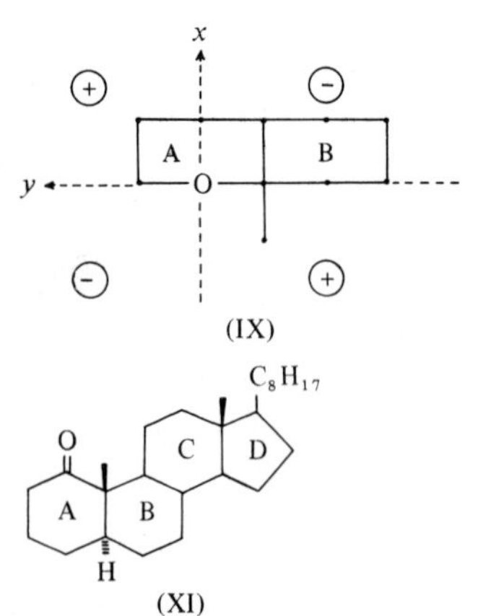

### 4.5.2 Conformational isomerism and configuration

At room temperature (2*R* : 5*R*)-*trans*-2-chloro-5-methylcyclohexanone gives near 300 nm a positive Cotton effect in methanol solution but a negative effect is iso-octane (Wellman *et al.*, 1963). In the mixed solvent ether–pentane–alcohol at room temperature the ketone exhibits a small negative CD band followed by a positive CD band of similar magnitude at higher frequency in the region of the carbonyl $\pi^* \leftarrow n$ absorption, and on lowering the temperature the positive CD band increases in area at the expense of the negative CD band, which is absent at liquid nitrogen temperatures (Wellman *et al.*, 1963; Moscowitz *et al.*, 1963). Account for these observations by using the octant (or quadrant) rule for chiral ketones.

**Solution**

(2*R* : 5*R*)-*trans*-2-chloro-5-methylcyclohexanone forms a diequatorial (XII) and a diaxial (XIII) conformational isomer. From the octant rule, (XII) is expected to give a positive Cotton effect and (XIII) a stronger negative Cotton effect at a slightly lower frequency, owing to the relatively large perturbation of the axial 2-chloro group. On account of the larger and oppositely signed Cotton effect, and the frequency displacement, a small fraction of the less stable diaxial conformer (XIII) is readily detected optically in an equilibrium mixture of the two isomers.

In solvents of low dielectric constant the Coulombic repulsion between the permanent dipole moments of the C—Cl and C=O bonds lowers the energy difference between the conformers, favouring the diaxial isomer (XIII). A solvent with a high dielectric constant favours an increase in the relative population of the diequatorial isomer (XII), as does a reduction in temperature. At liquid nitrogen temperatures the conformer population is entirely diequatorial (XII), and from temperature-dependent CD measurements it is found (Moscowitz *et al.*, 1963) that the diequatorial (XII) population is 97 ± 2% in methanol and 89 ± 3% in iso-octane at room temperature.

### 4.5.3 A regional rule for the d-electron transitions of tetragonal complexes

The *trans*-dichlorobis(diamine)cobalt(III) complexes, (XIV), (XV), and (XVI), which contain the $D_{4h}$ chromophore *trans*-$[CoN_4Cl_2]^+$, have the same configuration but different CD spectra in the visible region of the spectrum (figure 4.5.3a). The absorption and CD bands observed are components of the triply degenerate ${}^1T_{1g} \leftarrow {}^1A_{1g}$ transition of octahedral cobalt(III) complexes which, in the case of $[Co(NH_3)_6]^{3+}$, lies at 21 000 $cm^{-1}$. Identify the $D_{4h}$ components, ${}^1A_{2g} \leftarrow {}^1A_{1g}$ and ${}^1E_g \leftarrow {}^1A_{1g}$, of the octahedral ${}^1T_{1g} \leftarrow {}^1A_{1g}$ transition in the spectra of the complex ions, (XIV), (XV), and (XVI), and construct a regional rule (Mason, 1969) relating the position of an alkyl substituent to the sign of the CD induced in the region of the ${}^1A_{2g} \leftarrow {}^1A_{1g}$ ($D_{4h}$) absorption from the pseudoscalar property of the potential required. Do the nodal planes of the metal-ion orbitals, involved in the ${}^1A_{2g} \leftarrow {}^1A_{1g}$ ($D_{4h}$) transition, $d_{x^2-y^2} \leftarrow d_{xy}$, provide an adequate basis for a regional rule?

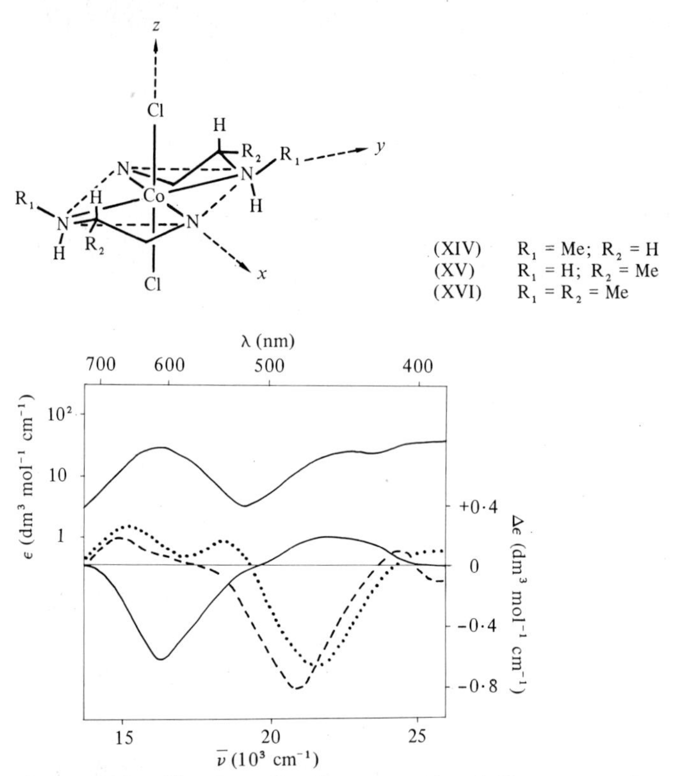


**Figure 4.5.3a.** The absorption (upper curve) and CD spectrum (lower curve) of the *trans*-dichloro-bis[(*S*)-(+)-propylenediamine]cobalt(III) ion (XV) (solid line), and the CD spectrum of the (−)-*trans,trans*-dichloro-bis[(*R*)-*N*-methylethylenediamine]cobalt(III) ion (XIV) (dotted line) and of the *trans*-dichloro-bis[(*R*)-$N_1$-methyl-(*S*)-(+)-propylene-diamine]-cobalt(III) ion (XVI) (dashed line).

**Solution**

The triply denegerate ${}^1T_{1g} \leftarrow {}^1A_{1g}$ ($O_h$) transition of the octahedral $[CoN_6]^{3+}$ chromophore breaks down in the tetragonal chromophore *trans*-$[CoN_4Cl_2]^+$ into a nondegenerate ${}^1A_{2g} \leftarrow {}^1A_{1g}$ and a doubly degenerate ${}^1E_g \leftarrow {}^1A_{1g}$ ($D_{4h}$) component. With the chloride ligands of *trans*-$[CoN_4Cl_2]^+$ coordinated along the $z$ axis, the nondegenerate ${}^1A_{2g} \leftarrow {}^1A_{1g}$ component is due to the $d_{x^2-y^2} \leftarrow d_{xy}$ excitation, and the ${}^1E_g \leftarrow {}^1A_{1g}$ component to the excitations

$d_{y^2-z^2} \leftarrow d_{yz}$ and $d_{z^2-x^2} \leftarrow d_{xz}$, where $d_{y^2-z^2}$ and $d_{z^2-x^2}$ represent appropriate linear combinations of $d_{z^2}$ and $d_{x^2-y^2}$.

The orbitals $d_{xy}$ and $d_{x^2-y^2}$ are unaffected to a fair approximation by the ligand replacement producing the *trans*-$[CoN_4Cl_2]^+$ from the $[CoN_6]^{3+}$ chromophore. The $^1A_{2g} \leftarrow {}^1A_{1g}$ transition of *trans*-$[CoN_4Cl_2]^+$ thus lies close to the frequency of the $^1T_{1g} \leftarrow {}^1A_{1g}$ transition of the $[CoN_6]^{3+}$ chromophore, namely 21 000 cm$^{-1}$.

The nodal planes of the $d_{xy}$ and $d_{x^2-y^2}$ orbitals give the coordinate function $[XY(X^2-Y^2)]$ for dissymmetric substituents perturbing the $^1A_{2g} \leftarrow {}^1A_{1g}$ transition of the *trans*-$[CoN_4Cl_2]^+$ chromophore. However, $[XY(X^2-Y^2)]$ is not a pseudoscalar function in the point group $D_{4h}$ of the *trans*-$[CoN_4Cl_2]^+$ chromophore (Schellman, 1966, 1968).

The simplest pseudoscalar function (Schellman, 1966, 1968) of the group $D_{4h}$ is $[XYZ(X^2-Y^2)]$, which provides a regional rule for the optical activity of dissymmetric tetragonal cobalt(III) complexes (Mason, 1969). The substituent coordinate function, $[XYZ(X^2-Y^2)]$, in which the tetragonal symmetry planes become nodal planes of the dissymmetric potential, gives the sign of the Cotton effect induced by the alkyl substituents of the *trans*-dichlorobis(diamine)cobalt(III) complexes (XIV), (XV), (XVI), and analogous tetragonal complexes (Mason, 1969) in the 21 000 cm$^{-1}$ region of the $^1A_{2g} \leftarrow {}^1A_{1g}$ ($D_{4h}$) transition (figure 4.5.3a).

In the graphical illustration of the tetragonal pseudoscalar regional rule (figure 4.5.3b), an alkyl substituent placed in a hatched region induces a positive Cotton effect, or in an unhatched region a negative Cotton effect, in the region of the $^1A_{2g} \leftarrow {}^1A_{1g}$ cobalt(III) absorption. The rule (figure 4.5.3b) accounts for the observation (figure 4.5.3a) that the *N*-methyl groups of the diamine ligands in the complexes (XIV) and (XVI) induce a Cotton effect of opposite sign to that given by the $CH_2$ or $CHCH_3$ groups of the diamine chain.

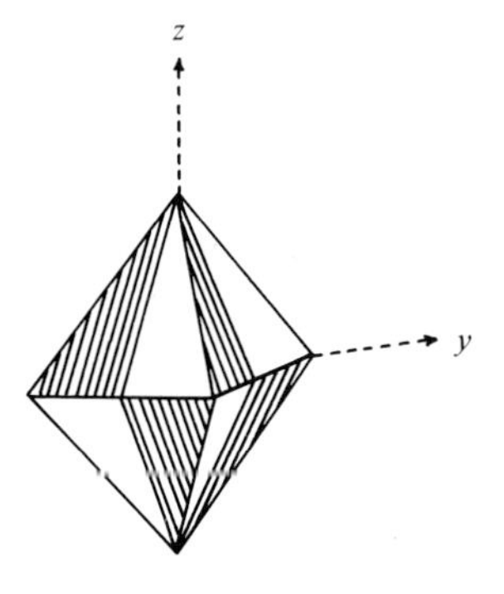


**Figure 4.5.3b.** The regional rule for the optical activity of tetragonal cobalt(III) complexes. Alkyl substitution in a hatched region induces a positive, and in an unhatched region a negative, Cotton effect in the frequency-range of the $^1A_{2g} \leftarrow {}^1A_{1g}$ ($D_{4h}$) transition $d_{x^2-y^2} \leftarrow d_{xy}$.

### 4.5.4 Symmetric and dissymmetric biphenyl

In the point-dipole-exciton approximation the rotational strengths of the coupling modes (I) and (II) of the benzene excitations in biphenyl (figure 4.5.0c) are given by

$$R_{\mathrm{I}} = -R_{\mathrm{II}} = \tfrac{1}{2}\pi d\bar{\nu}_{0n}\mu_{0n}^2 \sin 2\theta \,, \tag{4.5.4a}$$

where $d$ is the distance between the centres of the benzene rings in biphenyl, $2\theta$ is the dihedral angle between the planes of the two rings, and $\bar{\nu}_{0n}$ is the frequency in wavenumbers and $\bar{\mu}_{0n}$ the electric dipole moment of the benzene transition $\psi_n \leftarrow \psi_0$. Explain why perpendicular biphenyl, belonging to the group $D_{2d}$, is optically inactive, whilst $R_{\mathrm{I}}$ and $R_{\mathrm{II}}$ attain a maximum value when the dihedral angle equals $\frac{1}{2}\pi$.

What are the positions of substitution and the minimum number of (a) monoatomic and (b) polyatomic, substituents required in a chiral biphenyl with a dihedral angle of $\frac{1}{2}\pi$?

**Solution**

The biphenyl coupling modes (I) and (II) are necessarily degenerate when the dihedral angle $2\theta$ between the planes of the benzene rings (figure 4.5.0c) is $\frac{1}{2}\pi$. Although the rotational strengths of the two modes, $R_{\mathrm{I}}$ and $R_{\mathrm{II}}$ [equation (4.5.4a)], attain their maximum absolute value when $2\theta$ equals $\frac{1}{2}\pi$, they completely cancel one another, lying at the same frequency, and no optical activity is observable. In the point-dipole exciton approximation the frequency interval between the coupling modes (I) and (II) is given by

$$h(\nu_{\mathrm{I}} - \nu_{\mathrm{II}}) = \frac{2\mu_{0n}^2}{d^3}(\sin^2\theta - \cos^2\theta) \,, \tag{4.5.4b}$$

and the interval goes to zero when $\theta = \frac{1}{4}\pi$.

The secondary elements of symmetry in perpendicular biphenyl are the two mirror planes of the aromatic rings and a fourfold rotation-reflection axis. All substitution patterns in perpendicular biphenyl which eliminate these secondary symmetry elements give an optically active derivative. The minimum number of monoatomic substituents is two, one in the 2- or 3-position and the other in the 2′- or 3′-position. A single prochiral polyatomic substituent, e.g. of the type, RR′R″C—, in any position gives an optically active perpendicular biphenyl, but two such substituents in the 4- and 4′-positions produce an optically inactive derivative if one substituent has the (*S*) and the other the (*R*) configuration.

### 4.5.5 Coupled chromophores. I. The absolute configuration of an alkaloid

From the absorption and CD spectrum of the alkaloid (figure 4.5.5), demonstrate by means of the point-dipole exciton approximation that bulbocapnine (XVII) has the P(*z*) biphenyl configuration (Mason, 1967) or, equivalently, the (*S*) configuration at the asymmetric centre established by an x-ray diffraction study (Ashida *et al.*, 1963) of the corresponding methiodide. Show from the theory of the electronic spectra of benzene derivatives (Mason, 1961; Jaffé and Orchin, 1962; Murrell, 1963) that the coupling modes (I) and (II) hold in fair approximation for the lowest-energy and highest-energy accessible band systems of bulbocapnine (XVII).

O
O
NMe
H
HO
MeO
(XVII)

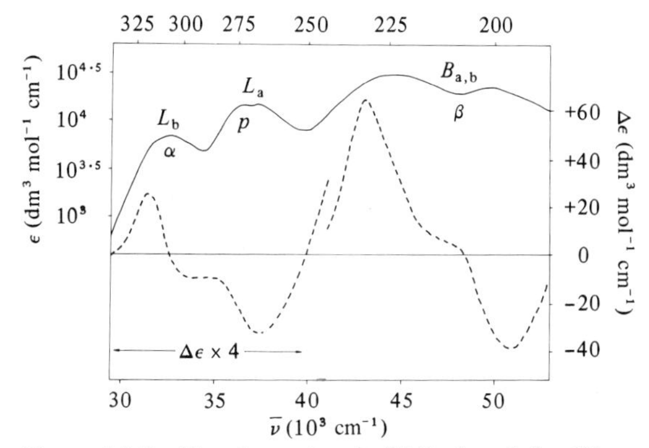


**Figure 4.5.5.** The absorption (solid line) and the CD spectrum (dashed line) of bulbocapnine (XVII) in water.

**Solution**

The lowest-energy absorption of benzene, the $^1L_b$ band, is due to a forbidden transition and the absorption intensity is weak. A substituent enhances the intensity of the benzene $^1L_b$ band and *ortho* or *meta* homo-disubstitution produces approximately the same intensity increase. A larger intensity enhancement, approximately fourfold, is given by *para* homo-disubstitution, whereas 1,2,3- or 1,3,5-homo-trisubstitution produces little or no intensity increase of the benzene $^1L_b$ (Mason, 1961; Jaffé and Orchin, 1962; Murrell, 1963). From these observations, Platt (1951) put forward the theory of spectroscopic moments whereby the $^1L_b$ absorption intensity of a polysubstituted benzene is given by the vector addition of the component transition moments induced by each substituent according to the pattern (XVIII). Platt's diagram (XVIII) also gives the polarisation direction of the resultant $^1L_b$ moment in a disubstituted or polysubstituted benzene derivative.

For each of the two catechol chromophores of bulbocapnine (XVII), the resultant $^1L_b$ transition moment lies along the twofold rotation symmetry axis of the chromophore (XIX) from the Platt diagram (XVIII). In bulbocapnine (XVII) the $^1L_b$ transition moments of the two catechol chromophores have the appropriate orientation to give the biphenyl coupling modes (I) and (II). From equation (4.5.4b) the mode (I) has the lower and mode (II) the higher frequency in the region of the $^1L_b$ absorption. In the CD spectrum of bulbocapnine (figure 4.5.5) the positive band at the lower frequency and the negative band at the higher frequency in the $^1L_b$ absorption region are thus due to the coupling modes (I) and (II) respectively. As these coupling modes in bulbocapnine have a positive and a negative rotational strength respectively, the aromatic moiety of the alkaloid (XVII) has the P($z$) biphenyl configuration (figure 4.5.0c) which corresponds to the ($S$) configuration at the asymmetric centre.

The CD associated with the higher-energy $^1L_a$ and $^1B_{a,b}$ bands of bulbocapnine (figure 4.5.5) supports this configurational assignment. Platt's (1951) diagram (XX) for the component transition moments induced by substitutents in the $^1L_a$ excitation of benzene indicates that the two catechol chromophores have resultant $^1L_a$ moments orientated in the direction of the long axis of the biphenyl system in the alkaloid (XVII). Of the two coupling modes of the $^1L_a$ moments of the catechol chromophores, one (XXI) is forbidden, the two moments cancelling identically, and no isotropic absorption or CD is associated with this mode. The other mode (XXII) is allowed and is responsible for all the $^1L_a$ absorption intensity of (XVII). Although the mode (XXII) is not optically active in the exciton approximation, CD is expected in the region of the $^1L_a$ band of bulbocapnine from the transfer of charge across the twisted inter-chromophore $\pi$ bond, an effect neglected in the present treatment.

The highest-energy absorption of bulbocapnine (XVII) is related to the allowed $^1B_{a,b}$ band of benzene. The benzene $^1B_{a,b}$ transition is doubly degenerate with one component directed between a pair of *para* carbon atoms (XXIII) and a second component, orthogonal to the first, bisecting parallel bonds (XXIV). The second component (XXIV) in the catechol chromophores of bulbocapnine (XVII) has the orientation (XIX) required to give rise to the biphenyl coupling modes (I) and (II). The CD pattern in the $^1B_{a,b}$ absorption region of bulbocapnine is thus expected to duplicate that found in the $^1L_b$ absorption region, as is observed (figure 4.5.5).

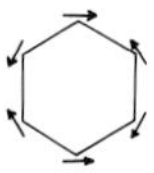

RO
RO

(XVIII) 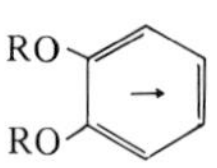 (XIX) 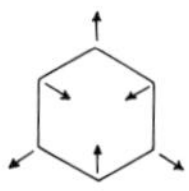 (XX)

(XXI) 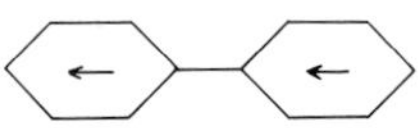 (XXII) 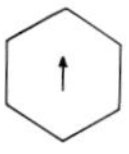 (XXIII) 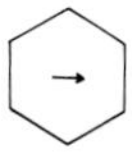 (XXIV)

Although they are a useful guide, the Platt diagrams (XVIII) and (XX) and the benzene $^1B_{a,b}$ polarisation diagrams (XXIII) and (XXIV) are not essential to the present analysis. The observed form of the CD spectrum of bulbocapnine, two CD bands of opposite sign and of comparable area associated with the $^1L_b$ absorption and also with the $^1B_{a,b}$ absorption (figure 4.5.5), indicates that the coupling modes (I) and (II) of moments directed along the $C_2$ axis of the catechol chromophore (XIX) are involved in the $^1L_b$ and in the $^1B_{a,b}$ transition of the alkaloid (XVII). Similarly the single CD band observed in the $^1L_a$ absorption region (figure 4.5.5) suggests that the component catechol chromophore moments are orientated in the direction of the long axis of the biphenyl moiety (XXII) of bulbocapnine (XVII).

**4.5.6 Coupled chromophores. II. The absolute configuration of a metal complex**

The ligand 1,10-phenanthroline (XXV) forms tris-complexes $[M(phen)_3]^{n+}$ with a range of metal ions and a number of these complexes have been resolved into optical isomers. An x-ray diffraction study (Templeton *et al.*, 1966) of $(-)_D$-$[Fe(phen)_3]^{2+}$ indicates that this isomer has the $M(C_3)$ or $\Lambda$ configuration (XXVI). Show that the same configurational assignment follows from an analysis of the CD spectrum of $(-)_D$-$[Fe(phen)_3]^{2+}$ (figure 4.5.6a) (Mason, 1968).

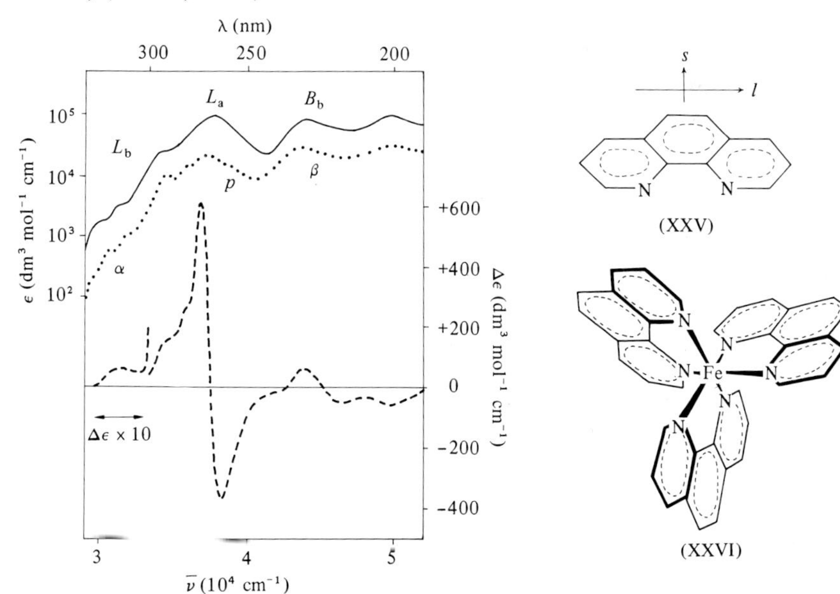


**Figure 4.5.6a.** The absorption (solid line) and the CD spectrum (dashed line) of the $(-)$-$[Fe(phen)_3]^{2+}$ ion (XXVI) in water, and the absorption spectrum of 1,10-phenanthroline (XXV) in ethanol (dotted line).

The absorption spectrum of $[Fe(phen)_3]^{2+}$ in the ultraviolet region resembles that of the free ligand (figure 4.5.6a), suggesting that the exciton approximation provides a valid method of analysis. For the complex (XXVI), work out the possible coupling modes of the ligand excitations which are short-axis ($s$) polarised in the free ligand (XXV), and of the excitations which are long-axis ($l$) polarised. Show that the coupling modes of the $s$-polarised ligand excitations are optically inactive to a first approximation, whereas the coupling modes of the $l$-polarised excitations are strongly optically active. Determine the signs of the rotational strengths and the energy order of the latter coupling modes, and show that these are consistent with the observed CD spectrum (figure 4.5.6a) for the M($C_3$) configuration (XXVI).

**Solution**

The Platt (1951) diagram for the spectroscopic moments induced by substituents or by atom replacements in phenanthrene indicates that the $^1L_b$ transition of 1,10-phenanthroline at 30000 $cm^{-1}$, like the $^1B_b$ transition at 43000 $cm^{-1}$, is polarised along the shorter in-plane axis of the ligand (XXV), whereas the moment of the $^1L_a$ transition at 37000 $cm^{-1}$ is oriented in the direction of the longer molecular axis. A similar conclusion emerges, together with an assignment of absolute configuration, from an analysis of the CD spectra of resolved tris(phenanthroline) metal-ion complexes (Mason, 1968).

For a given ligand transition there are three coupling modes in a tris(phenanthroline) complex $[M(phen)_3]^{n+}$ with $D_3$ symmetry, one which is symmetric with respect to a rotation by $\pm\frac{2}{3}\pi$ about the threefold symmetry axis of the complex, and two, forming a degenerate pair, which are antisymmetric with respect to the $C_3$ operation (figure 4.5.6b). In the case of a transition which is $s$-polarised in the free ligand (XXV) none of the three coupling modes involve an overall helical displacement of charge and all of these modes are optically inactive in the exciton approximation. The symmetric short-axis mode [figure 4.5.6b(i)] has a vanishing resultant and gives no absorption. The degenerate antisymmetric short-axis mode [figure 4.5.6b(ii)] has a nonzero resultant, producing isotropic light absorption but no optical activity.

For a transition which is $l$-polarised in the free ligand (XXV), all three coupling modes are strongly optically active. The symmetric long-axis mode [figure 4.5.6b(iii)] involves an overall helical charge displacement of the same chirality as that of the stereochemical configuration of the complex viewed along the $C_3$ axis. The degenerate antisymmetric long-axis mode [figure 4.5.6b(iv)] gives rise to an overall helical charge displacement of the opposite chirality.

With octahedral coordination, the plane of a chelate ring in a complex $[M(phen)_3]^{n+}$ lies at an angle of $\cos^{-1}[(\frac{2}{3})^{1/2}]$, namely 35°, to the threefold rotation axis of the complex, so that the Coulombic energy of the three component excitation dipoles in the symmetric long-axis coupling mode [figure 4.5.6b(iii)] is higher than that in the corresponding degenerate antisymmetric long-axis mode [figure 4.5.6b(iv)]. A given $l$-polarised ligand excitation thus gives in the spectrum of the $[M(phen)_3]^{n+}$ complex one CD band with a sign reflecting the stereochemical chirality of the complex along the principal threefold rotation axis of the molecule and another at lower frequency, of opposite sign, reflecting the stereochemical chirality of the complex viewed along one of the subsidiary twofold rotational axes of the ion.

The spectrum of the isomer (−)-$[Fe(phen)_3]^{2+}$ (figure 4.5.6a), shows a major positive CD band at 36800 $cm^{-1}$ and a large negative CD band at 38500 $cm^{-1}$ in the 37000 $cm^{-1}$ region of the $^1L_a$ absorption. These two CD bands arise from the degenerate antisymmetric [figure 4.5.6b(iv)] and nondegenerate symmetric [figure 4.5.6b(iii)] long-axis coupling mode respectively, and the signs of the bands indicate that the isomer (−)-$[Fe(phen)_3]^{2+}$ has the M($C_3$), or Λ-configuration (XXVI). The large magnitudes of the two CD bands show that the $^1L_a$ transition of 1,10-phenanthroline is largely $l$-polarised, and the weak CD band observed in the $^1L_b$ and $^1B_b$ absorption regions of the complex (figure 4.5.6a) suggest that the corresponding transitions in the free ligand are mainly $s$-polarised.

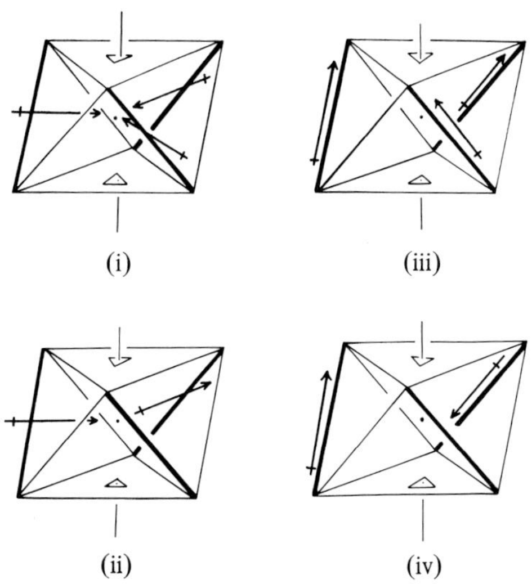


**Figure 4.5.6b.** The coupling modes of $\pi^* \leftarrow \pi$ ligand excitations in a $D_3$ complex $[M(phen)_3]^{n+}$ (XXVI). $s$-Polarised ligand excitations (XXV) give (i) a symmetric and (ii) an antisymmetric doubly degenerate mode, the former being forbidden having a vanishing resultant, and neither being optically active. $l$-Polarised ligand excitations give (iii) a symmetric and (iv) an antisymmetric doubly degenerate mode, both modes being allowed and optically active. Only one component of each of the two degenerate modes is illustrated.

### 4.5.7 The absolute configuration of an inherently dissymmetric chromophore

A number of derivatives of [4]-helicene (benzo[*c*]phenanthrene) (XXVII) have been resolved into optical isomers (Newman and Wheatley, 1948; Newman and Wise, 1956; Newman *et al.*, 1963), and the CD spectra of the (+)-1-fluoro-12-methyl and the (−)-1,12-dimethyl-5-acetic acid isomers have been recorded (Kemp and Mason, 1965, 1966). Show, from the CD spectrum of the isomer (figure 4.5.7a), that (−)-1,12-dimethylbenzo[*c*]phenanthrene-5-acetic acid has the M configuration, using the atomic orbital coefficients $C_{rj}$ of the conjugated carbon atoms $r$ in the Hückel $\pi$-orbitals $\psi_j$ of [4]-helicene (Coulson and Streitweiser, 1965; Heilbronner and Straub, 1966) listed in table 4.5.7.

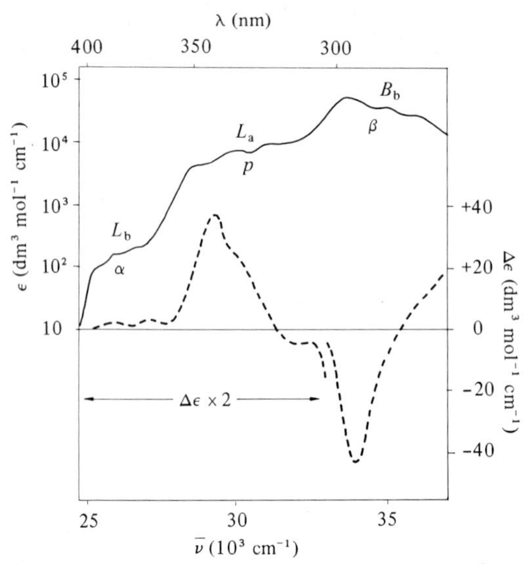


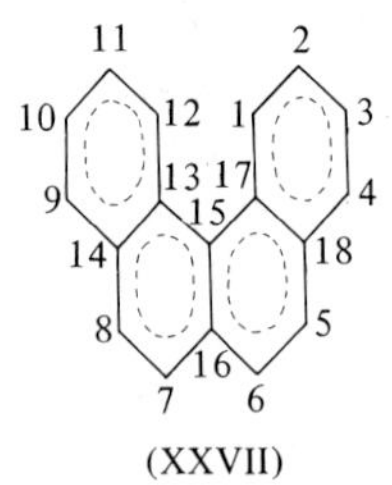


**Figure 4.5.7a.** The absorption (solid line) and the CD spectrum (dashed line) of (−)-1,12-dimethylbenzo[*c*]phenanthrene-5-acetic acid in cyclohexane.

**Table 4.5.7.**

| Atom, $r$ (XXVII) | $C_{r8}$ | $C_{r9}$ |
|---|---|---|
| 1 | −0·092 | +0·327 |
| 2 | −0·279 | +0·168 |
| 3 | −0·092 | −0·232 |
| 4 | +0·218 | −0·300 |
| 5 | −0·279 | +0·317 |
| 6 | −0·421 | +0·118 |
| 15 | 0·000 | −0·379 |
| 16 | 0·000 | −0·250 |
| 17 | +0·218 | +0·017 |
| 18 | +0·236 | +0·061 |

The atomic orbital coefficients of the atoms which are not listed in the above table are determined by symmetry. [4]-Helicene has a twofold rotational axis of symmetry, $C_2$, in the direction of the $C^{15}-C^{16}$ bond (XXVII), and the $\pi$-orbitals of the molecule are either symmetric ($A$) or antisymmetric ($B$) with respect to a rotation by $\pi$ about the $C_2$ axis. The highest occupied $\pi$-orbital, $\psi_9$, has nonzero coefficients at atoms 15 and 16 and the $2p_\pi$ orbitals of these atoms change sign under the $C_2$ rotation. Thus $\psi_9$ as a whole is antisymmetric to this operation, with $B$ symmetry, and two atoms which are related by the $C_2$ rotation have atomic orbital coefficients of the same sign and magnitude in the molecular orbital $\psi_9$. The next-highest occupied $\pi$-orbital, $\psi_8$, is symmetric under the $C_2$ operation, with $A$ symmetry as the coefficients on the atoms 15 and 16 in this molecular orbital are zero. Accordingly two atoms which are related by the $C_2$ rotation have atomic orbital coefficients of the same magnitude but of opposite sign in the molecular orbital $\psi_8$.

The lowest-energy absorption band of moderately high intensity in the electronic spectrum of an aromatic hydrocarbon is due to the promotion of an electron from the highest occupied to the lowest unoccupied $\pi$-orbital (Mason, 1961; Jaffé and Orchin, 1962; Murrell, 1963), $\psi_{10} \leftarrow \psi_9$ in the case of [4]-helicene, where the $\pi$-orbitals are numbered upwards from the most bonding orbital. This absorption is termed the $p$ band by Clar (1964), and the ${}^1L_a$ band by Platt (1949). The next-higher-energy excitations, $\psi_{10} \leftarrow \psi_8$ and $\psi_{11} \leftarrow \psi_9$ in the case of [4]-helicene, are degenerate in the Hückel approximation, but a consideration of the Coulombic interaction between the two transition charge distributions shows that these two excitations couple in phase to give a high-intensity absorption at a high frequency, and out of phase to give a forbidden band at low frequency. These are the $\beta$ band and $\alpha$ band respectively, in the terminology of Clar (1964), or the ${}^1B_b$ and ${}^1L_b$ band, respectively, in the nomenclature of Platt (1949). As in the case of [4]-helicene and its derivatives (figure 4.5.7a), the $p$ band generally appears on the frequency ordinate between the $\alpha$ and $\beta$ bands.

The atomic orbital coefficients of the lowest unoccupied, $\psi_{10}$, and subsequent antibonding $\pi$-orbital, $\psi_{11}$, are given by the alternant property of [4]-helicene. If alternate carbon atoms of [4]-helicene are starred, no two starred atoms, nor two unstarred atoms, are directly bonded. In such alternant hydrocarbons each antibonding $\pi$-orbital is paired with one of the bonding orbitals. The atomic orbital coefficient of a given atom has the same absolute value in the antibonding as in the paired bonding $\pi$-orbital, and the same sign if the atom is starred but the opposite sign if that atom is unstarred. In particular the highest occupied $\pi$-orbital of a neutral even alternant hydrocarbon is paired with the lowest unoccupied $\pi$-orbital, and in the case of [4]-helicene we have the relation,

$$\psi_{10} = \sum^{*} C_{r9}\varphi_r - \sum^{\circ} C_{s9}\varphi_s \qquad (4.5.7a)$$

between the coefficients of the $\pi$-orbitals $\psi_9$ and $\psi_{10}$. The starred sum in equation (4.5.7a) is taken over the $2p_\pi$ atomic orbitals $\varphi_r$ of the starred atoms $r$, and the circled sum over the corresponding atomic orbitals of the unstarred atoms $s$. A similar relation obtains between the atomic orbital coefficients of the $\pi$-orbitals $\psi_8$ and $\psi_{11}$ of [4]-helicene.

An electronic transition $\psi_j \leftarrow \psi_i$ of an aromatic molecule involves the displacement of a quantity of charge proportional to $(C_{ri}C_{sj})$ from atom $r$ to atom $s$, and the corresponding quantity $(C_{si}C_{rj})$ in the converse direction. The major transitional charge displacements are directed along the bonds of the $\pi$-system, and the quantity

$$(C_{ri}C_{sj} - C_{si}C_{rj}) \tag{4.5.7b}$$

represents the quantity of charge displaced along the bond between the atoms $r$ and $s$ in the direction $r \rightarrow s$.

A diagram based on the molecular structure (XXVII) giving the value and direction along each bond of the quantity (4.5.7b) for a particular electronic transition of [4]-helicene affords by inspection the left- or right-handed helical sense of the overall transitional charge displacement for each configurational isomer. Reference to the sign of the CD associated with the absorption band due to the electronic transition considered (figure 4.5.7a) then gives the particular configuration of the isomer studied. For the qualitative purpose of determining the absolute stereochemical configuration of a helicene, only the direction of the charge displacement along each bond, that is, the sign of quantity (4.5.7b), need be considered for transitions involving a helical charge displacement along and around the $C_2$ axis of the molecule, e.g. $\psi_{10} \leftarrow \psi_9$ giving the ${}^1L_a$ or $p$ band of [4]-helicene.

**Solution**

The evaluation of the sign and the magnitude of the bond charge displacements [quantity (4.5.7b)] for the ${}^1L_a$ transition $\psi_{10} \leftarrow \psi_9$ of [4]-helicene gives an overall charge-displacement diagram which is drawn up in figure 4.5.7b for the M configuration of the molecule. The form of the charge displacements depicted (figure 4.5.7b) corresponds for this configuration to a right-handed electronic motion along and around the twofold rotational symmetry axis of the molecule (the $z$ axis).

The analogous charge-displacement diagram for the ${}^1B_b$ transitions, $\psi_{10} \leftarrow \psi_8$ and $\psi_{11} \leftarrow \psi_9$, of [4]-helicene, again drawn up for the M configuration of the molecule (figure 4.5.7c), shows a dual electronic motion. There is a major left-handed helical charge displacement along and about the helix axis (the $x$ axis) and a minor right-handed helical charge displacement along the $y$ axis, perpendicular to the $C_2$ and the helix axis (Kemp and Mason, 1965, 1966).

The spectrum of (−)-1,12-dimethylbenzo[$c$]phenanthrene-5-acetic acid (figure 4.5.7a) shows a positive CD band in the region of the ${}^1L_a$ or

$p$-band absorption at 29000 $cm^{-1}$ and a negative CD band at 34000 $cm^{-1}$ in the region of the $^1B_b$ or $\beta$-band absorption. Accordingly, the spectrum shows that (−)-1,12-dimethylbenzo[$c$]phenanthrene-5-acetic acid has the M configuration illustrated (figures 4.5.7b and 4.5.7c).

The qualitative determination of the direction of the individual bond charge displacements [quantity (4.5.7b)] in an alternant aromatic hydrocarbon is particularly simple for a transition between a paired bonding and antibonding orbital, e.g. the $^1L_a$ transition $\psi_{10} \leftarrow \psi_9$ of [4]-helicene. Of two bonded carbon atoms $r$ and $s$, one is starred and the other is unstarred, so that, for the transition $\psi_{10} \leftarrow \psi_9$ between paired $\pi$-orbitals, the direction of the transitional charge displacement along the bond between the atoms $r$ and $s$ is given by the sign of the quantity $(2C_{r9}C_{s9})$ through the alternant pairing relation [equation (4.5.7a)]. The direction of the charge displacement is from the starred to the unstarred atom of the pair $r$ and $s$ if the sign of $(2C_{r9}C_{s9})$ is positive, or in the converse direction if that quantity is negative.

The qualitative procedure is reliable only if the transition considered is polarised along the $C_2$ symmetry axis of the helicene, as in the case of the $^1L_a$ transition $\psi_{10} \leftarrow \psi_9$ of [4]-helicene. The transition then involves only a single helical charge displacement, which is symmetry-determined along and around the twofold rotation axis. A transition polarised perpendicular to the $C_2$ rotation axis of a helicene generally involves one charge displacement along the helix axis of the helicene and another of opposite chirality along the direction perpendicular both to the $C_2$ and to the helix axis. The qualitative procedure of determining only the directions of the individual bond charge displacements is here unreliable, and it is generally necessary to evaluate the magnitudes of the two helical charge displacements involved in the transition or, at least, the chirality of their resultant.

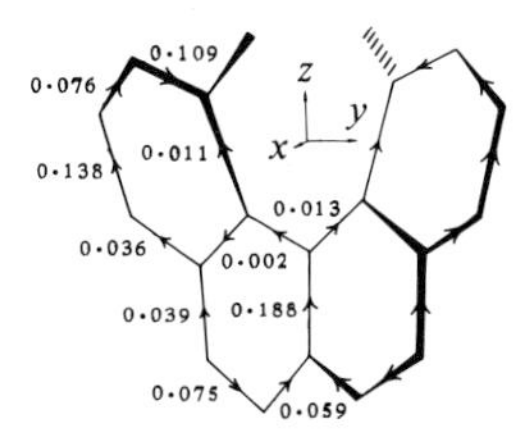


**Figure 4.5.7b.** Bond charge-displacement diagram for the $^1L_a$ transition $\psi_{10} \leftarrow \psi_9$ in the 1,12-dimethyl-[4]-helicene isomer with the M configuration relative to the helix axis ($x$). The overall charge displacement is right-handed along and about the twofold rotational symmetry axis of the molecule ($z$).

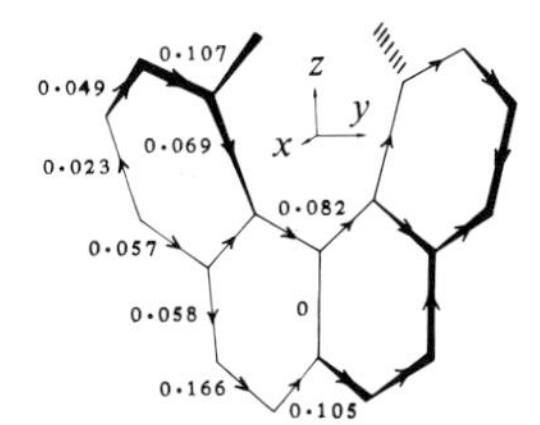


**Figure 4.5.7c.** Bond charge-displacement diagram for the $^1B_b$ transitions, $\psi_{10} \leftarrow \psi_8$ and $\psi_{11} \leftarrow \psi_9$, in the 1,12-dimethyl-[4]-helicene isomer with the M configuration. The major resultant charge displacement is left-handed along and around the helix axis ($x$), with a minor right-handed charge displacement along and about the $y$ axis.

**4.5.8 The rotatory dispersion of polypeptides**

The ORD curves of many optical isomers at wavelengths longer than those of the absorption bands of the molecule follow a single-term Drude relation (Drude, 1900),

$$[M]_\lambda = \frac{a_i\lambda_i^2}{\lambda^2-\lambda_i^2} \cdot \tag{4.5.8a}$$

In the sum (4.5.0a) there is a major contribution to the ORD of such molecules from a CD band lying near to the wavelength $\lambda_i$, and only minor contributions from the other CD bands of the molecule. The constant $a_i$ represents a weighted sum of the rotational strengths $R_i$ in the relation (4.5.0a).

The ORD curves of a number of polypeptides and proteins in the random-coil conformation follow relation (4.5.8a), giving a value for $\lambda_i$ near 210 nm from a plot of $[M]_\lambda^{-1}$ against $\lambda^2$. On assuming the α-helix conformation, polypeptides give a different ORD curve following the relation (Moffitt, 1956; Moffitt and Yang, 1956),

$$[M]_\lambda = \frac{a_0\lambda_0^2}{(\lambda^2-\lambda_0^2)} + \frac{b_0\lambda_0^4}{(\lambda^2-\lambda_0^2)^2} \cdot \tag{4.5.8b}$$

A plot of $[M](\lambda^2-\lambda_0^2)$ against $(\lambda^2-\lambda_0^2)^{-1}$ gives a straight line for a particular value of $\lambda_0$, which is found to be 212 ± 5 nm, and the constants $a_0$ and $b_0$ are obtained from the intercept and the slope of the line. For polypeptides derived from the L amino acids ($S$ configuration) the value of $b_0$ is $-630°$ per amino acid residue if the conformation is entirely α-helical and zero if completely random-coil. The percentage of a polypeptide in the α-helix conformation and the M or P configuration of the helix is given by the magnitude and the sign, respectively, of $b_0$.

Moffitt (1956) showed that, on adopting the α-helix conformation, a polypeptide acquires an additional source of optical activity from the coupling of the $\pi^* \leftarrow \pi$ transitions of the helically ordered amide chromophores of the polymer. For an infinite polypeptide there are two allowed coupling modes (figure 4.5.8). The parallel mode involves an overall charge displacement along and around the helix axis [figure 4.5.8(i)] and the perpendicular mode, which is doubly degenerate, gives rise to helical charge displacements along two mutually orthogonal directions perpendicular to the helix axis [figure 4.5.8(ii)].

In the polypeptide α-helix the electric dipole moment of the 200 nm $\pi^* \leftarrow \pi$ transition of the amide chromophore is directed perpendicular to the local radius vector from the helix axis, and lies at an angle of 50° to the line through the chromophore parallel to the helix axis. The transition moment is oriented at a large angle (~75°) to the line of the helix backbone, representing the averaged σ-bond frame of the α-helix (figure 4.5.8), so that the charge displacement of the parallel coupling mode is of opposite chirality to that of the σ-bond helix.

Determine the relative energies of the parallel and perpendicular coupling modes of the amide $\pi^* \leftarrow \pi$ excitations in the polypeptide $\alpha$-helix, and the signs of the rotational strengths of the two modes, $R_\parallel$ and $R_\perp$ respectively, for the M or the P configuration of the helix. From the sum rule, $R_\parallel$ and $R_\perp$ may be taken to have the same absolute magnitude. Show that a relation of the form (4.5.8b) follows from equation (4.5.0a) if $R_\parallel$ and $R_\perp$ make the only significant contributions to the rotatory dispersion, and that $b_0$ is negative for the P configuration of the polypeptide $\alpha$-helix.

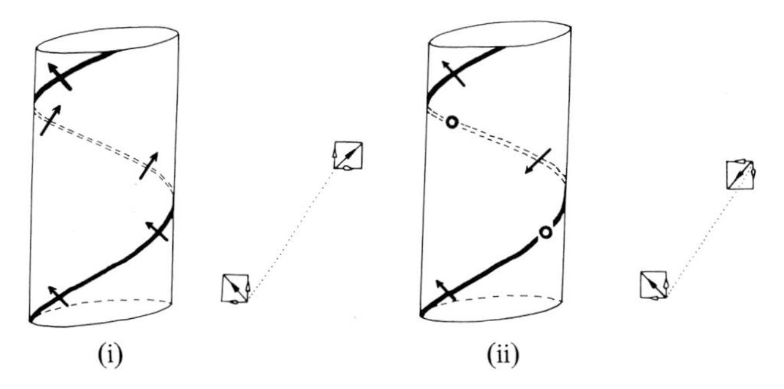

**Figure 4.5.8.** The allowed coupling modes of the moments of the 200 nm amide $\pi^* \leftarrow \pi$ excitation in the polypeptide $\alpha$-helix polarised (i) parallel and (ii) perpendicular to the helix axis. The third allowed mode, degenerate with (ii), has a resultant moment oriented perpendicular to that of (i) and of (ii).

**Solution**

If the ORD of a polypeptide in the $\alpha$-helix conformation is dominated by the contributions from the rotational strengths of the parallel and perpendicular coupling modes of the individual amide $\pi^* \leftarrow \pi$ excitation moments (figure 4.5.8), the molecular rotation is given by two terms of the sum (4.5.0a),

$$[M]_\nu = \frac{96\pi N}{hc}\left(\frac{R_\parallel \nu^2}{\nu_\parallel^2 - \nu^2} + \frac{R_\perp \nu^2}{\nu_\perp^2 - \nu^2}\right). \tag{4.5.8c}$$

Through the sum rule $R_\parallel = -R_\perp$ and the adoption of the approximation

$$\nu_\parallel \approx \nu_\perp \approx \nu_i\,, \tag{4.5.8d}$$

where $\nu_i$ is the mean frequency of the coupling modes, relation (4.5.8c) becomes

$$[M]_\nu = \frac{192\pi N}{hc}\left[\frac{R_\parallel \nu^2 \nu_i (\nu_\perp - \nu_\parallel)}{(\nu_i^2 - \nu^2)^2}\right]. \tag{4.5.8e}$$

The frequency dispersion of equation (4.5.8e) has the form required to account for the second term on the right-hand side of relation (4.5.8b), the contribution to $b_0$ from the coupling modes considered (figure 4.5.8) being (Moffitt, 1956; Moffitt and Yang, 1956):

$$b_i = \frac{192\pi N}{hc} \frac{(\nu_\perp - \nu_\parallel)R_\parallel}{\nu_i} . \tag{4.5.8f}$$

Moffitt considered additionally the contribution to $b_0$ in equation (4.5.8b) from analogous parallel and perpendicular coupling modes of the higher-energy amide $\pi^* \leftarrow \pi$ excitations near 160 nm, but the contribution of these modes to the ORD is of less importance owing to the magnitude of the frequency denominator entering into the appropriate dispersion relation analogous to (4.5.8e).

The sign of $b_0$ in equation (4.5.8b) is determined by that of the major contribution $b_i$ [equation (4.5.8f)], which refers to the coupling modes of the lower-energy $\pi^* \leftarrow \pi$ excitation near 200 nm (figure 4.5.8). As the moment of the 200 nm amide $\pi^* \leftarrow \pi$ transition lies at a large angle to the $\sigma$-bond backbone of the polypeptide $\alpha$-helix and is perpendicular to the radius vector from the helix axis, the parallel coupling mode [figure 4.5.8(i)] involves an overall helical charge displacement of an opposite chirality to that of the stereochemical configuration of the $\alpha$-helix. Thus $R_\parallel$ is negative for the P configuration of the polypeptide $\alpha$-helix [figure 4.5.8(i)].

The disposition of the amide chromophore in the polypeptide $\alpha$-helix, apart from absolute configuration, and the particular angle of 50° between the direction of the amide $\pi^* \leftarrow \pi$ transition moment and the line through the chromophore parallel to the helix axis, implies that the Coulombic energy of the parallel coupling mode [figure 4.5.8(i)] is less than that of the corresponding perpendicular mode [figure 4.5.8(ii)]. Accordingly $\nu_\perp - \nu_\parallel$ has a positive value, and $b_i$ [equation (4.5.8f)] and hence $b_0$ in equation (4.5.8b) is negative for a polypeptide $\alpha$-helix with the P configuration [figure 4.5.8(i)].

Coupling modes neglected by Moffitt in his early analysis (Moffitt, 1956) which vanish for an infinite helical polymer on the restrictive assumption that the dimensions of the polymer are negligible compared with the wavelength of light, were subsequently discussed by Moffitt *et al.* (1957). The additional coupling modes do not appear to be important for large polypeptides, and the Moffitt equation (4.5.8b) is empirically successful, but the additional modes are significant for polynucleotides and short-chain oligomers generally (Bradley *et al.*, 1963).

The CD bands corresponding to the parallel and perpendicular coupling modes (figure 4.5.8) of the 200 nm amide $\pi^* \leftarrow \pi$ transition in the polypeptide $\alpha$-helix are now accessible (Holworth *et al.*, 1962), together with the corresponding ORD (Blout *et al.*, 1962). The $\alpha$-helix of polypeptides composed of the L amino acids has been found, by X-ray diffraction studies (Kendrew *et al.*, 1961), to have the P configuration predicted by Moffitt (1956).

## References

Ashida, T., Pepinsky, R., Okaya, Y., 1963, *Acta Crystallogr. Sect. A,* **16**, A48.
Blout, E. R., Schmier, I., Simmons, N. S., 1962, *J. Am. Chem. Soc.,* **84**, 3193.
Bradley, D. F., Tinoco, I. Jr., Woody, R. W., 1963, *Biopolymers,* **1**, 239.
Brewster, J. H., 1967, *Topics in Stereochemistry,* Eds N. L. Allinger, E. L. Eliel (Interscience, Chichester, Sussex), volume 2, p.1.
Cahn, R. S., Ingold, C. K., Prelog, V., 1966, *Angew. Chem. Int. Ed. Engl.,* **5**, 385.
Clar, E., 1964, *Polycyclic Hydrocarbons,* volume 1 (Academic Press, London), p.47ff.
Coulson, C. A., Streitweiser, A. Jr., 1965, *Dictionary of π-electron Calculations* (Pergamon Press, Oxford).
Djerassi, C., Klyne, W., 1963, *J. Chem. Soc.,* 2390.
Drude, P., 1900, *Lehrbuch der Optik* (Hirzel, Leipzig); English translation, 1959, *Theory of Optics* (Dover, New York).
Heilbronner, E., Straub, P. A., 1966, *HMO–Hückel Molecular Orbitals* (Springer, Berlin).
Höhn, E. G., Weigang, O. E. Jr., 1968, *J. Chem. Phys.,* **48**, 1127.
Holworth, G., Gratzer, W. B., Doty, P., 1962, *J. Am. Chem. Soc.,* **84**, 3194.
IUPAC convention, 1970, *Inorg. Chem.,* **9**, 1.
Jaffé, H. H., Orchin, M., 1962, *Theory and Applications of Ultraviolet Spectroscopy* (John Wiley, Chichester, Sussex).
Kemp, C. M., Mason, S. F., 1965, *J. Chem. Soc., Chem. Commun.,* 559.
Kemp, C. M., Mason, S. F., 1966, *Tetrahedron,* **22**, 629.
Kendrew, J. C., Watson, H. C., Strandberg, B. E., Dickerson, R. E., Phillips, D. C., Shore, V. C., 1961, *Nature (London),* **190**, 666.
Mason, S. F., 1961, *Q. Rev. Chem. Soc.,* **15**, 287.
Mason, S. F., 1967, *Newer Physical Methods in Structural Chemistry,* Eds R. Bonnett, J. G. Davis (United Trade Press, London), p.149.
Mason, S. F., 1968, *Inorg. Chim. Acta, Rev.,* **2**, 89.
Mason, S. F., 1969, *J. Chem. Soc., Chem. Commun.,* 856.
Moffitt, W., 1956, *Proc. Nat. Acad. Sci. USA,* **42**, 736.
Moffitt, W., Fitts, D., Kirkwood, J. G., 1957, *Proc. Nat. Acad. Sci. USA,* **43**, 723.
Moffitt, W., Woodward, R. B., Moscowitz, A., Klyne, W., Djerassi, C., 1961, *J. Am. Chem. Soc.,* **83**, 4013.
Moffitt, W., Yang, J. T., 1956, *Proc. Nat. Acad. Sci. USA,* **42**, 596.
Moscowitz, A., Hansen, A. E., Forster, L. S., Rosenheck, K., 1964, *Biopolym. Symp.,* **1**, 75.
Moscowitz, A., Wellman, K. M., Djerassi, C., 1963, *Proc. Nat. Acad. Sci. USA,* **50**, 799.
Murrell, J. N., 1963, *Theory of the Electronic Spectra of Organic Molecules* (Methuen, London).
Newman, M. S., Mentzer, R. G., Slomp, G., 1963, *J. Am. Chem. Soc.,* **85**, 4018.
Newman, M. S., Wheatley, W. B., 1948, *J. Am. Chem. Soc.,* **70**, 1913.
Newman, M. S., Wise, R. M., 1956, *J. Am. Chem. Soc.,* **78**, 450.
Platt, J. R., 1949, *J. Chem. Phys.,* **17**, 484.
Platt, J. R., 1951, *J. Chem. Phys.,* **19**, 263.
Schellman, J. A., 1966, *J. Chem. Phys.,* **44**, 55.
Schellman, J. A., 1968, *Acc. Chem. Res.,* **1**, 144.
Templeton, D. H., Zalkin, A., Ueki, T., 1966, *Acta Crystallogr. Sect. A,* **21**, A154.
Wellman, K., Bunnenberg, E., Djerassi, C., 1963, *J. Am. Chem. Soc.,* **85**, 1870.

## Further reading

Mason, S. F., 1982, *Molecular Optical Activity and the Chiral Discrimination* (Cambridge University Press, Cambridge).

## 4.6 Photoelectron spectroscopy

**W C Price** King's College, University of London

### 4.6.0 Introduction

For a background in the field of photoelectron spectroscopy the reader is referred to reviews by Price (1981), Turner *et al.* (1970), and Siegbahn *et al.* (1967). The subject is concerned with the ejection of electrons from atomic systems according to the Einstein photoelectric equation

$$h\nu = E + \tfrac{1}{2}mv^2 , \tag{4.6.0a}$$

where $E$ is the initial binding energy (ionization energy) of the electron and $\frac{1}{2}mv^2$ the kinetic energy it acquires when ejected by a photon of frequency $\nu$. The use of two ranges of photon energies, one to eject outer-shell electrons and one to eject inner-shell electrons, conveniently divides the subject into the two branches ultraviolet photoelectron spectroscopy (UPS) and x-ray photoelectron spectroscopy (XPS). We shall treat these separately as they generally yield different types of information. Problems 4.6.1–4.6.8 deal with UPS, problems 4.6.9 and 4.6.10 with XPS.

*Ultraviolet photoelectron spectroscopy*

The photons usually used in ultraviolet photoelectron spectroscopy are atomic emission lines such as the 58·4 nm line of helium ($h\nu$ = 21·22 eV), the 30·4 nm line of ionized helium (40·83 eV), and similar lines of the other inert gases. These lines are capable of ejecting electrons from the valence shells of molecules, and the accurate measurement of the photoelectron energies by means of electrostatic deflection analysers enables the difference between the energies of the neutral ground state and that of the molecular ion to be obtained from the equation

$$I_0 + \Delta E_{\text{vib}} + \Delta E_{\text{rot}} = h\nu - \tfrac{1}{2}mv^2 , \tag{4.6.0b}$$

where $I_0$ is the 'adiabatic' ionization energy (i.e. the pure electronic energy change), and $\Delta E_{\text{vib}}$ and $\Delta E_{\text{rot}}$ are the changes in rotational and vibrational energy which accompany the photoionization.

The photoelectron spectrum of a molecule is a plot over the energy range of the numbers of photoelectrons emitted with a particular energy, i.e. an energy-distribution curve of the photoelectron emission resulting from a beam of monoenergetic photons. It is usually most convenient to plot this directly as photoelectron count against ionization energy, as measurements of photoelectron energy are affected by contact potentials in the analyser and it is invariably necessary to calibrate the scale by simultaneously recording the spectrum of a gas for which the ionization potentials are known accurately from spectroscopic data. Standards such as argon, nitrogen, methyl iodide, and helium are commonly used for this purpose.

The ultraviolet photoelectron spectra of two simple molecules are shown in figure 4.6.0. The figure also gives the potential energy curves of $H_2$ and $H_2^+$. When an electron is removed from the neutral $H_2$ molecule the nuclei find themselves suddenly in the potential field appropriate to the $H_2^+$ ion but still separated by the distance characteristic of the neutral molecule. The most probable change is thus a transition on the potential energy diagram from the internuclear separation of the ground state to a point on the potential energy curve of the ion vertically above this. This is the Franck–Condon principle and it determines to which vibrational level of the ion the most probable transition (strongest band) occurs. The energy corresponding to this change is called the vertical ionization potential, $I_v$. Transitions to vibrational levels on either side of $I_v$ are weaker, the one of lowest energy corresponding to the vibrationless state of the ion corresponding to $I_0$ [equation (4.6.0b)]. The photoelectron spectrum of $H_2$ is also plotted along the ordinate of the potential energy curve in figure 4.6.0(ii).

From the intensity distribution of the bands in the photoelectron spectrum it is possible to calculate the change in internuclear distance on ionization. Clearly when a bonding electron is removed, the part of the photoelectron spectrum corresponding to this will show wide vibrational structure with a frequency separation less than that of the ground-state vibration. On the other hand, the removal of relatively nonbonding

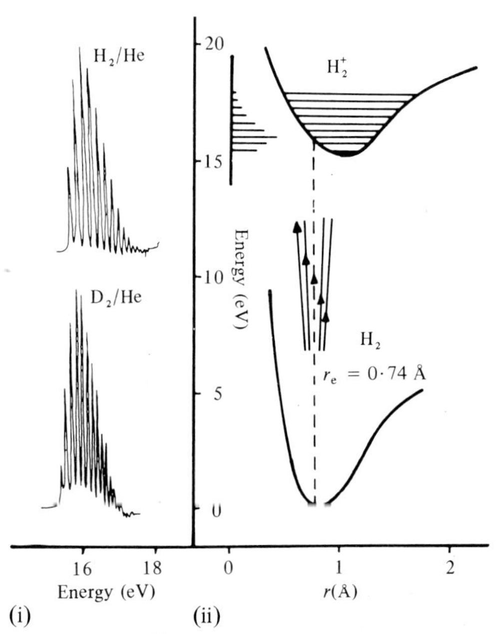


**Figure 4.6.0.** (i) Photoelectron spectra of $H_2$ and $D_2$; (ii) potential energy curves showing $H_2$ spectrum (schematic) plotted along ordinate.

electrons will give rise to photoelectron spectra rather similar to those of the monatomic gases, and little if any vibrational structure will accompany the main electronic band. The type of vibration associated with the pattern obtained when a bonding electron is removed can usually be identified as either a bending or a stretching mode and this can throw light on the function of the electron in the structure of the molecule, that is, either as angle-forming or distance-determining. From the band pattern it is frequently possible to calculate values of the changes in angle as well as changes in internuclear distance on ionization and so the geometry of the ionic states can be found if that of the neutral molecule is known.

Photoelectron data giving the ionization energies of electrons in the various orbitals are often used to check quantum theoretical calculations on molecules, but this is beyond the scope of the present section. The data will be used here to calculate changes in bond distance or bond angle on ionization and also to obtain dissociation energies in ionized states according to the equation

$$I_{\mathrm{M}} - I_{\mathrm{A}} = D_{\mathrm{M}} - D_{\mathrm{M}^+} , \qquad (4.6.0c)$$

where $I_{\mathrm{M}}$ is the ionization energy of the molecule, $I_{\mathrm{A}}$ the first ionization energy of the ionized atom produced when the molecular ion dissociates, $D_{\mathrm{M}}$ the dissociation energy of the molecule into ground-state atoms, and $D_{\mathrm{M}^+}$ the dissociation energy of the relevant state of the molecular ion. In the case of hydrogen a value for the dissociation energy in the ground state can also be obtained as follows. With $I_{\mathrm{v}}$ taken as 15·9 eV from figure 4.6.0(ii) and $I_{\mathrm{H}} = 13{\cdot}6$ eV, equation (4.6.0c) gives $D_{\mathrm{H_2}} - D_{\mathrm{H_2^+}} = 2{\cdot}3$ eV, which is the contribution of *one* electron to the bond in $H_2$ at the equilibrium distance. Doubling this value, we obtain $D_{\mathrm{H_2}} = 4{\cdot}6$ eV.

*x-Ray photoelectron spectroscopy*

x-Ray photoelectron spectroscopy (XPS) has mostly used the soft x-ray lines Mg $K\alpha$ or Al $K\alpha$ (1254 and 1487 eV respectively) as photon sources to examine electronic structures down to a depth of 1000 eV. Photoionization of the 1s shells of the light atoms Li to Ne and the $L$ and $M$ inner shells of heavier atoms is made possible by these higher-energy photons. Thus XPS has been concerned more with the core electrons than the valence electrons, since although photoelectrons of both types are ejected by an incident x-ray photon those from the valence shell will have such high velocities that the small energy differences associated with the structure of the valence electron shell are not so readily resolved as in UPS.

Although the binding energy of an electron in an inner shell is mainly controlled by the net positive charge within the shell, it is also affected to a lesser extent by the cloud of valence electrons lying above it. The XPS line of an element is found to vary over a few electron volts according to the state of chemical combination of the atom and the nature of its immediate neighbours. One of the major activities of the XPS field has been the correlation of the 'chemical shifts' exhibited by the core line of a given element in a large variety of chemical compounds with the effective atomic charges calculated from Pauling electronegativities or with thermodynamic data from chemical heats of formation or with a simple potential model.

**4.6.1**

Show that for photoionization from the nonvibrating ground state of a diatomic molecule the following formula gives the change in internuclear distance $r$ in the approximation of parabolic potential energy curves and the Franck–Condon principle:

$$(\Delta r)^2 = (5 \cdot 44 \times 10^5)(I_v - I_0)\mu^{-1}\omega^{-2} , \qquad (4.6.1a)$$

when $r$ is in Å, $I_v$, $I_0$ are in eV, $\mu$ the reduced mass is in atomic units (g mol$^{-1}$), and $\omega$ the mean spacing in the vibrational progression in the photoelectron band system between the onset and the maximum is expressed in cm$^{-1}$.

**Solution**

It can be seen from figure 4.6.0(ii) that $I_v - I_0$ is equal to the energy difference between the minimum of the potential energy curve of the molecular ion and that part of it lying vertically above the minimum of the corresponding curve of the neutral molecule. On the assumption of a parabolic potential energy curve (simple harmonic oscillator) this difference in energy is equal to $\frac{1}{2}k(\Delta r)^2$. Here $k$, the force constant appropriate to the state of the molecular ion, is equal to $4\pi^2 c^2 \omega^2 \mu / N_A$, where $c$ is the velocity of light and $N_A$ the Avogadro number. Hence we obtain

$$\tfrac{1}{2}k(\Delta r)^2 = 2\pi^2 c^2 \omega^2 \mu (\Delta r)^2 / N_A = 8065hc(I_v - I_0) ,$$

using the conversion factor 1 eV = 8065 cm$^{-1}$. With 1 Å = $10^{-8}$ cm, $h = 6 \cdot 626 \times 10^{-27}$ g cm$^2$ s$^{-1}$, $c = 2 \cdot 998 \times 10^{10}$ cm s$^{-1}$, and $N_A = 6 \cdot 022 \times 10^{23}$ mol$^{-1}$, this yields

$$(\Delta r)^2 = \frac{8065 \times (6 \cdot 626 \times 10^{-27}) \times (6 \cdot 022 \times 10^{23}) \times 10^{16}(I_v - I_0)}{2\pi^2 (2 \cdot 998 \times 10^{10})\omega^2 \mu} \text{Å}^2 ,$$

$$= (5 \cdot 44 \times 10^5)(I_v - I_0)\mu^{-1}\omega^{-2} \text{ Å}^2 ,$$

for $I$, $\mu$, and $\omega$ in the units specified above.

**4.6.2**

Apply equation (4.6.1a) to obtain a value for the change in internuclear distance in ionizing:

(a) $H_2(^1\Sigma_g^+)$ to $H_2^+(^2\Sigma_g^+)$, given that $I_0 = 15 \cdot 45$ eV and $I_v$ occurs at 15·98 eV on the third band of the vibrational progression,

(b) $N_2(^1\Sigma_g^+)$ to $N_2^+(A\,^2\Pi_u)$, given that $I_0 = 16 \cdot 69$ eV and $I_v$ occurs at 16·92 eV on the second band of the vibrational progression.

**Solution**

(a) For $H_2$,

$$\mu = \frac{m_1 m_2}{m_1+m_2} = \frac{1\cdot008^2}{2\times1\cdot008} = 0\cdot504\ ,$$

$$\omega = \tfrac{1}{2}(15\cdot98-15\cdot45)\times8065 = 2137\ \mathrm{cm}^{-1}\ ,$$

and hence equation (4.6.1a) yields

$$(\Delta r)^2 = \frac{(5\cdot44\times10^5)(15\cdot98-15\cdot45)}{(2137)^2\times0\cdot504} = 0\cdot125\ \text{Å}^2$$

and

$$\Delta r = 0\cdot35\ \text{Å}\ .$$

This is to be compared with the value 0·32 Å obtained by analysis of the rotational fine structure of the molecular spectrum (Herzberg, 1950) and also by theoretical calculation. The discrepancy is due mainly to the deviation of the potential energy curves from the parabolic form associated with the assumption of simple harmonic vibration.

(b)

$$\mu(N_2) = \frac{14\cdot007^2}{2\times14\cdot007} = 7\cdot004\ ,$$

$$\omega = (16\cdot92-16\cdot69)\times8065 = 1855\ \mathrm{cm}^{-1}\ .$$

Therefore

$$(\Delta r)^2 = \frac{(5\cdot44\times10^5)(16\cdot92-16\cdot69)}{(1855)^2\times7\cdot003} = 0\cdot0052\ \text{Å}^2\ ;$$

$$\Delta r = 0\cdot072\ \text{Å}\ .$$

This is to be compared with the value of 0·080 Å obtained from the rotational analysis of emission band systems of nitrogen.

**4.6.3**

In the second photoelectron band system of $H_2S$ (figure 4.6.3) which is due to the ionization of an electron from the $a_1$ orbital which lies along the bisector of the HSH angle, a series of bands is obtained which can be interpreted as a progression in the $\nu_2$ symmetrical bending vibration of the molecular ion. This arises from a change in the angle on ionization. Calculate the equilibrium angle in this state of $H_2S^+$, given that $I_0 = 12\cdot78$ eV, $I_v = 13\cdot33$ eV and that six bands occur between the onset and the maximum of the system. Since no vibrations corresponding to the stretching of the bonds appear it can be assumed that there is little change in the S–H distance from its value of 1·328 Å in the neutral ground state, for which the HSH angle is 92·2°.

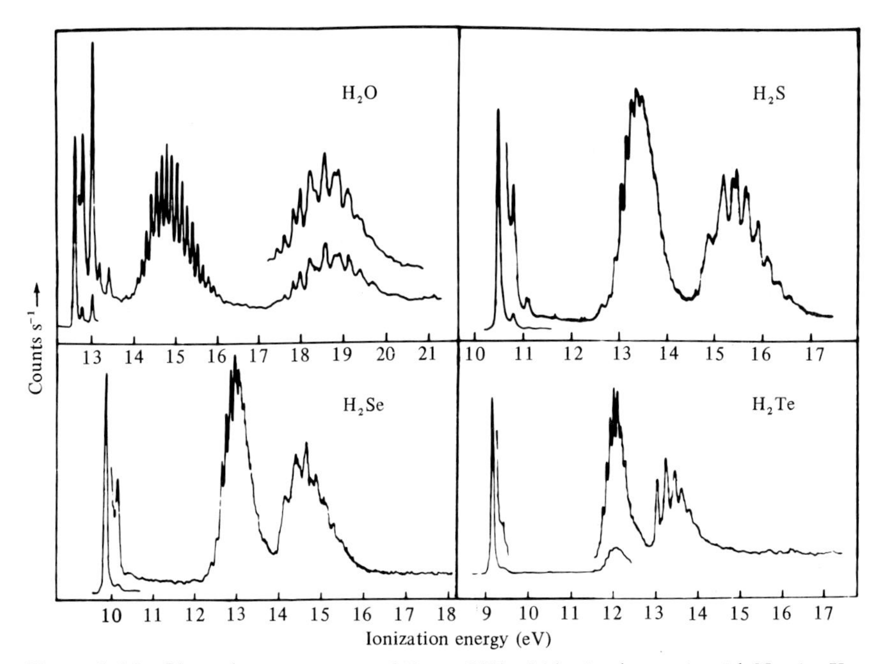


**Figure 4.6.3.** Photoelectron spectra of Group VI hydrides isoelectronic with Ne, Ar, Kr, and Xe, obtained with He (21·22 eV) radiation.

**Solution**

For a simple bending vibration the quantity $\Delta r$ in equation (4.6.1a) can be replaced by $l\Delta\theta$ where $l$ is the bond length in Å and $\theta$ the angle in rad. The mass of the sulphur atom, which moves little in this vibration, can be neglected when calculating $\mu$; thus $\mu \approx 0 \cdot 5$.

The mean value of $\omega$ is

$$\tfrac{1}{5}(13 \cdot 33 - 12 \cdot 78) \times 8065 = 887 \text{ cm}^{-1} .$$

Thus

$$(1 \cdot 328)^2 (\Delta\theta)^2 = \frac{(5 \cdot 44 \times 10^5)(13 \cdot 33 - 12 \cdot 78)}{(887)^2 \times 0 \cdot 5} ,$$

from which

$$\Delta\theta = 0 \cdot 657 \text{ rad} = 37 \cdot 6^\circ .$$

Since $\Delta\theta$ is an increase in the angle the new angle is $92 \cdot 2 + 37 \cdot 6 = 129 \cdot 8°$. This angle corresponds to the nonvibrating state of the ion. The state which is reached by the vertical ionization vibrates about this angle between the extremes $129 \cdot 8 \pm 37 \cdot 6$, that is, between $92 \cdot 2$ and $167 \cdot 4°$. In the outer extreme the molecule is very nearly linear and requires only about another $0 \cdot 1$ eV of vibrational energy to penetrate the barrier to linearity. When this happens the frequency is halved and doubling can be observed in the spectrum. Penetration of this barrier occurs at the onset in the case of the analogous band in $H_2O^+$. An analysis of the fine structure of the emission system of $H_2S^+$ confirms the value of the angle deduced by the above simple calculation.

**4.6.4**

The value of the lowest photoelectron band of oxygen is 12·070 eV. If the dissociation energy of oxygen into two normal $^3P_2$ atoms is 5·080 eV and the ionization energy of one of these atoms to the lowest $^4S_{3/2}$ state of $O^+$ is 13·614 eV, show that the electron removed is antibonding.

**Solution**

Using equation (4.6.0c) we obtain

$$12{\cdot}070 - 13{\cdot}614 = 5{\cdot}080 - D_{O_2^+} ,$$

from which

$$D_{O_2^+} = 6{\cdot}624 \text{ eV} .$$

Hence the dissociation energy has been *increased* by ionization, and the electron removed must have been antibonding.

**4.6.5**

The adiabatic ionization energies of the $\Pi$ and $\Sigma$ photoelectron bands of $HCl^+$ are 12·74 and 16·23 eV respectively (figure 4.6.5). If the first state dissociates into $H + Cl^+$, the ionization energy of Cl ($^3P_2 \leftarrow {}^2P_{3/2}$) being 13·01 eV, and the second into $H^+ + Cl$ ($I_H = 13{\cdot}60$ eV), calculate the dissociation energies of both these ionized states given that the neutral ground-state dissociation energy is 4·43 eV. Hence show that the first electron is nearly nonbonding and the second strongly bonding. Calculate also the wavelength of the $\nu_{00}$ band of the $\Sigma \rightarrow \Pi$ emission system of $HCl^+$.

**Solution**

The dissociation energy of the $\Pi$ system of $HCl^+$ is given by

$$D_{HCl^+} = D_{HCl} + I_{Cl} - I_{HCl} = 4{\cdot}43 + 13{\cdot}01 - 12{\cdot}74$$
$$= 4{\cdot}70 \text{ eV} ,$$

which indicates that the electron is slightly bonding. That of the $\Sigma$ system is similarly equal to $4{\cdot}43 + 13{\cdot}60 - 16{\cdot}23 = 1{\cdot}80$ eV, which is a considerable reduction compared with the ground-state value of 4·43 eV, and the electron removed must therefore have been strongly bonding. This is evident from the photoelectron spectrum (figure 4.6.5), which may also be used to obtain a value for the increased H–Cl separation in the $\Sigma$ state of $HCl^+$.

The wavelength of the $\nu_{00}$ band of the $\Sigma \rightarrow \Pi$ emission system is obtained by the difference in the adiabatic ionization energies: $16 \cdot 23 - 12 \cdot 74 = 3 \cdot 49$ eV. This is equivalent to a wavelength of 355·2 nm, i.e. in the near ultraviolet, where the bands are observed.

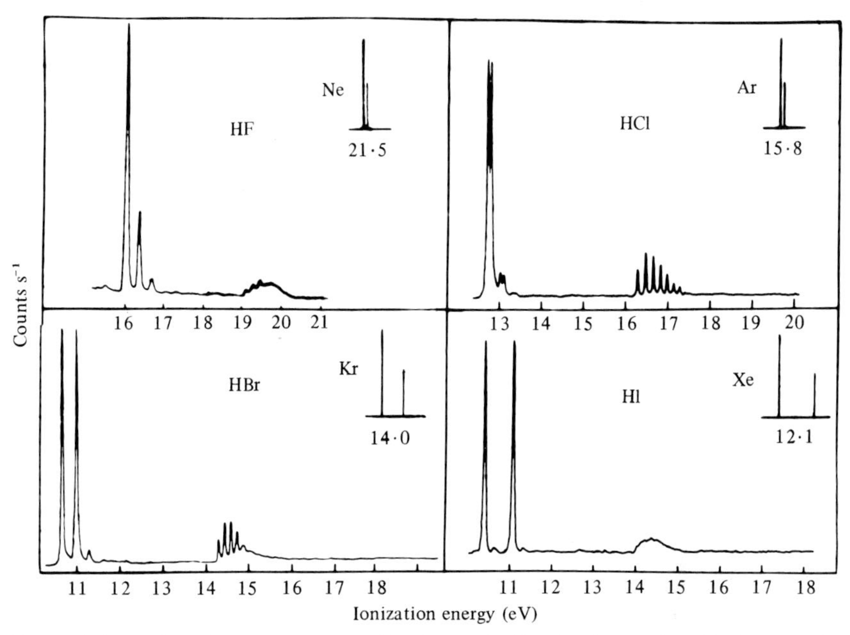


**Figure 4.6.5.** Photoelectron spectra of hydrogen halides isoelectronic with Ne, Ar, Kr, and Xe, obtained with He (21·22 eV) radiation.

**4.6.6**
The intensities of the two peaks in the photoelectron spectrum of the inert gases which correspond to the states of the ion $^2P_{3/2}$ and $^2P_{1/2}$ are in the ratio 2 : 1. Show that this is in agreement with their degeneracies.

**Solution**
The intensities depend upon the number of channels of escape open to the electrons, and in this case the degeneracies of the ionized states, which are equal to $2J + 1$, i.e. $(2 \times \frac{3}{2}) + 1 = 4$ for $^2P_{3/2}$ and $(2 \times \frac{1}{2}) + 1 = 2$ for $^2P_{1/2}$. For ionized states of different multiplicities, whether atomic or molecular, the intensities will be proportional to the multiplicity, e.g. the integrated intensities of the related singlet and triplet ionized states of $NO^+$ are in the ratio 1 : 3 and those of similar doublet and quartet states of $O_2^+$ are in the ratio 1 : 2.

**4.6.7**

Spurious bands are often produced in photoelectron spectroscopy if, for example, a small amount of hydrogen, giving Lyman $\alpha$ radiation [L($\alpha$)] at 121·567 nm, is present in the helium discharge. These are in fact useful for scale calibration (Lloyd, 1970). Consider the methyl iodide ionizations at 9·538 and 10·162 eV: what apparent ionization energies would these have on a He 21·22 eV scale if produced by L($\alpha$)?

**Solution**

The photoelectron energies produced by L($\alpha$) at 121·567 nm or 10·20 eV when photoionizing to the two first-ionized states of methyl iodide are $10{\cdot}20 - 9{\cdot}54 = 0{\cdot}66$ eV, and $10{\cdot}20 - 10{\cdot}16 = 0{\cdot}04$ eV, respectively. These photoelectron energies would be associated with ionization energies from a 21·22 eV He (I) photon which would be $21{\cdot}22 - 0{\cdot}66 = 20{\cdot}56$ eV, and $21{\cdot}22 - 0{\cdot}04 = 21{\cdot}18$ eV, respectively; they may in fact be used to calibrate the high-energy end of a He (I) spectrum.

**4.6.8**

The values of the ionization energies provided by photoelectron spectra are often useful in interpreting the vacuum-ultraviolet, energy-loss, or positive-ion-yield spectra of molecules. These spectra frequently show a few bands which might form a Rydberg series, but without enough members to establish the series convincingly. Firstly it is required that the band patterns, particularly those at higher energies, should resemble closely the pattern of the photoelectron band to which they converge. By subtracting the wavenumber values of the bands from the value of the ionization limit, term values can be found for the bands. From standard Rydberg term tables (see, for example, Sutcliffe, 1958) values for $R/(n-\delta)^2$ can be found to match these. If the quantum defect $\delta$ is fairly constant for successive members then the Rydberg series can be considered as established. The value of $\delta$ helps to assign the nature of the electron ionized.

In $NO_2$ the contour of the strong absorption system starting at 1290 Å correlates very closely with that of the photoelectron spectrum between 12·85 and 13·40 eV. Show that two other systems starting at 1089 and 1031·5 Å are members of a Rydberg series consecutive to the band starting at 1290 Å and find their quantum defects.

**Solution**

The ionization energy 12·85 eV is equivalent to a limit of 103 630 $cm^{-1}$. The three bands starting at 1290, 1089, and 1031·5 Å have frequencies of 77 520, 91 830, and 96 940 $cm^{-1}$ respectively. On subtraction from the limit, term values of 26 110, 11 800, and 6690 $cm^{-1}$ are obtained. From the Rydberg term tables it is seen that these term values correspond to Rydberg denominators of $(3{\cdot}0-0{\cdot}95)$, $(4{\cdot}0-0{\cdot}95)$, and $(5{\cdot}0-0{\cdot}95)$, thus showing that the bands form a Rydberg series. The value of the quantum defect by comparison with that of established series in similar molecules indicates that the excited orbitals are *n*s, from which it follows that the ground-state orbital being ionized is probably derived from a 2p atomic orbital.

**4.6.9**

Calculate the effective atomic charges on the nitrogen atoms in aniline and nitrobenzene using Pauling's valence bond model and show that they are consistent with N(1s) binding energies of 399·0 and 406 eV, respectively, in these two compounds.

**Solution**

According to Pauling the effective charge $q$ on an atom in a molecule is obtained as the sum of the partial ionic character of the bonds between the atom and its neighbours. Thus for the $j$th atom

$$q_i = Q_j \pm \sum_{i \neq j} n i_{jk}$$

where $Q_j$ is the formal charge on the $j$th atom, $i_{jk}$ is the partial ionic character of the bond between atoms $j$ and $k$, and $n$ is the bond number ($n = 1$ for a single bond, 2 for a double bond, etc.). The partial ionic character, $i_{jk}$, is obtained from the electronegativity difference $(x_j - x_k)$ between the bonding partners and is given by the relation

$$(\pm) i_{jk} = 1 - \exp[-0{\cdot}25(x_j - x_k)^2],$$

the positive sign being taken if $x_k - x_j$, and vice versa.

The values given in table 4.6.9 have been obtained by using Pauling electronegativities in the above formulae.

It is clear that the higher binding energy of N(1s) in nitrobenzene is consistent with the greater positive charge on this atom relative to the

**Table 4.6.9.**

| | Aniline | | | Nitrobenzene | | |
|---|---|---|---|---|---|---|
| | C— | N̈ | <H, H | C— | $\overset{+}{N}$ | =O, O⁻ |
| $x$ | 2·5 | 3·0 | 2·1 | 2·5 | 3·3 | 3·5<br>3·2 |
| $i$ | −0·06 | | −0·18<br>−0·18 | −0·15 | | +0·02<br>~0·00 |
| $Q$ | | 0 | | | +1·0 | |
| $q$ | | −0·42 | | | +0·89 | |

nitrogen atom in aniline. When similar calculations are carried out for other nitrogen-containing molecules and the binding energies of N(1s) plotted against calculated effective atomic charges an approximately linear graph is obtained (Siegbahn *et al.*, 1967).

**4.6.10**

Calculate the wavelengths of the photoelectrons ejected from a C(1s) orbital of binding energy 290 eV and from a C(2s) orbital of binding energy 23 eV by Mg $K\alpha$ (photon energy 1254 eV). Compare these with the wavelength of an electron ejected from a C(2p) orbital with an ionization energy of 10 eV by a 21 eV photon.

**Solution**

The wavelength of an electron which has an energy $\frac{1}{2}mv^2 = eV$ is given by

$$\lambda = h/mv = h(2m \times \tfrac{1}{2}mv^2)^{-\frac{1}{2}} = h(2meV)^{-\frac{1}{2}} = 12{\cdot}27\ V^{-\frac{1}{2}}\ ,$$

where $V$ is in volts and $\lambda$ in Å.

The kinetic energy of the C(1s) photoelectron is $1254 - 290 = 964$ eV. Hence $\lambda = 12{\cdot}27 \times (964)^{-\frac{1}{2}} = 0{\cdot}40$ Å. The kinetic energy of the C(2s) electron is $1254 - 23 = 1231$ eV and the wavelength of the photoelectron $= 12{\cdot}27 \times (1231)^{-\frac{1}{2}} = 0{\cdot}35$ Å. For a C(2p) electron with ionization energy 10 eV ejected by a 21 eV photon the photoelectron energy is $21 - 10 = 11$ eV and the associated wavelength is $12{\cdot}27 \times (11)^{-\frac{1}{2}} = 3{\cdot}7$ Å.

**References**

Herzberg, G., 1950, *Molecular Spectra and Molecular Structure*, second edition, *Volume I: Spectra of Diatomic Molecules* (Van Nostrand, Princeton, NJ).
Lloyd, D. R., 1970, *J. Sci. Instrum.*, **3**, 629.
Price, W. C., 1981, *Int. Rev. Phys. Chem.*, **1**, 1.
Siegbahn, K., *et al.* (10 authors), 1967, *ESCA, Atomic, Molecular, and Solid State Structure Studied by Electron Spectroscopy* (Almqvist and Wiksell, Stockholm).
Sutcliffe, L. H., 1958, *Spectrochim. Acta*, **12**, 179.
Turner, D. W., Baker, A. D., Baker, C., Brundle, C. R., 1970, *Molecular Photoelectron Spectroscopy* (John Wiley, Chichester, Sussex).

# 5 Nuclear spectroscopy

## 5.1 Nuclear magnetic resonance

**D J Greenslade** University of Essex
**R G Jones** University of Essex

**I Solid state n.m.r.** (by D J Greenslade)

**5.1.1**

(a) Calculate the intramolecular contribution to the second moment of the proton magnetic resonance absorption of (i) benzene, (ii) monodeuterobenzene, and (iii) 1,3,5-trideuterobenzene, in terms of the C–H bond length, given that the C–C bond length is 1·40 Å.

(b) Andrew and Eades (1953) measured the second moment of each of these compounds and found, at very low temperatures, the values 9·72, 7·94, and 3·57 $G^2$, respectively. Show that for benzene the intermolecular moment is 6·6 $G^2$.

(c) Use the results of parts (a) and (b) to deduce the length of the C–H bond in benzene. What assumptions have you made in your calculations? (1 G is equivalent to $10^{-4}$T.)

**Solution**

(a) The second moment, $M_2$, of a nuclear magnetic resonance absorption line is related to the internuclear vectors by the expression (Abragam, 1961, p.112):

$$\gamma_j^2 M_2 \equiv \gamma_j^2 \int_{-\infty}^{+\infty} \mathrm{f}(b) b^2 \, \mathrm{d}b = \tfrac{3}{4} a \gamma_j^2 \gamma_k^2 I_k (I_k + 1) \hbar^2 \sum_k \frac{(1 - 3\cos^2\theta_{jk})^2}{r_{jk}^6} , \quad (5.1.1\mathrm{a})$$

where $b$ is the field measured from the centre of the absorption line[1], and $\mathrm{f}(b)$ the height of the absorption line at $b$, $I_j$, and $I_k$ are the nuclear spins of nuclei $j$ and $k$, $j$ is one particular nucleus of the type being measured, $k$ any other nucleus, $r_{jk}$ the distance between them, $\gamma$ is the gyromagnetic ratio of a nucleus and $\theta_{jk}$ the angle between the internuclear vector and the magnetic field direction. The constant $a$ takes the value 1 if nuclei $j$ and $k$ are of the same type and $\frac{4}{9}$ if they are of different types. In the case of a powder or polycrystalline sample all orientations of the internuclear vector are possible and the angular term in equation (5.1.1a) must be replaced by its average:

$$\begin{aligned} \langle (3\cos^2\theta - 1)^2 \rangle &= \int_0^\pi (3\cos^2\theta - 1)^2 \sin\theta \, \mathrm{d}\theta \bigg/ \int_0^\pi \sin\theta \, \mathrm{d}\theta \\ &= \tfrac{1}{2} \int_0^\pi (6\cos^2\theta - 9\cos^4\theta - 1) \, \mathrm{d}(\cos\theta) \\ &= \tfrac{4}{5} . \end{aligned}$$

[1] See Editor's note on treatment of magnetic fields in Appendix 1.

The intramolecular second moment is simply calculated by taking one proton in the molecule and performing the summation over all the other nuclei in the molecule. Since the triangle formed by the centre of the benzene hexagon and two adjacent carbon atoms is equilateral, the circumscribing circle of the carbon atoms has radius $r_{C-C}$, the carbon–carbon bond length. The radius, $R$, of the circle through the protons is the sum of the carbon–hydrogen distance and the carbon–carbon distance. Again, from the properties of equilateral triangles (see figure 5.1.1) the *ortho*-proton is at a distance $R$, the *meta*-protons at $3^{1/2}R$, and the *para*-proton at $2R$. The deuterons will to a good approximation be similarly placed. The intramolecular second moment of benzene, $M_{2B}^{i}$, is then given in units of (magnetic field)$^2$ by

$$M_{2B}^{i} \doteq \tfrac{3}{5}\gamma_H^2 \tfrac{1}{2}(\tfrac{1}{2}+1)\hbar^2\left[\frac{2}{R^6}+\frac{2}{(3^{1/2}R)^6}+\frac{1}{(2R)^6}\right] . \qquad (5.1.1b)$$

For simplification, let

$$A = \tfrac{3}{5}\gamma_H^2 \tfrac{1}{2}(\tfrac{1}{2}+1)\frac{\hbar^2}{R^6} .$$

In the case of benzene (B) we are able to take a typical proton and sum over the other protons in the molecule, but in the case of monodeuterobenzene (M) we must average over the three possible situations. Firstly the deuteron may be *para* to the proton being considered (weight one), secondly *meta* (weight two), and thirdly *ortho* (weight two). In calculating the effect of the deuteron on the proton we must note that

$$\frac{I_D(I_D+1)}{I_H(I_H+1)} = \tfrac{8}{3} ,$$

and that $\gamma_H^2$ is replaced by $\frac{4}{9}\gamma_D^2$. Let $c = \gamma_D^2/\gamma_H^2 = 2{\cdot}356 \times 10^{-2}$, hence,

$$M_{2M}^{i} = \tfrac{1}{5}A[(2+\tfrac{2}{27}+\tfrac{4}{9}\tfrac{8}{3}\tfrac{1}{64}c)+2(2+\tfrac{1}{27}+\tfrac{4}{9}\tfrac{8}{3}\tfrac{1}{27}c+\tfrac{1}{64})+2(1+\tfrac{4}{9}\tfrac{8}{3}c+\tfrac{2}{27}+\tfrac{1}{64})] .$$

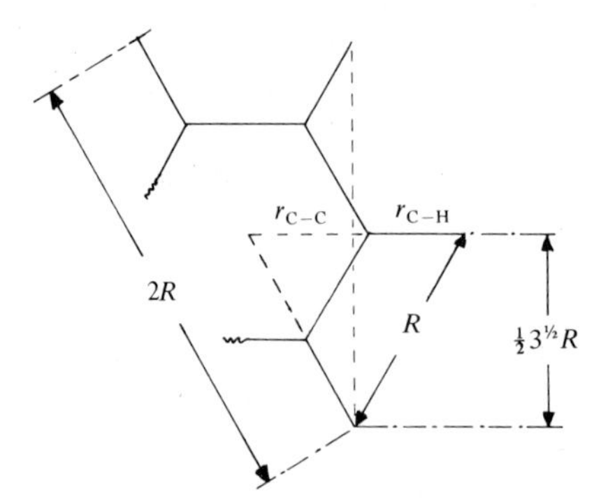


**Figure 5.1.1.**

For the trideuterobenzene (T) we may consider a typical proton:

$$M^{i}_{2T} = A[\tfrac{4}{9}\tfrac{8}{3}c(2+\tfrac{1}{64})+\tfrac{2}{27}] .$$

(b) The intermolecular second moments, $M^{e}_{2}$ are related by assuming that the lattice sums contain the same terms, but that in the deuterated species one-sixth (M) or one-half (T) of the terms relate to the deuterons, so that these are reduced by the factor $\tfrac{4}{9}\tfrac{8}{3}c$. Thus

$$\frac{M^{e}_{2T}}{M^{e}_{2B}} = \frac{(1+\tfrac{32}{27}c)}{2} = 0 \cdot 5140 .$$

From part (a) of the problem, we have

$$\frac{M^{i}_{2T}}{M^{i}_{2B}} = 0 \cdot 06240 .$$

Combining these results with the experimental results, we obtain

$$M_{2B} = M^{i}_{2B} + M^{e}_{2B} = 9 \cdot 72 \text{ G}^2 ;$$
$$M_{2T} = M^{i}_{2T} + M^{e}_{2T} = 3 \cdot 57 \text{ G}^2 .$$

Solving these equations, we deduce $M^{e}_{2B}$ to be $6 \cdot 6$ G$^2$.
(c) The intramolecular moment is simply obtained by subtracting the result of part (b) from the total moment, to give $M^{i}_{2B} = 3 \cdot 1$ G$^2$. Thus, with $R$ in Å, equation (5.1.1b) becomes

$$3 \cdot 1 = \tfrac{3}{5}(2 \cdot 675 \times 10^4)^2 \times \tfrac{3}{4}(1 \cdot 055 \times 10^{-27}) \times 2 \cdot 089 \times \frac{1}{R^6} ,$$

so that

$$R = r_{C-C} + r_{C-H} = 2 \cdot 49 \text{ Å} ,$$

and

$$r_{C-H} = 1 \cdot 09 \text{ Å} .$$

Throughout we have assumed a rigid lattice. Even at low temperatures there is motion of the protons and an averaging should have been carried out over this motion. In this calculation we need the carbon–carbon bond length, but it has been shown that line-narrowing and double-resonance methods enable the $^{13}$C–H dipolar interaction to be observed (Hester *et al.*, 1975; Mehring, 1976, p.145), and so direct determination of the carbon–hydrogen bond length is also possible.

5.1.2

(a) Calculate the effect on the intramolecular second moment, when the benzene molecule is rapidly rotated about an axis which makes an angle $\alpha$ to the plane of the molecule.
(b) Above 120 K Andrew and Eades (1953) found the second moment to decrease to $1 \cdot 61$ G$^2$ in the case of benzene and to $0 \cdot 44$ G$^2$ in the case of 1,3,5-trideuterobenzene. Suggest an explanation for their results.

(c) What is the effect of rotating the whole sample of benzene? Does this suggest an experimental method of narrowing the n.m.r. absorption lines exhibited by solids?

**Solution**

(a) If the vector $\boldsymbol{r}_{jk}$ joining a pair of protons $j$ and $k$ makes an angle $\theta_{jk}$ to the external magnetic field, $\boldsymbol{B}_0$ (see figure 5.1.2) then on rotation the second moment becomes

$$\overline{M}_2 \propto \frac{\overline{(3\cos^2\theta_{jk}-1)^2}}{r_{jk}^6} .$$

Use is now made of the spherical harmonic addition theorem (Morse and Feshbach, 1953):

$$(3\cos^2\theta_{jk}-1) = \tfrac{1}{2}(3\cos^2\theta'-1)(3\cos^2\alpha-1) .$$

For a polycrystalline sample we average, as in problem 5.1.1, over all possible $\theta'$. Since $\alpha$ is the same for all pairs,

$$M_2 = \tfrac{1}{4}A\sum_k \frac{1}{r_{jk}^6}(3\cos^2\alpha-1)^2 ,$$

where $A$ is a constant (see solution to problem 5.1.1).

(b) The separation of intermolecular and intramolecular moments is carried out as in problem 5.1.1:

$$\frac{M_{2\mathrm{T}}^{\mathrm{e}}}{M_{2\mathrm{B}}^{\mathrm{e}}} = 0\cdot514 , \qquad \frac{M_{2\mathrm{T}}^{\mathrm{i}}}{M_{2\mathrm{B}}^{\mathrm{i}}} = 0\cdot0624 ;$$

$$M_{2\mathrm{B}}^{\mathrm{i}}+M_{2\mathrm{B}}^{\mathrm{e}} = 1\cdot61\ \mathrm{G}^2 , \qquad M_{2\mathrm{T}}^{\mathrm{i}}+M_{2\mathrm{T}}^{\mathrm{e}} = 0\cdot44\ \mathrm{G}^2 ;$$

thus,

$$M_{2\mathrm{B}}^{\mathrm{i}} = 0\cdot85\ \mathrm{G}^2 ,$$

which is approximately a quarter of the value found for the rigid lattice. This suggests that $(3\cos^2\alpha-1)^2 = 1$, that is, $\cos\alpha = 0$, so that above 120 K the benzene molecule is rotating about its sixfold axis. Rotation does not imply uniform motion; rapid jumps achieve the same effect

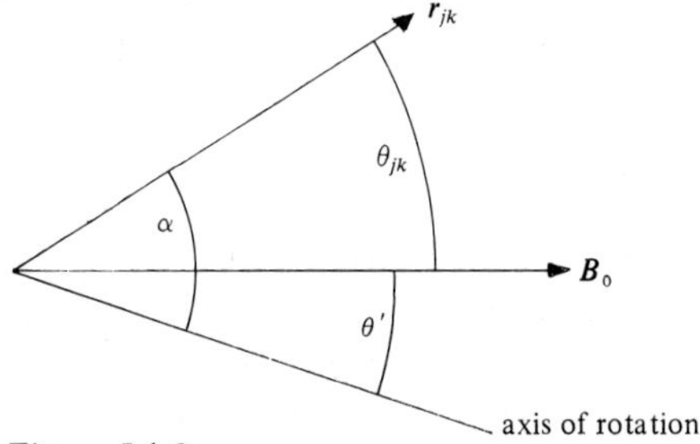


**Figure 5.1.2.**

provided that the magnetic resonance motional narrowing criterion,

$$(\overline{\Delta\omega_0^2})^{1/2}\tau_c \ll 1 ,$$

where $\overline{\Delta\omega_0^2}$ is the second moment of the resonance frequency shift and $\tau_c$ is the period of the motion, is met (Abragam, 1961, p.425).
(c) In the case of rotation of the whole sample, $\theta'$, the angle between the magnetic field and the axis of rotation, is constant and there is averaging over all possible values of $\alpha$. If $\theta' = \cos^{-1}(3^{-1/2})$, then the angular factor in the second moment is zero and the lines are narrowed. The motional narrowing criterion implies that the rotation must be sufficiently rapid. This method has been developed by Andrew (1972) who has achieved rotations at 10 kHz. An effective rotation of the nuclei can be achieved by applying a strong radio-frequency field and especially by applying a train of pulses of such a field. This method is discussed by Waugh (1970), Mansfield (1972), Haeberlen (1976), and Mehring (1976).

5.1.3
The hexafluorobenzene $^{19}F$ n.m.r. spectrum has a second moment at 77 K of $4{\cdot}90 \pm 0{\cdot}40$ $G^2$ at a frequency of 56·4 MHz (Albert *et al.*, 1972). Show that this is consistent with a rigid lattice by carrying out the following calculations.
(a) The magnetic field, $B$, experienced by a nucleus is

$$B = B_0[1-(\lambda_{xx}^2\sigma_{xx}+\lambda_{yy}^2\sigma_{yy}+\lambda_{zz}^2\sigma_{zz})] ,$$

where $\sigma_{xx}$, etc., are the principal values of the chemical shift tensor, and $\lambda_{xx}$ the direction cosines of its principal axes with respect to the external magnetic field vector $\boldsymbol{B}_0$. Derive an expression for the contribution to the second moment from the anisotropy of the chemical shift in the case of an axially symmetric tensor. Given that for the fluorine nuclei in hexafluorobenzene $\sigma_{xx} = \sigma_{yy} = -53$ and $\sigma_{zz} = 102$ p.p.m., calculate the contribution in the present case.
(b) Calculate the intramolecular contribution to the second moment from the carbon–carbon (1·40 Å) and the carbon–fluorine (1·33 Å) bond distances.
(c) Estimate the intermolecular second moment for a rigid lattice given that for benzene this quantity is 6·6 $G^2$, that the carbon–hydrogen bond length is 1·10 Å, and that the van der Waals radii of fluorine and hydrogen are 1·35 and 1·20 Å, respectively. The ratio of the nuclear moments is (56·4/60·0).

**Solution**
(a) The magnetic field experienced by a nucleus whose axially symmetric chemical shift tensor is at an angle $\theta$ to the external field is

$$B = B_0[1-(\sigma_\perp \sin^2\theta+\sigma_\parallel \cos^2\theta)] .$$

The average field for a collection of nuclei with randomly oriented tensors, as in a polycrystalline sample, is

$$\overline{B} = \tfrac{1}{2}\int_0^\pi B\sin\theta\,\mathrm{d}\theta = B_0[1-\tfrac{1}{3}(2\sigma_\perp+\sigma_\parallel)]\,,$$

so that $B = \overline{B} - b$, where

$$b = \tfrac{1}{3}B_0(\sigma_\parallel - \sigma_\perp)(3\cos^2\theta - 1)\,.$$

The strength of the absorption, $\mathrm{f}(b)$, at $b$ is proportional to the number of nuclei experiencing such a field shift, so that the second moment of the absorption,

$$\int_{-\infty}^{+\infty}\mathrm{f}(b)b^2\,\mathrm{d}b \bigg/ \int_{-\infty}^{+\infty}\mathrm{f}(b)\,\mathrm{d}b\,,$$

simply becomes

$$\int_0^\pi \tfrac{1}{3}B_0(\sigma_\parallel - \sigma_\perp)(3\cos^2\theta - 1)]^2\sin\theta\,\mathrm{d}\theta \bigg/ \int_0^\pi \sin\theta\,\mathrm{d}\theta\,,$$

since $\mathrm{f}(b)\,\mathrm{d}b$ is proportional to $\sin\theta\,\mathrm{d}\theta$, the solid angle element containing all nuclei whose tensors make an angle $\theta$ to the external field.

Hence, performing the integrations, we obtain

$$M_2^\sigma = \tfrac{4}{45}B_0^2(\sigma_\parallel - \sigma_\perp)^2\,. \qquad (5.1.3\mathrm{a})$$

This result suggests that the anisotropy might be deduced from the dependence of the second moment on the square of the field. This was the method used by Andrew and Tunstall (1963). The values given here, however, are those obtained by line-narrowing techniques (Mehring *et al.*, 1971), which are more accurate. Substituting these values into equation (5.1.3a) and noting that at 56·4 MHz the fluorine n.m.r. occurs at 14·09 kG, we obtain

$$M_2^\sigma = \tfrac{4}{45}(14{\cdot}09\times10^6)^2(155\times10^{-6})^2 = 0{\cdot}42\ \mathrm{G}^2\,.$$

(b) As in problem 5.1.1, the intramolecular moment is

$$M_2^\mathrm{i} = \tfrac{3}{5}\gamma_\mathrm{F}^2\tfrac{1}{2}(\tfrac{1}{2}+1)\hbar^2\left[\frac{2}{R^6}+\frac{2}{(3^{1/2}R^6)}+\frac{1}{(2R)^6}\right],$$

with $R = 1{\cdot}33 + 1{\cdot}40 = 2{\cdot}73$ Å; hence,

$$M_2^\mathrm{i} = 1{\cdot}6\ \mathrm{G}^2\,.$$

(c) The increase in intermolecular coordinates from benzene to hexafluorobenzene is assumed largely to be due to increased van der Waals radius:

$$M_2^\mathrm{e}(\mathrm{C_6F_6}) = \left(\frac{1{\cdot}20}{1{\cdot}35}\right)^6\left(\frac{56{\cdot}4}{60{\cdot}0}\right)^2 M_2^\mathrm{e}(\mathrm{C_6H_6}) = 2{\cdot}84\ \mathrm{G}^2\,.$$

The second moment for the rigid lattice is the sum of the various contributions (Andrew and Finch, 1957) and is, then,

$$2 \cdot 84 + 1 \cdot 6 + 0 \cdot 42 = 4 \cdot 86 \text{ G}^2 .$$

This is in good agreement with the experimental value.

**References to part I**

Abragam, A., 1961, *Principles of Nuclear Magnetism* (Oxford University Press, London).
Albert, S., Gutowsky, H. S., Ripmeester, J. A., 1972, *J. Chem. Phys.*, **56**, 2844.
Andrew, E. R., 1972, *Prog. Nucl. Magn. Reson. Spectrosc.*, **8**, 1.
Andrew, E. R., Eades, R. G., 1953, *Proc. R. Soc. London, Ser. A*, **218**, 537.
Andrew, E. R., Finch, N. D., 1957, *Proc. Phys. Soc. London, Sect. B*, **70**, 980.
Andrew, E. R., Tunstall, D. P., 1963, *Proc. Phys. Soc. London*, **81**, 986.
Haeberlen, U., 1976, *Adv. Magn. Reson.* supplement.
Hester, R. K., Cross, V. R., Ackermann, J. L., Waugh, J. S., 1975, *J. Chem. Phys.*, **63**, 3606.
Mansfield, P., 1972, *Prog. Nucl. Magn. Reson. Spectrosc.*, **8**, 41.
Mehring, M., 1976, *High Resolution NMR Spectroscopy in Solids* (Springer, Berlin).
Mehring, M., Griffin, R. G., Waugh, J. S., 1971, *J. Chem. Phys.*, **55**, 746.
Morse, P. M., Feshbach, H., 1953, *Methods of Theoretical Physics* (McGraw-Hill, New York), p. 1274.
Waugh, J. S., 1970, in *Magnetic Resonance*, Eds C. K. Coogan *et al.* (Plenum, New York), p.177.

**General reading**

A recent summary of work in n.m.r. of solids can be found in an issue of *Philosophical Transactions of the Royal Society of London, Series A*, 1981, volume 299, pages 475–689.

## II High-resolution n.m.r. (by R G Jones)

### Introduction

The problems which follow are based on the three principal analytical features of high-resolution n.m.r.:

*The relative positions of signals*: the structural environment of the protons is reflected in the chemical shifts. Correlation tables are available in the literature (Emsley *et al.*, 1965).

*The intensities of signals*: the relative intensities measured by use of the integral trace give the relative numbers of protons in different environments. This information can be converted into the real numbers of protons in different environments when the molecular formula is known.

*Multiplicities of signals*. Nuclei with magnetic moments interact to produce a mutual splitting of the n.m.r. signals. The multiplicities of the signals depend on the spin quantum numbers of the nuclei and the numbers of interacting nuclei in the simplest cases. The simple rule for the splitting created by $n$ protons (spin $\frac{1}{2}$) is

$$\text{number of lines} = (n+1) .$$

The relative intensities of the $(n+1)$ lines are given by the coefficients of the binomial expansion $(a+b)^n$ and are conveniently summarised in the Pascal triangle tabulated below.

| Number of interacting nuclei, $n$ | Number of lines, $(n+1)$ | Relative intensities |
|---|---|---|
| 1 | 2 | 1 1 |
| 2 | 3 | 1 2 1 |
| 3 | 4 | 1 3 3 1 |
| 4 | 5 | 1 4 6 4 1 |
| ⋮ | ⋮ | ⋰ ⋱ |

These rules are valid provided that two conditions are fulfilled:
(i) the chemical shift between the interacting groups must be large compared with the coupling constant, $J$;
(ii) the coupling constants characterizing the interaction between the nuclei of the groups must all be equal.

The magnitude of the coupling constants can be interpreted in terms of the stereochemistry of molecules and also the number of bonds separating the interacting nuclei. A comprehensive review of proton–proton coupling constants has been compiled by Bothner-By (1965).

Spectra very often appear far more complicated than predicted from the foregoing simple rules because either (i), (ii), or both are invalid. In these cases the analysis is more involved and can become very complicated. The problems which follow do include some of the less-involved cases of this kind, which have been clearly indicated and the relevant parts of the spectra have been expanded.

**Units and chart interpretation**

The units used in the problems are Hertz (Hz) both for chemical shifts and for coupling constants, and have dimensions $s^{-1}$. Many of the current texts use the units c/s (cycles per second) which are equivalent.

The spectra have all been recorded at 60 MHz and the charts are calibrated in parts per million (p.p.m.) units with the zero at tetramethylsilane (tms) as in the $\delta$ scale. An alternative scale which may be encountered in standard texts is the $\tau$ scale defined as

$$\tau = 10 - \delta ,$$

again in p.p.m. units; 1 p.p.m. = 60 Hz.

The figures are reduced to 36% of the size of the original charts, and the chart calibration is given in Hz and p.p.m. for each spectrum.

**5.1.4**

Analysis of a compound (relative molecular mass 137) gave H 5·1, C 61·2, N 10·2, and O 23·5 wt%. Identify the compound from its n.m.r. spectrum (figure 5.1.4) for a sample in solution in $CCl_4$.

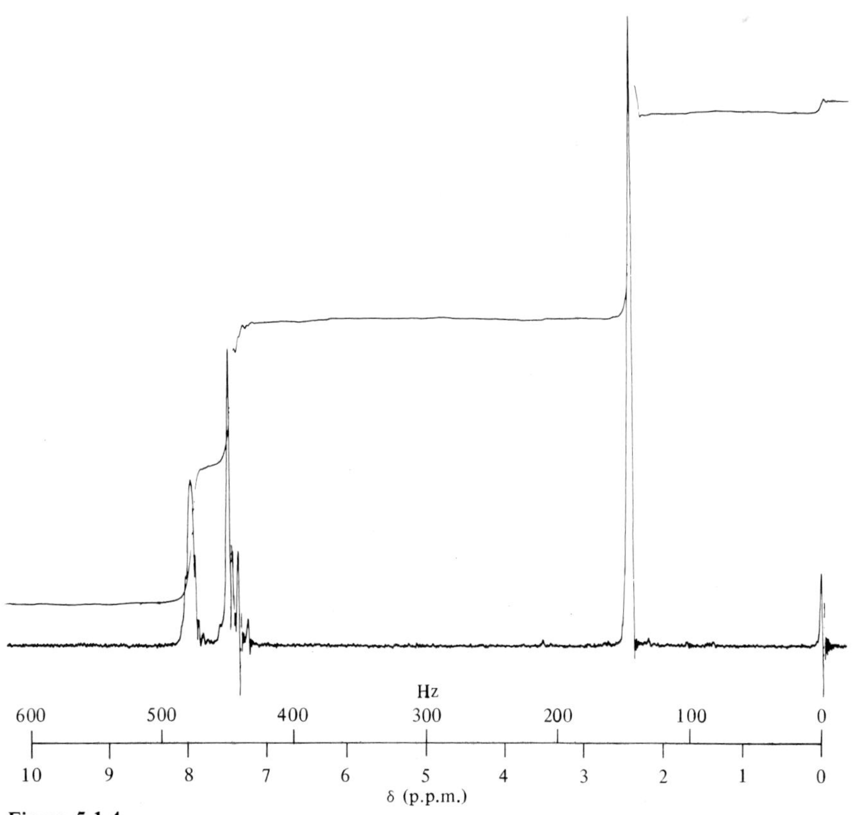


**Figure 5.1.4.**

**Solution**

The spectrum in figure 5.1.4 consists of two distinct multiplets at low field and a sharp single line at high field, which can all be assigned to a compound $C_7H_7NO_2$, since the solvent is $CCl_4$. No distinctive pattern of lines can be picked out, but the number of protons responsible for the observed signals can be calculated from the integral trace[2].

Total integral height = 64 mm.

Number of protons = 7.

Integral height per proton ≈ 9 mm.

The steps in the integral trace corresponding to the three regions of absorption at 7·95, 7·43, and 2·40 p.p.m.[3] are 18, 19, and 27 mm, respectively. It follows that the singlet at 2·40 p.p.m. must arise from a group of three equivalent protons, and each of the low-field multiplets from two protons.

The single peak at 2·40 p.p.m. can be assigned to methyl group protons. The chemical shift is very similar to that of toluene methyl protons at 2·34 p.p.m.

The multiplets at 7·95 and 7·43 p.p.m. can be assigned to four aromatic protons characterized principally by their chemical shifts, but also by the residual formula when the methyl group is subtracted:

$$C_7H_7NO_2 - CH_3 = C_6H_4NO_2 \, .$$

The compound $C_7H_7NO_2$ is therefore a nitrotoluene.

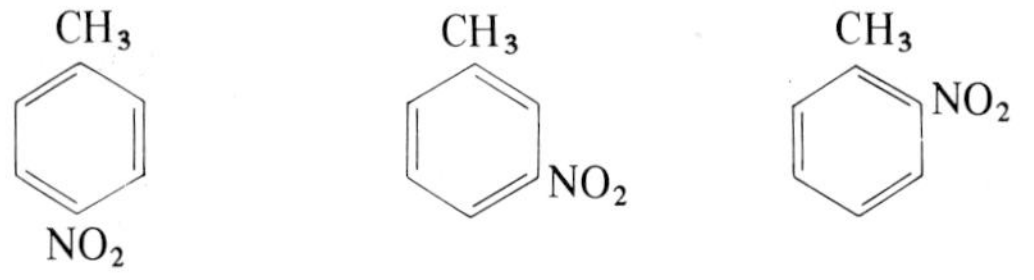

$CH_3$ $NO_2$ $CH_3$ $NO_2$ $CH_3$ $NO_2$

The 4-nitrotoluene has a plane of symmetry at right angles to the ring and therefore gives a more symmetrical spectrum than that in figure 5.1.4. The 2-nitrotoluene would be expected to have one proton at lower field than the others because the nitro group is adjacent to only one proton. The chemical shifts in such compounds as these can be calculated approximately from additive substituent effects (Mathieson, 1967).

Approximate shifts (p.p.m.) in the position of aromatic protons caused by substituents (relative to benzene protons at $\delta = 7{\cdot}27$ p.p.m.) are as follows:

| | *ortho* | *meta* | *para* |
|---|---|---|---|
| $CH_3$ | −0·15 | −0·1 | −0·1 |
| $NO_2$ | 1·0 | 0·3 | 0·4 |

[2] Integral trace heights are quoted to the nearest millimeter throughout these problems; errors on the original chart are ±2 mm.

[3] Errors ±2 Hz or ±0·03 p.p.m.

Numbering the ring substitutents from the methyl as 1, we obtain the following calculated values (p.p.m.) for 2-nitrotoluene:

3 proton, $\delta = 7{\cdot}27 + 1{\cdot}0 - 0{\cdot}1 = 8{\cdot}17$;
4 proton, $\delta = 7{\cdot}27 + 0{\cdot}3 - 0{\cdot}1 = 7{\cdot}47$;
5 proton, $\delta = 7{\cdot}27 + 0{\cdot}4 - 0{\cdot}1 = 7{\cdot}57$;
6 proton, $\delta = 7{\cdot}27 + 0{\cdot}3 - 0{\cdot}15 = 7{\cdot}42$.

The integral ratio for this part of the 2-nitrotoluene spectrum is predicted to be ~1 : 3. For 3-nitrotoluene, the values are:

2 proton, $\delta = 7{\cdot}27 + 1 - 0{\cdot}15 = 8{\cdot}12$;
4 proton, $\delta = 7{\cdot}27 + 1 - 0{\cdot}1 = 8{\cdot}17$;
5 proton, $\delta = 7{\cdot}27 + 0{\cdot}3 - 0{\cdot}1 = 7{\cdot}47$;
6 proton, $\delta = 7{\cdot}27 + 0{\cdot}4 - 0{\cdot}15 = 7{\cdot}52$.

The integral ratio for the aromatic region of the 3-nitrotoluene spectrum is predicted to be ~2 : 2, as observed in figure 5.1.4.

The assignment is summarized in table 5.1.4.

**Table 5.1.4.**

| δ (p.p.m.) | Relative intensity | Multiplicity | Assignment |
|---|---|---|---|
| 2·40 | 3 | singlet | $CH_3$ |
| ~7·4 | 2 | complex | aromatic |
| ~7·9 | 2 | complex | aromatic *ortho* to $NO_2$ |

Compound $C_7H_7NO_2$ is 3-nitrotoluene.

**5.1.5**

(a) Identify the two β-ketonic esters with molecular formulae $C_7H_{10}O_3$ (A) and $C_8H_{12}O_3$ (B) from the n.m.r. spectrum (figure 5.1.5) of a solution of the two in $CCl_4$.

(b) Estimate also the percentage of each compound present in the mixture.

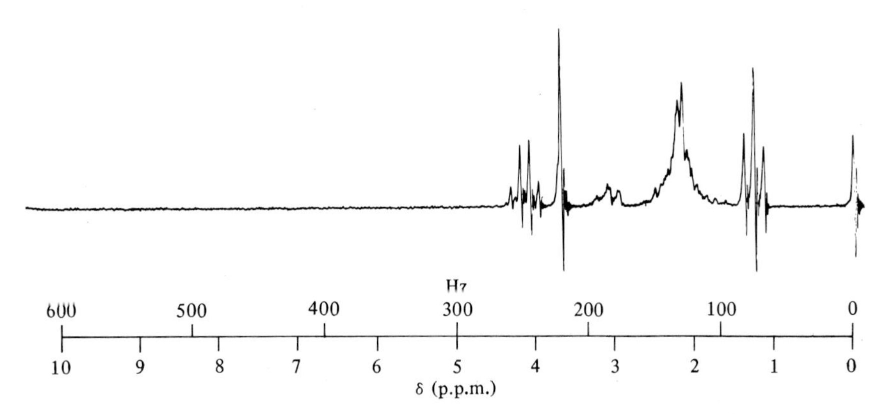


**Figure 5.1.5.** (See table 5.1.5 for integration data.)

**Solution**

(a) The spectrum in figure 5.1.5 can be divided into three regions. A triplet is very clearly defined at 1·25 p.p.m., and the same splitting is observed in a quartet to low field at 4·13 p.p.m. Slightly to higher field there is a sharp singlet at 3·7 p.p.m. (the shoulder on this and some low-intensity structure between the quartet peaks can be explained: see later).

The midfield region consists of two broad multiplets at ~2·2 p.p.m. and ~3·1 p.p.m.

The integral trace heights are given in table 5.1.5.

The information given with the problem defines a group CO·C·CO·OR or $C_3O_3R$. The group R can be deduced to be methyl, for one ester, from the singlet at 3·7 p.p.m. (observed at 3·65 p.p.m. in methyl acetate). The second ester certainly contains the ethyl group, which is characterized by the quartet at 4·13 p.p.m. (4·15 in ethyl acetate) and the triplet at 1·25 p.p.m.

The sample is therefore a mixture of a methyl and an ethyl ester: $-CO{\cdot}C{\cdot}CO{\cdot}OCH_3$ ($C_4O_3H_3$), and $-CO{\cdot}C{\cdot}CO{\cdot}OCH_2{\cdot}CH_3$ ($C_5H_5O_3$). The residues from the given molecular formulae are then $C_3H_7$, which could be n-propyl or isopropyl groups. However, the spectrum clearly excludes both because for n-propyl, the methyl should be clearly visible at ~1 p.p.m. as an approximate triplet (1 : 2 : 1), and for isopropyl, the methyl protons would appear as a doublet near 1 p.p.m. Clearly the valency of the $\beta$-keto ester group cannot be satisfied by either of the above groups.

The relevant part of the spectrum in figure 5.1.5 between 1·5 and 3·5 p.p.m. provides a clue to the identity of the remaining structural residue, $C_3H_7$. The two broad bands are in the ratio 1 : 7. This is slightly anomalous in that there are only seven protons to be accounted for. Since the low-field multiplet corresponds clearly to a proton in a unique position in the molecule the residue $C_3H_7$ may now be split into two parts, i.e. H and $C_3H_6$. The single proton must occupy a tertiary site, since when bonded to C=O the shift would be ⩾9 p.p.m. (as in aldehydes, RCHO).

The $C_3H_6$ group may be either $CH_3{-}CH{=}CH_2{-}$ or $-(CH_2)_3-$. The propene group cannot explain any of the relevant features of the spectrum in figure 5.1.5 but the trimethylene group does.

The structures of the unknown esters are therefore:

```
           O                              O
          //                             //
CH2—C                          CH2—C
 |     \                        |     \
 |      CH·CO·OCH3      and     |      CH·CO·OCH2·CH3
 |     /                        |     /
CH2—CH2                        CH2—CH2
    A                              B
```

**Table 5.1.5.**

| δ (p.p.m.) | 1·25 | 2·2 | 3·1 | 3·7 | 4·13 |
|---|---|---|---|---|---|
| Relative integral height | 14 | 43 | 6 | 8 | 8 |

(b) The total intensity of an n.m.r. absorption band from a set of nuclei in a molecule depends on $\gamma^2$, the square of the magnetogyric ratio of the nuclei; $B_1^2$, the intensity of the radio-frequency field; $N$, the number of nuclei in the set; and $x$, the mole fraction of the molecule present in the sample mixture.

In the present case the first two criteria are identical for all the protons in the mixture and can be neglected in the subsequent calculations as they are cancelled out in division operations. Thus the intensity $I$ can be expressed:

$$I \propto Nx \;. \tag{5.1.5a}$$

Given the assignment of the spectrum made in part (a), we can deduce $x_A$ and $x_B$, the mole fractions of A and B present in the mixture by comparing the observed intensities of the methyl and ethyl resonances with those of the $C_3H_7$ group. Thus if $I_1$ denotes the total intensity of the methyl and ethyl resonances, and $I_2$ that for the $C_3H_7$ group, we can use equation (5.1.5a) to write

$$\frac{I_1}{I_2} = \frac{3x_A + 5x_B}{7(x_A + x_B)} \;. \tag{5.1.5b}$$

But $x_A + x_B = 1$; and thus with substitution of the relative integral heights from table 5.1.5, equation (5.1.5b) becomes

$$\frac{8+8+14}{6+43} = \frac{5-2x_A}{7} \;;$$

thus

$$\begin{aligned} x_A &= \tfrac{1}{2}[5-(0\cdot61 \times 7)] \\ &= 0\cdot365 \;, \end{aligned}$$

and

$$x_B = 0\cdot635 \;.$$

The molecular weights of A and B, $M_A$ and $M_B$, can now be used to calculate $w_A$ and $w_B$, the percentage weight fractions of each molecule present:

$$\left.\begin{aligned} w_A &= \frac{100m_A}{m_A+m_B} = \frac{100M_Ax_A}{M_Ax_A+M_Bx_B} \;, \\ w_B &= \frac{100m_B}{m_A+m_B} = \frac{100M_Bx_B}{M_Ax_A+M_Bx_B} \;; \end{aligned}\right\} \tag{5.1.5c}$$

where $m_A m_B$ are the weights of A and B present in the mixture. Thus, with $M_A = 142$, $M_B = 156$, equations (5.1.5c) yield

$$w_A = 34\% \;; \qquad w_B = 66\% \;.$$

## 5.1.6

A compound has molecular formula $C_{14}H_{18}O$. Suggest possible structures from the n.m.r. spectrum (figure 5.1.6) of a sample in $CCl_4$. Ignore impurities at 270–190 Hz, 130–100 Hz, and 16 Hz.

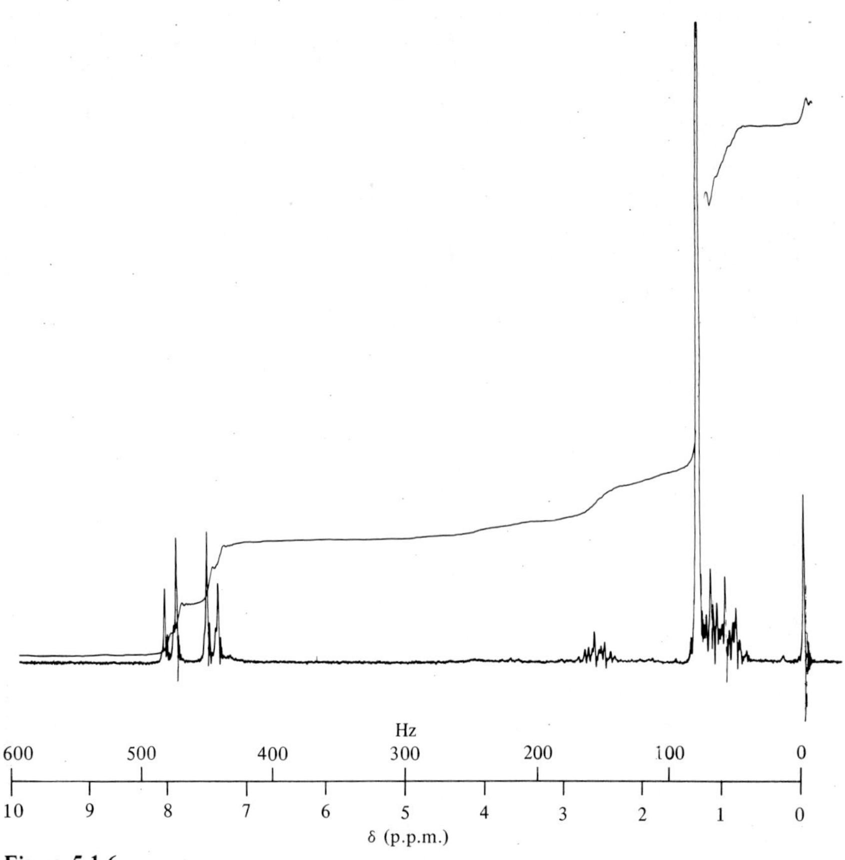


**Figure 5.1.6.**

**Solution**

The spectrum in figure 5.1.6 can be divided into five parts, as shown in table 5.1.6.

The total integral trace height is 65 mm, and since the number of protons is eighteen then one proton corresponds to ~4 mm of integral trace height on the chart. The numbers of protons which account for the five parts of the spectrum are two (at 7·96 p.p.m.), two (7·43 p.p.m.), one (~2·6 p.p.m.), and thirteen for the overlap of the complex multiplet and the sharp singlet.

The low-field signals centred at 7·96 and 7·43 p.p.m. have a form which is highly characteristic of a 1,4-substituted benzene. The coupling between mutually *meta* and *para* protons is small, and the n.m.r. spectrum approximates to the superposition of identical pairs of doublets arising from the interaction of mutually *ortho* protons. The separation of lines in the doublets (8–9 Hz) confirms this, and the chemical shifts provide further evidence in support of the 1,4-disubstituted benzene. The remaining residue of the molecule,

$$C_{14}H_{18}O - C_6H_4 = C_8H_{14}O \, ,$$

accounts for all the high-field signals in the spectrum. The dominant feature here is the very strong sharp line at 1·33 p.p.m. which obviously accounts for the major fraction of intensity in this region. The integral step-height of ~37 mm corresponds to ten protons, which is not a viable number of equivalent protons in the context of this problem. However, there is overlap with the complex multiplet centred at ~1·1 p.p.m., and a logical approach suggests that the sharp single line probably corresponds to nine equivalent protons and that therefore a tertiary butyl group may be present in the molecule. The possibility of three individual methyl groups with the same proton chemical shift is very remote.

$$C_8H_{14}O - C(CH_3)_3 = C_4H_5O \, .$$

The $C_4H_5O$ residue must then account for the two complex multiplets observed at ~1·1 and ~2·6 p.p.m. The integral indicates a single proton at ~2·6 p.p.m. and four protons at ~1·1 p.p.m. The chemical shift of the ~1·1 p.p.m. multiplet is characteristic of protons in a cyclopropyl ring and this latter structural unit accounts for the remaining five protons.

The structural units which have been deduced to explain the spectral features in figure 5.1.6 are:

[1,4-disubstituted benzene ring], $-C(CH_3)_3$, [cyclopropyl ring with H, H, H, H, H], CO.

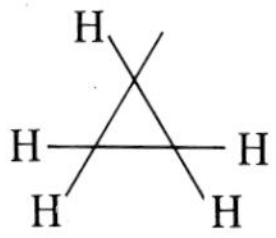

**Table 5.1.6.**

| Multiplicity | δ (p.p.m.) | Integral trace height (mm) |
|---|---|---|
| Complex multiplet | ~1·1 | 46 |
| Singlet | 1·33 | |
| Complex multiplet | ~2·6 | 4 |
| Doublet (some structure) | 7·43 | 8 |
| Doublet (mirror image of that at 7·43 p.p.m.) | 7·96 | 7 |

There are two structures which include these units:

(I) $Bu^t$–$C_6H_4$–CO–cyclopropyl and (II) $Bu^t$–CO–$C_6H_4$–cyclopropyl

The distinction between these two structures by use of n.m.r. is very difficult because the $\beta$-substituent effects of the phenyl and carbonyl groups are 0·35 and 0·2 p.p.m., respectively (Mathieson, 1967). The spectrum is actually that of 1-t-butyl, 4-cyclopropylketobenzene.

### 5.1.7

A compound has molecular formula $C_{11}H_{10}O_3$. From the n.m.r. spectrum of a sample of the compound in $CDCl_3$ (a) identify the compound (using figure 5.1.7a), and (b) identify and analyse the AB spin system and assign geometrical configuration (using figure 5.1.7b; sweep offset 370 Hz from tms).

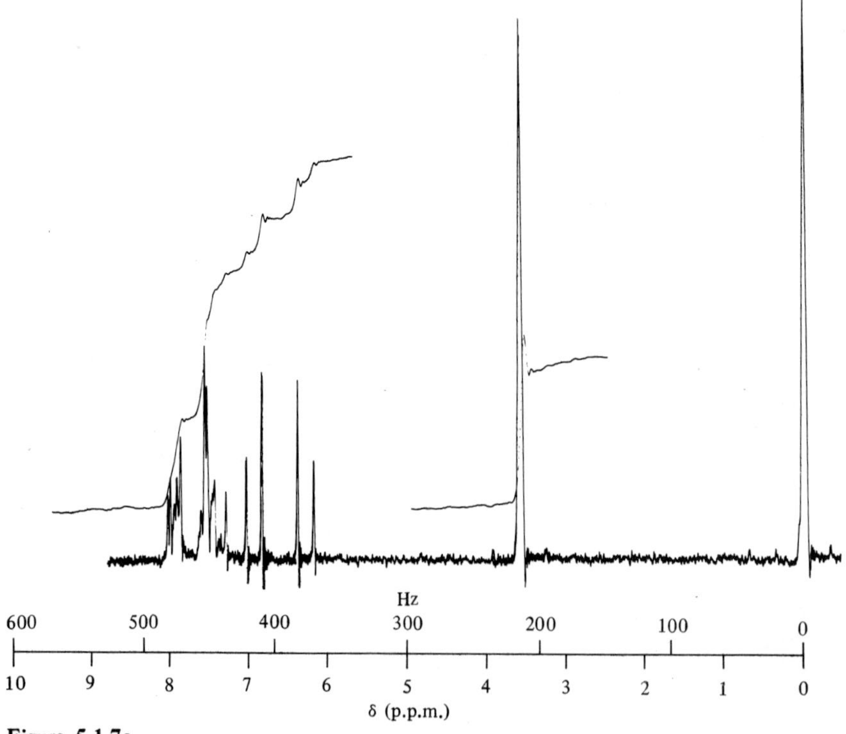


**Figure 5.1.7a.**

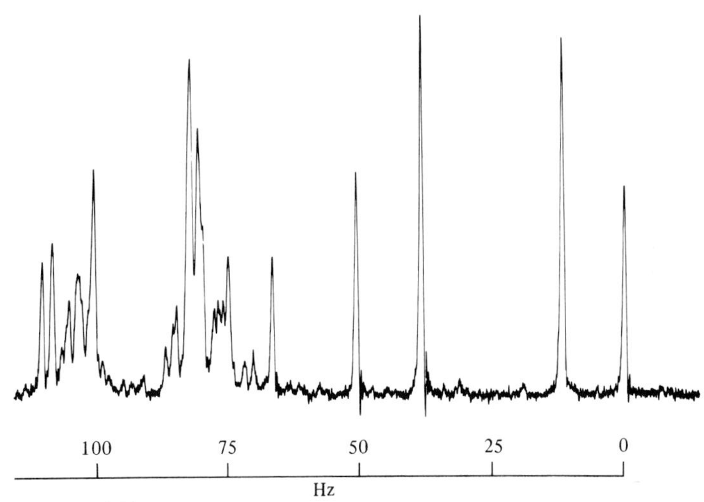


**Figure 5.1.7b.**

**Solution**

(a) The data available from figure 5.1.7a for the solution of this problem are summarized in table 5.1.7a. The intensity data are derived by dividing the total integral trace height by the known number of protons to obtain the integral trace height for one proton and then dividing the appropriately assigned trace heights by this intensity unit.

The chemical shift and intensity of the singlet at 3·6 p.p.m. are highly characteristic of a methyl group bonded directly to oxygen. Some typical shifts are shown below.

| | | | |
|---|---|---|---|
| $CH_3 \cdot OH$ | 3·38 | $CH_3 \cdot O \cdot C_6H_5$ | 3·73 |
| $CH_3 \cdot O \cdot CO \cdot R$ | 3·65 | $CH_3 \cdot O \cdot CO \cdot C_6H_5$ | 3·90 |

A conclusion can be tentatively drawn that the methyl group is bonded to oxygen of an ester group: $CH_3 \cdot O \cdot CO$– ($C_2H_3O_2$). The residual formula $C_9H_7O$ clearly cannot account for more than one phenyl ring, the presence of which is indicated both by the chemical shifts and by the

**Table 5.1.7a.**

| Multiplicity | δ (p.p.m.) | *J* (Hz) | Relative intensity |
|---|---|---|---|
| Singlet | $3 \cdot 6_0$ | | 3 |
| Doublet | $6 \cdot 2_5$ | ~12 | 1 |
| Doublet | $6 \cdot 9_3$ | ~12 | 1 |
| Complex | ~7·5 | | 3 |
| Complex | ~7·9 | | 2 |

overall relative intensity of the complex multiplets at ~7·5 and ~7·9 p.p.m. The structural formula can be divided into three parts: $C_6H_5$, $C_3H_2O$, and $C_2H_3O_2$.

The chemical shifts and relative intensities of the doublets at 6·9 and 6·2 p.p.m. are characteristic of two hydrogen atoms bonded to $sp^2$ hybridized carbon:

–CH=CH– or $\rangle C{=}C\langle^{H}_{H}$ .

*trans* or *cis* geminal

The remaining residue is CO which can be assumed to be a carbonyl group. The possible structures which can be assembled from the proposed structural units are shown below.

$C_6H_5{\cdot}CH{:}CH{\cdot}CO{\cdot}CO{\cdot}OCH_3$ (I)

$C_6H_5{\cdot}CO{\cdot}CH{:}CH{\cdot}CO{\cdot}OCH_3$ (II)

(*cis* or *trans*)

$(C_6H_5)(CO{\cdot}CO{\cdot}OCH_3)C{:}CH_2$ (III)

$(C_6H_5CO)(CO{\cdot}OCH_3)C{:}CH_2$ (IV)

The distinction between the α-keto ester structures [(I) and (III)] and the other ester structures [(II) and (IV)] can be made principally on the grounds of the aromatic proton chemical shifts, because the carbonyl is directly bonded to the phenyl group in (II) and (IV), and coupling constants.

The substituent data available (table 5.1.7b) predict little if any chemical shift difference between the aromatic protons of the α-keto ester structures (I) and (III), but ~0·3 p.p.m. difference between 2,6 and 3,4,5 for the other ester structures (II) and (IV) (~0·5 p.p.m. observed).

The coupling constants in substituted ethylenes are highly characteristic of the structures, depending on whether the protons are geminal, *cis*, or *trans* with respect to one another. The relevant part of the spectrum has been expanded in figure 5.1.7b.

**Table 5.1.7b.** Predicted approximate chemical shifts (p.p.m.) of aromatic protons in structures identified above (Mathieson, 1967).

| Proton | Structure type | |
|---|---|---|
| | (I), (III) | (II), (IV) |
| 2,6 | 7·5 | 7·9 |
| 3,5 | 7·5 | 7·6 |
| 4 | 7·5 | 7·6 |

(b) The four lines to high field of the spectrum (figure 5.1.7b) are those assigned to the two olefinic protons. The predicted first-order spectrum for the two nonequivalent protons consists of four lines of equal intensity and this is realized in those cases where the chemical shift between the proton resonances is very much greater than the coupling constant. In this and many similar cases the coupling constant cannot be neglected in comparison with the chemical shift, and the deviation from first order is manifested in two ways. The first is purely visual, in that the intensities of the lines in the doublets are no longer equal and the extreme lines lose intensity to those near the centre of the spectrum (figure 5.1.7c). This can be explicitly explained in analytical relationships involving the chemical shift and coupling constant. The second effect is the displacement of the chemical shifts of the protons from the mean chemical shift of the pairs of lines. The separation of the line pairs ($J$) does not change.

The analysis of this simple system is given in detail in standard texts. It is sufficient here to give the simplest method of analysis available. The product of the separations of the outer and inner lines (figure 5.1.7d) is equal to the square of the chemical shift (Pople *et al.*, 1959):

$$ab = (\nu_0 \delta_{AB})^2 .$$

From figure 5.1.7b, we obtain

$$(\nu_0 \delta_{AB})^2 = 50{\cdot}5 \times 26{\cdot}5 = 1338{\cdot}25 \text{ Hz}^2 ;$$

thus $|\nu_0 \delta_{AB}| = 36{\cdot}6$ Hz, with $|J| = 12{\cdot}0$ Hz .

The coupling constant is of prime importance in deciding the geometrical configuration of the molecule. Typical values of some relevance are for

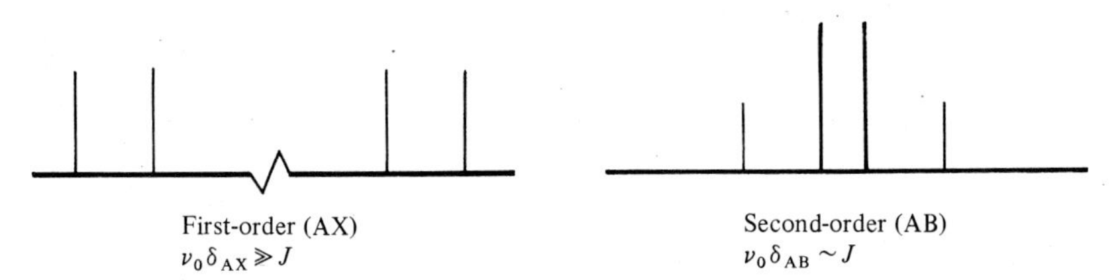


**Figure 5.1.7c.** Schematic comparison between first-order and second-order n.m.r. spectra for two protons with intramolecular coupling constant $J$. Here, $\nu_0 \delta$ is the chemical shift between the two nuclei, expressed as a frequency ($\nu_0$ is the operating frequency of the spectrometer).

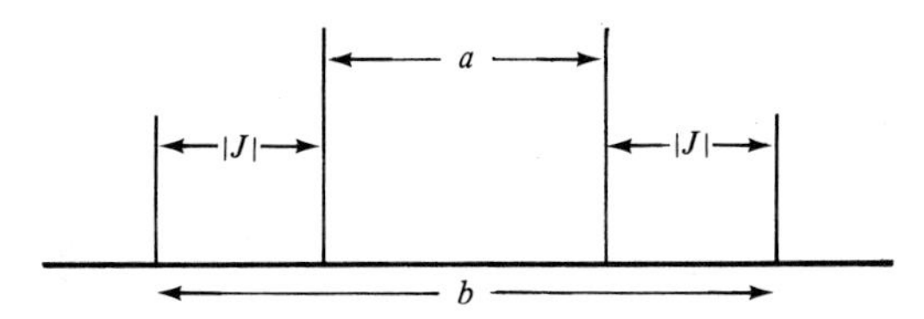


**Figure 5.1.7d.**

$H_5C_2O_2C \cdot CH\!:\!CH \cdot CO_2C_2H_5$, *cis* 11·9 and *trans* 15·5 Hz. This comparison of coupling constants favours the *cis* configuration:

$C_6H_5$ H
C—C
O C—H
O=C
O—$CH_3$ .

The geminal coupling is generally <3 Hz in purely organic molecules (Bothner-By, 1965).

### 5.1.8

A compound has molecular formula $C_{10}H_9O_3Cl$ and is related to that identified in problem 5.1.7. From the n.m.r. spectrum of a sample of the compound in $CDCl_3$ (a) identify the compound (using figure 5.1.8a), and (b) identify and analyse the simple AB system and assign the geometrical configuration (using figure 5.1.8b; sweep offset 406 Hz from tms).

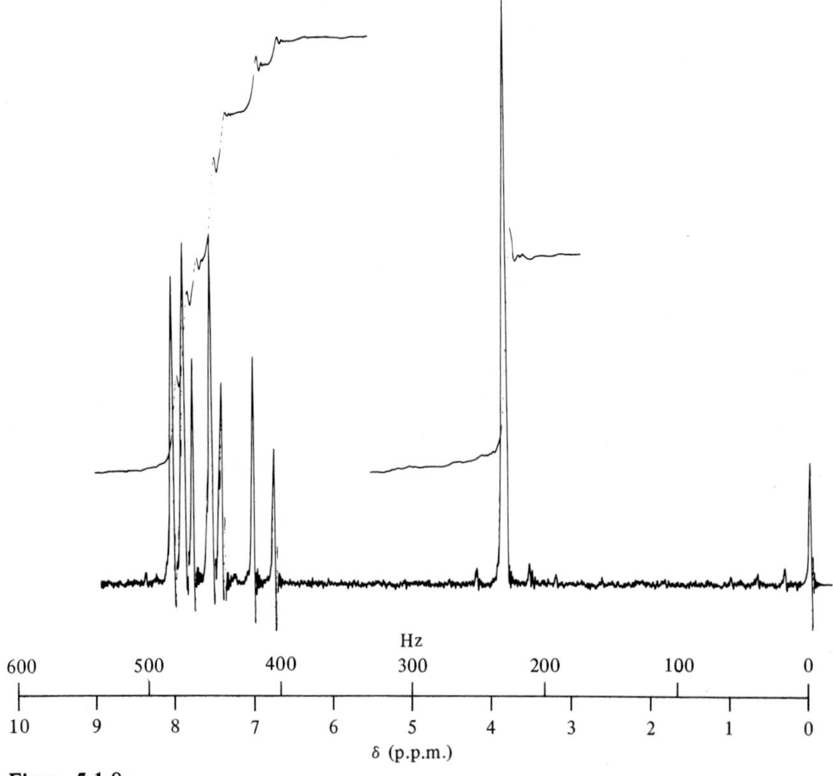


**Figure 5.1.8a.**

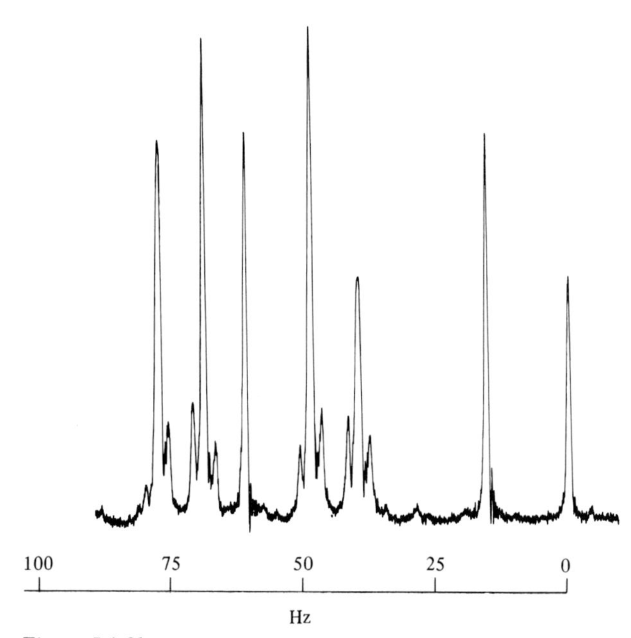


**Figure 5.1.8b.**

**Solution**

(a) The data available from figure 5.1.8a for the solution of this problem are summarized in table 5.1.8a. The relative intensity data are derived as indicated in problem 5.1.7(a).

The information given relates this molecule to the subject of problem 5.1.7 and from this the singlet at 3·8 p.p.m. can be assigned to the protons of a methyl group bonded to oxygen of an ester group $-CO{\cdot}OCH_3$. Note that this signal is 0·2 p.p.m. to lower field than the corresponding signal in problem 5.1.7.

It is possible to pick out from the low-field signals a pair of lines separated by 16 Hz with one unit of intensity at $6{\cdot}9_0$ p.p.m. This is seen more clearly in figure 5.1.8b. The total intensity in the low-field region is six units, one proton less than in problem 5.1.7.

**Table 5.1.8a.**

| Multiplicity | $\delta$ (p.p.m.) | $J$ (Hz) | Relative intensity |
|---|---|---|---|
| Singlet | $3{\cdot}8_3$ | | 3 |
| Doublet | $6{\cdot}9_0$ | ~16 | 1 |
| Doublet | $7{\cdot}5_0$ | ~9 | 2 |
| Two doublets overlapping | $7{\cdot}9_0$ | ~16<br>~9 | 3 |

The chlorine atom is therefore bonded to either the benzene ring or the olefinic carbon atoms. The number of lines observed in this region is seven, and these are clearly defined although some are broader than others. The simplicity of the spectrum compared with that in figure 5.1.7a suggests substitution in the *para*-position of the aromatic ring since this position alone provides the symmetry required. The two signals centred on $6{\cdot}9_0$ p.p.m. are separated by 16 Hz and are reminiscent of one half of the AB system identified in problem 5.1.7(b). The separation is much greater than any coupling constant known between protons in aromatic systems (10 Hz) and the signals can be assigned to an olefinic proton coupled to another nonequivalent olefinic proton.

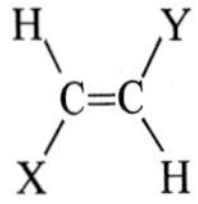

```
H     Y
 \   /
  C=C        (trans)
 /   \
X     H
```

The problem remaining is to identify in detail the signals arising from the four aromatic protons and two olefinic protons. This can best be solved with reference to figure 5.1.8b, where the relevant part of the spectrum has been expanded.
(b) The spectrum in figure 5.1.8b shows a great deal more detail than that in figure 5.1.8a. It can be seen that weaker lines exist which characterize the aromatic part of the spectrum and illustrate the shortcomings of the approximation describing the aromatic part as the superposition of two AB systems. The true description of such a system is given in the notation AA′BB′ or $[AB]_2$:

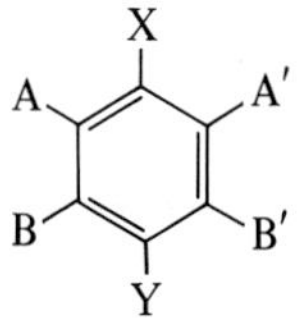

```
      X
A     |     A'
  \ /   \ /
   |     |
  / \   / \
B     |     B'
      Y
```

In this system the *ortho* and *para* A–B coupling constants ($J_{AB} = J_{A'B'}$ and $J_{AB'} = J_{A'B}$) are not identical, and the two A nuclei are said to be magnetically nonequivalent, as are the two B protons. The magnetic nonequivalence is denoted by the primed notation or by the square brackets as shown above.

The property of this system which is of prime importance here is that it is symmetrical about the mean of the A and B proton chemical shifts, which is the centre of the AA′BB′ spectrum. The lines in the spectrum can be assigned by use of this property; the assignment is given in table 5.1.8b.

The second pair of olefinic lines can be picked out by looking for the characteristic spacing of ~16 Hz and by taking into account the line-width and peak height. These are identified as line 9 and a line in the composite signal labelled 14 at $483{\cdot}5$ Hz from tms.

The low-field signal of the olefinic AB system distorts the symmetry of the aromatic proton spectrum, but it is still possible to distinguish the lines because of the different linewidth created by the near degeneracy of transition energies of the AA′BB′ system.

The value of the coupling constant measured for the olefinic protons as the separation of lines 1 and 2 in table 5.1.8b is 15·5 Hz. This value is compatible with the *trans* configuration of the olefinic protons:

$$\text{Cl–C}_6\text{H}_4\text{–C(=O)–CH=CH–CO·OCH}_3 \quad (\textit{trans})$$

**Table 5.1.8b.** Line data and assignment of portion of spectrum shown in figure 5.1.8b.

| Line | δ (p.p.m.) | ν ($H_3$)[a] | Peak height[b] (mm) | Assignment[c] |
|---|---|---|---|---|
| 1 | $6{\cdot}7_7$ | 406 | 32 | O |
| 2 | $7{\cdot}0_3$ | 421·5 | 50 | O |
| 3 | $7{\cdot}3_8$ | 443·5 | 11 | A |
| 4 | $7{\cdot}4_4$ | 446 | 31 | A |
| 5 | $7{\cdot}4_7$ | 448 | 13 | A |
| 6 | $7{\cdot}5_5$ | 453 | 14 | A |
| 7 | $7{\cdot}5_8$ | 455 | 65 | A |
| 8 | $7{\cdot}6_1$ | 457 | 10 | A |
| 9 | $7{\cdot}7_8$ | 467 | 50 | O |
| 10 | $7{\cdot}8_9$ | 473 | 10 | A |
| 11 | $7{\cdot}9_2$ | 475 | 63 | A |
| 12 | $7{\cdot}9_4$ | 477 | 14 | A |
| 13 | $8{\cdot}0_2$ | 481·5 | 12 | A |
| 14 | $8{\cdot}0_6$ | 483·5 | 50 | A, O |
| 15 | $8{\cdot}1_0$ | 486 | 5 | A |

[a] Chemical shift from tms in Hz for 60 MHz proton resonance.
[b] To nearest millimeter; this is not always reliable as a measure of intensity because of variations in the linewidth.
[c] A, aromatic proton signal; O, olefinic proton signal.

**5.1.9**

Identify the following high-resolution spin systems in standard notation (Pople *et al.*, 1959, pp.119–150).

(a)

$$\nu_1 = \nu_2 \neq \nu_3 = \nu_4\,; \qquad J_{13} = J_{24} \neq J_{14} = J_{23}\,; \qquad |\nu_1 - \nu_3| \gg J_{13} \text{ or } J_{14}\,.$$

(b)

$$\nu_1 \neq \nu_2 = \nu_3 \neq \nu_4\,; \qquad J_{12} = J_{13} \neq J_{14} \neq J_{23} \neq J_{24} = J_{34}\,;$$

$$|\nu_1 - \nu_2| \approx J_{12}\,; \qquad |\nu_1 - \nu_4| \gg J_{14}\,; \qquad |\nu_2 - \nu_4| \gg J_{24}\,.$$

In addition, (a) deduce the maximum number of resolvable lines in the complete spectrum for the special condition

$$(J_{13}+J_{14})[J_{13}+J_{14}+2(J_{12}+J_{34})] = (J_{13}-J_{14})^2 \ ;$$

(b) show that all the unique coupling constants but one can be measured directly from the spectrum, assuming that all the lines in the strongly coupled part are resolved. Indicate what information is available concerning relative signs of coupling constants and how the chemical shifts can be determined (Diehl *et al.*, 1967).

**Solution**

(a) The system can be represented diagrammatically as follows:

$$\begin{array}{ccc} 1 & -J_{AX}- & 3 \\ | & / & | \\ J_{AA'} & J_{AX'} & J_{XX'} \\ | & / & | \\ 2 & -J_{AX}- & 4 \end{array} \qquad \text{(with } C_2 \text{ or } C_s \text{ symmetry)} .$$

Thus, in conventional notation, it is AA′XX′ or $[AX]_2$.

Once the system has been identified the details of transition energies can be found in standard texts, where the following parameters have been invoked to abbreviate the analytical expressions

$$N = J_{13}+J_{14} \ , \qquad L = J_{13}-J_{14} \ ,$$
$$K = J_{12}+J_{34} \ , \qquad M = J_{12}-J_{34} \ .$$

Furthermore the $[AX]_2$ system has been identified as one where subspectral transformations are possible such that the A part of the $[AX]_2$ system is composed of two $a_2$ subspectra characterized by

$$\nu_a = \nu_A \pm \tfrac{1}{2}(J_{AX}+J_{AX'}) \ ;$$

or, according to the numbering system,

$$\nu_a = \nu_1 \pm \tfrac{1}{2}(J_{13}+J_{14}) = \nu_1 \pm \frac{N}{2} \ ;$$

and two ab subspectra characterized by the following transformations:

$$\left.\begin{array}{l} \nu_a = \nu_A+\tfrac{1}{2}(J_{AX}-J_{AX'}) \ , \\ \nu_b = \nu_A-\tfrac{1}{2}(J_{AX}-J_{AX'}) \ , \\ J_{ab} = J_{AA'}+J_{XX'} \ ; \end{array}\right\} \quad A \text{ symmetry species}$$

$$\left.\begin{array}{l} \nu_a = \nu_A+\tfrac{1}{2}(J_{AX}-J_{AX'}) \ , \\ \nu_b = \nu_A-\tfrac{1}{2}(J_{AX}-J_{AX'}) \ , \\ J_{ab} = J_{AA'}-J_{XX'} \ , \end{array}\right\} \quad B \text{ symmetry species}$$

where $A$ and $B$ are the irreducible representations of the $C_2$ point group (with which the n.m.r. group is isomorphic).

We note now the following

$$\nu_a - \nu_b = (J_{AX} - J_{AX'}) = L = (J_{13} - J_{14}) ,$$
$$(J_{13} + J_{14}) = N ,$$
$$(J_{12} + J_{34}) = K ;$$

so that

$$(J_{13} + J_{14})[J_{13} + J_{14} + 2(J_{12} + J_{34})] = N(N + 2K) .$$

The specified condition therefore states that

$$(\nu_a - \nu_b)^2 = N(N + 2|K|) .$$

Familiarity with the AB system allows recognition of this format since, stated in words, the product of the separations of the inner and outer lines is equal to the square of the chemical shift. Therefore, if the separation of the inner lines is taken to be $N$ and the separation of the outer lines to be $N + 2K$, the equality can be ascribed to the $A$ symmetry species ab subspectrum (figure 5.1.9).

It can be seen analytically that the two inner lines of the $A$ species ab subspectrum are coincident with the $a_2$ subspectra. A priori the stated condition does not impose the same conditions on the $B$ species ab subspectrum. The maximum number of lines observable in the A part of $[AX]_2$ is ten but in this case there will be only eight. The total number of observable lines will be sixteen, since the X part is identical to the A part.

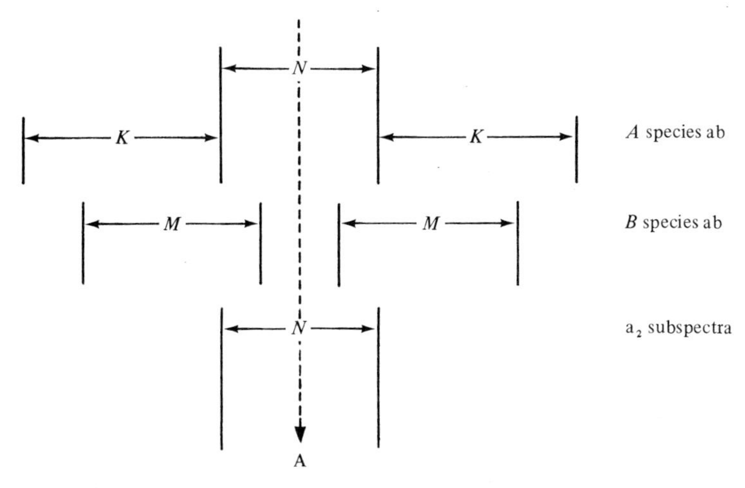


**Figure 5.1.9.** The A part of an $[AX]_2$ system. Only eight of the possible ten lines are observed for the condition $(\nu_a - \nu_b)^2 = N(N + 2|K|)$ (see text).

(b) The system can be represented thus.

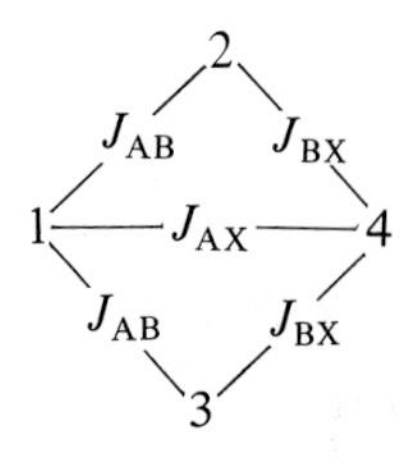


It can therefore be written $AB_2X$ in conventional notation.

The strongly coupled part of the $AB_2X$ system comprises two $ab_2$ subspectra characterized by $m_X = +\frac{1}{2}$ and $-\frac{1}{2}$, respectively, where $m_X$ denotes the magnetic quantum number of proton X, and by the following subspectral parameters:

$$\left.\begin{aligned} \nu_a &= \nu_A + \tfrac{1}{2}J_{AX} , \\ \nu_b &= \nu_B + \tfrac{1}{2}J_{BX} , \\ J_{ab} &= J_{AB} ; \end{aligned}\right\} \qquad m_X = +\tfrac{1}{2}$$

$$\left.\begin{aligned} \nu_{a'} &= \nu_A - \tfrac{1}{2}J_{AX} , \\ \nu_{b'} &= \nu_B - \tfrac{1}{2}J_{BX} , \\ J_{a'b'} &= J_{AB} . \end{aligned}\right\} \qquad m_X = -\tfrac{1}{2}$$

Each line of the $ab_2$ spectra can be labelled in the traditional way, so that the lines associated with A transitions become 1, 2, 3, 4 for $m_X = +\frac{1}{2}$, and 1′, 2′, 3′, 4′ for $m_X = -\frac{1}{2}$; lines associated with B transitions are labelled 5, 6, 7, 8 for $m_X = +\frac{1}{2}$, and 5′, 6′, 7′, 8′ for $m_X = -\frac{1}{2}$. The frequency $\nu_a$ is then given by the position of the line labelled 3, and $\nu_{a'}$ is given by line 3′. The difference between them gives $|J_{AX}|$, i.e. magnitude only.

The frequency $\nu_b$ is defined as the mean frequency of lines 5 and 7; $\nu_{b'}$ is given by $\frac{1}{2}(\nu_{5'}+\nu_{7'})$, so that $|J_{BX}|$ is given by $\frac{1}{2}|\nu_5+\nu_7-\nu_{5'}-\nu_{7'}|$.

The relative signs of $J_{AX}$ and $J_{BX}$ can be determined from the spectrum. $J_{AX}$ and $J_{BX}$ have the same sign if

$$\nu_3 < \nu_{3'}$$

and

$$(\nu_5+\nu_7) < (\nu_{5'}+\nu_{7'}) .$$

Their signs are opposite if

$$\nu_3 > \nu_{3'} , \text{ with } (\nu_5+\nu_7) < (\nu_{5'}+\nu_{7'}) ,$$

or if

$$\nu_3 < \nu_{3'} , \text{ with } (\nu_5+\nu_7) < (\nu_{5'}+\nu_{7'}) .$$

$J_{AB}$ can be determined independently from the two subspectra if the lines are well resolved, as

$$J_{AB} = \tfrac{1}{3}(\nu_1 - \nu_4 + \nu_6 - \nu_8) = \tfrac{1}{3}(\nu_{1'} - \nu_{4'} + \nu_{6'} - \nu_{8'}) \, .$$

$J_{BB}$ does not appear in the spectrum because the two B nuclei are magnetically equivalent (Gutowsky *et al.*, 1953). Thus, we can write

$$\nu_A - \nu_B = \tfrac{1}{2}[(\nu_3 + \nu_{3'}) - (\nu_5 + \nu_7 + \nu_{5'} + \nu_{7'})] \, .$$

**5.1.10**

Show that the AB part of an $ABX_2$ system is composed purely of ab subspectra, and derive the subspectral parameters (see Diehl *et al.*, 1967).

**Solution**

The problem is based upon the principles of subspectral analysis detailed in Diehl *et al.* (1967).

The first step is to determine the matrix elements of $\mathcal{H}$, the Hamiltonian for the $ABX_2$ system. This can be done systematically since the notation implies a first-order approximation such that the magnetic quantum numbers of the X nuclei, $m_X$, can be used to factorize the secular matrix, as well as the total spin ($I$) and total magnetic ($M$) quantum numbers:

| $M$ | DDT | | | DDS |
|---|---|---|---|---|
| 2 | 1 | | | |
| 1 | 2 | 1 | | 1 |
| 0 | 1 | 2 | 1 | 2 |
| −1 | | 1 | 2 | 1 |
| −2 | | | 1 | |
| $m_X$ | 1 | 0 | −1 | 0 |

Here D defines the state where $I = \frac{1}{2}$ in the case of single nucleus, T corresponds to the $I = 1$ (triplet) state for two spin $\frac{1}{2}$ nuclei, and S characterises the $I = 0$ (singlet) state. The degeneracies, $g$, of the different states can be easily calculated from equation (3) in the reference given above and turn out to be trivial ($g = 1$) in each case. In addition it is not necessary to define spin product wave functions [e.g. $\alpha\beta(\alpha\beta)$], in order to define the matrix elements. The diagonal elements of $\mathcal{H}$ are given by the sum

$$\sum_i \left( \nu_i m_i + \tfrac{1}{2} \sum_{l \neq i} J_{il} m_i m_l \right) ,$$

the quantum numbers $m_i$ and $m_l$ being the group values of the magnetic quantum numbers. So for the state $|\frac{1}{2}, \frac{1}{2}, 1\rangle$,

$$\langle \tfrac{1}{2}, \tfrac{1}{2}, 1 | \mathcal{H} | \tfrac{1}{2}, \tfrac{1}{2}, 1\rangle = \frac{\nu_A}{2} + \frac{\nu_B}{2} + \nu_X + \tfrac{1}{2}\left(\frac{J_{AB}}{2} + J_{AX} + J_{BX}\right) ,$$

$$\langle \tfrac{1}{2}, -\tfrac{1}{2}, 1 | \mathcal{H} | \tfrac{1}{2}, -\tfrac{1}{2}, 1\rangle = \frac{\nu_A}{2} - \frac{\nu_B}{2} + \nu_X + \tfrac{1}{2}\left(-\frac{J_{AB}}{2} + J_{AX} - J_{BX}\right) ,$$

$$\langle -\tfrac{1}{2}, \tfrac{1}{2}, 1 | \mathcal{H} | -\tfrac{1}{2}, \tfrac{1}{2}, 1\rangle = -\frac{\nu_A}{2} + \frac{\nu_B}{2} + \nu_X + \tfrac{1}{2}\left(-\frac{J_{AB}}{2} - J_{AX} + J_{BX}\right) ,$$

$$\langle -\tfrac{1}{2}, -\tfrac{1}{2}, 1 | \mathcal{H} | -\tfrac{1}{2}, -\tfrac{1}{2}, 1\rangle = -\frac{\nu_A}{2} - \frac{\nu_B}{2} + \nu_X + \tfrac{1}{2}\left(\frac{J_{AB}}{2} - J_{AX} - J_{BX}\right) .$$

Off-diagonal elements appear only between states with the same $M$, $m_X$, and $I$ quantum numbers and are nonzero only if the wave functions for two composite particles differ in their magnetic quantum numbers $m$, $i$ and $l$ [Diehl *et al.*, 1967, p.22, equation (12b)]; $i$ and $l$ are identified here as nuclei A and B in a trivial case:

$$\langle I_i^{m_i \pm 1}, I_l^{m_l \mp 1} | \mathcal{H} | I_i^{m_i}, I_l^{m_l}\rangle = \tfrac{1}{2} J_{il}\{[I_i(I_i + 1) - m_i(m_i \pm 1)][I_l(I_l + 1) - m_l(m_l \mp 1)]\}^{1/2} \langle \tfrac{1}{2}^{1/2}, \tfrac{1}{2}^{1/2} | \mathcal{H} | \tfrac{1}{2}^{-1/2}, \tfrac{1}{2}^{1/2}\rangle ,$$

$$= \tfrac{1}{2} J_{AB}\{[\tfrac{1}{2}(\tfrac{1}{2} + 1) - \tfrac{1}{2}(\tfrac{1}{2} - 1)][\tfrac{1}{2}(\tfrac{1}{2} + 1) - \tfrac{1}{2}(\tfrac{1}{2} - 1)]\}^{1/2} ,$$

$$= \tfrac{1}{2} J_{AB}[(\tfrac{3}{4} + \tfrac{1}{4})(\tfrac{3}{4} + \tfrac{1}{4})]^{1/2} = \tfrac{1}{2} J_{AB} .$$

Thus the 1 : 2 : 1 subsystem of energy levels characterized and distinguished from other subsystems by $I$ and $m_X$ quantum numbers, $(\frac{1}{2}, \frac{1}{2}, 1)$ and 1, respectively, has been defined explicitly in terms of the corresponding matrix elements of the Hamiltonian for the complete system. The three remaining 1 : 2 : 1 subsystems can be treated in the same way—details are not given here. The Hamiltonian matrix can be diagonalized to obtain eigenvalues and eigenvectors via the secular determinant.

However, we are concerned more with the possibility of subspectral transformations, i.e. we wish to establish the properties of the 1 : 2 : 1 subsystems in terms of ab subspectra with subspectral parameters $\nu_a$, $\nu_b$, and $J_{ab}$ expressed in terms of $\nu_A$, $\nu_B$, $J_{AB}$, $J_{AX}$, and $J_{BX}$.

The derivation of subspectral parameters has been clearly outlined in Diehl *et al.* (1967; section 2.2, p.11). The comparison is made between the matrix elements of the 1 : 2 : 1 subsystems and the ab system, as exemplified in table 5.1.10.

The two systems must have the same transition energies when diagonalized, if indeed the sub-Hamiltonian of the $ABX_2$ system gives rise to an ab type spectrum. However, the respective eigenvalues may be shifted relative to each other by a constant amount, $\epsilon$. These arguments allow four relations for the four unknowns $\nu_a$, $\nu_b$, $J_{ab}$, and $\epsilon$ to be derived:

$$A'_{11} + \epsilon = \tfrac{1}{2}(\nu_a + \nu_b + \tfrac{1}{2}J_{ab}) ,$$

$$C'_{11} + \epsilon = \tfrac{1}{2}(\tfrac{1}{2}J_{ab} - \nu_a - \nu_b) ,$$

$$B'_{11} + B'_{22} + 2\epsilon = -\tfrac{1}{2}J_{ab} ,$$

$$\epsilon^2 + \epsilon(B'_{11} + B'_{22}) + B'_{11}B'_{22} - (B'_{12})^2 = -\tfrac{3}{16}J_{ab}^2 - \tfrac{1}{4}(\nu_a - \nu_b)^2 .$$

Here $A'_{11}$, $B'_{11}$, $B'_{22}$, ... are elements of the matrices for the $ABX_2$ subsystem given in table 5.1.10.

The solutions are given in Diehl *et al.* (1967, p.12), and the transformations are as follows.

DDT, $m_X = \pm 1$:

$$\nu_a = \nu_A \pm J_{AX} ,$$
$$\nu_b = \nu_B \pm J_{BX} ,$$
$$J_{ab} = J_{AB} .$$

DDT, $m_X = 0$; and DDS, $m_X = 0$:

$$\nu_a = \nu_A ,$$
$$\nu_b = \nu_B ,$$
$$J_{ab} = J_{AB} .$$

**Table 5.1.10.** Comparison of matrix elements of the DDT, $m_X = 1$, 1 : 2 : 1 subsystem of $ABX_2$ with those of the ab system.

| | $ABX_2$ | ab |
|---|---|---|
| **A** | $[\tfrac{1}{2}(\nu_A+\nu_B)+\nu_X+\tfrac{1}{2}(\tfrac{1}{2}J_{AB}+J_{AX}+J_{BX})]$ | $[\tfrac{1}{2}(\nu_a+\nu_b+\tfrac{1}{2}J_{ab})]$ |
| **B** | $\begin{bmatrix} \tfrac{1}{2}(\nu_A-\nu_B)+\nu_X+\tfrac{1}{2}(J_{AX}-\tfrac{1}{2}J_{AB}-J_{BX}) & \tfrac{1}{2}J_{AB} \\ \tfrac{1}{2}J_{AB} & \tfrac{1}{2}(\nu_B-\nu_A)+\nu_X+\tfrac{1}{2}(J_{BX}-\tfrac{1}{2}J_{AB}-J_{AX}) \end{bmatrix}$ | $\begin{bmatrix} \tfrac{1}{2}(\nu_a-\nu_b-\tfrac{1}{2}J_{ab} & \tfrac{1}{2}J_{ab} \\ \tfrac{1}{2}J_{ab} & \tfrac{1}{2}(\nu_b-\nu_a-\tfrac{1}{2}J_{ab}) \end{bmatrix}$ |
| **C** | $[-\tfrac{1}{2}(\nu_A+\nu_B)+\nu_X+\tfrac{1}{2}(\tfrac{1}{2}J_{AB}-J_{AX}-J_{BX})]$ | $[-\tfrac{1}{2}(\nu_a+\nu_b-\tfrac{1}{2}J_{ab})]$ |

**5.1.11**

The twelve lines of a clearly resolved AMX spectrum are labelled 1 to 12 in order of decreasing frequency, so that 1, 2, 3, 4 are A lines; 5, 6, 7, 8 are M lines; and 9, 10, 11, 12 are X lines, for $|J_{MX}| > |J_{AX}| > |J_{AM}|$.

A double-resonance experiment was performed in which the line labelled 7 is irradiated with a second radio-frequency field, $B_2$, such that

$$|\Delta\nu_{1/2}| < \frac{\gamma B_2}{2\pi} \ll |J_{AM}| ,$$

where $\Delta\nu_{1/2}$ is the halfwidth of the transition, and $\gamma$ is the magnetogyric ratio of the proton, i.e. a 'tickling' experiment. Four lines in the spectrum were observed to split as a result of the experiment. Lines 3 and 12 appeared as sharp doublets, whereas lines 4 and 10 appeared as broader doublets.

Explain these observations qualitatively and use them to deduce the relative signs of the coupling constants in this system (see Harris and Lynden-Bell, 1969).

**Solution**

'Tickling' experiments are a direct way of deducing relative signs of coupling constants in first-order spectra by irradiating precisely at individual resonance-line frequencies with enough power to perturb the energy levels associated with that transition and no other. A consequence of the irradiation is that the n.m.r. Hamiltonian is modified and now includes the irradiating field, and the quantum mechanical problem must be solved once again, but only for the energy states involved in the specific transition. The perturbation leads to a mixing of the two states to an extent determined by the $B_2$ field, and therefore additional transitions become possible within the context of the normal selection rule, $\Delta M = \pm 1$ (see figure 5.1.11a).

The single-resonance transition labelled P in figure 5.1.11a is split into a broad doublet and is called 'progressive'. The single-resonance transition labelled R appears as a sharp doublet and is called 'regressive'.

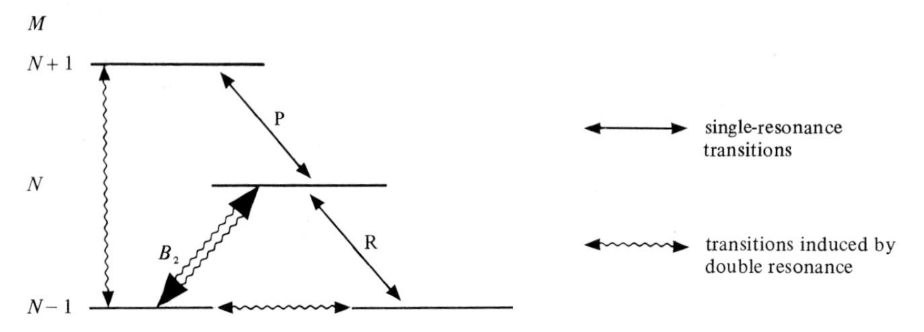


**Figure 5.1.11a.** Possible transitions in a double-resonance n.m.r. experiment.

*Relative signs of coupling constants*

One way of labelling the observed lines according to the spin states of the nuclei which explains the observed results is shown in figure 5.1.11b.

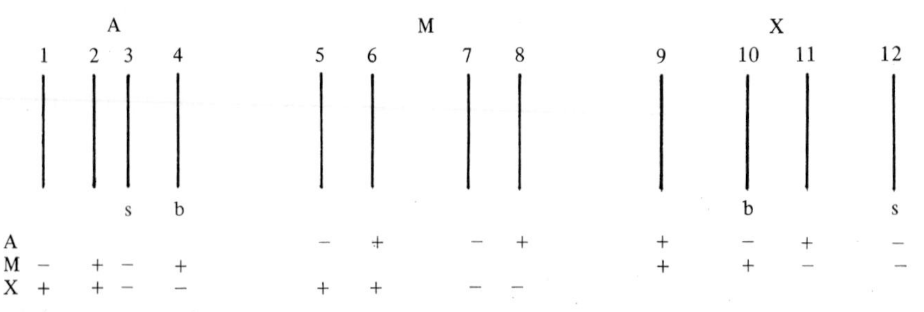


**Figure 5.1.11b.** Lines of the AMX spectrum labelled according to the nuclear spin states ($m = \pm\frac{1}{2}$) of the nuclei not undergoing transitions; s denotes sharp, b broad doublets obtained in the 'tickled' spectrum with transition 7 irradiated.

It is possible to see from consideration of the order of the spins that $J_{AM}$ has opposite sign to that of $J_{AX}$ or $J_{MX}$ (figure 5.1.11c).

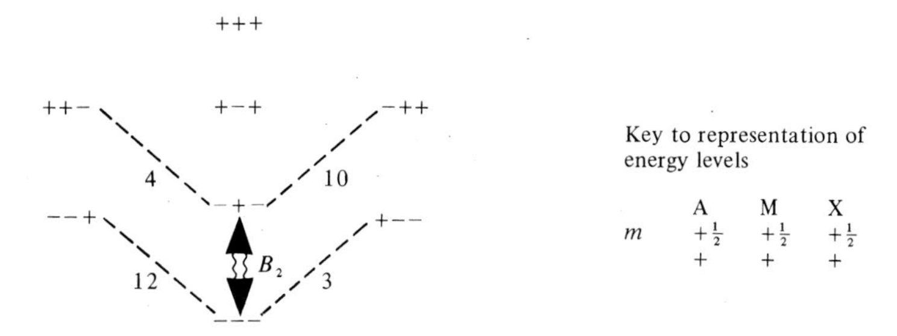


**Figure 5.1.11c.** Schematic of energy levels of the AMX nuclear spin states, showing 'tickled' transition and affected transitions numbered according to figure 5.1.11b.

**References to part II**

Bothner-By, A. A., 1965, *Adv. Magn. Reson.*, **1**, 195.

Diehl, P., Harris, R. K., Jones, R. G., 1967, *Prog. Nucl. Magn. Reson. Spectrosc.*, **3**, 1-61.

Emsley, J. W., Feeney, J., Sutcliffe, L. H., 1965, *High Resolution Nuclear Magnetic Resonance Spectroscopy* (Pergamon, Oxford).

Gutowsky, H. S., McCall, D. W., Slichter, C. P., 1953, *J. Chem. Phys.*, **21**, 279.

Harris, R. K., Lynden-Bell, R., 1969, *Nuclear Magnetic Resonance Spectroscopy* (Nelson, London).

Mathieson, D. W., 1967, *Nuclear Magnetic Resonance for Organic Chemists* (Academic Press, New York).

Pople, J. A., Schneider, W. G., Bernstein, H. J., 1959, *High Resolution Nuclear Magnetic Resonance* (McGraw-Hill, New York), p.119.

## 5.2 Nuclear quadrupole resonance

**E A C Lucken** Université de Genève

### 5.2.0 Introduction

The problems which follow are largely based on the analysis by Townes and Dailey (1949) of the magnitudes of nuclear quadrupole coupling constants (see also Lucken, 1969). This may be summarized as follows:
(a) Filled inner shells have overall spherical symmetry and contribute nothing to the field gradients.
(b) The same situation prevails for valence s electrons.
(c) The field gradients produced by the p, d, f ... electrons of the same principal quantum numbers have the relative magnitudes p:d:f ≈ 100:10:1. Thus, if an incomplete p shell is present, the field gradients from these electrons will dominate the coupling constant. Furthermore the field gradients produced by valence p electrons are much greater than those produced by neighbouring charges.
(d) In these circumstances the coupling constants may be analysed to yield the valence p electron populations $\rho_x$, $\rho_y$, $\rho_z$ at the atom in question according to the following equations,

$$e^2Q\frac{\partial^2 V}{\partial x^2} \equiv e^2Qq_{xx} = [\rho_x - \tfrac{1}{2}(\rho_y + \rho_z)]e^2Qq_0\,,$$

$$e^2Q\frac{\partial^2 V}{\partial y^2} \equiv e^2Qq_{yy} = [\rho_y - \tfrac{1}{2}(\rho_x + \rho_z)]e^2Qq_0\,,$$

$$e^2Q\frac{\partial^2 V}{\partial z^2} \equiv e^2Qq_{zz} = [\rho_z - \tfrac{1}{2}(\rho_x + \rho_y)]e^2Qq_0\,, \qquad (5.2.0a)$$

where $e^2Qq_0$ is the coupling constant of a valence p electron. As usual the field gradients are defined in the order $|q_{xx}| < |q_{yy}| < |q_{zz}|$ and the asymmetry parameter, $\eta$, by the relationship:

$$\eta \equiv \frac{e^2Qq_{xx} - e^2Qq_{yy}}{e^2Qq_{zz}}\,.$$

The asymmetry parameter is related to the orbital populations by the equation

$$e^2Qq_{zz}\eta = \tfrac{3}{2}(\rho_x - \rho_y)e^2Qq_0\,. \qquad (5.2.0b)$$

Values of $e^2Qq_0$ for the nuclei studied here will be found in table 5.2.0.

**Table 5.2.0.** Values of $e^2Qq_0$. For $^{14}N$ the value is derived from a variety of theoretical studies on simple molecules; the remainder come from direct measurements of atomic hyperfine coupling constants.

| Nucleus | Electron | $e^2Qq_0$ (MHz) | Reference |
|---|---|---|---|
| $^{14}N$ | 2p | −8 to −10 | Lucken (1969) |
| $^{23}Na$ | 3p | −5·16 | Perl *et al.* (1955) |
| $^{39}K$ | 4p | +5·6 | Buck *et al.* (1956) |
| $^{85}Rb$ | 5p | −48·8 | Senitzky and Rabi (1956) |
| $^{87}Rb$ | 5p | −23·6 | Senitzky and Rabi (1956) |
| $^{11}B$ | 2p | −5·390 | Wessel (1953) |
| $^{27}Al$ | 3p | −37·52 | Lew and Wessel (1953) |
| $^{69}Ga$ | 4p | −125·044940 | Daly and Holloway (1954) |
| $^{71}Ga$ | 4p | −78·79808 | Daly and Holloway (1954) |
| $^{115}In$ | 5p | −899·1048 | Kusch and Eck (1954) |
| $^{35}Cl$ | 3p | +109·746 | Jaccarino and King (1951) |
| $^{37}Cl$ | 3p | +86·510 | Jaccarino and King (1951) |
| $^{79}Br$ | 4p | −769·756 | King and Jaccarino (1954) |
| $^{81}Br$ | 4p | −643·032 | King and Jaccarino (1954) |
| $^{127}I$ | 5p | +2292·712 | Jaccarino *et al.* (1954) |

5.2.1

The halogen coupling constants of various diatomic halides are shown in table 5.2.1a. Calculate the ionic characters of the bonds assuming that in all cases the bonds use only the halogen valence $p_\sigma$ orbitals and that the remaining valence shell electrons are entirely centred on the halogen atom. Correlate these ionic characters with the electronegativities of the component atoms.

**Table 5.2.1a.** Halogen nuclear quadrupole coupling constants of diatomic halides.

| Compound | Nucleus | Coupling constant (MHz) | Compound | Nucleus | Coupling constant (MHz) |
|---|---|---|---|---|---|
| LiCl | $^{35}Cl$ | −3·07172 | NaI | $^{127}I$ | −259·87 |
| KCl | $^{35}Cl$ | −0·0585 | $Cl_2$ | $^{35}Cl$ | (−)108·95 |
| AgCl | $^{35}Cl$ | −37·32 | BrCl | $^{35}Cl$ | −103·6 |
| HCl | $^{35}Cl$ | −67·51 | FCl | $^{35}Cl$ | −146·0 |
| AgBr | $^{79}Br$ | 307 | ICl | $^{35}Cl$ | −82·5 |
| HBr | $^{79}Br$ | 535·44 | ICl | $^{127}I$ | −2944 |
| HI | $^{127}I$ | −1831·07 | BrCl | $^{79}Br$ | 876·8 |

**Solution**

For a halogen atom forming a pure $p_\sigma$ bond the populations of the valence p orbitals are

$\psi_x$, population 2;
$\psi_y$, population 2;
$\psi_z$, population $1+i$;

here $i$ is the ionic character of the $p_\sigma$ bond and the halogen is at the negative end of the dipole. Substitution of these populations in equation (5.2.0a) then yields

$$\frac{e^2Qq_{zz}}{e^2Qq_0} = -(1-i) . \tag{5.2.1a}$$

**Table 5.2.1b.** Ionic characters of gaseous diatomic halides.

| Halide | $i$ | $1-i$ | Halide | $i$ | $1-i$ |
|---|---|---|---|---|---|
| LiCl | 0·972 | 0·028 | NaI | 0·887 | 0·113 |
| KCl | 0·9995 | 0·0005 | $Cl_2$ | 0·007 | 0·993 |
| AgCl | 0·660 | 0·340 | BrCl | 0·056 | 0·944 |
| HCl | 0·385 | 0·615 | FCl | 0·33 | 0·67[a] |
| AgBr | 0·601 | 0·399 | ICl | 0·248 | 0·752 |
| HBr | 0·304 | 0·696 | ICl | 0·284 | 0·716[a] |
| HI | 0·201 | 0·799 | BrCl | 0·139 | 0·861[a] |

[a] Nucleus studied at *positive* end of dipole.

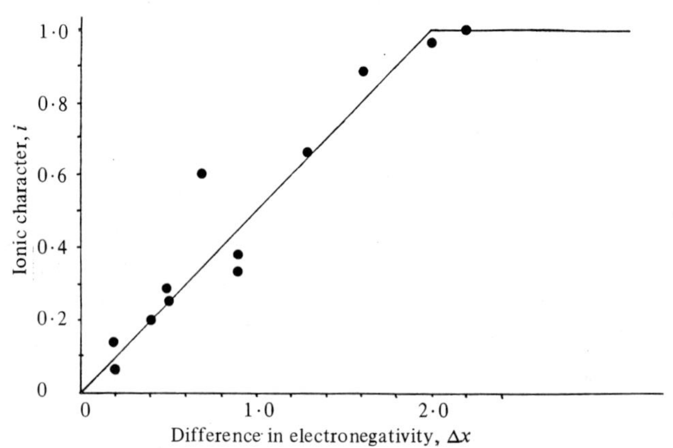


**Figure 5.2.1.** The relationship between the electronegativity differences (Pritchard and Skinner, 1955) of the component atoms and the ionic character deduced from the quadrupole coupling constant. The straight line indicates the rule of thumb proposed by Gordy (1955):

$$i = \begin{cases} \frac{1}{2}\Delta x, & 0 < \Delta x < 2{\cdot}0; \\ 1{\cdot}0, & \Delta x > 2{\cdot}0. \end{cases}$$

If the halogen atom is at the positive end of the dipole the corresponding $\psi_z$ population is $1-i$ and the corresponding equation for the coupling constant becomes

$$\frac{e^2Qq_{zz}}{e^2Qq_0} = -(1+i) . \qquad (5.2.1b)$$

Substitution of the coupling constants in these equations yields the values for the ionic characters which are listed in table 5.2.1b. Their relationship to the electronegativity differences of the components is shown in figure 5.2.1.

See Dailey and Townes (1955).

**5.2.2**

The quadrupole coupling constants of the Group III gaseous diatomic halides are shown in table 5.2.2a for nuclei at both ends of the bonds. Calculate the ionic characters of the bonds assuming, as in problem 5.2.1, that only the valence $p_\sigma$ orbitals are used in bond formation. Comment on the different ionic characters obtained for the two different nuclei in the same molecule and suggest ways in which the conflicting data might be reconciled.

**Table 5.2.2a.** Quadrupole coupling constants of Group III diatomic halides.

| Halide | Coupling constant (MHz) | |
|---|---|---|
| | halogen nucleus | metal nucleus |
| $^{27}Al^{35}Cl$ | −8·8 | −29·2 |
| $^{69}Ga^{35}Cl$ | −20 | −84·7 |
| $^{69}Ga^{79}Br$ | 134 | −74 |
| $^{69}Ga^{127}I$ | −549 | −66 |
| $^{115}In^{35}Cl$ | −18 | −655 |
| $^{115}In^{79}Br$ | 138 | −642 |

**Solution**

The first two columns in table 5.2.2b show the ionic characters for the two centres calculated by using, for the halogen atom, equation (5.2.1a) and for the metal atom, the equation

$$1-i = \frac{e^2Qq_{zz}}{e^2Qq_0} , \qquad (5.2.2a)$$

which follows directly from the orbital populations.

The ionic character given by the halogen coupling is in fair agreement with the predictions of the electronegativity differences, but those given by the metal atom couplings are much too low. They could be increased somewhat (~30%) by taking into account the reduction of the mean electron–nuclear distance brought about by the positive charge on the metal atom which thus increases the valence-electron coupling constant, $e^2Qq_0$. The probable source of the remaining discrepancy lies in polarization of the valence s electrons on the metal atom.

The orbitals on the metal atom then become:

$$\sigma_{\text{bonding}} = c^{1/2}\psi_s + (1-c)^{1/2}\psi_z, \quad \text{population } (1-i);$$
$$\sigma_{\text{nonbonding}} = (1-c)^{1/2}\psi_s - c^{1/2}\psi_z, \quad \text{population } 2.$$

The $\psi_x$ and $\psi_y$ orbitals as before are unoccupied. These orbitals and populations then yield the equation:

$$2c + (1-i)(1-c) = \frac{e^2Qq_{zz}}{e^2Qq_0} . \qquad (5.2.2b)$$

The values of $c$ calculated by equation (5.2.2b) with the use of the value of $i$ obtained from the halogen couplings (table 5.2.2a) is shown in the third column of table 5.2.2b. These values of $c$ would be reduced somewhat if the effect of the positive charge were included. A further modification of the above equations would arise, if a partial electron transfer of the halogen $p_\pi$ electrons to the vacant metal $p_\pi$ orbitals took place. This is unlikely in all the molecules concerned except perhaps in AlCl.

See Barret and Mandel (1958); Lide (1965).

**Table 5.2.2b.** Analysis of the quadrupole coupling data for Group III diatomic halides.

| Halide | $i$ from halogen coupling | $i$ from metal coupling | s-hybridization of metal valence orbital, $c$ |
|---|---|---|---|
| AlCl | 0·92 | 0·21 | 0·36 |
| GaCl | 0·82 | 0·32 | 0·27 |
| GaBr | 0·83 | 0·41 | 0·23 |
| GaI | 0·80 | 0·47 | 0·18 |
| InCl | 0·84 | 0·27 | 0·31 |
| InBr | 0·82 | 0·29 | 0·29 |

### 5.2.3

**Table 5.2.3a.** Quadrupole coupling constants and asymmetry parameters for organochlorine compounds.

| Compound | $e^2Qq_{zz}$(MHz) | $\eta$ |
|---|---|---|
| Vinyl chloride | 72·0 | 0·14 |
| Phosgene | 72·4 | 0·25 |
| 1,2,4,5-Tetrachlorobenzene | 72·6 | 0·116 |
| Chloranil | 73·6 | 0·21 |
| 3,4-Dichloropyridine | 69·4 | 0·086 |
| 2,6-Dichloropyridine | 67·7 | 0·118 |
| 2,4,6-Trichloro-*s*-triazine | 72·6 | 0·24 |

The quadrupole coupling constants and asymmetry parameters $\eta$ for the $^{35}Cl$ nucleus in various organochlorine compounds are shown in table 5.2.3a. Calculate from these data the ionic character $i$ and the partial double-bond character $\pi$ of the carbon–halogen bond in these compounds and comment on the values so obtained.

**Solution**

When one of the $p_\pi$ orbitals takes part in partial double-bond formation the valence orbitals and their populations then become (the halogen atom is assumed to bear the negative charge):

$\psi_x$, population $2-\pi$;
$\psi_y$, population $2$;
$\psi_z$, population $1+i$.

Substitution of these populations in equations (5.2.0a) and (5.2.0b) followed by straightforward algebraic manipulation then yields:

$$i = 1 + \left(\frac{e^2Qq_{zz}}{e^2Qq_0}\right)\left(1+\frac{\eta}{3}\right),$$

$$\pi = \tfrac{2}{3}\left|\frac{e^2Qq_{zz}}{e^2Qq_0}\right|\eta .$$

The application of these equations yields the results shown in table 5.2.3b.

**Table 5.2.3b.** Analysis of the ionic character $i$ and partial double-bond character $\pi$ of carbon–chlorine bonds from $^{35}Cl$ nuclear quadrupole coupling constants.

| Compound | $i$ | $\pi$ |
|---|---|---|
| Vinyl chloride | 0·314 | 0·062 |
| Phosgene | 0·285 | 0·111 |
| 1,2,4,5-Tetrachlorobenzene | 0·313 | 0·051 |
| Chloranil | 0·283 | 0·094 |
| 3,5-Dichloropyridine | 0·348 | 0·036 |
| 2,6-Dichloropyridine | 0·359 | 0·048 |
| 2,4,6-Trichloro-*s*-triazine | 0·284 | 0·110 |

The partial double-bond character clearly reflects the increased importance of the resonance structure (I) in phosgene compared with the analogous structure (II) in vinyl chloride which the relative electronegativities of oxygen and carbon would lead one to suspect. A similar remark applies to the pair 1,2,4,5-tetrachlorobenzene and chloranil. In the pair 3,5-dichloropyridine and 2,6-dichloropyridine the relative $\pi$ character may be explained by the possibility of resonance structures such as (III) for the latter compound whereas no such structures having the negative charge on the nitrogen atom can be formulated in the former molecule.

See Bersohn (1954).

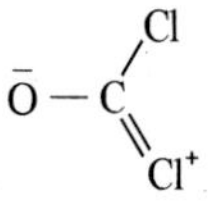

$\bar{O}-C(Cl)=Cl^+$ (I) $\quad CH_2^- - C(H)=Cl^+$ (II) $\quad$ Cl, $\bar{N}$, $Cl^+$ (III)

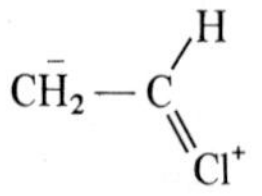

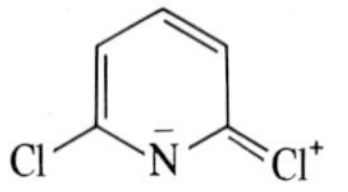

5.2.4

Derive the relationship between the valence orbital populations, the interorbital angle $R\widehat{X}R$ ($=\alpha$) and the quadrupole coupling constant for the symmetric pyramidal molecule $R_3X$, where X is a Group V atom.

**Solution**

The principal axis ($z$) of the field gradient coincides necessarily with the molecular threefold axis and the asymmetry parameter is identically zero. The asymmetry parameter is identically zero if the nucleus lies on an $n$-fold symmetry axis with $n \geqslant 3$ since then $x$ and $y$ vectors transform according to a doubly degenerate representation. The orbitals used in bonding are (see problem 1.2.2):

$$\phi_1 = \left(-\frac{\cos\alpha}{1-\cos\alpha}\right)^{1/2}\psi_s - \left(\tfrac{2}{3}\right)^{1/2}\psi_x + \left[\frac{1+2\cos\alpha}{3(1-\cos\alpha)}\right]^{1/2}\psi_z\,,$$

$$\phi_2 = \left(-\frac{\cos\alpha}{1-\cos\alpha}\right)^{1/2}\psi_s + 6^{-1/2}\psi_x + 2^{-1/2}\psi_y + \left[\frac{1+2\cos\alpha}{3(1-\cos\alpha)}\right]^{1/2}\psi_z\,,$$

$$\phi_3 = \left(-\frac{\cos\alpha}{1-\cos\alpha}\right)^{1/2}\psi_s + 6^{-1/2}\psi_x - 2^{-1/2}\psi_y + \left[\frac{1+2\cos\alpha}{3(1-\cos\alpha)}\right]^{1/2}\psi_z\,,$$

$$\phi_4 = \left(\frac{1+2\cos\alpha}{1-\cos\alpha}\right)^{1/2}\psi_s - \left(-\frac{3\cos\alpha}{1-\cos\alpha}\right)^{1/2}\psi_z\,.$$

The populations of the first three orbitals are by definition the same ($\equiv a$) and the fourth orbital contains two electrons. Substitution of the resultant populations in equation 5.2.0a yields

$$\frac{e^2Qq_{zz}}{e^2Qq_0} = -\frac{3\cos\alpha}{1-\cos\alpha}(2-a)\,. \qquad (5.2.4a)$$

The interorbital angle lies in the range $90° \leqslant \alpha \leqslant 120°$ and the angular term in equation (5.2.4a) then varies from zero through the value 0·75 for the tetrahedral angle to its maximum value of 1·0 for $\alpha = 120°$. The coupling constants are thus in this case extremely sensitive to the interbond angle.

5.2.5

The coupling constants of the linear trihalide anions are shown in table 5.2.5a. Analyse these data in terms of the orbital population at the various centres and comment on the bonding in these compounds.

**Table 5.2.5a.** Halogen quadrupole coupling constants of trihalide anions.

| Ion | Nucleus | Coupling constant (MHz) |
|---|---|---|
| $[I_{(1)}-I_{(2)}-I_{(3)}]^-$ (in $CsI_3$) | $^{127}I_{(1)}$ | 819·0 |
| | $^{127}I_{(2)}$ | 2477·5 |
| | $^{127}I_{(3)}$ | 1436·6 |
| $[Cl_{(1)}-I_{(2)}-Cl_{(3)}]^-$ (in $CsICl_2$) | $^{35}Cl_{(1)}$, $^{35}Cl_{(3)}$ | 39·72 |
| | $^{127}I_{(2)}$ | 3099 |

**Solution**

The populations of the various p orbitals in these three atoms may be written (the $z$ axis is taken to coincide with the molecular axis):

$\psi_x$, $\psi_y$, population 2·0 ,
$\psi_z$, population $a_i$ for atom $X_i$.

The coupling constants may then be analysed according to equation (5.2.0a) to yield the populations $a_i$ given in table 5.2.5b.

**Table 5.2.5b.** Populations, $a_i$, of the p orbitals in trihalide anions.

| Ion | Atom | | | Total $\sigma$ population |
|---|---|---|---|---|
| | $X_1$ | $X_2$ | $X_3$ | |
| $I_3^-$ | 1·64 | 0·94 | 1·37 | 3·95 |
| $ICl_2^-$ | 1·64 | 0·65 | 1·64 | 3·93 |

The bonding in these compounds, both symmetrical and unsymmetrical, is thus best described as a three-centre four-electron bond. This is equivalent to the resonance structure

$$Y^- \quad X-Y \longleftrightarrow Y-X \quad Y^-$$

This implies that for the totally symmetric isonuclear molecule the population of the centre orbital would be 1·0 while the outer atoms would have an orbital population of 1·5. The closeness of the sum of the orbital populations to 4·0 indicates that there is no appreciable d-orbital participation in the bonding.

See Cornwell and Yamasaki (1957); Yamasaki and Cornwell (1959).

**5.2.6**

The molecular structure of $Ga_2Cl_6$ is shown in figure 5.2.6 and the quadrupole coupling data are given in table 5.2.6. Use them to analyse the orbital populations in this molecule and hence comment on its electronic structure.

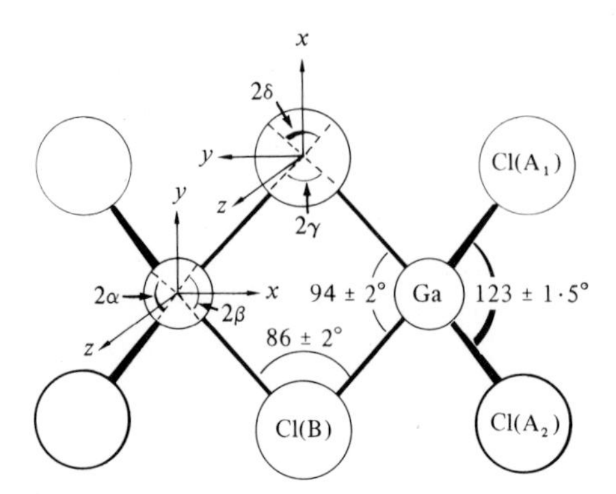


**Figure 5.2.6.** Molecular structure of $Ga_2Cl_6$. The values for the interbond angles are as determined experimentally. Angles $\alpha$, $\beta$, $\gamma$, $\delta$ represent interorbital angles.

**Table 5.2.6.** Quadrupole coupling data for crystalline $Ga_2Cl_6$.

| Nucleus | Coupling constant (MHz) | Asymmetry parameter |
|---|---|---|
| $^{71}Ga$ | 32·76 | 0·867 |
| $^{35}Cl(B)$ | 28·30 | 0·473 |
| $^{35}Cl(A_1)$ | 38·06 | 0·089 |
| $^{35}Cl(A_2)$ | 40·44 | 0·034 |

**Solution**

Using the systems of axes shown in figure 5.2.6 we can write the orbitals for the gallium atoms (cf. problem 5.2.4):

$$\psi_1 = 2^{-1/2}[(1-\cot^2\beta)^{1/2}\psi_s + \cot\beta\psi_x + \psi_y]\,, \text{ population } a;$$

$$\psi_2 = 2^{-1/2}[(1-\cot^2\beta)^{1/2}\psi_s + \cot\beta\psi_x - \psi_y]\,, \text{ population } a;$$

$$\psi_3 = 2^{-1/2}[(1-\cot^2\alpha)^{1/2}\psi_s - \cot\alpha\psi_x + \psi_z]\,, \text{ population } b;$$

$$\psi_4 = 2^{-1/2}[(1-\cot^2\alpha)^{1/2}\psi_s - \cot\alpha\psi_x - \psi_z]\,, \text{ population } b.$$

The *interorbital* angles $\alpha$ and $\beta$ are related by the equation

$$\cot^2\alpha + \cot^2\beta = 1\ . \tag{5.2.6a}$$

Similar orbitals are obtained for the bridging halogen atoms, with $\gamma$ and $\delta$ replacing $\alpha$ and $\beta$.

Substitution of the net p-orbital populations in equations (5.2.0a) and (5.2.0b), and application of equation (5.2.6a), shows that the asymmetry parameter depends only on the interorbital angle:

$$\eta = -3\cos 2\beta \qquad \text{for the Ga atom,} \tag{5.2.6b}$$

$$\eta = -3\cos 2\gamma \qquad \text{for the Cl atom.} \tag{5.2.6c}$$

Equations (5.2.6b) and (5.2.6c) yield respectively 107° for $2\beta$ and 99° for $2\gamma$, to be compared with the experimental *interbond* angles shown in figure 5.2.6. The relative values of the interorbital angles parallel those of the interbond angles, but they are not, of course, identical (since the interorbital angles cannot be less than 90° the interbond and interorbital angles can only be identical for a square disposition of the four central atoms). The bonding is probably best described in terms of bent bonds. The orbital populations may likewise be shown to be given by

$$b - a = \frac{e^2 Q q_{zz}}{e^2 Q q}\left(1 + \frac{\eta}{3}\right) . \tag{5.2.6d}$$

For the bridging chlorine atoms the value of $b$ may be taken as 2·0. The terminal chlorine atom populations may be analysed by equation (5.2.0a); this yields the halogen orbital populations:

$Cl(A_1)$, 1·65; $Cl(A_2)$, 1·63; Cl(B), 1·70.

Equation (5.2.6d) yields only the difference between the gallium orbital populations, 0·54, which differs considerably from the value of 0·06 indicated by the chlorine populations. Intermolecular field gradients may be responsible for some of the discrepancy, but much is probably due to the assignment of the value 2·0 to the bridging orbital $b$ populations as well as to the inadequacies of the simple theory used.

See Barnes and Segel (1956); Peterson and Bridenbaugh (1969).

**5.2.7**

The $^{14}N$ quadrupole coupling constants for pyridine and several 4-substituted derivatives are given in table 5.2.7a. Given that the field-gradient $z$ axis is the symmetry axis of the ring and that the $y$ axis is perpendicular to the molecular plane in all these compounds, analyse the data in terms of the nitrogen valence orbital populations. (Assume that the interorbital angle between the two $sp^2$ orbitals bonded to the carbon atoms is 120°.)

**Table 5.2.7a.** $^{14}N$ nuclear quadrupole coupling constants of the ring nitrogen atom in pyridine derivatives.

| | Coupling constant (MHz) | Asymmetry parameter |
|---|---|---|
| Pyridine | 4·60 | 0·396 |
| 4-Methylpyridine | 4·41 | 0·342 |
| 4-Chloropyridine | 4·56 | 0·363 |
| 4-Acetylpyridine | 4·76 | 0·435 |
| 4-Cyanopyridine | 4·77 | 0·437 |
| 4-Nitropyridine | 4·79 | 0·486 |

**Solution**

The expressions for the $\sigma$ orbitals of a pure $sp^2$ hybrid set are shown below together with symbols for their population. The lone-pair orbital containing two electrons lies along the molecular symmetry axis and the remaining orbitals have been labelled according to the assumptions set out in the problem.

$$\left.\begin{array}{ll} \phi_1 = 3^{-\frac{1}{2}}\psi_s + (\tfrac{2}{3})^{\frac{1}{2}}\psi_z\,, & \text{population } 2; \\ \phi_2 = 3^{-\frac{1}{2}}\psi_s - 6^{-\frac{1}{2}}\psi_z + 2^{-\frac{1}{2}}\psi_x\,, & \text{population } b; \\ \phi_3 = 3^{-\frac{1}{2}}\psi_s - 6^{-\frac{1}{2}}\psi_z - 2^{-\frac{1}{2}}\psi_x\,, & \text{population } b; \\ \phi_4 = \psi_y\,, & \text{population } a. \end{array}\right\} \qquad (5.2.7a)$$

Once again, substitution of the total p-orbital populations in equations (5.2.0a) and (5.2.0b) gives rise straightforwardly to the equations

$$b - a = \tfrac{2}{3}\eta\frac{e^2Qq_{zz}}{e^2Qq_0}\,, \qquad 2 - b = \tfrac{3}{2}\left(1 - \frac{\eta}{3}\right)\frac{e^2Qq_{zz}}{e^2Qq_0}\,.$$

The bond populations thus obtained for the compounds in question are shown in table 5.2.7b.

See Guibé and Lucken (1968); Lucken (1969).

**Table 5.2.7b.** Nitrogen bond populations of 4-substituted pyridine compounds.

| Compound | $a$ | $b$ |
|---|---|---|
| Pyridine | 1·201 | 1·335 |
| 4-Methylpyridine | 1·245 | 1·349 |
| 4-Chloropyridine | 1·209 | 1·332 |
| 4-Acetylpyridine | 1·169 | 1·326 |
| 4-Cyanopyridine | 1·168 | 1·321 |
| 4-Nitropyridine | 1·160 | 1·332 |

### 5.2.8

If the axes used to formulate the valence molecular orbitals do not coincide with the field-gradient axes, then the field gradients calculated according to equations (5.2.0a) and (5.2.0b) will not be the principal field gradients, but rather the values of the field gradients along the molecular orbital axes. In order to derive the principal values, the off-diagonal elements in the molecular orbital system must be known. If attention is as usual restricted to p orbitals then it becomes necessary to calculate the off-diagonal elements

$$\int \psi_i q_{jk} \psi_j \, d\tau\,,$$

where the indices $i$, $j$, $k$ represent one or other of the molecular orbital axes $x$, $y$, $z$. Calculate these off-diagonal elements in terms of $q_{zz}$ given the explicit forms below for the p orbitals and the field-gradient operators:

$$\psi_x = N\mathrm{f}(r)\sin\theta\cos\phi\ ,$$
$$\psi_y = N\mathrm{f}(r)\sin\theta\sin\phi\ ,$$
$$\psi_z = N\mathrm{f}(r)\cos\theta\ ,$$

where $N$ is a normalizing factor and $\mathrm{f}(r)$ is the radial function;

$$q_{zz} = \frac{3\cos^2\theta - 1}{r^3}\ ,$$

$$q_{xz} = \frac{3\sin\theta\cos\theta\cos\phi}{r^3}\ ,$$

$$q_{yz} = \frac{3\sin\theta\cos\theta\sin\phi}{r^3}\ ,$$

$$q_{xy} = \frac{3\sin^2\theta\sin\phi\cos\phi}{r^3}\ .$$

**Solution**

The off-diagonal integrals fall into four classes

(i) $\int \psi_i q_{ij} \psi_i \,\mathrm{d}\tau$ ,

(ii) $\int \psi_k q_{ij} \psi_k \,\mathrm{d}\tau$ ,

(iii) $\int \psi_i q_{ij} \psi_k \,\mathrm{d}\tau$ ,

(iv) $\int \psi_i q_{ij} \psi_j \,\mathrm{d}\tau$ .

Either explicit evaluation or simple symmetry considerations show that the first three classes of integrals are identically zero. Explicit evaluation of the integrals of type (iv) yields, for example,

$$\int \psi_z q_{xz} \psi_x \,\mathrm{d}\tau$$

$$= N^2 \int_\theta \int_\phi \int_r [\mathrm{f}(r)]^2 \sin\theta\cos\theta\cos\phi \left(\frac{3\sin\theta\cos\theta\cos\phi}{r^3}\right) \sin\theta r^2 \,\mathrm{d}\theta\,\mathrm{d}\phi\,\mathrm{d}r$$

$$= 3N^2\left\langle\frac{1}{r^3}\right\rangle \int_{\theta=0}^{\pi} \int_{\phi=0}^{2\pi} \sin^3\theta\cos^2\theta\cos^2\phi \,\mathrm{d}\theta\,\mathrm{d}\phi$$

$$= 3\pi N^2\left\langle\frac{1}{r^3}\right\rangle \int_0^{\pi} \sin^3\theta\cos^2\theta \,\mathrm{d}\theta = 3\pi N^2\left\langle\frac{1}{r^3}\right\rangle \times \tfrac{4}{15}$$

$$= \tfrac{12}{15}\pi N^2\left\langle\frac{1}{r^3}\right\rangle ,$$

where $\langle 1/r^3\rangle$ is the average value of $1/r^3$ for the particular radial function.

Again simple symmetry considerations show all type (iv) integrals to have the same value. Let us compare this with the basic $q_{zz}$ expectation value,

$$\int \psi_z q_{zz} \psi_z \, d\tau = N^2 \iiint [f(r)]^2 \cos^2\theta \left(\frac{3\cos^2\theta - 1}{r^3}\right) r^2 \sin\theta \, dr \, d\theta \, d\phi$$

$$= 2\pi N^2 \left\langle \frac{1}{r^3} \right\rangle \int_0^{\pi} (3\cos^4\theta \sin\theta - \cos^2\theta \sin\theta) \, d\theta$$

$$= 2\pi N^2 \left\langle \frac{1}{r^3} \right\rangle \left(\tfrac{6}{5} - \tfrac{2}{3}\right) = \tfrac{16}{15} \pi N^2 \left\langle \frac{1}{r^3} \right\rangle .$$

Thus

$$\int \psi_i q_{ij} \psi_j \, d\tau = \tfrac{3}{4} \int \psi_z q_{zz} \psi_z \, d\tau . \qquad (5.2.8a)$$

### 5.2.9

The nuclear quadrupole coupling constants of the $^{14}N$ nuclei in pyridazine measured in the vapour phase are

$$e^2 Q q_{zz} = -5 \cdot 65 \pm 0 \cdot 1 \text{ MHz}, \qquad \eta = 0 \cdot 16 \pm 0 \cdot 02 .$$

The directions of the principal axes are shown in figure 5.2.9. Analyse the quadrupole coupling data in terms of the populations of the various $\sigma$ and $\pi$ valence orbitals of the nitrogen atom assuming that the population of the lone-pair orbital is exactly $2 \cdot 0$ and that $e^2 Q q_0 = -9$ MHz. Compare these with the populations derived from the gas-phase coupling constants of pyridine, $e^2 Q q_{zz} = -4 \cdot 88$ MHz, $\eta = 0 \cdot 405$.

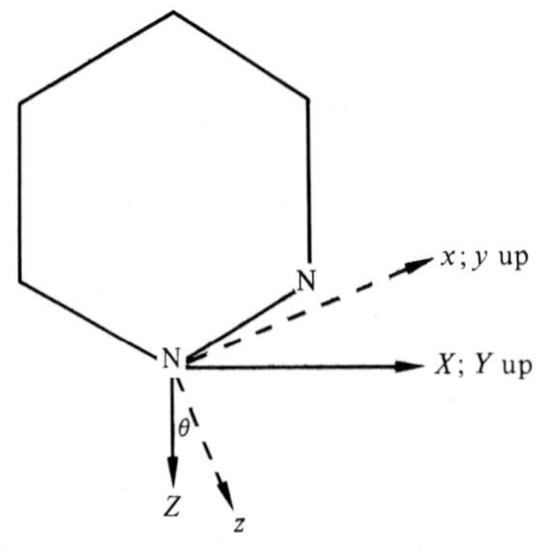


**Figure 5.2.9.** Axis systems in pyridazine. The $XYZ$ axes are those in which the $Z$ axis bisects the CNN angle, the $Y$ axis being perpendicular to the molecular plane. The $xyz$ axes are the experimental field-gradient axes, the $y$ axis being again perpendicular to the molecular plane and thus coincident with $Y$. The angle $\theta$ has the experimentally found value $9 \pm 1°$.

**Solution**

The $\sigma$ and $\pi$ valence orbitals for pyridazine are of course most naturally described in the $XYZ$ system according to equations (5.2.7a), for pyridine. The resultant equations may be used to analyse the data, provided that the field gradients are transformed to the $XYZ$ system.

The field gradients in the $XYZ$ system are related to the principal values by the usual transformation:

$$\left.\begin{aligned} e^2Qq_{XX} &= e^2Qq_{xx}\cos^2\theta + e^2Qq_{zz}\sin^2\theta\ ,\\ e^2Qq_{YY} &= e^2Qq_{yy}\ ,\\ e^2Qq_{ZZ} &= e^2Qq_{zz}\cos^2\theta + e^2Qq_{xx}\sin^2\theta\ ,\\ e^2Qq_{XZ} &= (e^2Qq_{xx} - e^2Qq_{zz})\sin\theta\cos\theta\ . \end{aligned}\right\} \tag{5.2.9a}$$

The asymmetry parameter in the $XYZ$ system, $\eta'$, thus becomes

$$\eta' = \frac{e^2Qq_{XX} - e^2Qq_{YY}}{e^2Qq_{ZZ}}\ . \tag{5.2.9b}$$

Let the populations of the two different $\sigma$ orbitals be represented by $b_1$ and $b_2$, where $b_1$ is the population of the $\sigma$ orbital bonded to the adjacent carbon atom and $b_2$ that of the $\sigma$ orbital bonded to the nitrogen atom. The substitution of these populations in equations (5.2.0a) and (5.2.0b) yields

$$b_1 + b_2 - 2a = \tfrac{4}{3}\eta'\frac{e^2Qq_{ZZ}}{e^2Qq_0}\ , \tag{5.2.9c}$$

$$4 - b_1 - b_2 = 3\left(1 - \frac{\eta'}{3}\right)\frac{e^2Qq_{ZZ}}{e^2Qq_0}\ . \tag{5.2.9d}$$

Equations (5.2.9c) and (5.2.9d) are not of course by themselves sufficient to determine the three unknowns. However, a third equation may be obtained by calculating the off-diagonal element $e^2Qq_{XZ}$, as in problem 5.2.8. This is made up of the contributions from orbitals $\phi_1$, $\phi_2$, and $\phi_3$ (orbital $\phi_4$ does not contribute since it is a pure $p_Y$ orbital and the $Y$ and $y$ axes coincide):

$$\int \phi_1 q_{XZ}\phi_1\,d\tau = 2\left(\tfrac{1}{3}\int\psi_s q_{XZ}\psi_s\,d\tau + \tfrac{2}{3}\int\psi_Z q_{XZ}\psi_Z\,d\tau + \tfrac{2}{3}2^{1/2}\int\psi_s q_{XZ}\psi_Z\,d\tau\right) = 0\ ,$$

$$\begin{aligned}\int \phi_2 q_{XZ}\phi_2\,d\tau = \frac{b_1}{2}\Big(&\tfrac{2}{3}\int\psi_s q_{XZ}\psi_s\,d\tau + \tfrac{1}{3}\int\psi_Z q_{XZ}\psi_Z\,d\tau + \int\psi_X q_{XZ}\psi_X\,d\tau\\ &- \tfrac{2}{3}2^{1/2}\int\psi_s q_{XZ}\psi_Z\,d\tau + 2(\tfrac{2}{3})^{1/2}\int\psi_s q_{XZ}\psi_X\,d\tau - 2(\tfrac{1}{3})^{1/2}\int\psi_X q_{XZ}\psi_Z\,d\tau\Big)\\ = -\tfrac{1}{4}3^{1/2}b_1q_0\ ,&\end{aligned}$$

by application of equation (5.2.8a). Similarly

$$\phi_3 q_{XZ} \phi_3 \, d\tau = \tfrac{1}{4} 3^{1/2} b_2 q_0 ,$$

and hence

$$e^2Qq_{XZ} = \tfrac{1}{4} 3^{1/2} (b_2 - b_1) e^2Qq_0 . \qquad (5.2.9e)$$

From the coupling constant data given above we have

$$\begin{aligned} e^2Qq_{xx} &= -\tfrac{1}{2} e^2Qq_{zz}(1-\eta) = 2\cdot 37 \text{ MHz} , \\ e^2Qq_{yy} &= -\tfrac{1}{2} e^2Qq_{zz}(1+\eta) = 3\cdot 28 \text{ MHz} , \\ e^2Qq_{zz} &= -5\cdot 65 \text{ MHz} ; \end{aligned}$$

and from equations (5.2.9a) and (5.2.9b)

$$\begin{aligned} e^2Qq_{ZZ} &= -5\cdot 65 \cos^2 9° + 2\cdot 37 \sin^2 9° = -5\cdot 45 \text{ MHz} , \\ e^2Qq_{XX} &= 2\cdot 37 \cos^2 9° - 5\cdot 65 \sin^2 9° = 2\cdot 17 \text{ MHz} , \\ e^2Qq_{YY} &= 3\cdot 28 \text{ MHz} , \\ e^2Qq_{XZ} &= (2\cdot 37 + 5\cdot 65) \sin 9° \cos 9° = 1\cdot 24 \text{ MHz} , \end{aligned}$$

$$\eta' = -\frac{2\cdot 17 - 3\cdot 28}{5\cdot 45} = 0\cdot 20 .$$

Now, equations (5.2.9c), (5.2.9d), and (5.2.9e) become

$$\begin{aligned} b_1 + b_2 - 2a &= 0\cdot 161 , \\ 4 - b_1 - b_2 &= 1\cdot 696 , \\ b_2 - b_1 &= -0\cdot 318 , \end{aligned}$$

whence

$$a = 1\cdot 07, \quad b_1 = 1\cdot 31, \quad b_2 = 0\cdot 99.$$

The corresponding values for pyridine are $a = 1\cdot 20$, $b = 1\cdot 34$. As expected the population in the N–N $\sigma$-bond in pyridazine is close to unity whereas that in the C–N bond is less than that in pyridine. Likewise the value of $a$ for pyridazine is less than that for pyridine.

**5.2.10**

For a nucleus X in a particular isolated molecule, RX, the coupling constant has a value $e^2Qq$ and the asymmetry parameter is zero. Calculate the effect on the quadrupole coupling constant of

(a) low-amplitude high-frequency simple harmonic vibration of the R—X bond, assuming a linear relationship between bond length and field gradient;

(b) low-amplitude high-frequency simple harmonic libration about an axis perpendicular to the R—X bond;

(c) rapid rotation about an axis at an angle $\theta$ to the R—X bond.

**Solution**

(a) The instantaneous coupling constant by hypothesis is given by

$$e^2Qq(t) = e^2Qq[1+ax(t)] ,$$

where $x$ is the displacement of the bond length from the equilibrium position.

For simple harmonic motion with frequency $\nu$, $x(t)$ is given by

$$x = x_{\max} \sin 2\pi\nu t .$$

The observed coupling is then given by the average value of $x$ over the motion, $\langle x \rangle$. This may readily be seen to be zero. Anharmonicity would, of course, introduce a dependence of $\langle x \rangle$ on the amplitude of vibration.

(b)

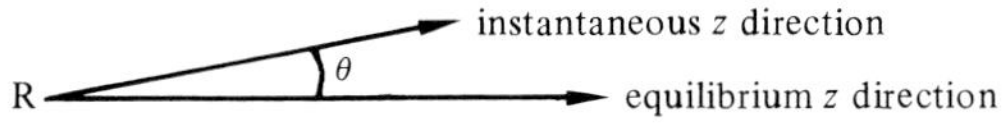

R — instantaneous $z$ direction — $\theta$ — equilibrium $z$ direction

The instantaneous field gradient along the $z$ direction is given by

$$q_{zz}(\theta) = \cos^2\theta q_{zz}(0) + \sin^2\theta q_{xx}(0) ,$$
$$q_{xx}(\theta) = \cos^2\theta q_{xx}(0) + \sin^2\theta q_{zz}(0) ,$$
$$q_{yy}(\theta) = q_{yy}(0) .$$

For the symmetric field gradient considered here we have

$$q_{xx}(0) = q_{yy}(0) = -\tfrac{1}{2}q_{zz}(0) ,$$
$$q_{zz}(\theta) = q_{zz}(0)(1-\tfrac{3}{2}\sin^2\theta) ,$$
$$q_{xx}(\theta) = q_{zz}(0)(\tfrac{3}{2}\sin^2\theta - \tfrac{1}{2}) , \qquad (5.2.10a)$$
$$q_{yy}(\theta) = -\tfrac{1}{2}q_{zz}(0) .$$

Hence the instantaneous asymmetry parameter is given by

$$\eta(\theta) = \tfrac{3}{2}\sin^2\theta . \qquad (5.2.10b)$$

For small displacements we can substitute $\theta$ for $\sin\theta$, and equations (5.2.10a) and (5.2.10b) become

$$q_{zz}(\theta) = q_{zz}(0)(1-\tfrac{3}{2}\theta^2) ,$$
$$\eta(\theta) = \tfrac{3}{2}\theta^2 .$$

For simple harmonic libration

$$\theta = \theta_{\max} \sin 2\pi\nu t ,$$

and the average value of $\theta^2$, taken over many vibrations, is $\frac{1}{2}\theta_{\max}^2$. Hence the average value of the coupling constant is given by

$$\langle e^2Qq_{zz} \rangle = e^2Qq_{zz}^0(1-\tfrac{3}{4}\theta_{\max}^2) , \qquad (5.2.10c)$$

with

$$\langle \eta \rangle = \tfrac{3}{4}\theta_{\max}^2 .$$

If the amplitude of vibration in a molecular solid depends on the temperature, equation (5.2.10c) predicts, as is usually observed, that the coupling constant will decrease with increasing temperature. This is the basis of the Bayer theory for the temperature dependence of pure quadrupole resonance frequencies.
(c) The situation in this case is rather similar to that in (b) except that now the angle $\theta$ is constant. The angle $\phi$, representing the angle of rotation about the $z$ axis, does not appear in this simple example since the asymmetry parameter is zero and the labelling of the axes is arbitrary. The rapid rotation about $z$ ensures that the average value of $\eta$ is zero; the average value of $e^2Qq$ is, however, given by

$$\langle e^2Qq\rangle = e^2Qq(\tfrac{3}{2}\cos^2\theta - \tfrac{1}{2}) . \qquad (5.2.10d)$$

See Ragle (1959); Bayer (1951); Kushida *et al.* (1956).

**5.2.11**

The pure quadrupole resonance frequencies for transitions $\Delta m = \pm 1$ for nuclei of spin 1, $\frac{3}{2}$, $\frac{5}{2}$, $\frac{7}{2}$, and $\frac{9}{2}$ are given by the following expressions:

$I = 1$

$$\nu_{0\leftrightarrow 1} = \frac{3e^2Qq}{4h}\left(1+\frac{\eta}{3}\right),$$

$$\nu_{0\leftrightarrow 1} = \frac{3e^2Qq}{4h}\left(1-\frac{\eta}{3}\right);$$

$I = \frac{3}{2}$

$$\nu_{\frac{3}{2}\leftrightarrow\frac{1}{2}} = \frac{e^2Qq}{2h}\left(1+\frac{\eta^2}{3}\right)^{\frac{1}{2}};$$

$I = \frac{5}{2}$

$$\nu_{\frac{5}{2}\leftrightarrow\frac{3}{2}} = \frac{6e^2Qq}{20h}(1-0{\cdot}2037\eta^2+0{\cdot}1622\eta^4),$$

$$\nu_{\frac{3}{2}\leftrightarrow\frac{1}{2}} = \frac{3e^2Qq}{20h}(1+0{\cdot}0926\eta^2-0{\cdot}6340\eta^4);$$

$I = \frac{7}{2}$

$$\nu_{\frac{7}{2}\leftrightarrow\frac{5}{2}} = \frac{3e^2Qq}{14h}(1-0{\cdot}1000\eta^2-0{\cdot}0180\eta^4),$$

$$\nu_{\frac{5}{2}\leftrightarrow\frac{3}{2}} = \frac{2e^2Qq}{14h}(1-0{\cdot}5667\eta^2+1{\cdot}8595\eta^4),$$

$$\nu_{\frac{3}{2}\leftrightarrow\frac{1}{2}} = \frac{e^2Qq}{14h}(1+3{\cdot}6333\eta^2-7{\cdot}2607\eta^4);$$

$I = \frac{9}{2}$

$$\nu_{\frac{9}{2}\leftrightarrow\frac{7}{2}} = \frac{4e^2Qq}{24h}(1-0{\cdot}0809\eta^2-0{\cdot}0043\eta^4)\,,$$

$$\nu_{\frac{7}{2}\leftrightarrow\frac{5}{2}} = \frac{3e^2Qq}{24h}(1-0{\cdot}1857\eta^2-0{\cdot}1233\eta^4)\,,$$

$$\nu_{\frac{5}{2}\leftrightarrow\frac{3}{2}} = \frac{2e^2Qq}{24h}(1-1{\cdot}3381\eta^2+11{\cdot}7224\eta^4)\,,$$

$$\nu_{\frac{3}{2}\leftrightarrow\frac{1}{2}} = \frac{e^2Qq}{24h}(1+9{\cdot}0333\eta^2-45{\cdot}691\eta^4)\,.$$

These expressions are exact for $I = 1$ and $I = \frac{3}{2}$. They are only approximate for nuclei of spin $\frac{5}{2}$, $\frac{7}{2}$, $\frac{9}{2}$, and for large values of the asymmetry parameter the Hamiltonian must be separately diagonalized. Pure quadrupole resonance frequencies, at intervals of 0·1 in $\eta$, calculated by such diagonalizations are shown in tables 5.2.11a, 5.2.11b, and 5.2.11c. Note that when $\eta$ is significant, transitions with $\Delta m = \pm 2$

**Table 5.2.11a.** Pure quadrupole resonance frequencies ($\Delta m = \pm 1$) for nuclei with spin $\frac{9}{2}$. The frequencies are in units of $e^2Qq/24h$.

| $\eta$ | $\nu_Q$ | | | |
|---|---|---|---|---|
| | $\frac{9}{2}\leftrightarrow\frac{7}{2}$ | $\frac{7}{2}\leftrightarrow\frac{5}{2}$ | $\frac{5}{2}\leftrightarrow\frac{3}{2}$ | $\frac{3}{2}\leftrightarrow\frac{1}{2}$ |
| 0·0 | 4·0000 | 3·0000 | 2·0000 | 1·0000 |
| 0·1 | 3·9967 | 2·9944 | 1·9754 | 1·0861 |
| 0·2 | 3·9870 | 2·9772 | 1·9212 | 1·3066 |
| 0·3 | 3·9707 | 2·9474 | 1·8713 | 1·5969 |
| 0·4 | 3·9478 | 2·9043 | 1·8489 | 1·9131 |
| 0·5 | 3·9179 | 2·8483 | 1·8659 | 2·2317 |
| 0·6 | 3·8810 | 2·7183 | 1·9252 | 2·5419 |
| 0·7 | 3·8366 | 2·7071 | 2·0240 | 2·8393 |
| 0·8 | 3·7845 | 2·6308 | 2·1562 | 3·1232 |
| 0·9 | 3·7243 | 2·5573 | 2·3147 | 3·3946 |
| 1·0 | 3·6555 | 2·4919 | 2·4919 | 3·6555 |

**Table 5.2.11b.** Pure quadrupole resonance frequencies for nuclei of spin $\frac{7}{2}$. The frequencies are in units of $e^2Qq/28h$.

| $\eta$ | $\nu_Q$ | | | $\eta$ | $\nu_Q$ | | |
|---|---|---|---|---|---|---|---|
| | $\frac{7}{2}\leftrightarrow\frac{5}{2}$ | $\frac{5}{2}\leftrightarrow\frac{3}{2}$ | $\frac{3}{2}\leftrightarrow\frac{1}{2}$ | | $\frac{7}{2}\leftrightarrow\frac{5}{2}$ | $\frac{5}{2}\leftrightarrow\frac{3}{2}$ | $\frac{3}{2}\leftrightarrow\frac{1}{2}$ |
| 0·0 | 6·0000 | 4·0000 | 2·0000 | 0·6 | 5·7732 | 3·6516 | 3·7173 |
| 0·1 | 5·9940 | 3·9780 | 2·0713 | 0·7 | 5·6881 | 3·6342 | 4·1321 |
| 0·2 | 5·9758 | 3·9198 | 2·2703 | 0·8 | 5·5894 | 3·6481 | 4·5461 |
| 0·3 | 5·9452 | 3·8424 | 2·5639 | 0·9 | 5·4778 | 3·6946 | 4·9546 |
| 0·4 | 5·9015 | 3·7634 | 2·9187 | 1·0 | 5·3549 | 3·7734 | 5·3549 |
| 0·5 | 5·8443 | 3·6967 | 3·3089 | | | | |

**Table 5.2.11c.** Pure quadrupole resonance frequencies for nuclei of spin $\frac{5}{2}$. The frequencies are in units of $e^2Qq/20h$.

| $\eta$ | $\nu_Q$ | | $\eta$ | $\nu_Q$ | |
|---|---|---|---|---|---|
| | $\frac{5}{2} \leftrightarrow \frac{3}{2}$ | $\frac{3}{2} \leftrightarrow \frac{1}{2}$ | | $\frac{5}{2} \leftrightarrow \frac{3}{2}$ | $\frac{3}{2} \leftrightarrow \frac{1}{2}$ |
| 0·0 | 6·0000 | 3·0000 | 0·6 | 5·6532 | 3·9992 |
| 0·1 | 5·9879 | 3·0326 | 0·7 | 5·5601 | 4·2984 |
| 0·2 | 5·9526 | 3·1282 | 0·8 | 5·4670 | 4·6166 |
| 0·3 | 5·8972 | 3·2809 | 0·9 | 5·3767 | 4·9489 |
| 0·4 | 5·8258 | 3·4827 | 1·0 | 5·2915 | 5·2915 |
| 0·5 | 5·7431 | 3·7249 | | | |

**Table 5.2.11d.** Pure quadrupole resonance frequencies (MHz).

| Compound | Nucleus | Spin | Frequency (MHz) |
|---|---|---|---|
| γ-Picoline | $^{14}N$ | 1 | 3·6883 |
| | | | 2·9325 |
| Benzonitrile | $^{14}N$ | 1 | 3·0183 |
| | | | 2·8098 |
| *N*-Chloropiperidine | $^{14}N$ | 1 | 5·9881 |
| | | | 3·3886 |
| | | | 2·5995 |
| *p*-Iodonitrobenzene | $^{127}I$ | $\frac{5}{2}$ | 565·82 |
| | | | 284·36 |
| Potassium niobate | $^{93}Nb$ | $\frac{9}{2}$ | 3·648 |
| | | | 3·030 |
| | | | 2·527 |
| | | | 2·085 |
| Triphenylbismuth | $^{209}Bi$ | $\frac{9}{2}$ | 29·785 |
| | | | 55·214 |
| | | | 83·516 |
| | | | 111·438 |
| Niobium pentachloride | $^{93}Nb$ | $\frac{9}{2}$ | 12·9032 |
| | | | 9·5612 |
| | | | 6·0621 |
| | | | 5·4228 |
| Lanthanum trifluoride | $^{139}La$ | $\frac{7}{2}$ | 3·414 |
| | | | 2·736 |
| | | | 2·229 |
| *o*-Iodobenzoic acid | $^{127}I$ | $\frac{5}{2}$ | 573·4 |
| | | | 294·6 |
| Antimony triiodide | $^{127}I$ | $\frac{5}{2}$ | 254·6 |
| | | | 174·5 |
| Bismuth trichloride | $^{209}Bi$ | $\frac{9}{2}$ | 51·740 |
| | | | 37·340 |
| | | | 31·876 |
| | | | 25·131 |

also have appreciable intensity. Their frequencies may be obtained by appropriate combinations of the above expressions or tabulated values. Observed pure quadrupole resonance frequencies are given in table 5.2.11d. Calculate the corresponding coupling constant and asymmetry parameter.

**Solution**

The quadrupole coupling constants and asymmetry parameters are shown in table 5.2.11e. Most of these may be obtained straightforwardly from the tables, the most convenient method being to plot the theoretical frequency ratios as a function of $\eta$ and compare with experiment to determine $\eta$ and hence the coupling constant. Care must however be exercised in assigning transitions when the asymmetry parameter is large and $I = \frac{7}{2}$ and $\frac{9}{2}$ (potassium niobate, lanthanum trifluoride, and bismuth trichloride). In these cases only the correct assignment gives a consistent value of $\eta$. The highest frequency can always be assigned to the $|m| = I \leftrightarrow |m| = I-1$ transition and the easiest procedure is to compare the theoretical and experimental ratios between this and the other frequencies. Note that when $I = 1$ the $\Delta m = \pm 2$ transition is relatively intense and has been included in the data for *N*-chloropiperidine.

**Table 5.2.11e.** Quadrupole coupling constants and asymmetry parameters.

| Compound | Nucleus | Spin | $e^2Qq_{zz}$ (MHz) | $\eta$ |
|---|---|---|---|---|
| $\gamma$-Picoline | $^{14}N$ | 1 | 4·4140 | 0·3424 |
| Benzonitrile | $^{14}N$ | 1 | 3·8854 | 0·1073 |
| *N*-Chloropiperidine | $^{14}N$ | 1 | 6·2512 | 0·8317 |
| *p*-Iodonitrobenzene | $^{127}I$ | $\frac{5}{2}$ | 1887·7 | 0·0630 |
| Potassium niobate | $^{93}Nb$ | $\frac{9}{2}$ | 23·12 | 0·80 |
| Triphenylbismuth | $^{209}Bi$ | $\frac{9}{2}$ | 669·06 | 0·09 |
| Niobium pentachloride | $^{93}Nb$ | $\frac{9}{2}$ | 78·08 | 0·3225 |
| Lanthanum trifluoride | $^{139}La$ | $\frac{7}{2}$ | 8·55 | 0·78 |
| *o*-Iodobenzoic acid | $^{127}I$ | $\frac{5}{2}$ | 1919 | 0·146 |
| Antimony triiodide | $^{127}I$ | $\frac{5}{2}$ | 895·8 | 0·565 |
| Bismuth trichloride | $^{209}Bi$ | $\frac{9}{2}$ | 318·8 | 0·555 |

**5.2.12**

(a) The $^{35}Cl$ pure quadrupole resonance frequencies of two crystalline modifications of phosphorus pentachloride are given in table 5.2.12a. Suggest possible structures for these two phases on the basis of these data.

(b) Suggest a structure for the 1 : 1 complex between $PCl_5$ and $SbCl_5$ on the basis of the $^{35}Cl$ resonance frequencies for this complex and for $SbCl_5$ (table 5.2.12b), and those for $PCl_5$ in table 5.2.12a.

(c) Comment on the crystal and molecular structures of $BI_3$ and $AlI_3$ on the basis of the $^{127}I$ resonance frequencies shown in table 5.2.12c.

**Table 5.2.12a.** $^{35}Cl$ resonance frequencies (MHz) for $PCl_5$.

| α phase | | β phase |
|---|---|---|
| 32·601 | 30·567 | 33·751 |
| 32·426 | 30·455 | 29·274 |
| 32·385 | 30·047 | 22·242 |
| 32·280 | 30·047 | |
| | 29·713 | |
| | 28·396 | |

**Table 5.2.12b.** $^{35}Cl$ resonance frequencies (MHz) for $SbCl_5$ and $PCl_5.SbCl_5$.

| $SbCl_5$ | $PCl_5.SbCl_5$ | |
|---|---|---|
| 30·4 | 32·478 | 25·60 |
| 28·3 | 32·275 | 24·96 |
| 27·9 | | 24·03 |

**Table 5.2.12c.** $^{127}I$ resonance frequencies (MHz) for $BI_3$ and $AlI_3$.

| $BI_3$ | | $AlI_3$ | |
|---|---|---|---|
| 111·320 | 201·380 | 131·844 | 263·920 |
| | | 131·371 | 263·228 |
| | | 112·314 | 215·614 |

**Solution**

(a) α Phase. The frequencies fall into two groups: four resonances in the range 32·28–32·60 MHz and six in the range 28·4–30·6 MHz. This suggests that two different species are present, one containing four (or perhaps two) chlorine atoms and the other six (or three) chlorine atoms. The resonance of course corresponds to the well-known structure of $PCl_4^+PCl_6^-$, the higher frequencies being associated with the $PCl_4^+$ ion.
β Phase. $PCl_5$ in the vapour phase has the trigonal bipyramidal structure. The β phase is a molecular crystal. The more axial chlorine atoms have the lower frequency (22·42 MHz), the two others are in the equatorial position, and the rather wide range of frequencies indicates considerable intermolecular interaction. The structure is probably isomorphous with that of $SbCl_5$ which is known to be a molecular crystal and whose frequencies are given in part (b).
(b) The two high frequencies at 32 MHz undoubtedly stem from the $PCl_4^+$ ion, and the lower frequencies can be attributed to $SbCl_6^-$.
(c) $BI_3$ has a planar molecule and crystallizes as a molecular crystal. The large asymmetry parameter is characteristic of pronounced double-bond

character between the iodine and the boron atoms, in agreement with Pauling's (1948) original suggestion.

In $AlI_3$ the frequencies fall in two groups, that at higher frequency exhibiting an almost negligible asymmetry parameter while that at lower frequency has a pronounced asymmetry. These two sets of resonances arise respectively from the outer and bridging iodine atoms in the dimeric structure $Al_2I_6$ (cf. $Ga_2Cl_6$ in problem 5.2.6).

**References**

Barnes, R. G., Segel, S. L., 1956, *J. Chem. Phys.,* **25**, 180.
Barret, A. H., Mandel, M., 1958, *Phys. Rev.,* **109**, 1572.
Bayer, H., 1951, *Z. Phys.,* **130**, 227.
Bersohn, R., 1954, *J. Chem. Phys.,* **22**, 2078.
Buck, P., Rabi, I. I., Senitzky, B., 1956, *Phys. Rev.,* **104**, 553.
Cornwell, C. D., Yamasaki, R. S., 1957, *J. Chem. Phys.,* **27**, 1060.
Dailey, B. P., Townes, C. H., 1955, *J. Chem. Phys.,* **23**, 118.
Daly, R. T., Jr., Holloway, J. H., 1954, *Phys. Rev.,* **96**, 539.
Gordy, W., 1955, *Discuss. Faraday Soc.,* **19**, 14.
Guibé, L., Lucken, E. A. C., 1968, *Mol. Phys.,* **14**, 79.
Jaccarino, V., King, J. G., 1951, *Phys. Rev.,* **63**, 471.
Jaccarino, V., King, J. G., Satten, R. H., Stroke H. H., 1954, *Phys. Rev.,* **94**, 1798.
King, J. G., Jaccarino, V., 1954, *Phys. Rev.,* **94**, 1610.
Kusch, P., Eck, T. G., 1954, *Phys. Rev.,* **94**, 1799.
Kushida, T., Benedek, G. B., Bloembergen, N., 1956, *Phys. Rev.,* **104**, 1364.
Lew, H., Wessel, G., 1953, *Phys. Rev.,* **90**, 1.
Lide, D. R., 1965, *J. Chem. Phys.,* **42**, 1013.
Lucken, E. A. C., 1969, *Nuclear Quadrupole Coupling Constants* (Academic Press, London).
Pauling, L., 1948, *The Nature of the Chemical Bond* (Cornell University Press, Ithaca, NY).
Perl, M., Rabi, I. I., Senitzky, B., 1955, *Phys. Rev.,* **98**, 611.
Peterson, G. M., Bridenbaugh, P. M., 1969, *J. Chem. Phys.,* **51**, 238.
Pritchard, H. O., Skinner, H. A., 1955, *Chem. Rev.,* **55**, 755.
Ragle, J. L., 1959, *J. Phys. Chem.,* **63**, 1395.
Senitzky, B., Rabi, I. I., 1956, *Phys. Rev.,* **103**, 315.
Townes, C. H., Dailey, B. P., 1949, *J. Chem. Phys.,* **17**, 782.
Wessel, G., 1953, *Phys. Rev.,* **92**, 1582.
Yamasaki, R. S., Cornwell, C. D., 1959, *J. Chem. Phys.,* **30**, 1265.

## 5.3 Mössbauer spectroscopy

**J J Zuckerman, N W G Debye¶** University of Oklahoma

### 5.3.1

Discuss critically the following statements:

(a) "Apart from mass and electric charge, nuclear properties are of little interest to the chemist."

(b) "Radioactive decay phenomena are independent of the chemical environment of the nucleus."

**Solution**

(a) Today a broad range of phenomena which arise from interactions of nuclei with their electron shells is known. These phenomena include:

(i) The small shift in optical spectra seen on isotopic substitution (said to result from the different dimensions of the nuclei).

(ii) The even smaller isomer shift in optical spectra produced by an excited (isomeric) nucleus [arising from the same cause as in (i)].

(iii) The chemical shift and the splitting of nuclear magnetic resonance lines (due to interactions between the nuclei, the valence electrons, and an applied magnetic field).

(iv) The dependence of the observed molecular rotational, nuclear magnetic resonance, or nuclear quadrupole resonance spectra on the symmetry of the chemical bonds around the nucleus under study (due to interaction between the nuclear electric quadrupole moment and electric field gradients).

(b) Despite the disparity in magnitude between the energies involved in nuclear decay processes and chemical binding forces, several examples are now known of the effect of chemical environment on nuclear transformations. These include:

(i) Changes in the lifetimes of radioactive decay by electron-capture processes on passing from one chemical situation to another (the probability of electron capture is proportional to the electron density at the nucleus; for example, the half-life of $^{7}Be$ for electron capture is 0·08% greater in $BeF_2$ than in beryllium metal).

(ii) Changes in the rate of isomeric nuclear transitions accompanied by strong internal electron conversion (the rate depends on the electronic states; for example, the half-life for internal conversion in $^{99}Tc^{m}$ is 0·27% greater in TcS than in $KTcO_3$).

(iii) Dependence of the mean lifetime of positrons and their mode of annihilation upon the electron density in the stopping material.

(iv) Dependence of the annihilation rate of the transient, bound species, $e^{+}e^{-}$, upon the oxidizing power of the medium, the presence of free radicals, etc.

¶ Present address: Department of Chemistry, Towson State University, Baltimore, Maryland.

(v) Dependence of the degree of anisotropy in cascade reactions on the medium—small for metals and ionic solids, large for liquids.

(vi) Dependence of the depolarization of muons upon the chemical state of the medium.

(vii) Dependence of the energies of $\gamma$ rays emitted by nuclei in excited states (the Mössbauer effect).

**5.3.2**

What considerations make possible the use of a nuclide in Mössbauer-effect studies?

**Solution**

The Mössbauer effect has been observed for at least thirty-three elements and predicted for an additional seventeen on the basis of the following factors:

(a) *The lifetime of the excited state.* The width of the resonance line is inversely proportional to the lifetime of the excited state; a very long lifetime may produce a linewidth so narrow that observation will be difficult if not impossible. Suitable lifetimes are in the range $10^{-6}-10^{-12}$ s.

(b) *The energy of the transition.* Acceptable $\gamma$-ray energies are found in the 5–2000 keV range. With high-energy $\gamma$ radiation the recoil effects are such that the recoil-free fraction will be low even at cryogenic temperatures, and observation of the Mössbauer event becomes unlikely. For very-low-energy $\gamma$ rays the problems of transmission through absorbers are aggravated. In addition it becomes a much more difficult matter to resolve the energy of interest from the general low-energy background.

In connection with both (a) and (b) above, it is of interest to examine the energy definition, $\Gamma/E$. Short-lifetime excited states will give rise to $\Gamma/E$ values too large and resonance selectivity would be lost. An extremely small $\Gamma/E$ ratio would reduce the probability of recoil-free emission and the possibility of observation would be slight. The most promising range for the $\Gamma/E$ ratio is $10^{-10}$ to $10^{-14}$.

(c) *The natural abundance of the ground-state absorber nuclei.* To avoid problems involved in isotopic enrichment, the natural abundance of the Mössbauer nuclide should be several percent. A high natural abundance of the precursor nuclide will also contribute to the ease of production of source material.

(d) *The lifetime of the precursor state.* It is desirable to have a reasonably long-lived precursor nuclide which, once prepared, can be utilized in a series of Mössbauer experiments.

(e) *The absence of interfering radiation.* Resolution of the Mössbauer $\gamma$ ray from other x and $\gamma$ radiation will be more difficult the closer this interference is to the energy of interest.

**5.3.3**

Which of the following materials are good host matrices for a Mössbauer nuclide to be used as a source material?

(i) Iron foil; (iii) tetraphenyltin;
(ii) tin(IV) chloride; (iv) $BaSnO_3$.

State the reasons for your decision.

**Solution**

The ideal host matrix is nonmagnetic material with the highest Debye temperature, where the Mössbauer nuclide occupies a site of cubic symmetry. These requirements ensure that there is no magnetic splitting of the source line, that there is a high recoil-free fraction, and that there is no quadrupole splitting of the source line. The source material should also be chemically inert. The characteristics of the radiation emitted should not be sensitive to the method of preparation or to the history of the particular source, which would render difficult the comparison of data with other investigators using the same type of source material. The radiation, for example, apart from strength, should not change with time. The host material itself should not give rise to interfering x rays or photoelectric or Compton scattering. Since broadening of the resonance line can come about because of the finite thickness of the source, a high specific activity material is desirable so that a thin source can be fabricated. Radioactive impurities will contribute to broadening or to new lines, and so the source material should be chemically pure in the Mössbauer element. Solid-state defects can also broaden resonance lines, so that the ideal source would consist of perfect crystals if crystalline, or a uniform glass if amorphous.

Thus iron foil is not a good source material since it shows magnetic splitting, is reactive in air and the structure is a function of the impurities. Tin(IV) chloride gives a singlet line but is liquid under ambient conditions and is reactive. Tetraphenyltin has a low recoil-free fraction at room temperature. $CaSnO_3$ is a good source, since it gives a singlet line which is narrow, has a high recoil-free fraction, and is inert.

**5.3.4**

In which of the following phases should the Mössbauer effect be observable?

(i) Polycrystalline solid; (v) chemisorbed gas;
(ii) amorphous solid; (vi) mobile liquid;
(iii) gas; (vii) thixotropic material.
(iv) single crystal;

**Solution**

Phases (i), (ii), (iv), (v), (vii).

The recoil-free fraction is given by:

$$f' = \exp\left[-\frac{\langle x^2 \rangle}{\lambda\!\!\!^-{}^2}\right],$$

where $\langle x^2 \rangle$ is the mean square amplitude of the nuclear vibration in the direction of the $\gamma$ ray, of wavelength $\lambda$, averaged over the lifetime of the Mössbauer event, and $\lambda\!\!\!^- = \lambda/2\pi$. Thus unbounded motion as in a gas or mobile liquid will cause $f'$ to vanish. Structural periodicity is not required for the observation of the Mössbauer effect.

**5.3.5**

Compounds of the formula $R_2Sn$, where R is an organic group, have been formulated as diorgano-derivatives of tin(II). The Mössbauer spectra of these materials show isomer shifts in the range $1{\cdot}3$–$1{\cdot}6$ mm s$^{-1}$ with respect to $SnO_2$. Discuss the nature of the compounds as revealed by the Mössbauer spectra.

**Solution**

The Mössbauer isomer shifts are in the range associated with tin(IV) compounds; the $R_2Sn$ materials are polymeric and should be reformulated as $[R_2Sn]_n$. Chemical evidence suggests the presence of Sn—Sn bonds, and this has been confirmed by crystal structure determination.

**5.3.6**

Study of both the $^{119}Sn^m$ ($E_\gamma = 23{\cdot}8$ keV) and $^{129}I$ ($E_\gamma = 26{\cdot}8$ keV) recoil-free fractions, $f'$, in $SnI_4$ as a function of temperature enables calculation of the Debye temperatures, $\Theta$, at the two lattice sites from the equation

$$f' \approx \exp(-6E_R T/k\Theta^2)$$

which holds for $T > \frac{1}{2}\Theta$. The $\Theta$ values are 166 K measured for $^{119}Sn$ and 85 K for $^{129}I$ (Bukshpan and Herber, 1967). A second isotope of iodine, $^{127}I$ ($E_\gamma = 57{\cdot}6$ keV), is also a Mössbauer nuclide. Assuming that the same $\Theta$ holds for the two iodine isotopes,

(a) derive the temperature dependence of the recoil-free fraction for the three nuclei $^{119}Sn$, $^{129}I$, and $^{127}I$;

(b) determine which of these nuclei may be expected to exhibit an observable Mössbauer effect at 100 K.

**Solution**

(a) The recoil energy, $E_R$, is given by

$$E_R = \frac{E_\gamma^2}{2mc^2},$$

where $E_\gamma$ is the $\gamma$-ray energy, $m$ is the nuclear mass, and $c$ is the speed of light.

$$E_R\ (^{119}Sn) = 4{\cdot}098 \times 10^{-22}\ J\ ,$$
$$E_R\ (^{127}I) = 2{\cdot}249 \times 10^{-21}\ J\ ,$$
$$E_R\ (^{129}I) = 4{\cdot}791 \times 10^{-22}\ J\ .$$

The recoil-free fractions are then given by:

$$f'(^{119}Sn) = \exp(-0{\cdot}00647T)\ ,$$
$$f'(^{127}I) = \exp(-0{\cdot}1352T)\ ,$$
$$f'(^{129}I) = \exp(-0{\cdot}0288T)\ .$$

(b)

$$f'(^{119}Sn) \approx 0{\cdot}5 \rightarrow \text{observable}\ ,$$
$$f'(^{127}I) = \exp(-13{\cdot}52) \ll 1 \rightarrow \text{not observable}\ ,$$
$$f'(^{129}I) \approx 0{\cdot}06 \rightarrow \text{observable}\ .$$

**5.3.7**

Comment on the following experimental findings.

(a) Changing substituent groups in a compound containing the Mössbauer nuclides $^{40}K$, $^{57}Fe$, $^{119}Sn$, and $^{129}I$ increases all the isomer-shift values. Treating the same compound in which the $^{127}I$ isotope had been substituted resulted in increases in all the isomer shifts, but a decrease for $^{127}I$.

(b) Iodine-129 undergoes $\beta$-decay to the excited Mössbauer level of $^{129}Xe$. The square planar compound $^{129}I\ Cl_4^-$ gives rise to a $^{129}Xe$ spectrum ($E_\gamma$ = 39·6 keV) which is a doublet (Perlow and Perlow, 1964).

(c) The $^{57}Fe$ spectrum of $Fe_3(CO)_{12}$ consists of three lines of equal intensity (Herber *et al.*, 1963).

**Solution**

(a) The isomer shift is directly related to the difference in radius between the ground and excited nuclear states, $\Delta R$. The sign of this change is opposite for $^{127}I$ and $^{129}I$.

(b) The spectrum arises from neutral $XeCl_4$ which has yet to be prepared chemically. The doublet spectrum rules out a tetrahedral structure, and is consistent with the square planar structure of the precursor $ICl_4^-$.

(c) Among possible structures are: linear $(OC)_4Fe(OC)_2Fe(OC)_2Fe(CO)_4$ (I), and $(OC)_3Fe(OC)_3Fe(OC)_3Fe(CO)_3$ (II) designated as 4.2.2.4 and 3.3.3.3, and the cyclic forms:

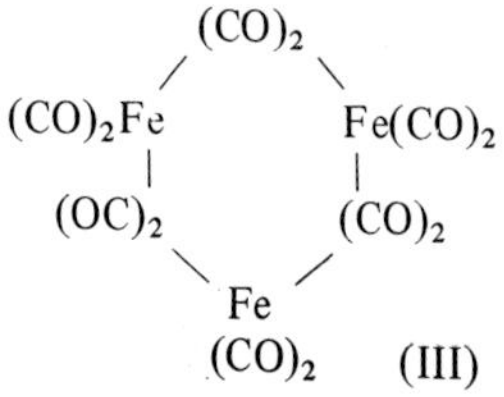

```
           (CO)2                                  (CO)4
          /     \                                  Fe
(CO)2Fe           Fe(CO)2                        /    \
   |                 |                   (OC)3Fe        Fe(CO)3
 (OC)2             (CO)2                         \    /
          \     /                                 (CO)2
            Fe
          (CO)2        (III)                               (IV)
```

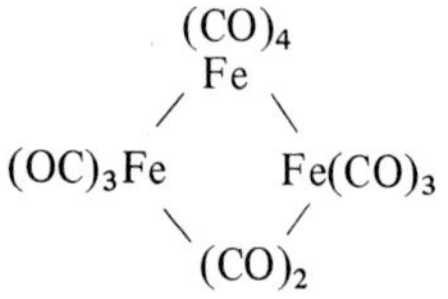

Considering the spectrum as the result of the superposition of a singlet and a doublet would rule out structure (III) in which the three iron atoms are equivalent. The structure currently favored on the basis of x-ray structural determinations of related compounds is (IV).

**5.3.8**
The two Mössbauer-active iodine nuclei have the following properties

$$^{127}\mathrm{I}:\ E_\gamma = 58\ \mathrm{keV},\ \text{and}\ \frac{\Delta R}{R} = +4 \times 10^{-5}\ ;$$

$$^{129}\mathrm{I}:\ E_\gamma = 28\ \mathrm{keV},\ \text{and}\ \frac{\Delta R}{R} = -5 \times 10^{-5}\ .$$

The spectrum of $Na_3H_2{}^{127}IO_6$ consists of a line with an isomer shift of $+1 \cdot 2$ mm $s^{-1}$ with respect to a $Zn^{127}Te$ source. Calculate the corresponding parameter for $Na_3H_2{}^{129}IO_6$ relative to $Zn^{129}I$.

**Solution**
The isomer shift, $\delta$, is given by

$$\delta = \text{const.} \frac{\Delta R}{R} [|\Psi_a(0)|^2 - |\Psi_s(0)|^2]\ ,$$

where the constant and electric field density terms are assumed to be invariant with isotopic substitution. Thus

$$\frac{\delta^{129}}{\delta^{127}} = \left(\frac{\Delta R}{R}\right)^{129} \Bigg/ \left(\frac{\Delta R}{R}\right)^{127}\ ,$$

and $\delta^{129} = -1 \cdot 5$ mm $s^{-1}$ .

**5.3.9**
Consider a treatment in which the atoms or ligands attached to a Mössbauer atom in an array of octahedral symmetry are approximated by point charges, $Q_i$, arranged at distances $R_i$ from the center. Taking examples of the type $Mab_5$, $Ma_2b_4$ and $Ma_3b_3$ in which the Mössbauer atom is in an axially symmetric field, the quadrupole splitting ($q$) is given by

$$q = \left|\frac{eQV_{zz}}{2}\right|\ ,$$

where $V_{zz}$ is the largest element of the diagonalized electric field gradient (e.f.g.) tensor whose elements are:

$$V_{xx} = \sum_i \frac{Q_i}{R_i^3}(1 - 3\sin^2\theta_i \cos^2\phi_i)\ ,$$

$$V_{yy} = \sum_i \frac{Q_i}{R_i^3}(1 - 3\sin^2\theta \sin^2\phi_i)\ ,$$

$$V_{zz} = \sum_i \frac{Q_i}{R_i^3}(1-3\cos^2\theta_i)\,,$$

$$V_{xy} = V_{yz} = -\sum_i \frac{Q_i}{R_i^3}(3\sin^2\theta_i \sin\phi_i \cos\phi_i)\,,$$

$$V_{xz} = V_{zx} = -\sum_i \frac{Q_i}{R_i^3}(3\sin\theta_i \cos\theta_i \cos\phi_i)\,,$$

$$V_{yz} = V_{xy} = -\sum_i \frac{Q_i}{R_i^3}(3\sin\theta_i \cos\theta_i \sin\phi_i)\,,$$

where the summations are over the six point charges. On the basis of this model, calculate the relative values of the quadrupole splittings expected for $Mab_5$, *cis*- and *trans*-$Ma_2b_4$, and for *facial*-$Ma_3b_3$.

**Solution**

The spherical coordinates of the six point charges surrounding the central Mössbauer atom are:

| charge | $\theta$ | $\phi$ | $r$ |
|---|---|---|---|
| $Q_1$ | 90° | 0° | $R_1$ |
| $Q_2$ | 90° | 180° | $R_2$ |
| $Q_3$ | 90° | 90° | $R_3$ |
| $Q_4$ | 90° | 270° | $R_4$ |
| $Q_5$ | 0° | 0° | $R_5$ |
| $Q_6$ | 180° | 0° | $R_6$ |

The coordinate system must be chosen so that the cross terms (for example, $V_{xy}$, $V_{xz}$, $V_{yz}$) vanish in order to diagonalize the e.f.g. tensor. Performing the summations over the point charges yields:

$$V_{xx} = \sum_{i=1}^{6} \frac{Q_i}{R_i^3} - 3\sum_{i=1}^{2} \frac{Q_i}{R_i^3}\,,$$

$$V_{yy} = \sum_{i=1}^{6} \frac{Q_i}{R_i^3} - 3\sum_{i=3}^{4} \frac{Q_i}{R_i^3}\,,$$

$$V_{zz} = \sum_{i=1}^{6} \frac{Q_i}{R_i^3} - 3\sum_{i=5}^{6} \frac{Q_i}{R_i^3}\,.$$

**Table 5.3.9.**

| Structure | $V_{xx}$ | $V_{yy}$ | $V_{zz}$ | Relative $q$ |
|---|---|---|---|---|
| $MaB_5$ | $P_a - P_b$ | $P_a - P_b$ | $2(P_b - P_a)$ | 2 |
| *cis*-$Ma_2b_4$ | $P_b - P_a$ | $P_b - P_a$ | $2(P_a - P_b)$ | 2 |
| *trans*-$Ma_2b_4$ | $2(P_a - P_b)$ | $2(P_a - P_b)$ | $4(P_a - P_b)$ | 4 |
| *facial*-$Ma_3b_3$ | 0 | 0 | 0 | 0 |

These equations may be evaluated for the given geometries, and the resultant e.f.g. elements are given in table 5.3.9 in terms of a parameter $P_i$ defined by

$$P_i \equiv \frac{Q_i}{R_i^3} \; .$$

**5.3.10**

(a) The $^{57}$Fe Mössbauer transition is from an $I = \frac{1}{2}$ to an $I = \frac{3}{2}$ state. Derive the splitting pattern for iron in a magnetic field.

(b) In the Mössbauer spectrum of metallic iron the resonances are found at $-5\cdot07$, $-2\cdot87$, $-0\cdot58$, $+1\cdot10$, $+3\cdot34$, and $+5\cdot58$ mm s$^{-1}$ relative to a sodium nitroprusside standard. What does this imply about the site symmetry of the iron atoms in the sample? What are the magnitudes of the magnetic hyperfine splitting constants of the $I = \frac{3}{2}$ and $I = \frac{1}{2}$ states?

(c) In what way can this phenomenon be used to measure low temperatures?

**Solution**

(a) For a nuclear state of spin $I$, there are $2I+1$ allowed values of the magnetic quantum number, $m_I$. In a magnetic field the permitted energy levels are $-gM_\mu Hm_I$, where $H$ is the magnetic field strength, $g$ the gyromagnetic ratio and $M_\mu$ the nuclear magneton. For the transition $I = \frac{1}{2}$ to $I = \frac{3}{2}$, the magnetic field gives rise to the splitting of these levels into two and four energy levels respectively. The selection rules for magnetic dipole transitions give that $\Delta m = 0, \pm 1$. For the $^{57}$Fe nucleus the six allowed transitions are given in order of increasing energy by the sequence:

(1) $-\frac{1}{2} \rightarrow -\frac{3}{2}$ : $E_t$
(2) $-\frac{1}{2} \rightarrow -\frac{1}{2}$ : $E_t + \beta$
(3) $-\frac{1}{2} \rightarrow +\frac{1}{2}$ : $E_t + 2\beta$
(4) $+\frac{1}{2} \rightarrow -\frac{1}{2}$ : $E_t + \beta + \alpha$
(5) $+\frac{1}{2} \rightarrow +\frac{1}{2}$ : $E_t + 2\beta + \alpha$
(6) $+\frac{1}{2} \rightarrow +\frac{3}{2}$ : $E_t + 3\beta + \alpha$ .

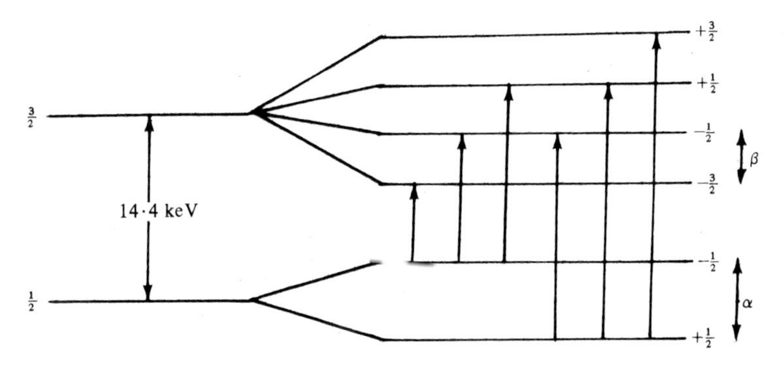


(b) The energies of the six lines, assuming cubic symmetry, are given above. The difference in energy of the two sets of outer lines should be $\beta$; and between the two sets of lines second and fourth from either end, $\alpha$.

The observation of such expected symmetry in the Mössbauer spectrum implies that the site symmetry is indeed close to cubic. The differences in the experimental transitions are found to be:

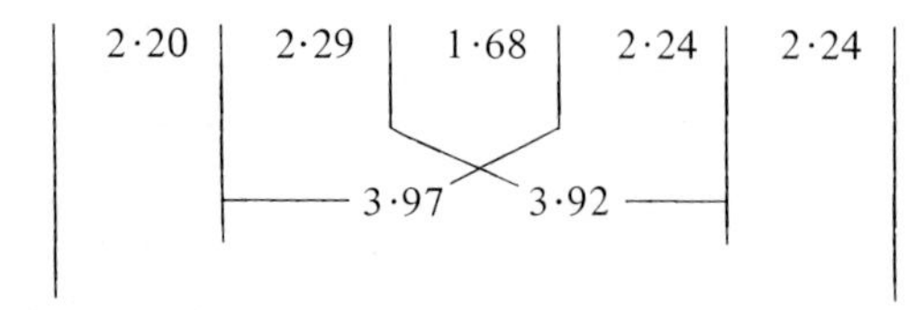


and $\alpha$ and $\beta$ are given by:

$$\beta = \tfrac{1}{4}[2\cdot20+2\cdot29+2\cdot24+2\cdot25] = 2\cdot24 \text{ mm s}^{-1} ,$$
$$\alpha = \tfrac{1}{2}[3\cdot97+3\cdot92] = 3\cdot94 \text{ mm s}^{-1} .$$

(c) Note that the energies $\alpha$ and $\beta$ can be comparable to $kT$ at very low, but accessible, temperatures (fractions of a Kelvin). At these temperatures the populations of the six states become determined by the Boltzmann distribution function, and the intensities of the six lines become skewed in favor of the lower energy transitions.

**5.3.11**

The $^{57}Fe$ Mössbauer spectrum of $Fe(NH_4)_2(SO_4)_2.6H_2O$ exhibits a simple doublet with quadrupole splitting and isomer shift characteristic of $Fe^{2+}$. The spectrum of $^{57}Fe$ produced in the decay of $^{57}Co$-substituted $Fe(NH_4)_2(SO_4)_2.6H_2O$ consists of a similar doublet plus a second doublet with parameters characteristic of $Fe^{3+}$. Similar substitution in conductors does not change the observed spectra. Comment on these results.

**Solution**

The electron capture of $^{57}Co$ during the $^{57}Co \rightarrow {}^{57}Fe$ decay is usually followed by Auger electron emission resulting in highly ionized atomic states. These anomalous charge states generally decay very rapidly by capturing electrons from neighboring atoms, but if the Auger cascade and subsequent electron transfer processes are slow compared to the half-life of the 14·4 keV $^{57}Fe$ excited state, then the anomalous states may be observed in the Mössbauer spectrum. Such conditions may be attained by incorporating the $^{57}Co$ in insulators such as $Fe(NH_4)_2(SO_4)_2.6H_2O$, but in conductors the lifetimes of the anomalous charge states are, as expected, too short to be observed.

**References**

Bukshpan, S., Herber, R. H., 1967, *J. Chem. Phys.,* **46,** 3375.
Herber, R. H., Kingston, W. R., Wertheim, G., 1963, *Inorg. Chem.,* **2,** 153.
Perlow, G. J., Perlow, M. R., 1964, *J. Chem. Phys.,* **41,** 1157.

# 6 Mass spectrometry

## 6.1 Mass spectrometry

**J R Gilbert** University of Essex

### 6.1.0

A systematic method of identifying a molecule from its mass spectrum has not yet been developed. However, there are certain guidelines that can be followed:

(1) Identify the mass of the molecular ion, $M$, and determine the possible elemental formulae which could represent the molecular ion. The measurement of isotope ratios is of invaluable help in determining the elemental compositions of ions. In many cases the molecular ion will be fairly obvious; if this is not so we look to see if ions of lower mass represent the loss of reasonable fragments from the molecular ions, e.g. if the next lowest mass ion corresponds to a loss of 8 mass units then the ion under consideration cannot be a molecular ion as it is most unlikely that an ion will easily lose eight hydrogen atoms.

(2) Determine the total number of rings plus double bonds.

(3) Identify both the most abundant fragment ions and the corresponding neutral fragments.

(4) Postulate molecular structures and test against reference spectra.

All the spectra given in these problems are of compounds which contain two or more of the following elements: C, H, N, O, Cl, Br, and I. The natural isotopic abundances, which will be useful in the problems, are given in table 6.1.0. The spectra are presented in a normalised form, which makes the most intense peak (called the base peak) in any spectrum equal to a height of 100 units. The peaks are identified by $m/e$, the ratio of the mass (relative to $^{12}C$ = 12) of the ion to its charge. It has recently become customary to express the mass ratio as $m/z$, where $z$ is the number of elementary charges on the ion. The notation $m/e$ has been retained, as all the ions discussed here are assumed to carry a single charge $e$.

**Table 6.1.0.**

| Element | Mass | Relative abundance | Element | Mass | Relative abundance |
|---|---|---|---|---|---|
| H | 1 | 100 | Cl | 35 | 100 |
| C | 12 | 100 | Cl | 37 | 32·7 |
| C | 13 | 1·08 | Br | 79 | 100 |
| N | 14 | 100 | Br | 81 | 97·5 |
| O | 16 | 100 | I | 127 | 100 |
| O | 18 | 0·2 | | | |

### 6.1.1

Deduce the structures of the compounds (a)–(h) giving the following 70 eV mass spectra (base peaks are denoted by bold type):

| (a) *m/e* | Relative intensity | *m/e* | Relative intensity | *m/e* | Relative intensity |
|---|---|---|---|---|---|
| 12 | 0·1 | 40 | 2·3 | 58 | 0·5 |
| 13 | 0·1 | 41 | 42·0 | 59 | 0·0 |
| 14 | 0·6 | 42 | 59·9 | 60 | 0·0 |
| 15 | 4·9 | **43** | **100·0** | 61 | 0·1 |
| 16 | 0·1 | 44 | 3·3 | 62 | 0·5 |
| 25 | 0·3 | 45 | 0·0 | 63 | 0·1 |
| 26 | 4·2 | 48 | 0·0 | 64 | 0·0 |
| 27 | 35·3 | 49 | 0·2 | 65 | 0·0 |
| 28 | 7·0 | 50 | 0·5 | 66 | 0·0 |
| 29 | 25·0 | 51 | 0·9 | 67 | 0·9 |
| 30 | 0·6 | 52 | 0·5 | 68 | 0·5 |
| 31 | 0·3 | 53 | 1·0 | 69 | 0·9 |
| 36 | 0·0 | 54 | 0·2 | 70 | 0·2 |
| 37 | 0·8 | 55 | 5·2 | 71 | 0·7 |
| 38 | 2·5 | 56 | 4·1 | 72 | 9·2 |
| 39 | 15·2 | 57 | 13·6 | 73 | 0·5 |

| (b) *m/e* | Relative intensity | *m/e* | Relative intensity | *m/e* | Relative intensity |
|---|---|---|---|---|---|
| 26 | 0·5 | 46 | 3·0 | 74 | 1·0 |
| 27 | 5·2 | 49 | 0·5 | 75 | 0·8 |
| 28 | 0·2 | 50 | 8·0 | 76 | 1·2 |
| 29 | 0·5 | 51 | 10·2 | 77 | 3·5 |
| 37 | 2·5 | 52 | 2·0 | 78 | 0·6 |
| 38 | 4·0 | 53 | 0·9 | 85 | 0·7 |
| 39 | 16·2 | 61 | 2·5 | 86 | 0·5 |
| 40 | 2·5 | 62 | 5·5 | 87 | 0·3 |
| 41 | 2·4 | 63 | 8·3 | 89 | 5·2 |
| 42 | 0·6 | 64 | 2·0 | 90 | 6·1 |
| 43 | 2·1 | 65 | 14·3 | **91** | **100·0** |
| 44 | 0·9 | 66 | 1·4 | 92 | 76·8 |
| 45 | 6·3 | 73 | 0·5 | 93 | 6·1 |

| (c) *m/e* | Relative intensity | *m/e* | Relative intensity | *m/e* | Relative intensity |
|---|---|---|---|---|---|
| 14 | 4·0 | 43 | 1·0 | 74 | 9·1 |
| 15 | 13·0 | 44 | 0·2 | 75 | 6·2 |
| 25 | 1·3 | 49 | 1·9 | 76 | 7·0 |
| 26 | 7·1 | 50 | 18·3 | 77 | 16·0 |
| 27 | 12·2 | 51 | 25·5 | 78 | 40·4 |
| 28 | 1·3 | 52 | 6·6 | 79 | 2·6 |
| 29 | 1·6 | 53 | 1·0 | 87 | 1·6 |
| 36 | 0·6 | 58 | 1·0 | 88 | 0·3 |
| 37 | 4·0 | 61 | 2·3 | 98 | 1·0 |
| 38 | 6·4 | 62 | 4·0 | 101 | 1·2 |
| 39 | 13·2 | 63 | 8·5 | 102 | 7·2 |
| 40 | 0·9 | 64 | 1·2 | 103 | 40·0 |
| 41 | 1·1 | 65 | 2·5 | **104** | **100·0** |
| 42 | 0·5 | 73 | 1·8 | 105 | 8·8 |

| (d) *m/e* | Relative intensity | *m/e* | Relative intensity | *m/e* | Relative intensity |
|---|---|---|---|---|---|
| 14 | 1·8 | 28 | 5·0 | 42 | 3·0 |
| 15 | 4·8 | 29 | 27·0 | 43 | 8·5 |
| 18 | 0·3 | 30 | 5·5 | 44 | 2·0 |
| 19 | 3·2 | **31** | **100·0** | 45 | 33·0 |
| 25 | 2·2 | 32 | 1·2 | 46 | 15·0 |
| 26 | 9·3 | 33 | 0·2 | 47 | 0·4 |
| 27 | 23·0 | 41 | 0·8 | | |

| (e) *m/e* | Relative intensity | *m/e* | Relative intensity | *m/e* | Relative intensity |
|---|---|---|---|---|---|
| 26 | 5·7 | 41 | 20·9 | 55 | 1·0 |
| 27 | 30·0 | 42 | 9·1 | 56 | 0·4 |
| 28 | 5·1 | **43** | **100·0** | 57 | 0·7 |
| 29 | 25·0 | 44 | 2·4 | 58 | 7·2 |
| 31 | 0·8 | 45 | 1·0 | 71 | 10·1 |
| 37 | 2·1 | 50 | 1·1 | 72 | 0·4 |
| 38 | 5·2 | 51 | 0·9 | 86 | 9·3 |
| 39 | 19·7 | 53 | 0·9 | 87 | 0·5 |
| 40 | 2·6 | | | | |

| (f) *m/e* | Relative intensity | *m/e* | Relative intensity | *m/e* | Relative intensity |
|---|---|---|---|---|---|
| 14 | 1·1 | 39 | 10·5 | 66 | 0·9 |
| 15 | 0·5 | 46 | 2·4 | 73 | 2·3 |
| 16 | 2·1 | 49 | 2·7 | 74 | 9·2 |
| 25 | 0·8 | 50 | 25·0 | 75 | 5·4 |
| 26 | 4·0 | 51 | 58·3 | 76 | 4·4 |
| 27 | 8·5 | 52 | 3·1 | **77** | **100·0** |
| 28 | 5·1 | 61 | 1·3 | 78 | 7·1 |
| 29 | 1·3 | 62 | 1·7 | 93 | 9·4 |
| 30 | 11·2 | 63 | 2·7 | 123 | 35·4 |
| 37 | 5·2 | 64 | 1·3 | 124 | 2·4 |
| 38 | 6·2 | 65 | 8·5 | | |

| (g) *m/e* | Relative intensity | *m/e* | Relative intensity | *m/e* | Relative intensity |
|---|---|---|---|---|---|
| 35 | 0·3 | 53 | 1·0 | 77 | 49·0 |
| 36 | 0·7 | 54 | 0·2 | 78 | 2·1 |
| 37 | 3·3 | 55 | 0·5 | 85 | 1·5 |
| 38 | 8·6 | 56 | 6·1 | 86 | 1·5 |
| 39 | 2·0 | 57 | 1·7 | 111 | 0·8 |
| 40 | 0·1 | 72 | 0·5 | **112** | **100·0** |
| 49 | 2·4 | 73 | 2·9 | 113 | 6·8 |
| 50 | 13·7 | 74 | 5·6 | 114 | 32·1 |
| 51 | 16·5 | 75 | 5·1 | 115 | 2·1 |
| 52 | 1·8 | 76 | 4·2 | | |

| (h) *m/e* | Relative intensity | *m/e* | Relative intensity | *m/e* | Relative intensity |
|---|---|---|---|---|---|
| 12 | 0·3 | **29** | **100·0** | 95 | 3·1 |
| 13 | 0·8 | 30 | 2·2 | 104 | 0·5 |
| 14 | 2·2 | 79 | 6·1 | 105 | 0·5 |
| 15 | 2·0 | 80 | 2·3 | 106 | 0·3 |
| 24 | 0·8 | 81 | 6·0 | 107 | 0·4 |
| 25 | 4·5 | 91 | 0·9 | 108 | 27·4 |
| 26 | 27·4 | 92 | 0·5 | 109 | 0·7 |
| 27 | 80·6 | 93 | 4·5 | 110 | 26·1 |
| 28 | 18·4 | 94 | 0·6 | 111 | 0·5 |

**Solution**

(a) The large peaks at *m/e* 29, 43, 57 (due to $C_nH_{2n+1}^+$ ions) and 72 suggest that this is a spectrum of an aliphatic hydrocarbon with a molecular ion of mass 72, i.e. pentane. Aldehydes and ketones with molecular weight 72 can be discounted because of the absence of peaks at *m/e* 31, 45 and 59 which are usually present in the spectra of oxygenated compounds. The $^{13}C$ isotope peak at *m/e* 73 confirms the absence of oxygen; its abundance is 5·4% of the molecular ion which is consistent with a molecular ion containing five carbon atoms. The compound must therefore be pentane, $C_5H_{12}$.

Pentane exists in three isomeric forms, shown below:

$$CH_3CH_2CH_2CH_2CH_3 \quad \text{(I)}$$

$$CH_3CH(CH_3)CH_2CH_3 \quad \text{(II)}$$

$$CH_3-C(CH_3)_2-CH_3 \quad \text{(III)}$$

The base peak at *m/e* 43, which is due to loss of $C_2H_5$ ($M-29$) from the molecular ion,

$$C_5H_{12}^{+\cdot} \rightarrow C_3H_7^{+\cdot} + C_2H_5 \ ,$$

rules out structure (III). An isomer of structure (III) is also ruled out by the size of the peak at *m/e* 57. The ion of *m/e* 57 is formed by loss of a methyl group from the molecular ion, and an isomer with such a highly branched structure would easily lose a methyl group:

$$C_5H_{12}^{+\cdot} \rightarrow C_4H_9^{+\cdot} + CH_3$$

giving a peak which would probably be the base peak in the spectrum. It is not possible to say definitively which of the remaining two structures is responsible for the spectrum, and to be certain a comparison with known spectra should be made. Unknown (a) is n-pentane (I).

(b) The molecular weight of this compound is 92. The isotope peak at *m/e* 93, which is 7·8% of the molecular ion, suggests that we are dealing with the spectrum of a compound containing seven carbon atoms. These seven carbon atoms will account for 84 mass units which leaves only 8 unaccounted for, hence, the molecular formula must be $C_7H_8$.

For an elemental formula $C_aH_bN_cO_d$ the total number of rings and double bonds, $N$, is given by:

$$N = (a - \tfrac{1}{2}b + c + 1) .$$

Applying this equation to $C_7H_8$ gives $N = (7 - \frac{1}{2} \times 8 + 1) = 4$. The ions at $m/e$ 39, 51, 65, 77, and 91 suggest that the molecule is an alkyl benzene which is in agreement with the finding of 4 rings and double bonds. The alkyl benzene is in this case toluene (methyl benzene).

The base peak at $m/e$ 91 is due to the formation of the resonance-stabilised tropylium ion,

$CH_3$ (on benzene ring)$^{+\cdot}$ → (tropylium ion, +) + H

(c) The molecular weight of this compound is 104. The peaks at $m/e$ 39, 51, 65, and 77 suggest that the compound probably contains a phenyl group. The intensity of the $^{13}C$ isotope peak at $m/e$ 105 indicates that the compound contains eight carbon atoms. Hence, compound (c) has an elemental formula $C_8H_8$. The total number of rings and double bonds is 5, which is consistent with a phenyl group and a side chain, containing 1 double bond, of $C_2H_3$. Hence this compound is vinyl benzene (styrene).

(d) This compound appears to have a molecular weight of 46. The presence of oxygen, which is suggested by the peaks at $m/e$ 31 and 45, and the $^{13}C$ isotope peak, which is consistent with the presence of two carbon atoms, means that the elemental formula of this compound must be $C_2H_6O$. There are two possible compounds with this formula, viz. dimethyl ether and ethanol. The base peak in the spectrum is at $m/e$ 31 which is formed by loss of a methyl group from the molecular ion ($M - 15$). Both dimethyl ether and ethanol can easily lose a methyl group:

$$CH_3OCH_3^{+\cdot} \rightarrow CH_3O^{+\cdot} + CH_3 ,$$
$$CH_3CH_2OH^{+\cdot} \rightarrow CH_2OH^{+\cdot} + CH_3 .$$

However, the peaks at $m/e$ 29 ($M - 17$) and 27 are consistent with the presence of an ethyl group,

$$C_2H_5OH^{+\cdot} \xrightarrow{-OH} C_2H_5^{+\cdot} \xrightarrow{-H_2} C_2H_3^{+\cdot} ,$$

thus identifying the compound as ethanol. The peak at $m/e$ 19 is of interest as it frequently occurs in the spectra of alcohols. It is a rearrangement ion due to $H_3O^+$,

$$C_2H_5OH^{\cdot+} \longrightarrow H_3O^{\cdot+} + C_2H_3 .$$

(e) The peaks at *m/e* 29, 43, 57, and 71 taken in conjunction with a molecular ion of *m/e* 86 might lead one to suppose that this compound is an aliphatic hydrocarbon, i.e. an isomer of hexane. However, the peak at *m/e* 58 makes this unlikely as this peak would be due to the $C_4H_{10}^{\cdot+}$ (butane) ion and such rearrangement ions are not found in the mass spectra of aliphatic hydrocarbons. The intensity of the $^{13}C$ isotope peak at *m/e* 87 is consistent with a $C_5$ compound. The only possible elemental formula which gives a molecular ion with five carbon atoms and a molecular weight of 86 is $C_5H_{10}O$. The presence of oxygen in the compound is confirmed by the small peaks at *m/e* 31 and 45. The similarity to an aliphatic hydrocarbon suggests that this molecule is either an aldehyde or ketone. The similarity of aldehydes and ketones to aliphatic hydrocarbons is largely due to the CO and $(CH_2)_2$ groups having the same nominal mass.

Absence of a large $M-1$ peak rules out the possibility of this compound being an aldehyde

$$RCHO^{\cdot+} \longrightarrow RCO^{\cdot+} + H .$$

There are two possible $C_5$ ketones, $CH_3COC_3H_7$ and $C_2H_5COC_2H_5$. The prominent peak at *m/e* 58 is due to a rearrangement process involving fragmentation $\beta$ to the carbonyl group and transfer of a $\gamma$-hydrogen atom:

[Structure: cyclic arrangement $R^2$–C(=O)–$CH_2$–$CH_2$–CH($R^1$)···H···O, radical cation] $\longrightarrow$ [$R^2$–C(OH)=$CH_2$, radical cation] $+ CH_2{=}CHR^1$.

In ketones the hydrogen atoms are transferred almost exclusively from the $\gamma$-position which means that this compound must be methyl n-propyl ketone. This is confirmed by the ions at *m/e* 43, 71, and 57; the former pair of ions being formed by fusion of the molecular ion at the bond $\alpha$ to the carbonyl group,

$$CH_3 \vdots CO \vdots C_3H_7^{\cdot+} \longrightarrow \underset{m/e\ 71}{C_3H_7CO^{\cdot+}} + CH_3$$

$$\longrightarrow \underset{m/e\ 43}{CH_3CO^{\cdot+}} + C_3H_7 ,$$

while the ion at *m/e* 57 is due to loss of an ethyl group ($M-29$) from the molecular ion

$$CH_3COCH_2CH_2CH_3^{\neg +} \longrightarrow CH_3COCH_2^{\neg +} + C_2H_5 \, .$$

(f) The molecular weight of this compound can be seen to be 123. A simple rule which helps with the recognition of elements in a molecular ion, states that all organic molecules having an odd molecular weight must contain an odd number of nitrogen atoms. The rule depends on the fact that the most abundant isotope of an element with even valency has an even mass while the most abundant isotope of most elements of odd valency has an odd mass. Nitrogen is an exception to this, in that it has an odd valency and its most abundant isotope is of even mass. It follows that compound (f) must therefore contain an odd number of nitrogen atoms. The peaks at *m/e* 39, 51, 65, and 77 suggest the presence of a phenyl group, which leaves a group of 46, i.e. 123 − 77, mass units unaccounted for and this group must contain an odd number of nitrogen atoms. The prominent peak at *m/e* 93, which is due to loss of 30 mass units from the molecular ion, is typical of aromatic nitro-compounds which undergo loss of NO:

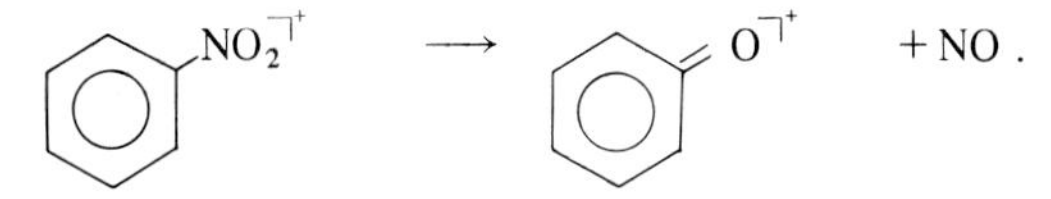


Hence this compound must be nitrobenzene.
(g) The most significant feature of this spectrum is the high-abundance peaks, two mass units apart, at *m/e* 114 and 112. These peaks have intensities in the ratio of 1 : 3, which suggests the presence of one chlorine atom. The high intensity of these two peaks also suggests that we are dealing with a stable molecular ion, i.e. probably an aromatic compound. This is confirmed by both the presence of an intense peak at *m/e* 77, due to the $C_6H_5^+$ ion, and the intensity of the $^{13}C$ isotope peak at *m/e* 115 which is consistent with a $C_6$ compound. Hence the unknown is chlorobenzene.
(h) The spectrum of this compound also shows two intense peaks, two mass units apart, at *m/e* 108 and 110. However, the ratio of the intensities is 1 : 1 which is indicative of the presence of one bromine atom. Similar pairs of peaks also occur at *m/e* 93, 95, and 79, 81. The last two confirm the presence of bromine. The peaks at *m/e* 93 and 95 are due to loss of 15 mass units which indicates the presence of a methyl group. Absence of peaks at *m/e* 31, 45, and 59 means that this compound does not contain oxygen; hence the base peak at *m/e* 29 must be due to one of the following ions. $C_2H_5$, $CH_3N$, or $N_2H$. The last two can be discounted on chemical grounds, since molecules with the constitution $CH_3NBr$ or $N_2HBr$ are unknown. Hence the compound is ethyl bromide.

### 6.1.2

The compound X gave the following mass spectrum.

| *m/e* | Relative intensity | *m/e* | Relative intensity | *m/e* | Relative intensity |
|---|---|---|---|---|---|
| 15 | 2·2 | 40 | 5·7 | 56 | 4·1 |
| 25 | 0·9 | 41 | 88·5 | 57 | 63·9 |
| 26 | 11·5 | 42 | 5·4 | 58 | 2·8 |
| 27 | 61·5 | 43 | 12·3 | 127 | 27·9 |
| 28 | 23·8 | 51 | 4·4 | 128 | 5·1 |
| **29** | **100·0** | 52 | 1·6 | 141 | 3·6 |
| 30 | 2·2 | 53 | 4·3 | 155 | 5·7 |
| 37 | 0·5 | 54 | 1·4 | 184 | 17·2 |
| 38 | 6·6 | 55 | 10·0 | 185 | 0·8 |
| 39 | 29·2 | | | | |

Metastable peaks were observed at *m/e* values of 15·3, 24·1, 25·1, 29·4, 35·1, and 37·1.

Using this spectrum identify compound X and show the decomposition reactions which give metastable peaks.

**Solution**

This compound has a molecular weight of 184 and the prominent peak at *m/e* 127 indicates the presence of iodine in the compound. As *m/e* 127 is $M-57$ this compound is probably a butyl iodide. The $^{13}C$ isotope peak at *m/e* 185 is consistent with a $C_4$ compound, and the peaks at *m/e* 57, 43, and 29 are typical of hydrocarbon radical ions. Hence this compound is $C_4H_9I$. Butyl iodide can exist as four isomers:

| $CH_3CH_2CH_2CH_2I$ | $(CH_3)_2CHCH_2I$ | $CH_3CH_2CHICH_3$ | $(CH_3)_3CI$ |
|---|---|---|---|
| (I) | (II) | (III) | (IV) |

The presence of a peak at *m/e* 155, due to $(X-C_2H_5)$ rules out structures (II) and (IV). To be certain which of the remaining two isomers is the unknown a comparison with a known spectrum should be made. Unknown X is n-butyl iodide.

Metastable peaks appear in a mass spectrum when ions which leave the ion source with mass $M_1$ decompose during their flight to produce an ion of lower mass $M_2$. If the decomposition takes place in the field-free region between the ion accelerator and the magnetic separator, and involves little or no release of kinetic energy, then the ions of mass $M_2$ are collected as mass $M^*$, which is given by

$$M^* \cong \frac{M_2^2}{M_1} .$$

Use of the above formula shows that the observed metastable peaks are produced by the reactions in table 6.1.2.

Table 6.1.2.

| $M^*$ | Reaction |
|---|---|
| 15·3 | $C_4H_7^+$ (55) ⟶ $C_2H_5^+$ (29) + $C_2H_2$ |
| 24·1 | $C_2H_4^+$ (28) ⟶ $C_2H_2^+$ (26) + $H_2$ |
| 25·1 | $C_2H_5^+$ (29) ⟶ $C_2H_3^+$ (27) + $H_2$ |
| 29·4 | $C_4H_9^+$ (57) ⟶ $C_3H_5^+$ (41) + $CH_4$ |
| 35·1 | $C_3H_3^+$ (39) ⟶ $C_3H^+$ (37) + $H_2$ |
| 37·1 | $C_3H_5^+$ (41) ⟶ $C_3H_3^+$ (39) + $H_2$ |

**6.1.3**

The $C_5H_6^+$ ion found in the mass spectra of phenol and aniline may have a cyclic or open-chain structure. The appearance energies of this ion from phenol and aniline are 12·50 and 12·56 eV respectively. Given that the ionisation energies of cyclopentadiene and pent-3-en-1-yne are 8·86 and 9·14 eV respectively, use the thermochemical data in table 6.1.3 to decide whether the most likely structure is cyclic or open-chain in each case (note: 1 eV = 96·5 kJ mol$^{-1}$). Comment on any assumptions you make and discuss the validity of this approach to the elucidation of ion structure.

Table 6.1.3

| Molecule | $\Delta H^{\circ}_{f,298}$ (kJ mol$^{-1}$) | Molecule | $\Delta H^{\circ}_{f,298}$ (kJ mol$^{-1}$) |
|---|---|---|---|
| Carbon monoxide | −110·5 | Aniline | 88 |
| Hydrogen cyanide | 130·5 | Cyclopentadiene | 134 |
| Phenol | −96·7 | Pent-3-en-1-yne | 264 |

**Solution**

For the reaction

$$M + e^- \rightarrow A^+ + B + 2e^- ,$$

the appearance energy $V_a$ of $A^+$ may be expressed:

$$V_a(A^+) = \Delta H_f(A^+) + \Delta H_f(B) - \Delta H_f(M) + E , \tag{6.1.3a}$$

where the term $E$ represents the excess energy in the products. Then, if the heats of formation of M and B are known, and the excess energy $E$ can be estimated, a measurement of $V_a(A^+)$ can be used to calculate the heat of formation of the ion $A^+$. If the ionisation energies $I$ of two or more molecules $A_1$ and $A_2$, where $A_1^+$ and $A_2^+$ may be electronically identical or isomeric with $A^+$, are measured, then from the equation

$$I(A_i) = \Delta H_f(A_i^+) - \Delta H_f(A_i) + E' \tag{6.1.3b}$$

it is possible to calculate the heats of formation of the ions $A_1^+$ and $A_2^+$.

A comparison of these heats of formation with that of $A^+$ should then indicate the probable structure of $A^+$. It is usually assumed that the excess energies $E$ and $E'$ are negligible.

The reactions of interest in this problem are

$$[C_6H_5\text{—OH}]^{+\cdot} \rightarrow [C_5H_6]^{+\cdot} + CO$$

and

$$[C_6H_5\text{—NH}_2]^{+\cdot} \rightarrow [C_5H_6]^{+\cdot} + HCN\ .$$

Using equation (6.1.3a) we obtain

$$\Delta H_f(C_5H_6^+) = V_a(C_5H_6^+) - \Delta H_f(CO) + \Delta H_f(C_6H_5OH)$$
$$= 1206{\cdot}3 - (-110{\cdot}5) + (-96{\cdot}7) = 1220{\cdot}1 \text{ kJ mol}^{-1}$$

from phenol, and

$$\Delta H_f(C_5H_6^+) = V_a(C_5H_6^+) - \Delta H_f(HCN) + \Delta H_f(C_6H_5NH_2)$$
$$= 1212{\cdot}0 - 130{\cdot}5 + 88 = 1169{\cdot}5 \text{ kJ mol}^{-1}$$

from aniline.

The heats of formation of the $C_5H_6^+$ ion from cyclopentadiene and pent-3-en-1-yne are calculated by means of equation (6.1.3b):

$$\Delta H_f(C_5H_6^+) = I(C_5H_6) + \Delta H_f\left(\text{cyclopentadiene} \text{ or } \equiv\!-\!=\!-\right)$$
$$= 855 + 134 = 989 \text{ kJ mol}^{-1} \text{ from cyclopentadiene,}$$
$$= 882 + 264 = 1146 \text{ kJ mol}^{-1} \text{ from pent-3-en-1-yne.}$$

The calculations suggest that the open-chain pent-3-en-1-yne ion is the most likely structure for the $C_5H_6^+$ ion formed in the fragmentation of both phenol and aniline molecular ions (Occolowitz and White, 1968).

The reliability of the above technique for elucidating the structure of ions depends on (a) the accuracy with which $I$ and $V_a$ can be measured, (b) the errors introduced by assuming that $E$ and $E'$ are negligible, and (c) the errors introduced by the 'kinetic shift'. These can be accounted for as follows.

(i) The limit of accuracy in the measurement of $V_a$ and $I$ is probably $\pm 0{\cdot}2$ eV ($\sim\pm 19$ kJ mol$^{-1}$) and in many cases may be larger.

(ii) The term $E'$ in equation (6.1.3b) arises because the ionisation process is a vertical transition and the ions are formed with excess vibrational energy. This excess vibrational energy is usually $0{\cdot}2$ to $0{\cdot}4$ eV and in some cases may be as high as $0{\cdot}8$ eV. In a reaction of the type $M^+ \rightarrow A^+ + B$, if the back reaction requires zero or little activation energy, then the appearance energy of $A^+$ will be the same as the heat of reaction, and $E$ in equation (6.1.3a) will have the same origin as $E'$ in equation (6.1.3b). This will often be the case for bond-fission reactions. However, rearrangement reactions, like the reactions considered in this problem, will almost certainly have an activation energy for the back reaction.

(iii) A second error in the interpretation of appearance energies arises from the kinetic shift. The kinetic shift effect arises because the appearance energy of fragment ions corresponds not to the energy required to reach the transition state for dissociation, but instead to the energy required to make the dissociation occur with a rate constant of $\sim 10^5$–$10^6$ $s^{-1}$. The error introduced by the kinetic shift can be at least 1·5 eV (Cooks *et al.*, 1969).

**General references**

Beynon, J. H., 1960, *Mass Spectrometry and Its Applications to Organic Chemistry* (Elsevier, Amsterdam).
Beynon, J. H., Saunders, R. A., Williams, A. E., 1968, *The Mass Spectra of Organic Molecules* (Elsevier, Amsterdam).
McLafferty, F. W., 1966, *Interpretation of Mass Spectra* (Benjamin, New York)
Budzikiewicz, H., Djerassi, C., Williams, D. H., 1964, *Interpretation of Mass Spectra of Organic Compounds* (Holden-Day, San Francisco).

**References to problem 6.1.3**

Chupka, W. A., 1959, *J. Chem. Phys.*, **30**, 191.
Cooks, R. G., Howe, I., Williams, D. H., 1969, *Org. Mass Spectrom.*, **2**, 137.
Occolowitz, J. L., White, G. L., 1968, *Aust. J. Chem.*, **21**, 997.

# 7 Structure and energy

## 7.1 Bond energies and lattice energies

**H A Skinner** University of Manchester

### 7.1.1

Tetrafluorohydrazine dissociates on heating to form difluoroamino radicals,

$$N_2F_4\ (g) \rightleftharpoons 2NF_2\ (g)\ . \tag{7.1.1a}$$

The concentration of $NF_2$ in the equilibrium mixture was determined spectroscopically from

$$c_{NF_2} = \frac{A}{\epsilon L}\ , \tag{7.1.1b}$$

where $\epsilon$ is molar absorbancy index, $L$ path length, and $A$ absorbance, which was measured at 260 nm. Measurements of absorbance between 343 and 413 K under conditions such that the degree of dissociation was very small (partial pressure of $N_2F_4 \gg$ partial pressure of $NF_2$) gave

$$\lg A = -\frac{2{\cdot}31 \times 10^3}{T} + 6{\cdot}62\ . \tag{7.1.1c}$$

(a) Assuming $L$ = const, and that $\epsilon$ and $\Delta E^\circ$ are temperature independent, calculate $\Delta E^\circ$ and $\Delta H^\circ$ for the dissociation process (7.1.1a) at 298 K.
(b) Given that $\Delta H^\circ_{f,298}$ ($N_2F_4$, g) = $-7{\cdot}1$ kJ mol$^{-1}$, use $\Delta H^\circ$ from (a) to calculate $\Delta H^\circ_{f,298}$ ($NF_2$, g).
(c) Given that the enthalpies of the reactions

$$S(c) + 3F_2\ (g) \rightarrow SF_6\ (g) \tag{7.1.1d}$$

and

$$S(c) + 2NF_3\ (g) \rightarrow SF_6\ (g) + N_2\ (g) \tag{7.1.1e}$$

at 298 K are $\Delta H^\circ = -1220{\cdot}9$ and $-955{\cdot}2$ kJ mol$^{-1}$ respectively, calculate $\Delta H^\circ_f$ ($NF_3$, g).
(d) Use $\Delta H^\circ_f$ ($NF_2$, g) from (b), $\Delta H^\circ_f$ ($NF_3$, g) from (c), and $\Delta H^\circ_f$ (F, g) = $79{\cdot}1$ kJ mol$^{-1}$ to calculate the bond dissociation energy, $D^\circ_{298}$ ($NF_2$–F), i.e. $\Delta H^\circ$ for the process

$$NF_3\ (g) \rightarrow NF_2\ (g) + F\ (g)\ . \tag{7.1.1f}$$

($R = 8{\cdot}314$ J mol$^{-1}$ K$^{-1}$).

**Solution**

(a) For the equilibrium, $N_2F_4 \rightleftharpoons 2NF_2$,

$$K_c = \frac{c^2_{NF_2}}{c_{N_2F_4}} = \frac{A^2}{\epsilon^2 L^2 c_{N_2F_4}} \tag{7.1.1g}$$

and

$$\ln K_c = 2 \ln A - \ln(\epsilon^2 L^2 c_{N_2F_4})\ . \tag{7.1.1h}$$

Since $\partial\epsilon/\partial T = 0$, $L$ = const, and $c_{N_2F_4}$ for small degrees of dissociation is virtually constant,

$$\frac{\partial \ln K_c}{\partial T} = 2\frac{\partial \ln A}{\partial T} , \qquad (7.1.1i)$$

$$\ln A = 2\cdot 303 \lg A = 2\cdot 303\left(-\frac{2\cdot 31 \times 10^3}{T} + 6\cdot 62\right) , \qquad (7.1.1j)$$

so that

$$\frac{\partial \ln A}{\partial T} = \frac{2\cdot 303 \times 2\cdot 31 \times 10^3}{T^2} = \frac{5\cdot 32 \times 10^3}{T^2} . \qquad (7.1.1k)$$

From equation (7.1.1i)

$$\frac{\partial \ln K_c}{\partial T} = \frac{10\cdot 64 \times 10^3}{T^2} = \frac{\Delta E^\circ}{RT^2} , \qquad (7.1.1l)$$

whence on putting $R = 8\cdot 314$ J mol$^{-1}$ K$^{-1}$, we obtain

$$\Delta E^\circ = 10\cdot 64 \times 10^3 R = 88\cdot 45 \text{ kJ mol}^{-1} .$$

For $N_2F_4 \rightarrow 2NF_2$, we have $\Delta n = 1$ and $\Delta H = \Delta E + RT$; hence

$$\Delta H^\circ_{298} = \Delta E^\circ + 2\cdot 47 = 90\cdot 92 \text{ kJ mol}^{-1} .$$

(b) For $N_2F_4 \rightarrow 2NF_2$,

$$\Delta H^\circ_{298} = 2\Delta H^\circ_{f,298}(NF_2) - \Delta H^\circ_{f,298}(N_2F_4) ,$$

$$90\cdot 92 = 2\Delta H^\circ_{f,298}(NF_2) + 7\cdot 1 ,$$

so

$$\Delta H^\circ_{f,298}(NF_2, g) = 41\cdot 9 \text{ kJ mol}^{-1} .$$

(c) For $S(c) + 3F_2(g) \rightarrow SF_6(g)$,

$$\Delta H^\circ_{298} = -1220\cdot 9 = \Delta H^\circ_{f,298}(SF_6, g) .$$

For $S(c) + 2NF_3(g) \rightarrow SF_6(g) + N_2(g)$,

$$\begin{aligned}\Delta H^\circ_{298} &= -955\cdot 2 \\ &= \Delta H^\circ_{f,298}(SF_6, g) - 2\Delta H^\circ_{f,298}(NF_3, g) \\ &= -1220\cdot 9 - 2\Delta H^\circ_{f,298}(NF_3, g) ,\end{aligned}$$

whence

$$\Delta H^\circ_{f,298}(NF_3, g) = -132\cdot 85 \text{ kJ mol}^{-1} .$$

(d) For $NF_3(g) \rightarrow NF_2(g) + F(g)$ we have

$$\begin{aligned} D^\circ_{298}(NF_2{-}F) &= \Delta H^\circ_{298} \\ &= \Delta H^\circ_{f,298}(F, g) + \Delta H^\circ_{f,298}(NF_2, g) - \Delta H^\circ_{f,298}(NF_3, g) \\ &= 79{\cdot}1 + 41{\cdot}9 + 132{\cdot}85 = 253{\cdot}85 \text{ kJ mol}^{-1} . \end{aligned}$$

The original experimental data on the dissociation of tetrafluorohydrazine were reported by Johnson and Colburn (1961).

7.1.2

The enthalpies of *atomization*, $\Delta H^\circ_{a,298}$, of saturated hydrocarbons in the gaseous state can be approximated by summing contributions from each of the bonds in the molecule. It is necessary to attribute different contributions from primary (p), secondary (s), and tertiary (t) C—H bonds.
(a) Given the following experimental data for $\Delta H^\circ_{a,298}$ (in kJ $mol^{-1}$), evaluate the enthalpy contributions to $\Delta H^\circ_{a,298}$ made by CC, CH(p), CH(s), and CH(t) bonds.

| | | | |
|---|---|---|---|
| $C_2H_6$ | 2822·44 | i-$C_4H_{10}$ | 5174·14 |
| n-$C_4H_{10}$ | 5165·78 | neo-$C_5H_{12}$ | 6359·39 |

(b) Use your results to calculate $\Delta H^\circ_{a,298}$ of n-hexane, cyclohexane, 2-methylpentane, 3-methylpentane, and 2,2-dimethylbutane. (The experimental $\Delta H^\circ_{a,298}$ values are 7508·69, 7029·25, 7515·80, 7513·13, and 7527·06 kJ $mol^{-1}$ respectively.)
(c) Show that agreement between calculated and observed $\Delta H^\circ_{a,298}$ values is improved if correction is made for steric repulsion between pairs of C—H bonds in 1, 4 mutually gauche positions

(i.e. for pairs $>\overset{1}{C}$—$\overset{2}{C}$—$\overset{3}{C}$—$\overset{4}{C}<$ with H···H between C1 and C4).

**Solution**

(a) From the application of the given additivity scheme we have:

(i) $\Delta H^\circ_{a,298}(C_2H_6) = 2822{\cdot}44 = [CC] + 6[p]$ ,

(ii) $\Delta H^\circ_{a,298}(\text{n-}C_4) = 5165{\cdot}78 = 3[CC] + 6[p] + 4[s]$ ,

(iii) $\Delta H^\circ_{a,298}(\text{i-}C_4) = 5174{\cdot}14 = 3[CC] + 9[p] + [t]$ ,

(iv) $\Delta H^\circ_{a,298}(\text{neo-}C_5) = 6359{\cdot}39 = 4[CC] + 12[p]$ ,

where [p], [s], and [t] indicate primary, secondary, and tertiary C—H bonds.

From (iv) $-[2 \times \text{(i)}]$ we have

$$[\mathrm{CC}] = 357{\cdot}255 \text{ kJ mol}^{-1} .$$

From (iii) $- [(\tfrac{3}{4}) \times \text{(iv)}]$,

$$[\mathrm{t}] = 404{\cdot}598 \text{ kJ mol}^{-1} .$$

From (ii) $-$ (i) and $[\mathrm{CC}] = 357{\cdot}255$,

$$[\mathrm{s}] = 407{\cdot}208 \text{ kJ mol}^{-1} .$$

From (i), using $[\mathrm{CC}] = 357{\cdot}255$, we find

$$[\mathrm{p}] = 410{\cdot}864 \text{ kJ mol}^{-1} .$$

(b) For n-hexane:

$$\Delta H^{\circ}_{\mathrm{a},298}(\text{calc}) = 5[\mathrm{CC}] + 6[\mathrm{p}] + 8[\mathrm{s}] = 7509{\cdot}12 \text{ kJ mol}^{-1} ;$$

for cyclohexane:

$$\Delta H^{\circ}_{\mathrm{a},298}(\text{calc}) = 6[\mathrm{CC}] + 12[\mathrm{s}] = 7030{\cdot}03 \text{ kJ mol}^{-1} ;$$

for 2-methylpentane:

$$\Delta H^{\circ}_{\mathrm{a},298}(\text{calc}) = 5[\mathrm{CC}] + 9[\mathrm{p}] + 4[\mathrm{s}] + [\mathrm{t}] = 7517{\cdot}48 \text{ kJ mol}^{-1} ;$$

for 3-methylpentane $\Delta H^{\circ}_{\mathrm{a},298}(\text{calc})$ is the same as for 2-methylpentane; and for 2,2-dimethylbutane:

$$\Delta H^{\circ}_{\mathrm{a},298}(\text{calc}) = 5[\mathrm{CC}] + 12[\mathrm{p}] + 2[\mathrm{s}] = 7531{\cdot}06 \text{ kJ mol}^{-1} .$$

A comparison with the experimental data is given in table 7.1.2.

**Table 7.1.2.**

| | $\Delta H^{\circ}_{\mathrm{a},298}$ (kJ mol$^{-1}$) | | | |
|---|---|---|---|---|
| | (calc) | (obs) | (calc) − (obs) | (calc*) − (obs) |
| n-Hexane | 7509·12 | 7508·69 | +0·43 | +0·43 |
| Cyclohexane | 7030·03 | 7029·25 | +0·78 | +0·78 |
| 2-Methylpentane | 7517·48 | 7515·80 | +1·68 | −0·32 |
| 3-Methylpentane | 7517·48 | 7513·13 | +4·35 | +0·35 |
| 2,2-Dimethylbutane | 7531·06 | 7527·06 | +4·00 | 0·00 |

(c) Steric 1, 4 gauche C—H interactions are present in the branched hexanes, and corrections for these amount to $\Delta H_{\mathrm{S}}$ (in 2-methylpentane) and $2\Delta H_{\mathrm{S}}$ in 3-methylpentane and 2,2-dimethylbutane. If $\Delta H_{3}$ is set at 2 kJ mol$^{-1}$, the differences (calc) − (obs) are reduced to the values given in the column (calc*) − (obs). The latter values in all cases lie within the experimental errors of measurement of $\Delta H^{\circ}_{\mathrm{a},298}$.

For further examples of the steric correction, see Skinner (1962), and Skinner and Pilcher (1963).

### 7.1.3

The standard enthalpies of formation of the n-alkanes and cycloalkanes ($C_3$ to $C_9$) in the ideal gas state are given in table 7.1.3a. Assuming that the cyclohexane molecule is strain free, use these data to evaluate the magnitude of the strain energies in each of the cycloalkanes listed.

**Table 7.1.3a.**

| n-Alkanes | $\Delta H^\circ_{f,298}$ (kJ mol$^{-1}$) | Cycloalkanes | $\Delta H^\circ_{f,298}$ (kJ mol$^{-1}$) |
|---|---|---|---|
| $C_3$ | −103·85 | $C_3$ | 53·30 |
| $C_4$ | −126·15 | $C_4$ | 26·65 |
| $C_5$ | −146·44 | $C_5$ | −77·24 |
| $C_6$ | −165·94 | $C_6$ | −123·14 |
| $C_7$ | −187·78 | $C_7$ | −119·33 |
| $C_8$ | −208·45 | $C_8$ | −125·77 |
| $C_9$ | −229·03 | $C_9$ | −133·0 |

**Solution**

Consider the hydrogenation reaction

$$\underset{CH_2-CH_2}{[CH_2]_{n-2}}(g) + H_2(g) \rightarrow CH_3\cdot[CH_2]_{n-2}\cdot CH_3(g)\,, \qquad (7.1.3a)$$

for which

$$\Delta H^\circ_{n,298} = \Delta H^\circ_{f,298}(C_n\,,\text{ n-alkane}) - \Delta H^\circ_{f,298}(C_n\,,\text{ cycloalkane}). \qquad (7.1.3b)$$

$\Delta H^\circ_{n,298}$ may be divided into contributions from:
(i) breaking a C–C bond in the $C_n$ cycloalkane, $\Delta H_n(\text{i})$,
(ii) breaking H–H in $H_2$, $\Delta H(\text{ii})$, and
(iii) forming two terminal primary C–H bonds in the $C_n$ $n$-alkane, $\Delta H_n(\text{iii})$,
so that

$$\Delta H^\circ_{n,298} = \Delta H_n(\text{i}) + \Delta H(\text{ii}) + \Delta H_n(\text{iii})\,. \qquad (7.1.3c)$$

$\Delta H(\text{ii})$ is a constant; $\Delta H_n(\text{iii})$ is considered virtually constant for values of $n \geqslant 3$; $\Delta H_n(\text{i})$ varies with $n$ according to the strain energies in cycloalkane rings.

Attributing differences between $\Delta H_n(\text{i})$ and the cyclohexane value, $\Delta H_6(\text{i})$, to strain contributions, $\Delta H_{S(n)}$, we can write

$$\Delta H_n(\text{i}) = \Delta H_6(\text{i}) - \Delta H_{S(n)}\,. \qquad (7.1.3d)$$

From equation (7.1.3c) we have

$$\Delta H^\circ_6 = \Delta H_6(\text{i}) + \Delta H(\text{ii}) + \Delta H_6(\text{iii})\,, \qquad (7.1.3e)$$

and

$$\Delta H^\circ_{n,298} = \Delta H_n(\text{i}) + \Delta H(\text{ii}) + \Delta H_n(\text{iii})\,, \qquad (7.1.3f)$$

so that, by subtraction,

$$\Delta H_6^\circ - \Delta H_{n,298}^\circ = \Delta H_6(\mathrm{i}) - \Delta H_n(\mathrm{i}) . \tag{7.1.3g}$$

Combining equation (7.1.3g) with equation (7.1.3d) we obtain

$$\Delta H_6^\circ - \Delta H_{n,298}^\circ = \Delta H_{S(n)} . \tag{7.1.3h}$$

The given data can now be used to calculate the values $\Delta H_n^\circ$ from (7.1.3b). The $\Delta H_{S(n)}$ values are then calculated from (7.1.3h) (see table 7.1.3b).

**Table 7.1.3b.**

| $n$ | $\Delta H_{n,298}^\circ$ (kJ mol$^{-1}$) | $\Delta H_{S(n)}$ (kJ mol$^{-1}$) | $n$ | $\Delta H_{n,298}^\circ$ (kJ mol$^{-1}$) | $\Delta H_{S(n)}$ (kJ mol$^{-1}$) |
|---|---|---|---|---|---|
| 3 | −157·15 | 114·35 | 7 | −68·45 | 25·65 |
| 4 | −152·80 | 110·00 | 8 | −82·68 | 39·88 |
| 5 | −69·20 | 26·40 | 9 | −96·03 | 53·23 |
| 6 | −42·80 | 0·00[a] | | | |

[a] Assumed.

## 7.1.4

The standard enthalpies of formation of gaseous silane and disilane were determined from measurements of the enthalpies of decomposition to elements:

$$\Delta H_{\mathrm{f},298}^\circ(\mathrm{SiH_4, g}) = 34\cdot3\ \mathrm{kJ\ mol^{-1}} ,$$

$$\Delta H_{\mathrm{f},298}^\circ(\mathrm{Si_2H_6, g}) = 80\cdot3\ \mathrm{kJ\ mol^{-1}} .$$

Electron impact studies gave the following appearance potentials $V_a$ of the ions $SiH_3^+$ and $SiH_2^+$ from the parent molecules listed in table 7.1.4.

**Table 7.1.4.**

| Parent | $V_a$ ($SiH_3^+$) (eV) | $V_a$ ($SiH_2^+$) (eV) |
|---|---|---|
| $SiH_4$ | 12·4 | 11·9 |
| $Si_2H_6$ | 11·85 | 11·9 |
| $CH_3SiH_3$ | - | 11·5 |

The measured appearance potentials set an *upper* limit to the enthalpy of the relevant ionization process ($V_a \geqslant \Delta H$). Given that $\Delta H_{\mathrm{f},298}^\circ(\mathrm{CH_4, g}) = -74\cdot9$ kJ mol$^{-1}$, $\Delta H_{\mathrm{f},298}^\circ(\mathrm{H, g}) = 218\cdot0$ kJ mol$^{-1}$, and $\Delta H_{\mathrm{f},298}^\circ(\mathrm{CH_3, g}) = 145\cdot6$ kJ mol$^{-1}$, calculate (a) $\Delta H_{\mathrm{f},298}^\circ(\mathrm{SiH_3^+})$, $\Delta H_{\mathrm{f},298}^\circ(\mathrm{SiH_3})$, $\Delta H_{\mathrm{f},298}^\circ(\mathrm{SiH_2^+})$, and $\Delta H_{\mathrm{f},298}^\circ(\mathrm{CH_3SiH_3})$, (b) the dissociation energies $D_{298}^\circ(\mathrm{H{-}SiH_3})$, $D_{298}^\circ(\mathrm{H_3Si{-}SiH_3})$, $D_{298}^\circ(\mathrm{H_3C{-}SiH_3})$, and $D_{298}^\circ(\mathrm{H_2Si^+{-}H})$.

(1 eV = 96484·6 J mol$^{-1}$) .

**Solution**

(a) From $V_a(SiH_3^+)$ we have

$$SiH_4(g) + e^- \rightarrow SiH_3^+(g) + H(g) + 2e^-.$$
$$V_a = 12{\cdot}4 \text{ eV} = 1196{\cdot}4 \text{ kJ mol}^{-1},$$
$$\Delta H_1 = \Delta H^\circ_{f,298}(SiH_3^+, g) + \Delta H^\circ_{f,298}(H, g) - \Delta H^\circ_{f,298}(SiH_4, g),$$
$$\Delta H_1 = \Delta H^\circ_{f,298}(SiH_3^+, g) + 218{\cdot}0 - 34{\cdot}3 \text{ kJ mol}^{-1}.$$

But $\Delta H_1 \leqslant 1196{\cdot}4 \text{ kJ mol}^{-1}$, and therefore

$$\Delta H^\circ_{f,298}(SiH_3^+) \leqslant 1012{\cdot}7 \text{ kJ mol}^{-1}.$$

Also we can write

$$Si_2H_6(g) + e^- \rightarrow SiH_3^+(g) + SiH_3(g) + 2e^-.$$
$$V_a = 11{\cdot}85 \text{ eV} = 1143{\cdot}3 \text{ kJ mol}^{-1},$$
$$\Delta H_2 = \Delta H^\circ_{f,298}(SiH_3^+, g) + \Delta H^\circ_{f,298}(SiH_3, g) - \Delta H^\circ_{f,298}(Si_2H_6, g),$$
$$\Delta H_2 = \Delta H^\circ_{f,298}(SiH_3^+, g) + \Delta H^\circ_{f,298}(SiH_3, g) - 80{\cdot}3 \text{ kJ mol}^{-1}.$$

Since $\Delta H_2 \leqslant 1143{\cdot}3 \text{ kJ mol}^{-1}$ and $\Delta H^\circ_{f,298}(SiH_3^+, g) \leqslant 1012{\cdot}7 \text{ kJ mol}^{-1}$,

$$\Delta H^\circ_{f,298}(SiH_3) \approx 210{\cdot}9 \text{ kJ mol}^{-1}.$$

From $V_a(SiH_2^+)$ we have

$$SiH_4(g) + e^- \rightarrow SiH_2^+(g) + H_2(g) + 2e^-.$$
$$V_a = 11{\cdot}9 \text{ eV} = 1148{\cdot}2 \text{ kJ mol}^{-1},$$
$$\Delta H_3 = \Delta H^\circ_{f,298}(SiH_2^+, g) - \Delta H^\circ_{f,298}(SiH_4, g),$$
$$\Delta H_3 = \Delta H^\circ_{f,298}(SiH_2^+, g) - 34{\cdot}3 \text{ kJ mol}^{-1}.$$

Since $\Delta H_3 \leqslant 1148{\cdot}2 \text{ kJ mol}^{-1}$,

$$\Delta H^\circ_{f,298}(SiH_2^+) \leqslant 1182{\cdot}5 \text{ kJ mol}^{-1}.$$

Again, we can also write

$$Si_2H_6(g) + e^- \rightarrow SiH_2^+(g) + SiH_4(g) + 2e^-.$$
$$V_a = 11{\cdot}9 \text{ eV} = 1148{\cdot}2 \text{ kJ mol}^{-1},$$
$$\Delta H_4 = \Delta H^\circ_{f,298}(SiH_2^+, g) + \Delta H^\circ_{f,298}(SiH_4, g) - \Delta H^\circ_{f,298}(Si_2H_6, g),$$
$$\Delta H_4 = \Delta H^\circ_{f,298}(SiH_2^+, g) + 34{\cdot}3 - 80{\cdot}3 \text{ kJ mol}^{-1}.$$

Since $\Delta H_4 \leqslant 1148{\cdot}2 \text{ kJ mol}^{-1}$,

$$\Delta H^\circ_{f,298}(SiH_2^+) \leqslant 1194{\cdot}2 \text{ kJ mol}^{-1}.$$

Finally, we have

$$CH_3SiH_3(g) + e^- \rightarrow SiH_2^+(g) + CH_4(g) + 2e^-.$$
$$V_a = 11{\cdot}5 \text{ eV} = 1109{\cdot}6 \text{ kJ mol}^{-1},$$
$$\Delta H_5 = \Delta H^\circ_{f,298}(SiH_2^+, g) + \Delta H^\circ_{f,298}(CH_4, g) - \Delta H^\circ_{f,298}(CH_3SiH_3, g),$$
$$\Delta H_5 = \Delta H^\circ_{f,298}(SiH_2^+, g) - 74{\cdot}9 - \Delta H^\circ_{f,298}(CH_3SiH_3, g) \text{ kJ mol}^{-1}.$$

Choosing the lower value for $\Delta H^\circ_{f,298}(SiH_2^+, g)$, with $\Delta H_5 \leqslant 1109{\cdot}6 \text{ kJ mol}^{-1}$,

$$\Delta H^\circ_{f,298}(CH_3SiH_3) \approx -2 \text{ kJ mol}^{-1}.$$

(b) We can now calculate the dissociation energies:

For the process $SiH_4(g) \rightarrow SiH_3(g) + H(g)$,

$$D^\circ_{298}(SiH_3{-}H) = \Delta H_6 = \Delta H^\circ_{f,298}(SiH_3, g) + \Delta H^\circ_{f,298}(H, g) - \Delta H^\circ_{f,298}(SiH_4, g)$$

$$\approx 210{\cdot}9 + 218{\cdot}0 - 34{\cdot}3 \text{ kJ mol}^{-1},$$

whence

$$D^\circ_{298}(SiH_3{-}H) \approx 394{\cdot}6 \text{ kJ mol}^{-1}.$$

For the process $Si_2H_6(g) \rightarrow 2SiH_3(g)$,

$$D^\circ_{298}(H_3Si{-}SiH_3) = \Delta H_7 = 2\Delta H^\circ_{f,298}(SiH_3, g) - \Delta H^\circ_{f,298}(Si_2H_6, g)$$

$$\approx 421{\cdot}8 - 80{\cdot}3 \text{ kJ mol}^{-1},$$

whence

$$D^\circ_{298}(H_3Si{-}SiH_3) \approx 341{\cdot}5 \text{ kJ mol}^{-1}.$$

For the process $CH_3SiH_3(g) \rightarrow CH_3(g) + SiH_3(g)$,

$$D^\circ_{298}(H_3C{-}SiH_3) = \Delta H_8 = \Delta H^\circ_{f,298}(CH_3, g) + \Delta H^\circ_{f,298}(SiH_3, g) - \Delta H^\circ_{f,298}(CH_3SiH_3, g)$$

$$\approx 145{\cdot}6 + 210{\cdot}9 + 2{\cdot}0 \text{ kJ mol}^{-1},$$

whence

$$D^\circ_{298}(H_3C{-}SiH_3) \approx 358{\cdot}5 \text{ kJ mol}^{-1}.$$

For the process $SiH_3^+(g) \rightarrow SiH_2^+(g) + H(g)$,

$$D^\circ_{298}(H_2Si^+{-}H) = \Delta H_9 = \Delta H^\circ_{f,298}(SiH_2^+, g) + \Delta H^\circ_{f,298}(H, g) - \Delta H^\circ_{f,298}(SiH_3^+, g).$$

Since $\Delta H^\circ_{f,298}(SiH_2^+, g) \leqslant 1182{\cdot}5$ kJ mol$^{-1}$, $\Delta H^\circ_{f,298}(H, g) = 218{\cdot}0$ kJ mol$^{-1}$, and $\Delta H^\circ_{f,298}(SiH_3^+, g) \leqslant 1012{\cdot}7$ kJ mol$^{-1}$,

$$D^\circ_{298}(H_2Si^+{-}H) \approx 387{\cdot}8 \text{ kJ mol}^{-1}.$$

The electron impact data are taken from studies by Potzinger and Lampe (1970), Steele *et al.* (1962), and Steele and Stone (1962).

### 7.1.5[(1)]

High-temperature gaseous equilibria in the Mg–Cu–F system have been examined by mass-spectroscopic analysis of the vapours effusing from a tungsten Knudsen cell containing a mixture of powdered $MgF_2$ and Cu. At cell temperatures above 1300 K, the ions $Mg^+$, $MgF^+$, $MgF_2^+$, $Cu^+$, and $CuF^+$ were detected in the vapour mass spectrum.

Using ionizing electron energies 5 eV above the respective appearance potentials, the relative ion intensities of $Mg^+$, $MgF^+$, $Cu^+$, and $CuF^+$ at selected temperatures in the range 1400–1600 K were measured, and are listed in table 7.1.5a (Hildenbrand, 1968a).

[(1)] Editors' note: the thermochemical calorie (cal) is retained as the unit of energy in this problem for consistency with the original data used, and to facilitate the calculations (1 cal = 4·184 J).

(a) From the data given in table 7.1.5a evaluate the equilibrium constants, $K_p$, at each listed temperature, for the process

$$MgF(g) + Cu(g) \rightleftharpoons Mg(g) + CuF(g) \ . \qquad (7.1.5a)$$

(b) Calculate the enthalpy of reaction (7.1.5a) at 298 K by the 'third law' method, using your calculated $K_p$ values and the values of the Gibbs energy function for reactants and products given in table 7.1.5b.

**Table 7.1.5a.** Measured relative ion intensities $I$.

| $T$ (K) | $I(Mg^+)$ | $I(MgF^+)$ | $I(Cu^+)$ | $I(CuF^+)$ |
|---|---|---|---|---|
| 1433 | 0·105 | 0·055 | 8·25 | 0·080 |
| 1489 | 0·480 | 0·255 | 24·0 | 0·270 |
| 1524 | 0·650 | 0·345 | 31·0 | 0·420 |
| 1563 | 1·20 | 0·800 | 57·0 | 0·850 |
| 1590 | 1·25 | 0·600 | 27·0 | 0·375 |

**Table 7.1.5b.** Values of $\dfrac{G_T - H^\circ_{298}}{T}$ (cal K$^{-1}$ mol$^{-1}$). Source: Stull (1965).

| | $T$ (K) | | | | $T$ (K) | | |
|---|---|---|---|---|---|---|---|
| | 1400 | 1500 | 1600 | | 1400 | 1500 | 1600 |
| Mg(g) | 39·278 | 39·550 | 39·809 | Cu(g) | 43·518 | 43·790 | 44·049 |
| CuF(g) | 60·595 | 61·077 | 61·537 | MgF(g) | 59·173 | 59·649 | 60·104 |

(c) Given that $D^\circ_{298}$ (Mg—F, g) = 110 ± 1 kcal mol$^{-1}$, calculate the value of $D^\circ_{298}$ (Cu—F, g).
(d) The standard enthalpy of formation of $CuF_2$(c) is −129·7 kcal mol$^{-1}$ in the revised NBS tables (Wagman *et al.*, 1969), and −131·2 kcal mol$^{-1}$ in the JANAF compilation (Stull, 1965). If $\Delta H^\circ_{f,298}(CuF_2, c)$ is chosen to be −130 ± 2 kcal mol$^{-1}$, and the value 62·5 ± 0·5 kcal mol$^{-1}$ is accepted for the enthalpy of sublimation at 298 K (Kent *et al.*, 1966), calculate $D^\circ_{298}$(FCu−F, g). Assume the values $\Delta H^\circ_{f,298}$(F, g) = 18·9 kcal mol$^{-1}$; $\Delta H^\circ_{f,298}$(Cu, g) = 81·0 ± 0·5 kcal mol$^{-1}$.

**Solution**

(a) The relative ion intensities at a given temperature are proportional to the partial pressures of the parent molecules in the gaseous equilibrium mixture (Hildenbrand and Murad, 1965). Hence, for reaction (7.1.5a) we can write

$$K_p = \frac{[Mg][CuF]}{[Cu][MgF]} \equiv \frac{I(Mg^+)I(CuF^+)}{I(Cu^+)I(MgF^+)} \ , \qquad (7.1.5b)$$

from which the calculated $K_p$ values are as follows.

| $T$(K) | 1433 | 1489 | 1524 | 1563 | 1590 |
|---|---|---|---|---|---|
| $10^2 K_p$ | 1·85 | 2·12 | 2·55 | 2·24 | 2·89 |

(b) Enthalpies of reaction, $\Delta H_r^\circ$, are related to equilibrium constants by

$$\Delta G^\circ_{r,T} = -RT \ln K_p = \Delta H^\circ_{r,T} - T\Delta S^\circ_{r,T}\,, \qquad (7.1.5c)$$

and can be calculated from $K_p$ values provided that values are available for the standard entropies of all reactants and products. For ideal gases, entropies have been calculated by statistical mechanics from spectroscopically determined molecular constants for a large number of diatomic molecules. The evaluation of $\Delta H^\circ_{r,T}$ by this procedure is known as the 'third law' method.

For evaluation of $\Delta H^\circ_{r,298}$ it is convenient to make use of tabulated values of the Gibbs energy function, $(G^\circ_T - H^\circ_{298})/T$, instead of standard entropy values. We write

$$-\frac{\Delta G^\circ_{r,T}}{T} = \Delta\left(-\frac{G^\circ_T - H^\circ_{298}}{T}\right) - \frac{\Delta H^\circ_{r,298}}{T}\,, \qquad (7.1.5d)$$

whence

$$\Delta H^\circ_{r,298} = -RT \ln K_p + T\sum\left(-\frac{G^\circ_T - H^\circ_{298}}{T}\right)_{\text{reactants}} - T\sum\left(-\frac{G^\circ_T - H^\circ_{298}}{T}\right)_{\text{products}}\,. \qquad (7.1.5e)$$

From the data in table 7.1.5b, we have

| $T$(K) | 1400 | 1500 | 1600 |
|---|---|---|---|
| $\Delta\left(-\frac{G^\circ_T - H^\circ_{298}}{T}\right)$ (cal K$^{-1}$ mol$^{-1}$) | −2·818 | −2·812 | −2·807 |

Applying equation (7.1.5e), we obtain the results in table 7.1.5c.

(c) For reaction (7.1.5a), we write

$$\Delta H^\circ_{r,298} = D^\circ_{298}(\text{Mg—F}) - D^\circ_{298}(\text{Cu—F}),$$
$$7\cdot1(\pm0\cdot3) = 110(\pm1) - D^\circ_{298}(\text{Cu—F})\ \text{kcal mol}^{-1},$$

whence

$$D^\circ_{298}(\text{Cu—F, g}) = 102\cdot9 \pm 1\ \text{kcal mol}^{-1}\,.$$

(d)

$$\Delta H^\circ_{f,298}(CuF_2, c) = -130 \pm 2\ \text{kcal mol}^{-1},$$
$$\Delta H^\circ_{\text{sub},298}(CuF_2, c) = 62\cdot5 \pm 0\cdot5\ \text{kcal mol}^{-1},$$

whence

$$\Delta H^\circ_{f,298}(CuF_2, g) = -67\cdot5 \pm 2\ \text{kcal mol}^{-1}.$$

For the process

$$\mathrm{CuF_2(g)} \rightarrow \mathrm{Cu(g)} + 2\mathrm{F(g)}\ ,$$
$$\Delta H^{\circ}_{\mathrm{r},298} = \Delta H^{\circ}_{\mathrm{f},298}(\mathrm{Cu, g}) + 2\Delta H^{\circ}_{\mathrm{f},298}(\mathrm{F, g}) - \Delta H^{\circ}_{\mathrm{f},298}(\mathrm{CuF_2, g})$$
$$= (81\cdot0 \pm 0\cdot5) + 2(18\cdot9) + (67\cdot5 \pm 2) = 186\cdot3 \pm 2\ \mathrm{kcal\ mol^{-1}}.$$

But

$$\Delta H^{\circ}_{\mathrm{r},298} = D^{\circ}_{298}(\mathrm{Cu{-}F}) + D^{\circ}_{298}(\mathrm{FCu{-}F}),$$
$$186\cdot3(\pm2) = 102\cdot9(\pm1) + D^{\circ}_{298}(\mathrm{FCu{-}F})\ \mathrm{kcal\ mol^{-1}},$$

whence

$$D^{\circ}_{298}(\mathrm{FCu{-}F}) = 83\cdot4 \pm 2\cdot3\ \mathrm{kcal\ mol^{-1}}\ .$$

A discussion on the factors influencing the magnitudes of $D$(Cu—F) and $D$(FCu—F), and reasons for the difference between the two is given by Hildenbrand (1968b).

**Table 7.1.5c.** Values of $\Delta H^{\circ}_{298}$ for reaction (7.1.5a).

| $T$(K) | $10^2 K_p$ | $\Delta\left(-\frac{G^{\circ}_T - H^{\circ}_{298}}{T}\right)$ (cal K$^{-1}$ mol$^{-1}$) | $-RT\ln K_p$ (kcal mol$^{-1}$) | $\Delta H^{\circ}_{\mathrm{r},298}$ (kcal mol$^{-1}$) |
|---|---|---|---|---|
| 1433 | 1·85 | −2·816 | 11·37 | 7·33 |
| 1489 | 2·12 | −2·813 | 11·41 | 7·22 |
| 1524 | 2·55 | −2·811 | 11·12 | 6·84 |
| 1563 | 2·24 | −2·809 | 11·80 | 7·41 |
| 1590 | 2·89 | −2·808 | 11·20 | 6·73 |
| | | | Mean value | 7·10 |
| | | Standard deviation of the mean | | ±0·14 |
| | | Uncertainty interval[a] | | ±0·28 |

[a] The absolute uncertainty is considered by Hildenbrand (1968a) to be substantially larger.

7.1.6

A polar molecule AB is considered to consist of polarizable ions, $A^+$ and $B^-$, held together by electrostatic forces.

(a) Show that the potential energy, $V(r)$, of this ion-pair model is given, approximately, by

$$V(r) = -\frac{e^2}{r} - \frac{e^2(\alpha_A + \alpha_B)}{2r^4} + A\exp\left(-\frac{r}{\rho}\right), \qquad (7.1.6a)$$

where $\alpha_A$, $\alpha_B$ are the polarizabilities of $A^+$, $B^-$; the term $A\exp(-r/\rho)$ allows for the repulsion energy of the ions in contact.

(b) Given that the equilibrium internuclear separation in AB is $r_e$, and that the force-constant of the molecule is $k_e$, derive expressions for

(i) the dipole moment, $\mu$;

(ii) the constant $\rho$ in the repulsion term $A\exp(-r/\rho)$; and

(iii) the dissociation energy, $D^{\circ}_0$(A—B) of the AB molecule, starting from the potential function (7.1.6a).

(c) Assuming that the molecules LiF, NaF, KF, and CsF are adequately described as ion-pair molecules obeying (7.1.6a), calculate the values of $\mu$, $\rho$, and $D_0^\circ$ for each of these molecules, given the molecular constants listed in table 7.1.6a, and atomic constants listed in table 7.1.6b.

**Table 7.1.6a.** Molecular constants.

| Molecule | $10^8 r_e$ (cm) | $\omega_e$ (cm$^{-1}$) | $10^{-5} k_e$ (dyn cm$^{-1}$) |
|---|---|---|---|
| LiF | 1·563 | 914·3 | 2·504 |
| NaF | 1·926 | 536·1 | 1·7614 |
| KF | 2·172 | 426·0 | 1·3671 |
| CsF | 2·345 | 353·0 | 1·2204 |

**Table 7.1.6b.** Atomic constants.

| Atom | $10^{24}\alpha_A$ (cm$^3$) | $10^{24}\alpha_B$ (cm$^3$) | $I_A$ (kJ mol$^{-1}$)[a] | $A_B$ (kJ mol$^{-1}$)[b] |
|---|---|---|---|---|
| Li | 0·029 | - | 520·28 | - |
| Na | 0·181 | - | 495·85 | - |
| K | 0·84 | - | 418·86 | - |
| Cs | 2·44 | - | 575·81 | - |
| F | - | 1·05 | - | 332·46 |

[a] $I_A$ = ionization potential of A.
[b] $A_B$ = electron affinity of B.

$e = 1{\cdot}602 \times 10^{-19}$ C
$h = 6{\cdot}6262 \times 10^{-34}$ J s
$c = 2{\cdot}9979 \times 10^{10}$ cm s$^{-1}$
$N_A = 6{\cdot}02205 \times 10^{23}$ mol$^{-1}$

1 dyn cm = 1 erg = $10^{-7}$ J
1 debye = 1 D = $3{\cdot}336 \times 10^{-28}$ C cm.

**Solution**

The model ion pair $A^+B^-$ is represented in figure 7.1.6.

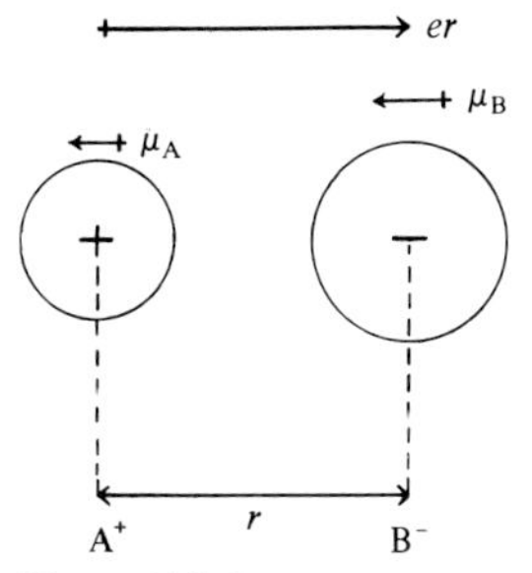


**Figure 7.1.6.**

The dipole moment would be $er$ if both ions were nonpolarizable, but the field from $A^+$ induces a dipole $\mu_B$ in $B^-$, and likewise the field from $B^-$ induces a dipole $\mu_A$ in $A^+$. These induced dipoles are opposite in direction to the principal moment, so that the net dipole moment is

$$\mu = er - \mu_A - \mu_B = er - \alpha_A E_A - \alpha_B E_B \,, \tag{7.1.6b}$$

where $E_A$ is the field strength from ion $B^-$ at A, and $E_B$ is the field strength from ion $A^+$ at B.

(a) The potential energy, $V(r)$, of the ion pair is made up from four major contributions:

(1) the coulombic attraction between the ions, $-e^2/r$;

(2) the attractions of the ions for the dipoles they induce in one another, $-e\mu_A/r^2 - e\mu_B/r^2$;

(3) the work done to induce the ion dipoles in the fields $E_A$ and $E_B$, $\mu_A^2/2\alpha_A + \mu_B^2/2\alpha_B$;

(4) the repulsion of the two ions in contact, $A\exp(-r/\rho)$.

Since $E_A = E_B = e/r^2$, addition of contributions (2) and (3) gives $-e^2(\alpha_A + \alpha_B)/2r^4$. Hence

$$V(r) = -\frac{e^2}{r} - \frac{e^2(\alpha_A + \alpha_B)}{2r^4} + A\exp\left(-\frac{r}{\rho}\right). \tag{7.1.6c}$$

[Several minor contributions to $V(r)$ are omitted in this treatment. For a more complete evaluation of $V(r)$ see Rittner (1951).]

(b) (i) Substituting $E_A = E_B = e/r_e^2$ in equation (7.1.6b) gives

$$\mu = er_e\left[1 - \frac{(\alpha_A + \alpha_B)}{r_e^3}\right]. \tag{7.1.6d}$$

(ii) At the equilibrium internuclear separation $r_e$,

$$\left(\frac{dV}{dr}\right)_{r=r_e} = 0 \,, \tag{7.1.6e}$$

and

$$\left(\frac{d^2V}{dr^2}\right)_{r=r_e} = k_e \,. \tag{7.1.6f}$$

From equations (7.1.6c) and (7.1.6e), we have

$$\left(\frac{dV}{dr}\right)_{r=r_e} = 0 = \frac{e^2}{r_e^2} + \frac{2e^2(\alpha_A + \alpha_B)}{r_e^5} - \frac{A}{\rho}\exp\left(-\frac{r_e}{\rho}\right), \tag{7.1.6g}$$

whence

$$A = \frac{\rho e^2}{r_e^2}\left[1 + \frac{2(\alpha_A + \alpha_B)}{r_e^3}\right]\exp\left(\frac{r_e}{\rho}\right). \tag{7.1.6h}$$

From equation (7.1.6f), we have

$$\left(\frac{d^2V}{dr^2}\right)_{r=r_e} = k_e = -\frac{2e^2}{r_e^3} - \frac{10e^2(\alpha_A + \alpha_B)}{r_e^6} + \frac{A}{\rho^2}\exp\left(-\frac{r_e}{\rho}\right). \tag{7.1.6i}$$

Combining equations (7.1.6h) and (7.1.6i), we obtain

$$\left[k_e + \frac{2e^2}{r_e^3} + \frac{10e^2(\alpha_A + \alpha_B)}{r_e^6}\right] = \frac{e^2}{\rho r_e^2}\left[1 + \frac{2(\alpha_A + \alpha_B)}{r_e^3}\right], \tag{7.1.6j}$$

from which

$$\rho = \frac{e^2}{r_e^2}\left[1 + \frac{2(\alpha_A + \alpha_B)}{r_e^3}\right] \Big/ \left[k_e + \frac{2e^2}{r_e^3} + \frac{10e^2(\alpha_A + \alpha_B)}{r_e^6}\right]. \tag{7.1.6k}$$

(iii) The dissociation energy $D_0^\circ(\mathrm{A—B})$ relates to the process

$$\mathrm{AB(g)} \rightarrow \mathrm{A(g)} + \mathrm{B(g)} ,$$

and is given by

$$D_0^\circ(\mathrm{A—B}) = -V(r_e) - \tfrac{1}{2}hc\omega_e - (I_A - A_B) , \tag{7.1.6l}$$

where $\frac{1}{2}hc\omega_e$ is the zero-point vibrational energy of AB, $I_A$ is the ionization potential of A, and $A_B$ is the electron affinity of B.

Now from equations (7.1.6c) and (7.1.6h) we find

$$V(r_e) = -\frac{e^2}{r_e} - \frac{e^2(\alpha_A + \alpha_B)}{2r_e^4} + \frac{\rho e^2}{r_e^2}\left[1 + \frac{2(\alpha_A + \alpha_B)}{r_e^3}\right]$$

$$= -\frac{e^2}{r_e}\left[1 - \frac{\rho}{r_e} + \frac{(\alpha_A + \alpha_B)}{2r_e^3}\left(1 - \frac{4\rho}{r_e}\right)\right]. \tag{7.1.6m}$$

so that

$$D_0^\circ = \frac{e^2}{r_e}\left[1 - \frac{\rho}{r_e} + \frac{(\alpha_A + \alpha_B)}{2r_e^3}\left(1 - \frac{4\rho}{r_e}\right)\right] - \tfrac{1}{2}hc\omega_e - (I_A - A_B) . \tag{7.1.6n}$$

(c) Values of $\mu$ calculated from equation (7.1.6d) are:

| | $10^{24}(\alpha_A + \alpha_B)$ (cm$^3$) | $10^{24}r_e^3$ (cm$^3$) | $\frac{(\alpha_A + \alpha_B)}{r_e^3}$ | $\mu$ (D) |
|---|---|---|---|---|
| LiF | 1·079 | 3·818 | 0·2826 | 5·39 |
| NaF | 1·231 | 7·1445 | 0·1723 | 7·66 |
| KF | 1·89 | 10·240 | 0·1846 | 8·51 |
| CsF | 3·49 | 12·895 | 0·2706 | 8·22 |

Values of $\rho$ calculated from equation (7.1.6k) are:

| | $10^4 \frac{e^2}{r_e^2}$ (dyn) | $\frac{e^2}{r_e^3}$ (dyn cm$^{-1}$) | $\frac{10e^2(\alpha_A + \alpha_B)}{r_e^6}$ (dyn cm$^{-1}$) | $10^8\rho$ (cm) |
|---|---|---|---|---|
| LiF | 9·4429 | 60415 | 170722 | 0·2732 |
| NaF | 6·2188 | 32289 | 55634 | 0·2822 |
| KF | 4·8918 | 22526 | 41574 | 0·2999 |
| CsF | 4·1951 | 17889 | 48416 | 0·3125 |

Values of $V(r_e)$ calculated from equation (7.1.6m) are:

| | $10^{12}\frac{e^2}{r_e}$ (erg) | $\frac{\rho}{r_e}$ | $10^{12}V(r_e)$ (erg) | $N_A V(r_e)$ (kJ mol$^{-1}$) |
|---|---|---|---|---|
| LiF | 14·7595 | 0·1748 | −12·81 | −771·4 |
| NaF | 11·9776 | 0·1464 | −10·65 | −641·3 |
| KF | 10·623 | 0·1381 | −9·595 | −577·8 |
| CsF | 9·8374 | 0·1337 | −9·141 | −550·5 |

Values of $D_0^\circ$ calculated from equation (7.1.6n) are:

| | $10^{14}\frac{1}{2}hc\omega_e$ (erg) | $\frac{1}{2}N_A hc\omega_e$ (kJ mol$^{-1}$) | $(I_A - A_B)$ (kJ mol$^{-1}$) | $D_0^\circ$ (kJ mol$^{-1}$) |
|---|---|---|---|---|
| LiF | 9·08 | 5·468 | 187·82 | 578·1 |
| NaF | 5·32 | 3·204 | 163·39 | 474·7 |
| KF | 4·23 | 2·547 | 86·40 | 488·9 |
| CsF | 3·51 | 2·114 | 43·35 | 505·0 |

The experimental data given in tables 7.1.6a and 7.1.6b are taken from Chao (1970) and Rittner (1951).

The best available experimental values for $D_0^\circ$, as quoted by Chao, are 137·8, 113·2, 117·5, and 121·7 kcal mol$^{-1}$ (576·6, 473·6, 491·6, and 509·2 kJ mol$^{-1}$) respectively in LiF, NaF, KF, and CsF.

The agreement of calculated dipole moments with experiment is less satisfactory. Experimental values of $\mu$ (de Wijn, 1966) are given as 6·33, 8·15, 8·60, and 7·87 D for LiF, NaF, KF, and CsF, respectively.

### 7.1.7

Photoionization studies have established the threshold energy (at 0 K) for the process

$$CH_4(g) + h\nu \rightarrow CH_3^+(g) + H(g) + e^- \tag{7.1.7a}$$

at 1381·6 kJ mol$^{-1}$, and at 1456·0 kJ mol$^{-1}$ for the process

$$CH_3(g) + h\nu \rightarrow CH_2^+(g) + H(g) + e^- . \tag{7.1.7b}$$

The ionization potential of $CH_3$, as determined spectroscopically, is 949·8 kJ mol$^{-1}$, and of $CH_2$ is 1002·9 kJ mol$^{-1}$.

(a) Use these data to evaluate the bond dissociation energies $D_0^\circ(CH_3{-}H)$ and $D_0^\circ(CH_2{-}H)$.

(b) Given the enthalpies of formation, $\Delta H_{f,0}^\circ(CH_4, g) = -66{\cdot}82$ kJ mol$^{-1}$ and $\Delta H_{f,0}^\circ(H, g) = 216{\cdot}0$ kJ mol$^{-1}$, calculate the values of $\Delta H_{f,0}^\circ(CH_3, g)$ and $\Delta H_{f,0}^\circ(CH_2, g)$ from your values for $D_0^\circ(CH_3{-}H)$ and $D_0^\circ(CH_2{-}H)$.

(c) Using your value for $\Delta H^{\circ}_{f,0}(CH_3)$, and the thermochemical data listed in table 7.1.7, evaluate the bond dissociation energies $D^{\circ}_0(CH_3—Cl)$, $D^{\circ}_0(CH_3—Br)$, and $D^{\circ}_0(CH_3—I)$.

**Table 7.1.7.** Enthalpies of formation.

| Substance | $\Delta H^{\circ}_{f,0}$ (kJ mol$^{-1}$) | Substance | $\Delta H^{\circ}_{f,0}$ (kJ mol$^{-1}$) |
|---|---|---|---|
| $CH_3Cl$ (g) | −72·92 | Cl (g) | 120·0 |
| $CH_3Br$ (g) | −19·75 | Br (g) | 117·9 |
| $CH_3I$ (g) | 22·51 | I (g) | 107·2 |

(d) Diazomethane reacts readily with the hydrogen halides

$$CH_2N_2(g) + HX(g) \rightarrow CH_3X(g) + N_2(g) \qquad (X = Cl, Br, I)\,. \qquad (7.1.7c)$$

The reactions are exothermic, and sufficiently so in the case X = I to cause thermal decomposition of a fraction of the $CH_3I$ produced. $CH_3$ radicals have not, however, been detected in the reactions with HBr, and HCl. Use these observations to deduce lower and upper limits to the enthalpy of formation of diazomethane [$\Delta H^{\circ}_{f,0}(HI, g) = 28\cdot7$ kJ mol$^{-1}$; $\Delta H^{\circ}_{f,0}(HBr, g) = -28\cdot6$ kJ mol$^{-1}$], and an upper limit to $D^{\circ}_0(CH_2—N_2)$.

The experimental investigations on reaction (7.1.7c) are reported by Hassler and Setzer (1965) and by Bauer *et al.* (1965).

The photoionization results quoted are taken from Chupka (1968) and Chupka and Lifshitz (1968). The thermochemical data are taken from the NBS Tables (Wagman *et al.*, 1965).

**Solution**

(a) From equation (7.1.7a), and the ionization potential of $CH_3$ we have

$$CH_4 + 1381\cdot6 \text{ kJ mol}^{-1} \rightarrow CH_3^+ + H + e^-\,,$$

$$CH_3 + 949\cdot8 \text{ kJ mol}^{-1} \rightarrow CH_3^+ + e^-\,.$$

By subtraction

$$CH_4 + 431\cdot8 \text{ kJ mol}^{-1} \rightarrow CH_3 + H\,, \qquad (7.1.7d)$$

whence

$$D^{\circ}_0(CH_3—H) = 431\cdot8 \text{ kJ mol}^{-1}\,.$$

From equation (7.1.7b), and the ionization potential of $CH_2$ we have

$$CH_3 + 1456\cdot0 \text{ kJ mol}^{-1} \rightarrow CH_2^+ + H + e^-\,,$$

$$CH_2 + 1002\cdot9 \text{ kJ mol}^{-1} \rightarrow CH_2^+ + e^-\,.$$

By subtraction

$$CH_3 + 453\cdot1 \text{ kJ mol}^{-1} \rightarrow CH_2 + H\,, \qquad (7.1.7e)$$

whence

$$D_0^\circ(\text{CH}_2\text{—H}) = 453{\cdot}1 \text{ kJ mol}^{-1} .$$

(b) From equation (7.1.7d) we have

$$\Delta H_0^\circ = D_0^\circ = 431{\cdot}8 \text{ kJ mol}^{-1} = \Delta H_{\text{f},0}^\circ(\text{CH}_3) + \Delta H_{\text{f},0}^\circ(\text{H}) - \Delta H_{\text{f},0}^\circ(\text{CH}_4)$$
$$= \Delta H_{\text{f},0}^\circ(\text{CH}_3) + 216{\cdot}0 - (-66{\cdot}82) \text{ kJ mol}^{-1} ,$$

whence

$$\Delta H_{\text{f},0}^\circ(\text{CH}_3, \text{g}) = 149{\cdot}0 \text{ kJ mol}^{-1} .$$

From equation (7.1.7e) we have

$$\Delta H_0^\circ = D_0^\circ = 453{\cdot}1 \text{ kJ mol}^{-1} = \Delta H_{\text{f},0}^\circ(\text{CH}_2) + \Delta H_{\text{f},0}^\circ(\text{H}) - \Delta H_{\text{f},0}^\circ(\text{CH}_3)$$
$$= \Delta H_{\text{f},0}^\circ(\text{CH}_2) + 216{\cdot}0 - 149{\cdot}0 \text{ kJ mol}^{-1} ,$$

whence

$$\Delta H_{\text{f},0}^\circ(\text{CH}_2, \text{g}) = 386{\cdot}1 \text{ kJ mol}^{-1} .$$

(c) Using $\Delta H_{\text{f},0}^\circ(\text{CH}_3) = 149{\cdot}0 \text{ kJ mol}^{-1}$ we obtain

$$D_0^\circ(\text{CH}_3\text{—Cl}) = \Delta H_{\text{f},0}^\circ(\text{CH}_3) + \Delta H_{\text{f},0}^\circ(\text{Cl}) - \Delta H_{\text{f},0}^\circ(\text{CH}_3\text{Cl})$$
$$= 149{\cdot}0 + 120{\cdot}0 - (-72{\cdot}92) = 341{\cdot}9 \text{ kJ mol}^{-1} .$$

$$D_0^\circ(\text{CH}_3\text{—Br}) = \Delta H_{\text{f},0}^\circ(\text{CH}_3) + \Delta H_{\text{f},0}^\circ(\text{Br}) - \Delta H_{\text{f},0}^\circ(\text{CH}_3\text{Br})$$
$$= 149{\cdot}0 + 117{\cdot}9 - (-19{\cdot}75) = 286{\cdot}7 \text{ kJ mol}^{-1} .$$

$$D_0^\circ(\text{CH}_3\text{—I}) = \Delta H_{\text{f},0}^\circ(\text{CH}_3) + \Delta H_{\text{f},0}^\circ(\text{I}) - \Delta H_{\text{f},0}^\circ(\text{CH}_3\text{I})$$
$$= 149{\cdot}0 + 107{\cdot}2 - 22{\cdot}51 = 233{\cdot}7 \text{ kJ mol}^{-1} .$$

(d) For reaction (7.1.7c) with X = Br, we can write

$$\Delta H_0^\circ = -19{\cdot}75 + 28{\cdot}6 - \Delta H_{\text{f},0}^\circ(\text{CH}_2\text{N}_2) \text{ kJ mol}^{-1}$$
$$= -\Delta H_{\text{f},0}^\circ(\text{CH}_2\text{N}_2) + 8{\cdot}85 \text{ kJ mol}^{-1} .$$

The enthalpy of reaction in this case is insufficient to cause thermal disruption of $CH_3Br$, i.e. to break the C—Br bond. Hence we may conclude that the exothermicity of reaction (7.1.7c) is less than $D_0^\circ(\text{CH}_3\text{–Br})$ (286·7 kJ mol$^{-1}$). On this basis,

$$\Delta H_{\text{f},0}^\circ(\text{CH}_2\text{N}_2) < 295{\cdot}6 \text{ kJ mol}^{-1} .$$

For reaction (7.1.7c) with X = I, we have

$$\Delta H_0^\circ = 22{\cdot}51 - 28{\cdot}7 - \Delta H_{\text{f},0}^\circ(\text{CH}_2\text{N}_2) \text{ kJ mol}^{-1}$$
$$= -\Delta H_{\text{f},0}^\circ(\text{CH}_2\text{N}_2) - 6{\cdot}19 \text{ kJ mol}^{-1} .$$

In this case, the reaction is sufficiently exothermic to rupture the C—I bond in $CH_3I$: $D_0^\circ(CH_3—I) = 233\cdot7$ kJ mol$^{-1}$. Hence

$$\Delta H_{f,0}^\circ(CH_2N_2) > 227\cdot5 \text{ kJ mol}^{-1} .$$

Accepting that $\Delta H_{f,0}^\circ(CH_2N_2) > 228$ kJ mol$^{-1}$,

$$D_0^\circ(CH_2—N_2) = \Delta H_{f,0}^\circ(CH_2) - \Delta H_{f,0}^\circ(CH_2N_2) \leqslant 386 - 228 \text{ kJ mol}^{-1}$$
$$\leqslant 158 \text{ kJ mol}^{-1} .$$

7.1.8

Pauling made use of an empirical relationship which can be expressed as

$$D(A—B) \approx \tfrac{1}{2}[D(A—A)+D(B—B)]+96\cdot48(x_A - x_B)^2 \text{ kJ mol}^{-1} \quad (7.1.8a)$$

to construct his well-known table of electronegativity values for atoms. A—B, B—B, and A—A are single electron-pair covalent bonds, and $x_A$, $x_B$ are the respective electronegativities of A and B.

(a) Apply equation (7.1.8a) to evaluate electronegativities of F, OH, $NH_2$, $CH_3$, Cl, SH, $PH_2$, and $SiH_3$ relative to $x_H = 2\cdot1$, using the enthalpies of hydrogenation, $\Delta H_{298}^\circ$, listed in table 7.1.8a, which have been calculated from standard enthalpies of formation in the NBS Tables (Wagman *et al.*, 1968).

**Table 7.1.8a.** Enthalpies of hydrogenation at 25°C.

| Reaction | $\Delta H_{298}^\circ$ (kJ mol$^{-1}$) |
|---|---|
| $H_2(g) + F_2(g) \rightarrow 2HF(g)$ | −543·9 |
| $H_2(g) + HO—OH(g) \rightarrow 2H_2O(g)$ | −347·3 |
| $H_2(g) + H_2N—NH_2(g) \rightarrow 2NH_3(g)$ | −187·4 |
| $H_2(g) + H_3C—CH_3(g) \rightarrow 2CH_4(g)$ | −65·3 |
| $H_2(g) + Cl_2(g) \rightarrow 2HCl(g)$ | −184·5 |
| $H_2(g) + HS—SH(g) \rightarrow 2H_2S(g)$ | −51·9 |
| $H_2(g) + H_2P—PH_2(g) \rightarrow 2H_3P(g)$ | −10·0 |
| $H_2(g) + H_3Si—SiH_3(g) \rightarrow 2SiH_4(g)$ | −11·7 |

(b) Kistiakowsky and coworkers (1936) measured the enthalpies of hydrogenation in the vapour phase of cyclohexene and of benzene to form cyclohexane, from which they derived a value of 36 kcal mol$^{-1}$ (151 kJ mol$^{-1}$) for the net resonance energy of benzene, attributed to delocalization of the six $\pi$ electrons.

Use the Kistiakowsky method, and the enthalpies of hydrogenation given in table 7.1.8b to evaluate the net resonance energies in naphthalene, anthracene, cyclopentadiene, pyrrole, furan, and pyridine. The $\Delta H_{298}^\circ$ values given in table 7.1.8b are calculated from standard enthalpies of formation recommended by Cox and Pilcher (1970).

**Table 7.1.8b.** Enthalpies of hydrogenation at 25°C.

| Reaction | $\Delta H^\circ_{298}$ (kJ mol$^{-1}$) |
|---|---|
| (1) Cyclohexene (g) → cyclohexane (g) | −118·8 |
| (2) Naphthalene (g) → *trans*-$C_{10}H_{18}$(g) | −333·0 |
| (3) Anthracene (g) → *trans,syn,trans*-$C_{14}H_{24}$(g) | −474·0 |
| (4) Cyclopentene (g) → cyclopentane (g) | −111·7 |
| (5) Cyclopentadiene (g) → cyclopentane (g) | −210·9 |
| (6) Pyrrole (g) → pyrrolidine (g) | −111·7 |
| (7) Furan (g) → tetrahydrofuran (g) | −149·4 |
| (8) $Bu^nN{=}CHPr^i$(g) → $Bu^nNHBu^i$(g) | −83·7[a] |
| (9) Pyridine (g) → piperidine (g) | −193·7 |

[a] Assume this value to be typical for a localized —CH=N— bond.

**Solution**

(a) Let us rewrite the Pauling equation (7.1.8a) to apply to the hydrogenation of R—R at 298 K:

$$D^\circ_{298}(\mathrm{R{-}H}) = \tfrac{1}{2}[D^\circ_{298}(\mathrm{R{-}R}) + D^\circ_{298}(\mathrm{H{-}H})] + 96{\cdot}48(x_\mathrm{R} - x_\mathrm{H})^2 \ . \quad (7.1.8b)$$

But

$$\Delta H^\circ_{298}(\text{hydrogenation}) = D^\circ_{298}(\mathrm{R{-}R}) + D^\circ_{298}(\mathrm{H{-}H}) - 2D^\circ_{298}(\mathrm{R{-}H}) \ , \quad (7.1.8c)$$

so that equation (7.1.8b) becomes

$$-\tfrac{1}{2}\Delta H^\circ_{298} = 96{\cdot}48(x_\mathrm{R} - x_\mathrm{H})^2 = \Delta \quad (7.1.8d)$$

or

$$(x_\mathrm{R} - x_\mathrm{H}) = 0{\cdot}102\Delta^{1/2} \ . \quad (7.1.8e)$$

From table 7.1.8a, putting $x_\mathrm{H} = 2{\cdot}1$, we obtain the results in table 7.1.8c.

**Table 7.1.8c.**

| Reaction | $\Delta$ | $0{\cdot}102\Delta^{1/2}$ | R | $x_\mathrm{R}$ |
|---|---|---|---|---|
| $H_2 + F_2$ | 272·0 | 1·68 | F | 3·8 |
| $H_2$ + HO—OH | 173·7 | 1·34 | OH | 3·44 |
| $H_2 + H_2N{-}NH_2$ | 93·7 | 0·99 | $NH_2$ | 3·1 |
| $H_2 + H_3C{-}CH_3$ | 32·7 | 0·58 | $CH_3$ | 2·7 |
| $H_2 + Cl_2$ | 92·3 | 0·98 | Cl | 3·1 |
| $H_2$ + HS—SH | 26·0 | 0·52 | SH | 2·6 |
| $H_2 + H_2P{-}PH_2$ | 5·0 | 0·23 | $PH_2$ | 2·3 |
| $H_2 + H_3Si{-}SiH_3$ | 5·9 | 0·25 | $SiH_3$ | 2·35 |

(b) The Kistiakowsky method compares $\Delta H^\circ_{298}$ (hydrogenation) of benzene with 3 × $\Delta H^\circ_{298}$ (hydrogenation) of cyclohexene. The latter is considered to give the enthalpy of hydrogenation of the hypothetical nonconjugated cyclohexatriene.

Applying this procedure we obtain

for naphthalene

| | |
|---|---|
| $\Delta H^\circ_{298}$ | $= -333\cdot0$ kJ mol$^{-1}$, |
| $5 \times \Delta H^\circ_{298}$(cyclohexene) | $= -594\cdot0$ kJ mol$^{-1}$, |
| Resonance energy | $= 261\cdot0$ kJ mol$^{-1}$; |

for anthracene

| | |
|---|---|
| $\Delta H^\circ_{298}$ | $= -474\cdot0$ kJ mol$^{-1}$, |
| $7 \times \Delta H^\circ_{298}$(cyclohexene) | $= -831\cdot6$ kJ mol$^{-1}$, |
| Resonance energy | $= 357\cdot6$ kJ mol$^{-1}$; |

for cyclopentadiene

| | |
|---|---|
| $\Delta H^\circ_{298}$ | $= -210\cdot9$ kJ mol$^{-1}$, |
| $2 \times \Delta H^\circ_{298}$(cyclopentene) | $= -223\cdot4$ kJ mol$^{-1}$, |
| Resonance energy | $= 12\cdot5$ kJ mol$^{-1}$; |

for pyrrole

| | |
|---|---|
| $\Delta H^\circ_{298}$ | $= -111\cdot7$ kJ mol$^{-1}$, |
| $2 \times \Delta H^\circ_{298}$(cyclopentene) | $= -223\cdot4$ kJ mol$^{-1}$, |
| Resonance energy | $= 111\cdot7$ kJ mol$^{-1}$, |

for furan

| | |
|---|---|
| $\Delta H^\circ_{298}$ | $= -149\cdot4$ kJ mol$^{-1}$, |
| $2 \times \Delta H^\circ_{298}$(cyclopentene) | $= -223\cdot4$ kJ mol$^{-1}$, |
| Resonance energy | $= 74\cdot0$ kJ mol$^{-1}$; |

for pyridine

| | |
|---|---|
| $\Delta H^\circ_{298}$ | $= -193\cdot7$ kJ mol$^{-1}$, |
| $2 \times \Delta H^\circ_{298}$(cyclohexene) | $= -237\cdot6$ kJ mol$^{-1}$, |
| $\Delta H^\circ_{298}$(—CH=N—) | $= -83\cdot7$ kJ mol$^{-1}$, |
| Resonance energy | $= 127\cdot6$ kJ mol$^{-1}$. |

**7.1.9**

Crystalline MgO has the rock-salt structure, with $r_0$(Mg—O) = 2·106 Å. The compressibility, $\beta$, of the crystal is $0\cdot44 \times 10^{-2}$ dyn$^{-1}$ cm$^{2}$. The characteristic Debye frequency for the solid oxide is $\nu_{max} \approx 17\cdot1 \times 10^{12}$ s$^{-1}$.
(a) Assuming an ionic model, calculate the lattice energy per molecule, $u(r_0)$, using a potential function of the Born–Mayer type; include terms $Cr^{-6}$ and $Dr^{-8}$ to allow for dipole–dipole and dipole–quadrupole interactions, and represent the ionic repulsion by a term $B\exp(-r/\rho)$. [For MgO, $C = 47 \times 10^{-60}$ erg cm$^6$ molecule$^{-1}$, $D = 17 \times 10^{-76}$ erg cm$^8$ molecule$^{-1}$. 1 dyn $\equiv 10^{-5}$N; 1 erg $\equiv 10^{-7}$J.]
(b) From the calculated lattice energy, and using the Born–Haber cycle, calculate the electron affinity of the oxygen atom (O → $O^{2-}$) given the

following subsidiary data:

| | |
|---|---|
| $\Delta H^\circ_{f,0}(\mathrm{MgO, c})$ | $= -597{\cdot}1$ kJ mol$^{-1}$, |
| $\Delta H^\circ_{sub,0}(\mathrm{Mg, c} \rightarrow \mathrm{Mg, g})$ | $= 146{\cdot}4$ kJ mol$^{-1}$, |
| $I(\mathrm{Mg} \rightarrow \mathrm{Mg^{2+}})$ | $= 2187{\cdot}8$ kJ mol$^{-1}$, |
| $D^\circ_0(\mathrm{O_2} \rightarrow 2\mathrm{O})$ | $= 493{\cdot}7$ kJ mol$^{-1}$. |

**Solution**

(a) The Born–Mayer model for a crystal lattice composed of ions $\mathrm{M}^{z+}$, $\mathrm{X}^{z-}$, gives

$$u(r) = -\frac{Az^2e^2}{r} + B\exp\left(-\frac{r}{\rho}\right) - \frac{C}{r^6} - \frac{D}{r^8} + \frac{9}{4}h\nu_{\max}\,, \qquad (7.1.9a)$$

where $u(r)$ is the lattice energy per MX molecule at internuclear separation $r$, and $A$ is the Madelung constant for the crystal type. The first term measures the Coulomb energy, and the last term takes care of the zero-point lattice energy.

The rock-salt unit cell has a volume $(2r)^3$, and contains the substance of four molecules. Hence the volume per molecule is

$$v = \frac{(2r)^3}{4} = 2r^3\,, \qquad (7.1.9b)$$

and the molar volume is

$$Nv = V = 2Nr^3\,. \qquad (7.1.9c)$$

The molar lattice energy is

$$U(r) = Nu(r)\,. \qquad (7.1.9d)$$

The thermodynamic condition for equilibrium, $(\mathrm{d}U/\mathrm{d}V)_{V=V_0} = 0$, is equivalent to the condition

$$\left(\frac{\mathrm{d}u}{\mathrm{d}r}\right)_{r=r_0} = 0\,, \qquad (7.1.9e)$$

from which

$$\frac{B}{\rho}\exp\left(-\frac{r_0}{\rho}\right) = \left(\frac{z^2Ae^2}{r_0^2} + \frac{6C}{r_0^7} + \frac{8D}{r_0^9}\right). \qquad (7.1.9f)$$

A second equilibrium condition (Born and Mayer, 1932) is that

$$\left(\frac{\mathrm{d}^2U}{\mathrm{d}V^2}\right)_{r=r_0} = \frac{1}{\beta V_0}\,. \qquad (7.1.9g)$$

Now

$$\frac{\mathrm{d}U}{\mathrm{d}V} = \frac{\mathrm{d}U}{\mathrm{d}r}\frac{\mathrm{d}r}{\mathrm{d}V} = \frac{1}{6Nr^2}\frac{\mathrm{d}U}{\mathrm{d}r} = \frac{1}{6r^2}\frac{\mathrm{d}u}{\mathrm{d}r}$$

and

$$\frac{d^2U}{dV^2} = \frac{d}{dV}\left(\frac{dU}{dV}\right) = \frac{1}{6Nr^2}\frac{d}{dr}\left(\frac{dU}{dV}\right) = \frac{1}{6Nr^2}\frac{d}{dr}\left(\frac{1}{6r^2}\frac{du}{dr}\right)$$

$$= \frac{1}{6Nr^2}\left(\frac{1}{6r^2}\frac{d^2u}{dr^2} - \frac{2}{6r^3}\frac{du}{dr}\right).$$

At $r = r_0$, $du/dr = 0$, so that

$$\left(\frac{d^2U}{dV^2}\right)_{r=r_0} = \frac{1}{\beta V_0} = \frac{1}{36Nr_0^4}\left(\frac{d^2u}{dr^2}\right)_{r=r_0}, \tag{7.1.9h}$$

$$\left(\frac{d^2u}{dr^2}\right)_{r=r_0} = -\frac{2z^2Ae^2}{r_0^3} + \frac{B}{\rho^2}\exp\left(-\frac{r_0}{\rho}\right) - \frac{42C}{r_0^8} - \frac{72D}{r_0^{10}},$$

which, with equation (7.1.9f), becomes

$$\left(\frac{d^2u}{dr^2}\right)_{r=r_0} = -\frac{2z^2Ae^2}{r_0^3} - \frac{42C}{r_0^8} - \frac{72D}{r_0^{10}} + \frac{1}{\rho r_0}\left(\frac{z^2Ae^2}{r_0} + \frac{6C}{r_0^6} + \frac{8D}{r_0^8}\right). \tag{7.1.9i}$$

From equation (7.1.9h) we obtain

$$\left(\frac{d^2u}{dr^2}\right)_{r=r_0} = \frac{36Nr_0^4}{\beta V_0} = \frac{36Nr_0^4}{\beta 2Nr_0^3} = \frac{18r_0}{\beta},$$

so that

$$\frac{18r_0}{\beta} + \frac{2z^2Ae^2}{r_0^3} + \frac{42C}{r_0^8} + \frac{72D}{r_0^{10}} = \frac{1}{\rho r_0}\left(\frac{z^2Ae^2}{r_0} + \frac{6C}{r_0^6} + \frac{8D}{r_0^8}\right),$$

from which we obtain

$$\frac{\rho}{r_0} = \left(\frac{z^2Ae^2}{r_0} + \frac{6C}{r_0^6} + \frac{8D}{r_0^8}\right) \Big/ \left(\frac{18r_0^3}{\beta} + \frac{2z^2Ae^2}{r_0} + \frac{42C}{r_0^6} + \frac{72D}{r_0^8}\right). \tag{7.1.9j}$$

Combining equations (7.1.9a) and (7.1.9f), we obtain the lattice energy per molecule in the convenient form

$$u(r_0) = -\frac{z^2Ae^2}{r_0}\left(1 - \frac{\rho}{r_0}\right) - \frac{C}{r_0^6}\left(1 - \frac{6\rho}{r_0}\right) - \frac{D}{r_0^8}\left(1 - \frac{8\rho}{r_0}\right) + \frac{9}{4}h\nu_{max}. \tag{7.1.9k}$$

[See for example Ladd and Lee (1958, 1959).]

For MgO, $z = 2$ and $A = 1{\cdot}748$. Using equations (7.1.9j) and (7.1.9k), we can calculate $\rho/r_0$ and $u(r_0)$ for MgO.

Calculation of $\rho/r_0$ (energy terms in units of $10^{-12}$ erg):

$$\frac{4Ae^2}{r_0} = 76{\cdot}59, \quad \frac{6C}{r_0^6} = 3{\cdot}23, \quad \frac{8D}{r_0^8} = 0{\cdot}35;$$

$$\frac{18r_0^3}{\beta} = 382{\cdot}12, \quad \frac{8Ae^2}{r_0} = 153{\cdot}17, \quad \frac{42C}{r_0^6} = 22{\cdot}62, \quad \frac{72D}{r_0^8} = 3{\cdot}16.$$

Hence

$$\frac{\rho}{r_0} = \frac{76\cdot59+3\cdot23+0\cdot35}{382\cdot12+153\cdot17+22\cdot62+3\cdot16} = \frac{80\cdot17}{561\cdot07} = 0\cdot1429 .$$

Calculation of $u(r_0)$ in units of $10^{-12}$ erg ($h = 6\cdot6262 \times 10^{-27}$ erg s):

$$1-\frac{\rho}{r_0} = 0\cdot8571 , \qquad 1-\frac{6\rho}{r_0} = 0\cdot1426 , \qquad 1-\frac{8\rho}{r_0} = -0\cdot1432 ,$$

$$\tfrac{9}{4} h\nu_{\mathrm{max}} = 0\cdot255 .$$

Hence

$$u(r_0) = -76\cdot59 \times 0\cdot8571 - \tfrac{1}{6} \times 3\cdot23 \times 0\cdot1426 - \tfrac{1}{8} \times 0\cdot35 \times (-0\cdot1432)$$
$$+0\cdot255 = -65\cdot645 - 0\cdot077 + 0\cdot006 + 0\cdot255$$
$$= -65\cdot46 \times 10^{-12} \text{ erg} ,$$

or (with $N_A = 6\cdot02205 \times 10^{23}$ mol$^{-1}$)

$$U_0 = -3942\cdot0 \text{ kJ mol}^{-1} .$$

(b) The Born–Haber cycle, carried out in steps at 0 K, is represented schematically below:

$$\begin{array}{lll}
Mg^{2+}\,(g) & +\ O^{2-}\,(g) & \xrightarrow{(U_0)}\ MgO(c) \\
\uparrow (I) & \uparrow (A) & \\
Mg\,(g) & +\ O\,(g) & \\
\uparrow \Delta H^\circ_{\mathrm{sub},0} & \uparrow (\tfrac{1}{2}D^\circ_0) & \nearrow \Delta H^\circ_{\mathrm{f},0} \\
Mg(c) & +\ \tfrac{1}{2}O_2\,(g) &
\end{array} ,$$

from which it follows that

$$U_0 = \Delta H^\circ_{\mathrm{f},0} - \Delta H^\circ_{\mathrm{sub},0} - I(\mathrm{Mg} \to \mathrm{Mg}^{2+}) - \tfrac{1}{2}D^\circ_0 - A(\mathrm{O} \to \mathrm{O}^{2-}) , \qquad (7.1.91)$$

where $A$ is electron affinity, and $I$ is ionization energy. Substituting $U_0 = -3942\cdot0$ kJ mol$^{-1}$ and the given values for $\Delta H^\circ_{\mathrm{f},0}$, $\Delta H^\circ_{\mathrm{sub},0}$, $I$, and $D^\circ_0$, we obtain

$$A(\mathrm{O} \to \mathrm{O}^{2-}) = 3942\cdot0 - 597\cdot1 - 146\cdot4 - 2187\cdot8 - 246\cdot9$$
$$= 763\cdot8 \text{ kJ mol}^{-1}.$$

[This calculation may be compared with that of Ladd and Lee (1960).]

**References**

Bauer, S. H., Marshall, D., Baer, T., 1965, *J. Am. Chem. Soc.,* **87**, 5514.
Born, M., Mayer, J. E., 1932, *Z. Phys.,* **75**, 1.
Chao, J., 1970, *Thermochim. Acta,* **1**, 71.
Chupka, W. A., 1968, *J. Chem. Phys.,* **48**, 2337.
Chupka, W. A., Lifshitz, C., 1968, *J. Chem. Phys.,* **48**, 1109.

Cox, J. D., Pilcher, G., 1970, *Thermochemistry of Organic and Organometallic Compounds* (Academic Press, London and New York).
de Wijn, H. W., 1966, *J. Chem. Phys.*, **44**, 810.
Hassler, J. C., Setzer, D. W., 1965, *J. Am. Chem. Soc.*, **87**, 3793.
Hildenbrand, D. L., 1968a, *J. Chem. Phys.*, **48**, 2457.
Hildenbrand, D. L., 1968b, *J. Chem. Phys.*, **48**, 3657.
Hildenbrand, D. L., Murad, E., 1965, *J. Chem. Phys.*, **43**, 1400.
Johnson, F. A., Colburn, C. B., 1961, *J. Am. Chem. Soc.*, **83**, 3043.
Kent, R. A., McDonald, J. D., Margrave, J. L., 1966, *J. Phys. Chem.*, **70**, 874.
Kistiakowsky, G. B., Ruhoff, J. R., Smith, H. A., Vaughan, W. E., 1936, *J. Am. Chem. Soc.*, **58**, 137, 146.
Ladd, M. F. C., Lee, W. H., 1958, *Trans. Faraday Soc.*, **54**, 34.
Ladd, M. F. C., Lee, W. H., 1959, *J. Inorg. Nucl. Chem.*, **11**, 264.
Ladd, M. F. C., Lee, W. H., 1960, *Acta Crystallogr.*, **13**, 959.
Potzinger, P., Lampe, F. W., 1970, *J. Phys. Chem.*, **74**, 719.
Rittner, E. S., 1951, *J. Chem. Phys.*, **19**, 1030.
Skinner, H. A., 1962, *J. Chem. Soc.*, 4396.
Skinner, H. A., Pilcher, G., 1963, *Q. Rev. Chem. Soc.*, **17**, 264.
Steele, W. C., Nichols, L. D., Stone, F. G. A., 1962, *J. Am. Chem. Soc.*, **84**, 4441.
Steele, W. C., Stone, F. G. A., 1962, *J. Am. Chem. Soc.*, **84**, 3599.
Stull, D. R. (Project Director), 1965, *JANAF Thermochemical Tables* (Dow Chemical Co., Michigan).
Wagman, D. D., Evans, W. H., Parker, V. B., Halow, I., Bailey, S. M., Schumm, R. H., 1965, *NBS Technical Note 270-1* (National Bureau of Standards, Washington, DC) 1968, *NBS Technical Note 270-3* (National Bureau of Standards, Washington, DC) 1969, *NBS Technical Note 270-4* (National Bureau of Standards, Washington, DC)

## 7.2 Wave functions and bonding

**C A Coulson** (deceased) formerly of the University of Oxford

### 7.2.1 Rydberg levels and molecular ionization potentials

The data in table 7.2.1a (taken from Price, 1935) are a set of successive absorption frequencies for acetylene. It is believed that they belong to one series. Show that they fit a Rydberg-type formula

$$\nu_n = \nu_\infty - \frac{R}{(n-\delta)^2},$$

where $R$ is the Rydberg constant, and $\delta$ is a quantum defect. Hence show that the corresponding ionization potential is approximately 11·35 eV.

What is the molecular-orbital (MO) description of the electron being ionized?

**Table 7.2.1a.** Ultraviolet absorption frequencies ($\mathrm{cm}^{-1}$) in acetylene.

| | | | |
|---|---|---|---|
| 74516 | 86665 | 89464 | 90560 |
| 83140 | 88431 | obscured | 90860 |

**Solution**

The first job is to estimate $\nu_\infty$. Do this by plotting the given $\nu$ against the number of the band (figure 7.2.1a). This shows that $\nu$ tends to about 92000 $\mathrm{cm}^{-1}$ as $n$ increases. The actual best-fit value is 92076 $\mathrm{cm}^{-1}$. Now plot $(\nu_\infty - \nu_n)^{-1/2}$ against the number of the band. Since we do not know the starting value, call it $n = n_0$. We then have values for $n_0$, $n_0 + 1$, $n_0 + 2$, $n_0 + 3$, $n_0 + 4$, $n_0 + 6$, and $n_0 + 7$. If a Rydberg-type formula holds, we should get a straight line whose slope is $R^{-1/2}$ and whose ordinate gives us the value of $n - \delta$ for each line (figure 7.2.1b).

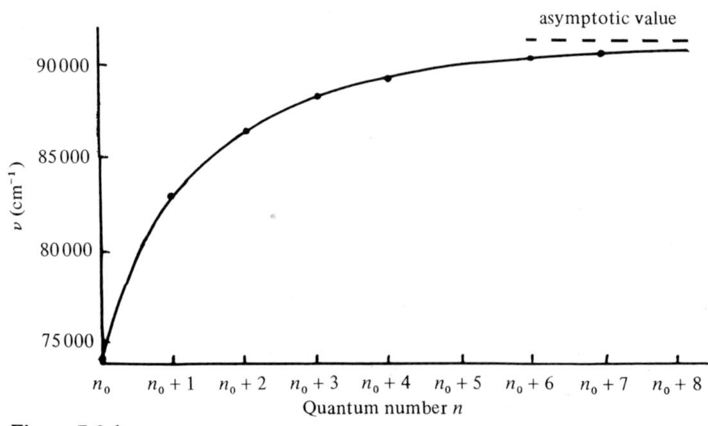


**Figure 7.2.1a.**

In acetylene all the atomic orbitals with principal quantum numbers $\leqslant 2$ for carbon are used in bonding. We therefore expect the lowest value of $n$ to be 3, corresponding to the least energy of excitation. (By analogy with other molecules we may take the electron involved in the spectrum to be a largely carbon–carbon bond electron.) If we put $n_0 = 3$, it follows from figure 7.2.1b that $\delta$ is about 0·50. The reason for this is that if we complete the straight line it crosses the horizontal axis at $n_0 - 2{\cdot}5$, and this must not have a negative value. Thus the Rydberg series has the form

$$\nu_n = 92076 - \frac{R}{(n - 0{\cdot}50)^2} \qquad n = 3, 4, \ldots .$$

The ionization potential is $h\nu_\infty$, i.e. 11·35 eV.

The MO description of acetylene is (Greene *et al.*, 1971):

$$(1\sigma_g)^2(1\sigma_u)^2(2\sigma_g)^2(2\sigma_u)^2(3\sigma_g)^2(1\pi_u)^4\,, \quad {}^1\Sigma_g^+ .$$

Here

$1\sigma_g$, $1\sigma_u$ are inner-shell orbitals of the carbons of the form $1s_A \pm 1s_B$, where $1s_A$ denotes the functional form of the 1s orbital on carbon atom A;

$2\sigma_g$, $2\sigma_u$, $3\sigma_g$ collectively provide the $\sigma$-bonds for $H{-}C_A$, $C_A{-}C_B$, and $C_B{-}H$;

$1\pi_u^4$ provides the two $\pi$-bonds that convert the C—C single bond into a triple bond.

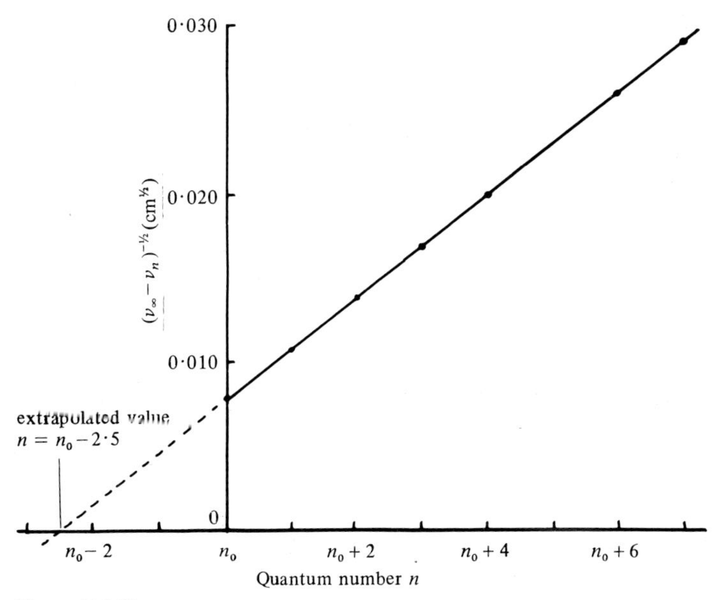


**Figure 7.2.1b.**

These latter electrons are the easiest to remove, so that the electron being ionized is in a $1\pi_u$ molecular orbital. Table 7.2.1b gives the necessary numerical results.

**Table 7.2.1b.** (all values of $\nu$ in $cm^{-1}$).

| $n$ | $\nu_n$ | $\nu_\infty - \nu_n$ | $(\nu_\infty - \nu_n)^{-1/2}$ | $n$ | $\nu_n$ | $\nu_\infty - \nu_n$ | $(\nu_\infty - \nu_n)^{-1/2}$ |
|---|---|---|---|---|---|---|---|
| $n_0$ | 74516 | 17560 | 0·0075 | $n_0+4$ | 89464 | 2612 | 0·0196 |
| $n_0+1$ | 83140 | 8936 | 0·0106 | $n_0+5$ | - | - | - |
| $n_0+2$ | 86665 | 5411 | 0·0136 | $n_0+6$ | 90560 | 1516 | 0·0257 |
| $n_0+3$ | 88431 | 3645 | 0·0166 | $n_0+7$ | 90860 | 1216 | 0·0287 |

**7.2.2 The free-electron model**

One very highly simplified model of a molecule (or an atom) is found by supposing that the moving particle (normally an electron) is confined within a region $0 \leqslant x \leqslant l$, but is otherwise free. This is a one-dimensional potential well of length $l$ and infinitely high walls. Show that the allowed wave functions are

$$\psi_n = \left(\frac{2}{l}\right)^{1/2} \sin\frac{n\pi x}{l}, \qquad n = 1, 2, \ldots, \tag{7.2.2a}$$

with energies

$$E_n = \frac{n^2h^2}{8ml^2} + V,$$

where $m$ is the mass of the electron and $V$ is the constant potential along the bottom of the well. We can usually choose $V = 0$. Why?

Verify that the functions (7.2.2a) are orthonormal; i.e. show that

$$\int \psi_{n_1}\psi_{n_2}\,dx = \begin{cases} 0 & \text{if } n_1 \neq n_2 \\ 1 & \text{if } n_1 = n_2 . \end{cases}$$

(a) Consider a system of $2N$ electrons, each having a spin, and moving as free noninteracting particles in the region $0 \leqslant x \leqslant l$. According to the aufbau principle, in the ground state they will occupy the lowest $N$ levels. Show that the wavelength of the first, i.e. lowest energy, absorption band is given by

$$\nu = \frac{(2N+1)h}{8ml^2}$$

$$\lambda = \frac{329\cdot 7}{2N+1} l^2 \text{ Å}. \tag{7.2.2b}$$

(b) Justify the use of this formula for the $\pi$-electrons of a polyene chain, and use it to estimate $\lambda$ for (i) ethylene, (ii) butadiene, and (iii) octatetraene.

The major difficulty is in the appropriate choice of $l$. First take $l$ equal to 1, 3, and 7 times the mean carbon–carbon aromatic bond length 1·40 Å. Notice that this gives rather poor agreement for ethylene and butadiene, the experimental values being (i) 1625, (ii) 2100, (iii) 2900 Å. For this reason it is now usual to add on half a bond length at each end of the chain. What is the justification for this value? Show that it improves the agreement with experiment. What value of $l$ would be necessary, for ethylene, to give exact agreement?

(c) The same method may be applied to the $\pi$-electrons of benzene by closing the linear segment on itself to form a circle. If $\phi$ measures the angle round an axis perpendicular to the ring, show that the electronic orbitals are

$$\psi_n = \left(\frac{1}{2\pi}\right)^{1/2} \exp(in\phi) \qquad n = 0, \pm 1, \pm 2, \ldots, \tag{7.2.2c}$$

with energies

$$E_n = \frac{n^2 h^2}{8\pi^2 m \rho^2} = \frac{n^2 h^2}{2ml^2}, \tag{7.2.2d}$$

where $\rho$ is the radius and $l$ the circumference of the circle.

Calculate the value of $\lambda$ for benzene, taking $l$ to be the circumference of a circle of radius 1·4 Å (experimental value $\lambda = 2600$ Å).

**Solution**

$\psi$ must satisfy the wave equation

$$\frac{d^2\psi}{dx^2} + \frac{8\pi^2 m}{h^2}(E - V)\psi = 0 .$$

Since $V$ is constant in $0 \leqslant x \leqslant l$, we may put

$$\frac{8\pi^2 m}{h^2}(E - V) = \text{const} = k^2 . \tag{7.2.2e}$$

The general solution of equation (7.2.2e) is

$$\psi = A \sin kx + B \cos kx ,$$

where $A$ and $B$ are constants to be fixed by the initial conditions that $\psi = 0$ at $x = 0$ and $x = l$. (Since the particle cannot get out of the well, $\psi$ must be identically zero for $x < 0$ and $x > l$. Moreover, $\psi$ is continuous, so that $\psi = 0$ at $x = 0, l$.) The condition at $x = 0$ implies that $B = 0$. The condition at $x = l$ now implies that $A \sin kl = 0$. We cannot have $A = 0$ also, since then there is no wave function left! So $\sin kl = 0$, $kl = n\pi$, where $n = 1, 2, \ldots$ . Hence, for equation (7.2.2e)

$$E_n - V = \frac{h^2 k^2}{8\pi^2 m} = \frac{n^2 h^2}{8ml^2}, \tag{7.2.2f}$$

as required. Furthermore

$$\psi_n = A \sin\frac{n\pi x}{l} .$$

$A$ is found from the normalization condition

$$\int_0^l \psi_n^2 \, \mathrm{d}x = 1 ,$$

which gives $A = (2/l)^{1/2}$.

This automatically shows that the wave functions are normalized. Orthogonality follows since

$$\begin{aligned}\int_0^l \psi_{n_1} \psi_{n_2} \, \mathrm{d}x &= \frac{2}{l}\int_0^l \sin\frac{n_1\pi x}{l} \sin\frac{n_2\pi x}{l} \, \mathrm{d}x \\ &= \frac{1}{l}\int_0^l \left[\cos\frac{(n_1-n_2)\pi x}{l} - \cos\frac{(n_1+n_2)\pi x}{l}\right] \mathrm{d}x \\ &= \frac{1}{l}\left[\frac{\sin[(n_1-n_2)\pi x/l]}{(n_1-n_2)\pi/l} - \frac{\sin[(n_1+n_2)\pi x/l]}{(n_1+n_2)\pi/l}\right]_0^l \\ &= 0 .\end{aligned}$$

We can usually take $V = 0$ because $V$ is a constant and equation (7.2.2e) shows that a change in the value of $V$ merely shifts all the energy values equally.

(a) According to the aufbau principle all the $N$ lowest energy levels will be doubly filled in the ground state. Thus the highest occupied level is $E_N$, the lowest unoccupied level is $E_{N+1}$, and the excitation energy is

$$E_{N+1} - E_N = \frac{(2N+1)h^2}{8ml^2} .$$

This energy difference is $h\nu$, so that

$$\nu = \frac{(2N+1)h}{8ml^2}$$

and

$$\lambda = \frac{c}{\nu} = \frac{8ml^2c}{(2N+1)h} = \frac{329\cdot 7}{2N+1} l^2 \text{ Å} . \qquad (7.2.2b)$$

(b) In a polyene chain the $\pi$-electrons are 'reasonably' free to move along the chain, so that the analysis applies only to them, and not to the $\sigma$-electrons, which may be taken to be more nearly confined to localised C—C and C—H $\sigma$-bonds. For ethylene $2N = 2$; for butadiene $2N = 4$, and for octatetraene $2N = 8$. The corresponding values of $\lambda$ are 215, 1164, and 3520 Å. Only the third of these is at all satisfactory.

In figure 7.2.2 we show the C—C bond in ethylene. If we think of the bond length as the sum of two atomic radii as shown by the central dotted line, then each atom will 'project' an equal length on the far side of the bond, so that the effective size of the molecule is found by adding one-half of a bond length at each end.

The new values are 860, 2070, and 4590 Å. Agreement with experiment is improved for the two shorter molecules, but is worsened for the longest one.

To get agreement for ethylene we need a value of $l$ such that from equation (7.2.2b):

$$\frac{329\cdot 7l^2}{3} = 1625\ \text{Å}^2\ .$$

i.e. $l = 3\cdot 85$ Å.

(c) In terms of the angle $\phi$, distance round the circumference is given by $s = \rho\phi$, and the wave equation

$$\frac{d^2\psi}{ds^2}+\frac{8\pi^2 m}{h^2}(E-V)\psi = 0$$

becomes

$$\frac{d^2\psi}{d\phi^2}+\frac{8\pi^2 m\rho^2}{h^2}(E-V)\psi = 0\ .$$

Putting $8\pi^2 m\rho^2(E-V)/h^2 = k^2$, we get the solution

$$\psi = A\exp(ik\phi)+B\exp(-ik\phi)\ .$$

$A$ and $B$ are now arbitrary constants (apart from normalization, of course). But $\psi$ is single-valued if it returns to its original value when $\phi$ increases by $2\pi$, so that $\exp(2\pi ik) = 1$ and $2\pi ik = 2n\pi i$; i.e. $k = n$ where $n = 0, \pm 1, \pm 2, \ldots$ .

The solutions are thus $\exp(in\phi)$ with

$$E_n - V = \frac{h^2k^2}{8\pi^2 m\rho^2} = \frac{n^2h^2}{8\pi^2 m\rho^2} = \frac{n^2h^2}{2ml^2}$$

as required. Since $\psi$ is now complex, normalization requires

$$\int_{s=0}^{l}\psi^*\psi\, ds = 1\ .$$

But $\psi^*\psi = |A|^2$, leading to the required normalization if we choose $A$ to be real.

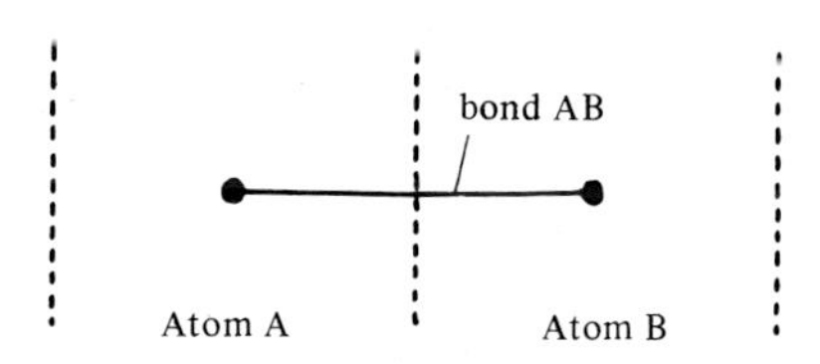


**Figure 7.2.2.**

For benzene, with this model

$$h\nu = E_2 - E_1 ,$$

since we have six $\pi$-electrons, and $\psi_0$ is nondegenerate, holding two electrons while $\psi_1$ is doubly degenerate, holding four. Thus

$$\lambda = \frac{c}{\nu} = \frac{hc}{E_2 - E_1} ,$$

leading to

$$\lambda = 2100 \text{ Å} ,$$

which is not far from the true value.

Note that in the cyclic system we may put $n = 0$ in equation (7.2.2d) to get the lowest energy; but in the open chain we have to start with $n = 1$ in equation (7.2.2f).

For further study of this model, see Bayliss (1952), Griffith (1953), and various papers in the *Journal of Chemical Physics* by J. R. Platt, K. Ruedenberg, and collaborators (e.g. Ruedenberg, 1958).

### 7.2.3 Molecular-orbital treatment of $H_2^+$

If we use the simple LCAO (linear combination of atomic orbitals) approximation, we write the molecular orbital (MO) as

$$\chi = N(\phi_a + \phi_b) , \qquad (7.2.3a)$$

where

$$\phi_a = \left(\frac{1}{\pi}\right)^{1/2} \exp(-r_a) , \qquad (7.2.3b)$$

and $N$ is a normalizing factor (see Coulson, 1961, for details). A and B are the two nuclei, and $r_a$, $r_b$ are the distances of the electron from A and B respectively. Verify that the atomic orbital (7.2.3b) is correctly normalized, in atomic units. Show also that it satisfies the wave equation for an isolated H atom:

$$\nabla^2 \phi_a + 2\left(E + \frac{1}{r_a}\right)\phi_a = 0 , \qquad (7.2.3c)$$

where $E = -\frac{1}{2}$, and atomic units are used throughout.

Next show that

$$N^{-2} = 2(1+S) , \qquad (7.2.3d)$$

where

$$S = \int \phi_a \phi_b \, d\tau$$

is the familiar overlap integral between the atomic orbitals $\phi_a$ and $\phi_b$.

Write down the molecular Hamiltonian $\mathcal{H}$, and work out the value of

$$E_\chi = \frac{\int \chi^* \mathcal{H} \chi \, d\tau}{\int \chi^* \chi \, d\tau} .$$

Use equation (7.2.3c), work in terms of the atomic orbitals, and show that

$$E_\chi = -\frac{1}{2} - \frac{\int \frac{1}{r_b} \phi_a^2 \, d\tau + \int \frac{1}{r_b} \phi_a \phi_b \, d\tau}{1+S} . \qquad (7.2.3e)$$

Evaluate $E_\chi$ when the internuclear distance is $R = 0, 1, 2, 3, 4a_0$, using the numerical values of the integrals given in table 7.2.4a. Add the coulombic repulsion of the nuclei, and verify that the total energy, with fixed nuclei, has a minimum at about $R = 1 \cdot 42$ Å ($a_0 = 0 \cdot 529$ Å $= 0 \cdot 0529$ nm), with $D_e = 1 \cdot 63$ eV. (1 a.u. = $27 \cdot 20$ eV.) The true values with a correct wave function are $2 \cdot 78$ eV at $R = 2 \cdot 00 a_0 = 1 \cdot 06$ Å.

**Solution**

To show that $\phi_a$ in equation (7.2.3b) is normalized we have to show that $\int \phi_a^2 \, d\tau = 1$. Since $\phi_a$ is spherically symmetrical about nucleus A we can take the volume element to be a tiny spherical shell lying between $r_a$ and $r_a + dr_a$. Then $d\tau = 4\pi r_a^2 \, dr_a$, and

$$\int \phi_a^2 \, d\tau = \int_0^\infty \frac{1}{\pi} \exp(-2r_a) 4\pi r_a^2 \, dr_a = 4 \int_0^\infty r_a^2 \exp(-2r_a) \, dr_a = 1 .$$

Note that

$$\int_0^\infty r^n \exp(-\alpha r) \, dr = \frac{n!}{\alpha^{n+1}} .$$

The full expression for $\nabla^2$ in spherical polar coordinates $(r, \theta, \phi)$ is

$$\nabla^2 = \frac{\partial^2}{\partial r^2} + \frac{2}{r}\frac{\partial}{\partial r} + \frac{1}{r^2}\left(\frac{1}{\sin\theta}\frac{\partial}{\partial\theta}\sin\theta\frac{\partial}{\partial\theta} + \frac{1}{\sin^2\theta}\frac{\partial^2}{\partial\phi^2}\right) .$$

So if $\phi_a$ depends only on $r_a$, and we use A as origin, we obtain:

$$\nabla^2 \phi_a = \frac{d^2\phi_a}{dr_a^2} + \frac{2}{r_a}\frac{d\phi_a}{dr_a} = \frac{1}{\pi^{1/2}}\left(1 - \frac{2}{r_a}\right)\exp(-r_a) .$$

Thus

$$\nabla^2 \phi_a + 2\left(E + \frac{1}{r_a}\right)\phi_a = 0 ,$$

with $E = -\frac{1}{2}$, as required. Notice that in atomic units $\hbar = h/2\pi = 1$, $m = 1$, and so the wave equation becomes

$$\nabla^2 \psi + 2(E - V)\psi = 0 .$$

Next we have to normalize $\chi$, i.e. put $\int \chi^2 \, d\tau = 1$, so that

$$1 = N^2 \int (\phi_a^2 + 2\phi_a \phi_b + \phi_b^2) \, d\tau .$$

Thus

$$N^{-2} = 1 + 2S + 1 = 2(1+S)$$

as required. For the $H_2^+$ ion, since $V = -1/r_a - 1/r_b$, we have in atomic units

$$\mathcal{H} = -\frac{1}{2}\nabla^2 - \frac{1}{r_a} - \frac{1}{r_b} .$$

Then

$$E_\chi = \frac{\int \chi^* \mathcal{H} \chi \, d\tau}{\int \chi^* \chi \, d\tau} = \int \chi^* \mathcal{H} \chi \, d\tau$$

$$= N^2 \int (\phi_a + \phi_b) \mathcal{H} (\phi_a + \phi_b) \, d\tau$$

$$= N^2 (\mathcal{H}_{aa} + \mathcal{H}_{ab} + \mathcal{H}_{ba} + \mathcal{H}_{bb}) ,$$

in an obvious notation. Now

$$\mathcal{H}_{aa} = \int \phi_a \left( -\frac{1}{2}\nabla^2 - \frac{1}{r_a} - \frac{1}{r_b} \right) \phi_a \, d\tau$$

$$= \int \phi_a \left( -\frac{1}{2}\nabla^2 - \frac{1}{r_a} \right) \phi_a \, d\tau - \int \frac{1}{r_b} \phi_a^2 \, d\tau$$

$$= -\frac{1}{2} \int \phi_a^2 \, d\tau - \int \frac{1}{r_b} \phi_a^2 \, d\tau = -\frac{1}{2} - \int \frac{1}{r_b} \phi_a^2 \, d\tau ,$$

if we use the wave equation for $\phi_a$.

Similarly,

$$\mathcal{H}_{ab} = \int \phi_a \left( -\frac{1}{2}\nabla^2 - \frac{1}{r_a} - \frac{1}{r_b} \right) \phi_b \, d\tau$$

$$= \int \phi_a \left( -\frac{1}{2}\nabla^2 - \frac{1}{r_b} \right) \phi_b \, d\tau - \int \frac{1}{r_a} \phi_a \phi_b \, d\tau$$

$$= -\frac{1}{2} \int \phi_a \phi_b \, d\tau - \int \frac{1}{r_a} \phi_a \phi_b \, d\tau .$$

Also, by symmetry, $\mathcal{H}_{aa} = \mathcal{H}_{bb}$, and $\mathcal{H}_{ab} = \mathcal{H}_{ba}$. Thus, after collecting terms, we obtain the required expression for $E$.

Table 7.2.3 shows the required numerical values.

**Table 7.2.3.** Values of the energy terms (a.u.).

| | $R(a_0)$ | | | | |
|---|---|---|---|---|---|
| | 0 | 1 | 2 | 3 | 4 |
| $E$ | $-1{\cdot}500$ | $-1{\cdot}288$ | $-1{\cdot}054$ | $-0{\cdot}892$ | $-0{\cdot}787$ |
| Coulombic repulsion, $1/R$ | $\infty$ | $1{\cdot}000$ | $0{\cdot}500$ | $0{\cdot}333$ | $0{\cdot}250$ |
| $E(H_2^+)$ | $\infty$ | $-0{\cdot}288$ | $-0{\cdot}554$ | $-0{\cdot}559$ | $-0{\cdot}537$ |

One way of finding the minimum, which clearly lies between $R = 2a_0$ and $R = 3a_0$, is by means of a graph, but this does not give a very good result since the curve is very flat near its minimum. Suppose, therefore, that we put a parabola through the last three points. We can write

$$E = -0{\cdot}559+0{\cdot}005\frac{(R-3)(R-4)}{2}+0{\cdot}022\frac{(R-2)(R-3)}{2},$$

since this fits exactly at $R = 2, 3, 4a_0$. Hence, near $R = 3a_0$,

$$E = -0{\cdot}463-0{\cdot}0725R+0{\cdot}0135R^2 .$$

This is a minimum when $\mathrm{d}E/\mathrm{d}R = 0$, i.e. $0{\cdot}0725 = 0{\cdot}0270R$, so that $R = 2{\cdot}685a_0 = 1{\cdot}41$ Å. Moreover $E_{\min} = -0{\cdot}5599$, leading to $D_e = 0{\cdot}0599$ a.u. $= 1{\cdot}63$ eV. Ideally we ought to take points closer to $R = 2{\cdot}685a_0$ to get a better value of $D_e$.

### 7.2.4 Comparison of Heitler–London and molecular-orbital treatments of $H_2$

(a) The MO wave function is

$$\psi_{\mathrm{MO}} = \chi(1)\chi(2),$$

where $\chi(1)$ is given by equation (7.2.3a). All spin factors are omitted in this problem since they will be the same in all our functions, and will therefore carry through all the calculations.

Now work out

$$E_{\mathrm{MO}} = \frac{\int \psi_{\mathrm{MO}}^* \mathcal{H} \psi_{\mathrm{MO}}\, \mathrm{d}\tau}{\int \psi_{\mathrm{MO}}^* \psi_{\mathrm{MO}}\, \mathrm{d}\tau},$$

where $\mathcal{H}$ is now the two-electron Hamiltonian. It is convenient to write

$$\mathcal{H} = \mathcal{H}(1)+\mathcal{H}(2)+\frac{1}{r_{12}} \qquad \text{(in atomic units)},$$

where

$$\mathcal{H}(1) = -\frac{1}{2}\nabla_1^2-\frac{1}{r_{a_1}}-\frac{1}{r_{b_1}}$$

would be the Hamiltonian for electron 1 if electron 2 were not present at all.

Deduce that

$$E_{\mathrm{MO}} = 2E_{\chi} + \iint \chi(1)\chi(2)\frac{1}{r_{12}}\chi(1)\chi(2)\,\mathrm{d}\tau_1\,\mathrm{d}\tau_2\ ,$$

where $E_{\chi}$ is given by equation (7.2.3e).

Evaluate $E_{\mathrm{MO}}$ for $R = 0, 1, 2, 3, 4a_0$, and deduce that the estimated $D_e$ is about 2·2 eV at $R_e$ just less than $2a_0$.

(b) The unnormalized Heitler–London wave function is

$$\psi_{\mathrm{HL}} = \phi_a(1)\phi_b(2) + \phi_b(1)\phi_a(2)\ .$$

Use this expression, and the fact that the two atoms are equivalent, to show that, similarly to part (a),

$$E_{\mathrm{HL}} = -1 + \frac{Q+J}{1+S^2}\ ,$$

where

$S$ = overlap integral, $\int\phi_a\phi_b\,\mathrm{d}\tau$ ;

$Q$ = coulomb integral, $\displaystyle\iint\frac{\phi_a^2(1)\phi_b^2(2)}{r_{12}}\mathrm{d}\tau_1\,\mathrm{d}\tau_2 - 2\int\frac{\phi_a^2(1)}{r_{b_1}}\mathrm{d}\tau_1$ ;

$J$ = exchange integral, $\displaystyle\iint\frac{\phi_a(1)\phi_b(1)\phi_a(2)\phi_b(2)}{r_{12}}\mathrm{d}\tau_1\,\mathrm{d}\tau_2 - 2S\int\frac{\phi_a(1)\phi_b(1)}{r_{a_1}}\mathrm{d}\tau_1$ .

The electronic energy is now known in terms of certain definite integrals. The values of these are listed in table 7.2.4a.

**Table 7.2.4a.** Values of the Heitler–London integrals (a.u.).

| Integral | $R(a_0)$ | | | | |
|---|---|---|---|---|---|
| | 0 | 1 | 2 | 3 | 4 |
| $S$ | 1·000 | 0·8584 | 0·5865 | 0·3485 | 0·1893 |
| $\int\frac{\phi_a^2(1)}{r_{b_1}}\mathrm{d}\tau_1$ | 1·000 | 0·7293 | 0·4725 | 0·3300 | 0·2496 |
| $\int\frac{\phi_a(1)\phi_b(1)}{r_{a_1}}\mathrm{d}\tau_1$ | 1·000 | 0·7358 | 0·4060 | 0·1996 | 0·0916 |
| $\iint\frac{\phi_a^2(1)\phi_b^2(2)}{r_{12}}\mathrm{d}\tau_1\,\mathrm{d}\tau_2$ | 0·6250 | 0·5545 | 0·4260 | 0·3198 | 0·2476 |
| $\iint\frac{\phi_a^2(1)\phi_a(2)\phi_b(2)}{r_{12}}\mathrm{d}\tau_1\,\mathrm{d}\tau_2$ | 0·6250 | 0·5070 | 0·3080 | 0·1607 | 0·0770 |
| $\iint\frac{\phi_a(1)\phi_b(1)\phi_a(2)\phi_b(2)}{r_{12}}\mathrm{d}\tau_1\,\mathrm{d}\tau_2$ | 0·6250 | 0·4367 | 0·1842 | 0·0585 | 0·0156 |

Evaluate the electronic energy for $R = 0, 1, 2, 3, 4a_0$, add the Coulombic energy of the nuclei ($1/R$), and plot $E(R)$ against $R$. Estimate the dissociation energy. This is effectively what Heitler and London (1927) did.

(c) Compare the dissociation energies in $\psi_{\text{MO}}$ and $\psi_{\text{HL}}$ with the true value 4·75 eV.

Notice that the single-centre integral $\int[\phi_a^2(1)\phi_a^2(2)/r_{12}]\,d\tau_1\,d\tau_2$ is a particular case of each of the last three integrals in the table, corresponding to $R = 0$, so that $\phi_a \equiv \phi_b$.

**Solution**

(a) $\psi_{\text{MO}}$ is normalized if each $\chi$ is normalized, since

$$\iint \psi_{\text{MO}}^2\,d\tau_1\,d\tau_2 = \iint \chi^2(1)\chi^2(2)\,d\tau_1\,d\tau_2$$
$$= \left(\int \chi^2(1)\,d\tau_1\right)\left(\int \chi^2(2)\,d\tau_2\right) = 1\,.$$

So

$$E_{\text{MO}} = \iint \psi_{\text{MO}}\,\mathcal{H}\,\psi_{\text{MO}}\,d\tau_1\,d\tau_2$$
$$= \iint \chi(1)\chi(2)\left[\mathcal{H}(1)+\mathcal{H}(2)+\frac{1}{r_{12}}\right]\chi(1)\chi(2)\,d\tau_1\,d\tau_2\,.$$

The first term in this expression gives

$$\iint \chi(1)\chi(2)\mathcal{H}(1)\chi(1)\chi(2)\,d\tau_1\,d\tau_2 = \left(\int \chi(1)\mathcal{H}(1)\chi(1)\,d\tau_1\right)\times\left(\int \chi^2(2)\,d\tau_2\right)$$
$$= \int \chi(1)\mathcal{H}(1)\chi(1)\,d\tau_1 = E_\chi\,.$$

The second term gives a further $E_\chi$, so that

$$E_{\text{MO}} = 2E_\chi + \iint \chi(1)\chi(2)\frac{1}{r_{12}}\chi(1)\chi(2)\,d\tau_1\,d\tau_2\,,$$

as required.

To evaluate the final integral we expand in terms of atomic orbitals and obtain

$$N^4\iint[\phi_a(1)+\phi_b(1)]^2\frac{1}{r_{12}}[\phi_a(2)+\phi_b(2)]^2\,d\tau_1\,d\tau_2\,.$$

On expansion there are nine terms, which are often equal in pairs; for example

$$\iint \frac{\phi_a^2(1)\phi_a(2)\phi_b(2)}{r_{12}}\,d\tau_1\,d\tau_2 = \iint \frac{\phi_b^2(1)\phi_b(2)\phi_a(2)}{r_{12}}\,d\tau_1\,d\tau_2\,,$$

since the two nuclei are similar. On gathering similar terms together we obtain

$$E_{\mathrm{MO}} = \frac{1}{2(1+S)^2}\left[\iint \frac{\phi_a^2(1)\phi_a^2(2)}{r_{12}}\,\mathrm{d}\tau_1\,\mathrm{d}\tau_2 + 4\iint \frac{\phi_a^2(1)\phi_a(2)\phi_b(2)}{r_{12}}\,\mathrm{d}\tau_1\,\mathrm{d}\tau_2 \right.$$

$$\left. + \iint \frac{\phi_a^2(1)\phi_b^2(2)}{r_{12}}\,\mathrm{d}\tau_1\,\mathrm{d}\tau_2 + 2\iint \frac{\phi_a(1)\phi_b(1)\phi_a(2)\phi_b(2)}{r_{12}}\,\mathrm{d}\tau_1\,\mathrm{d}\tau_2\right].$$

On putting in the numerical values the results listed in table 7.2.4b are obtained.

**Table 7.2.4b.** Values of the energy terms (a.u.).

| Term | $R(a_0)$ | | | | |
|---|---|---|---|---|---|
| | 0 | 1 | 2 | 3 | 4 |
| $2E_X$ | −3·000 | −2·576 | −2·108 | −1·784 | −1·574 |
| Last integral | 0·625 | 0·591 | 0·527 | 0·469 | 0·428 |
| $E_{\mathrm{MO}}$ | −2·375 | −1·985 | −1·581 | −1·315 | −1·146 |
| $E_{\mathrm{coulomb}}$ | ∞ | 1·000 | 0·500 | 0·333 | 0·250 |
| Total MO energy | ∞ | −0·985 | −1·081 | −0·983 | −0·896 |

The minimum occurs at just less than $2a_0$, and the dissociation energy is approximately 0·082 a.u. = 2·2 eV.

(b)

$$E_{\mathrm{HL}} = \frac{\iint \psi_{\mathrm{HL}}\mathcal{H}\psi_{\mathrm{HL}}\,\mathrm{d}\tau_1\,\mathrm{d}\tau_2}{\iint \psi_{\mathrm{HL}}^2\,\mathrm{d}\tau_1\,\mathrm{d}\tau_2}.$$

The denominator is

$$\iint [\phi_a(1)\phi_b(2)+\phi_b(1)\phi_a(2)][\phi_a(1)\phi_b(2)+\phi_b(1)\phi_a(2)]\,\mathrm{d}\tau_1\,\mathrm{d}\tau_2.$$

Take the whole of the first bracket with the first product in the second bracket. We are to integrate over both sets of electron coordinates, so that no electron 'labels' will remain at the end. We *could* call the electrons 2 and 1 instead of 1 and 2. If we do so, the integral reduces to precisely the same as the integral of the product of the whole of the first bracket with the second term in the second bracket. So the denominator is

$$2\iint [\phi_a(1)\phi_b(2)+\phi_b(1)\phi_a(2)]\phi_a(1)\phi_b(2)\,\mathrm{d}\tau_1\,\mathrm{d}\tau_2.$$

Exactly the same argument applies to the numerator, leading to

$$E_{\mathrm{HL}} = \frac{\iint [\phi_a(1)\phi_b(2)+\phi_b(1)\phi_a(2)]\mathcal{H}\phi_a(1)\phi_b(2)\,\mathrm{d}\tau_1\,\mathrm{d}\tau_2}{\iint [\phi_a(1)\phi_b(2)+\phi_b(1)\phi_a(2)]\phi_a(1)\phi_b(2)\,\mathrm{d}\tau_1\,\mathrm{d}\tau_2}.$$

The denominator is now simply $1+S^2$. For the numerator, use the fact that $\phi_a$ satisfies an equation

$$\left(-\frac{1}{2}\nabla^2 - \frac{1}{r_{a_1}}\right)\phi_a(1) = -\tfrac{1}{2}\phi_a(1) ,$$

and similarly for $\phi_b$, to show that

$$\mathcal{H}\phi_a(1)\phi_b(2) = \left(-1 - \frac{1}{r_{b_1}} - \frac{1}{r_{a_2}} + \frac{1}{r_{12}}\right)\phi_a(1)\phi_b(2) .$$

By substituting this in the earlier expression for $E_{HL}$ the required result is obtained.

Numerical substitution gives the values in table 7.2.4c. The minimum occurs at about $1\cdot65a_0 = 0\cdot87$ Å, and it leads to $D_e = 0\cdot116$ a.u. = $3\cdot14$ eV. But this value could only be found reliably by taking a set of points between $R = a_0$ and $R = 2a_0$.

(c) Neither value is very good. Both are reasonable, with the Heitler–London method scoring a small advantage over the MO method.

**Table 7.2.4c.** Heitler–London energy terms (a.u.) for $H_2^+$.

| Term | $R(a_0)$ | | | | |
|---|---|---|---|---|---|
| | 0 | 1 | 2 | 3 | 4 |
| $\dfrac{Q+J}{1+S^2}$ | $-1\cdot375$ | $-0\cdot996$ | $-0\cdot604$ | $-0\cdot375$ | $-0\cdot261$ |
| $E_{HL}$ | $-2\cdot375$ | $-1\cdot996$ | $-1\cdot604$ | $-1\cdot375$ | $-1\cdot261$ |
| Total HL energy | $\infty$ | $-0\cdot996$ | $-1\cdot104$ | $-1\cdot042$ | $-1\cdot011$ |

For a more complete account of this, see Coulson (1937).

### 7.2.5 Koopmans' theorem

Koopmans' theorem (1933) states that a good approximation to the ionization potential $I$ of an atom or a molecule is given by the magnitude of the corresponding Hartree–Fock parameter. This is the eigenvalue of the Hartree–Fock one-electron wave equation for the orbital in question. With a closed-shell single-determinant wave function and the remaining orbitals assumed unchanged, the theorem would be exact. For any other situation it is not exact, even though it is widely used.

For a simple illustration of the effect of relaxation, or readjustment, of the remaining orbitals when one electron has been removed, consider the variational wave function for helium in its ground state (spin neglected, since it plays no part in our discussion),

$$\psi(\text{He}) = \phi(c, 1)\phi(c, 2) ,$$

where $\phi(c, 1)$ is a normalized atomic orbital for electron 1, with orbital exponent $c$, so that

$$\phi(c, 1) = \left(\frac{c^3}{\pi}\right)^{1/2} \exp(-cr_1) .$$

The best value of $c$ for helium is $c = 2 - \frac{5}{16} = \frac{27}{16}$. However, when electron 2 is removed, the remaining electron will have a wave function

$$\psi(\mathrm{He}^+) = \phi(c, 1) ,$$

where now $c = 2$. The relaxation in the wave function after ionization is shown by the difference between $\frac{27}{16}$ and 2.

Show that the Hartree–Fock self-consistent-field equation for the helium atom is, in atomic units,

$$\left(-\frac{1}{2}\nabla_1^2 - \frac{2}{r_1}\right)\phi(1) + \left(\int \frac{\phi^2(2)}{r_{12}}\,\mathrm{d}\tau_2\right)\phi(1) = \epsilon\phi(1) .$$

Deduce that the Hartree–Fock parameter $\epsilon$ is given by

$$\epsilon = \int \phi(1)\left(-\frac{1}{2}\nabla_1^2 - \frac{2}{r_1}\right)\phi(1)\,\mathrm{d}\tau_1 + \iint \frac{\phi^2(1)\phi^2(2)}{r_{12}}\,\mathrm{d}\tau_1\,\mathrm{d}\tau_2 .$$

Next, show that, if there is no relaxation after ionization,

$$E(\mathrm{He}) = 2\int \phi(1)\left(-\frac{1}{2}\nabla_1^2 - \frac{2}{r_1}\right)\phi(1)\,\mathrm{d}\tau_1 + \iint \frac{\phi^2(1)\phi^2(2)}{r_{12}}\,\mathrm{d}\tau_1\,\mathrm{d}\tau_2 ,$$

$$E(\mathrm{He}^+) = \int \phi(1)\left(-\frac{1}{2}\nabla_1^2 - \frac{2}{r_1}\right)\phi(1)\,\mathrm{d}\tau_1 .$$

Use this result to verify Koopmans' theorem for the ionisation potential, $I_{\mathrm{He}}$:

$$I_{\mathrm{He}} = E(\mathrm{He}^+) - E(\mathrm{He}) = -\epsilon .$$

Next, assuming that all these formulae hold for the atomic orbitals $\phi(c, 1)$, show that

$$E(\mathrm{He}) = -\left(\tfrac{27}{16}\right)^2 , \qquad E(\mathrm{He}^+) = \tfrac{1}{2}c^2 - 2c .$$

Assuming no relaxation, show that this gives

$$I_{\mathrm{He}} = -\epsilon = -\tfrac{1}{2}c^2 + \tfrac{11}{8}c = \tfrac{459}{512} .$$

Assuming relaxation, show that this leads to

$$I_{\mathrm{He}} = \left(\tfrac{27}{16}\right)^2 - 2 = \tfrac{217}{256} ,$$

which is about $1 \cdot 3$ eV less than the Koopmans' value. Why would the use of Koopmans' theorem be expected to lead to an estimated ionization potential which is too big?

*Note*. Assume that

$$\iint \frac{\phi^2(c, 1)\phi^2(c, 2)}{r_{12}}\,\mathrm{d}\tau_1\,\mathrm{d}\tau_2 = \tfrac{5}{8}c .$$

**Solution**

Since there is no exchange between the two electrons, the effect of electron 2 on the motion of electron 1 is to provide an additional potential energy $\int[\phi^2(2)/r_{12}]\,d\tau_2$. As the nuclear charge is +2, the one-electron wave equation is therefore

$$\left(-\frac{1}{2}\nabla_1^2-\frac{2}{r_1}\right)\phi(1)+\left(\int\frac{\phi^2(2)}{r_{12}}\,d\tau_2\right)\phi(1) = \epsilon\phi(1)\,. \tag{7.2.5a}$$

This equation can also be obtained by requiring that the total electronic energy be stationary (have a minimum) with respect to all small changes in $\phi$.

Multiplying equation (7.2.5a) by $\phi(1)$ and integrating leads (since $\phi$ is supposed to be normalized) to

$$\epsilon = \int\phi(1)\left(-\frac{1}{2}\nabla_1^2-\frac{2}{r_1}\right)\phi(1)\,d\tau_1+\iint\frac{\phi^2(1)\phi^2(2)}{r_{12}}\,d\tau_1\,d\tau_2\,. \tag{7.2.5b}$$

The full Hamiltonian is

$$\mathcal{H} = \left(-\frac{1}{2}\nabla_1^2-\frac{2}{r_1}\right)+\left(-\frac{1}{2}\nabla_2^2-\frac{2}{r_2}\right)+\frac{1}{r_{12}}\,.$$

The total energy with a wave function $\psi$ (supposed real) is

$$E = \frac{\iint\psi\mathcal{H}\psi\,d\tau_1\,d\tau_2}{\iint\psi^2\,d\tau_1\,d\tau_2}\,.$$

Putting $\psi = \phi(1)\phi(2)$ we obtain

$$E(\text{He}) = \int\phi(1)\left(-\frac{1}{2}\nabla_1^2-\frac{2}{r_1}\right)\phi(1)\,d\tau_1+\int\phi(2)\left(-\frac{1}{2}\nabla_2^2-\frac{2}{r_2}\right)\phi(2)\,d\tau_2 + \iint\frac{\phi^2(1)\phi^2(2)}{r_{12}}\,d\tau_1\,d\tau_2\,.$$

The first two terms on the right-hand side are equal, since both are pure numbers, so that

$$E(\text{He}) = 2\int\phi(1)\left(-\frac{1}{2}\nabla_1^2-\frac{2}{r_1}\right)\phi(1)\,d\tau_1+\iint\frac{\phi^2(1)\phi^2(2)}{r_{12}}\,d\tau_1\,d\tau_2\,. \tag{7.2.5c}$$

But with a single electron, as in $He^+$:

$$E(\text{He}^+) = \int\phi(1)\left(-\frac{1}{2}\nabla_1^2-\frac{2}{r_1}\right)\phi(1)\,d\tau_1\,. \tag{7.2.5d}$$

From equations (7.2.5b), (7.2.5c), and (7.2.5d) it follows that

$$-\epsilon = E(\text{He}^+)-E(\text{He}) = I_{\text{He}}\,.$$

On putting $\phi(1)$ in equation (7.2.5d) equal to $\phi(c, 1) = (c^3/\pi)^{1/2}\exp(-cr_1)$, it soon follows from this equation that

$$E(\mathrm{He}^+) = \tfrac{1}{2}c^2 - 2c\ . \tag{7.2.5e}$$

Similarly, from equation (7.2.5c):

$$E(\mathrm{He}) = 2(\tfrac{1}{2}c^2 - 2c) + \tfrac{5}{8}c = c^2 - \tfrac{27}{8}c\ . \tag{7.2.5f}$$

If we use the same value of $c$ in equations (7.2.5e) and (7.2.5f), so that we assume no relaxation after the removal of one electron, then

$$I_{\mathrm{He}} = -\tfrac{1}{2}c^2 + \tfrac{11}{8}c\ , \tag{7.2.5g}$$

as required. It follows from equation (7.2.5f) that for He the best value of $c$, which is such that the expression for $E$ is stationary, is given by $2c - \frac{27}{8} = 0$, i.e. $c = \frac{27}{16} = 2 - \frac{5}{16}$. Putting this value in equation (7.2.5g) gives $\frac{459}{512}$, as stated. But if for $\mathrm{He}^+$ we take $c = 2$, so that $E(\mathrm{He}^+) = -2$, then $I_{\mathrm{He}} = (\frac{27}{16})^2 - 2 = \frac{217}{256}$. This value is $\frac{459}{512} - \frac{217}{256} = \frac{25}{512}$ a.u. $= 1 \cdot 3$ eV less than the Koopmans' value.

In general the use of Koopmans' theorem will lead to too big an ionization potential, since relaxation, which is not allowed for in the theorem, will lower the energy of the ion, and thus reduce the amount of energy needed to create it from the neutral system. Table 7.2.5 (Lorquet, 1960) shows a few values of $I$ for some simple molecules.

**Table 7.2.5.**

| | $I$ (eV) | | |
|---|---|---|---|
| | OH | HF | $NH_3$ |
| By Koopmans' theorem | 16·2 | 17·3 | 14·0 |
| Including relaxation | 13·2 | 15·8 | 10·2 |

**7.2.6 The Hellmann–Feynman theorem**

The Hellmann–Feynman theorem (Hellmann, 1937; Feynman, 1939) states that if the Hamiltonian $\mathcal{H}$ contains some parameter $\alpha$, so that the eigenvalues $E$ and eigenfunctions $\psi$ depend on $\alpha$, then

$$\frac{\partial E}{\partial \alpha} = \left(\psi, \frac{\partial \mathcal{H}}{\partial \alpha}\psi\right) = \int \psi^* \frac{\partial \mathcal{H}}{\partial \alpha}\psi\, \mathrm{d}\tau\ .$$

The theorem holds for Hartree–Fock wave functions, and, of course, for exact wave functions. It is often used, however, with approximate wave functions. We must then anticipate some errors.

A particularly important application is to a polyatomic molecule, in which $\alpha$ is some geometrical coordinate such as a bond angle or bond length. The present problem deals with a diatomic molecule AB, with nuclear charges $Z_a$, $Z_b$, and bond length $R$ (see figure 7.2.6).

By taking the parameter $\alpha$ to be the $z$-coordinate of B, we compute $\partial E/\partial z$, which is equal to the force on nucleus B in the negative $z$-direction. Alternatively, by evaluating $\partial E/\partial z$ over a range of values of $R$ we can integrate to give

$$E(R) = \int_{-\infty}^{R} \frac{\partial E}{\partial z}\,\mathrm{d}z\ ,$$

and thus obtain the potential-energy curve of the molecule.
(a) The valence-bond (VB) and molecular-orbital (MO) wave functions for the ground state of $H_2$ are given, as usual, in terms of the two normalized atomic orbitals $\phi_a$, $\phi_b$ together with their overlap integral $S$, by the expressions

$$\psi_{VB}(1, 2) = N[\phi_a(1)\phi_b(2)+\phi_b(1)\phi_a(2)]\ ,$$

$$\psi_{MO}(1, 2) = N'\chi(1)\chi(2)\ ,$$

where the spin factor has been omitted, since it will play no part in the rest of our calculations, and

$$\chi(1) = \phi_a(1)+\phi_b(1)\ .$$

Show that normalization requires

$$N^{-2} = 2(1+S^2)\ , \quad (N')^{-1} = 2(1+S)\ .$$

(b) Calculate the total electron density, as it might be measured by x-ray diffraction, showing that in each case it takes the form

$$\rho = \lambda(\phi_a^2+\phi_b^2)+2\mu\phi_a\phi_b\ ,$$

where

$$\lambda+\mu S = 1\ .$$

Obtain $\lambda_{VB}$ and $\lambda_{MO}$ in terms of $S$. If we call $2\mu\phi_a\phi_b$ the bond charge, show that this is greater for $\psi_{MO}$ than for $\psi_{VB}$. It follows that the atomic charges $\lambda_{MO}$ and $\lambda_{VB}$ are in the opposite order.
(c) Using the notation of figure 7.2.6, employ the Hellmann–Feynman theorem to determine the force due to the total electronic charge cloud plus the charge on nucleus A, acting on nucleus B, for $\psi_{VB}$ and $\psi_{MO}$.

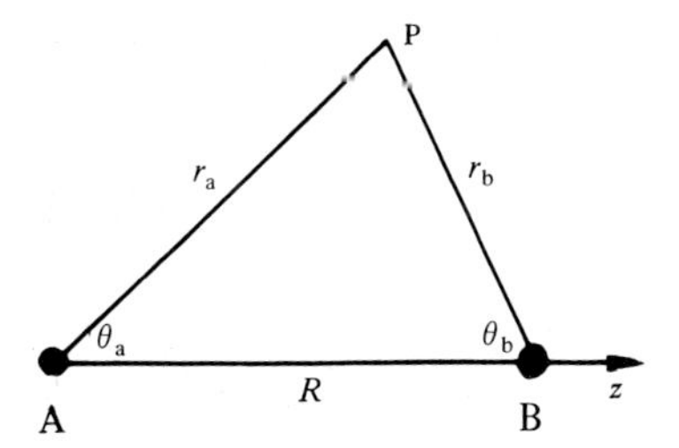

**Figure 7.2.6.**

Interpret physically the various force terms that appear.

(d) Put $\phi_a = (\zeta^3/\pi)^{1/2}\exp(-\zeta r_a)$, and similarly for $\phi_b$, where the effective nuclear charge $\zeta$ is taken to be 1·2. Can you calculate, by simple electrostatics, without any large amount of work, the value of the integral

$$\int \frac{\phi_a^2 \cos\theta_b}{r_b^2} d\tau \ ?$$

Next, using the numerical values in table 7.2.6a, plot the force acting on **B** as a function of $R$ for the two wave functions. What are the equilibrium distances predicted for the two wave functions? Recall that at $R = R_e$ there is a delicate balance of forces acting on either proton: the attraction of the overlap charge and the repulsion of the only partially screened other nucleus. At $R_e$ these two terms are equal and opposite, but they have different values for $\psi_{VB}$ and $\psi_{MO}$. Can you interpret this difference in terms of the results in (b)?

**Table 7.2.6a.**

| $\frac{R}{a_0}$ | $2R^2\int \frac{\phi_a\phi_b\cos\theta_b}{r_b^2} d\tau$ (a.u.) | $S$ | $R^2\int \frac{\phi_a^2\cos\theta_b}{r_b^2} d\tau$ (a.u.) |
|---|---|---|---|
| 1·2 | 0·665 | 0·742 | 0·549 |
| 1·3 | 0·737 | 0·708 | 0·603 |
| 1·4 | 0·801 | 0·675 | 0·652 |
| 1·5 | 0·858 | 0·642 | 0·697 |
| 1·6 | 0·908 | 0·608 | 0·738 |
| 1·7 | 0·951 | 0·576 | 0·773 |
| 1·8 | 0·984 | 0·544 | 0·805 |

**Solution**

(a) Normalization of $\psi_{VB}$ requires $\int\int \psi_{VB}^2 \, d\tau_1 \, d\tau_2 = 1$, so that

$$N^{-2} = \int [\phi_a(1)\phi_b(2) + \phi_b(1)\phi_a(2)]^2 \, d\tau = 2(1+S^2) \, ,$$

as in problem 7.2.4. Similarly normalization of $\psi_{MO}$ requires

$$(N')^{-2} = \int\int [\chi(1)\chi(2)]^2 \, d\tau_1 \, d\tau_2 = [2(1+S)]^2 \, .$$

(b) With either wave function, the electron density of electron 1 is $\int [\psi(1, 2)]^2 \, d\tau_2$, so that the total density is

$$\rho = 2\int [\psi(1, 2)]^2 \, d\tau_2 \, .$$

Thus for the MO wave function

$$\rho = 2(N')^2[\chi(1)]^2\int[\chi(2)]^2\,\mathrm{d}\tau_2$$

$$= 2N'[\chi(1)]^2 = \frac{1}{1+S}[\phi_a^2+\phi_b^2+2\phi_a\phi_b]\ .$$

Check that $\int\rho\,\mathrm{d}\tau = 2$, since this is the total number of electrons. Hence

$$\lambda_{MO} = \frac{1}{1+S}\ ,\quad \mu_{MO} = \frac{1}{1+S}\ ,\quad \lambda+\mu S = 1\ .$$

Similarly, with $\psi_{VB}$,

$$\rho = 2N^2\int[\phi_a(1)\phi_b(2)+\phi_b(1)\phi_a(2)]^2\,\mathrm{d}\tau_2$$

$$= \frac{1}{1+S^2}[\phi_a^2(1)+\phi_b^2(1)+2S\phi_a(1)\phi_b(1)]\ .$$

Check again that $\int\rho\,\mathrm{d}\tau = 2$. We then have

$$\lambda_{VB} = \frac{1}{1+S^2}\ ,\qquad \mu_{VB} = \frac{S}{1+S^2}\ ,\qquad \lambda+\mu S = 1\ .$$

Since $S<1$, it follows that $\lambda_{VB} > \lambda_{MO}$, $\mu_{VB} < \mu_{MO}$, indicating that the bond charge has a larger value in $\psi_{MO}$ than in $\psi_{VB}$. Numerical values of the coefficients $\lambda_{MO}$, $\lambda_{VB}$, $\mu_{MO}$, and $\mu_{VB}$ are given as a function of $R$ in table 7.2.6b.

(c) If we put $\alpha$ of the Hellmann–Feynman theorem equal to $z_B$, then

$$\frac{\partial E}{\partial z_B} = \int\psi\frac{\partial\mathcal{H}}{\partial z_B}\psi\,\mathrm{d}\tau\ ,$$

where, in atomic units,

$$\mathcal{H} = -\tfrac{1}{2}(\nabla_1^2+\nabla_2^2)-\left(\frac{1}{r_{a_1}}+\frac{1}{r_{a_2}}+\frac{1}{r_{b_1}}+\frac{1}{r_{b_2}}\right)+\frac{1}{r_{12}}+\frac{1}{R}\ .$$

In calculating $\partial\mathcal{H}/\partial z_B$ we keep A fixed, and also the two electrons, but move nucleus B. Only the terms $1/r_{b_1}$, $1/r_{b_2}$, and $1/R$ change in this process. But

$$\frac{\partial}{\partial z_B}\left(\frac{1}{r_{b_1}}\right) = -\frac{1}{r_{b_1}^2}\frac{\partial r_{b_1}}{\partial z_B}$$

$$= -\frac{1}{r_{b_1}^2}\frac{\partial}{\partial z_B}[(x_1^2+y_1^2)+(z_1-z_B)^2]^{1/2}$$

$$= \frac{z_1-z_B}{r_{b_1}^3} = -\frac{\cos\theta_{b_1}}{r_{b_1}^2}\ .$$

The electronic contribution to $\partial E/\partial z_{\mathrm{B}}$ is thus

$$2\iint \frac{\cos\theta_{b_1}}{r_{b_1}^2}[\psi(1,2)]^2\,d\tau_1\,d\tau_2 = \int \frac{\cos\theta_b}{r_b^2}\rho\,d\tau\,.$$

If we think of $\rho\,d\tau$ as the charge in tiny volume element $d\tau$ around the point P, then $\rho\,d\tau/r_b^2$ is the attractive force exerted by this charge on the nucleus B. The factor $\cos\theta_b$ gives the component along BA. The integral $\int(\cos\theta_b/r_b^2)\rho\,d\tau$ is therefore exactly the force that would be obtained by a classical calculation in which the charge density was $\rho$.

The term $1/R$ in $\mathcal{H}$ gives rise to the repulsion $1/R^2$ from nucleus A acting on nucleus B. The net force on B away from A is thus

$$F_{\mathrm{B}} = \frac{1}{R^2} - \int \frac{\cos\theta_b}{r_b^2}\rho\,d\tau$$

$$= \frac{1}{R^2} - \int (\lambda\phi_a^2 + \lambda\phi_b^2 + 2\mu\phi_a\phi_b)\frac{\cos\theta_b}{r_b^2}\,d\tau\,.$$

The second term in the integral contributes nothing. Hence

$$F_{\mathrm{B}} = \frac{1}{R^2} - \lambda\int \phi_a^2\frac{\cos\theta_b}{r_b^2}\,d\tau - 2\mu\int \frac{\phi_a\phi_b\cos\theta_b}{r_b^2}\,d\tau\,.$$

The first term is the nuclear repulsion; the second term is the attraction from the atomic charge on A; the third term is the attraction from the bond charge. The first two terms together could be called the repulsion from the only partially screened nucleus A.

(d) $\int\phi_a^2(\cos\theta_b/r_b^2)\,d\tau$ is the force acting on B in the direction of A due to the inverse-square law of attraction from the charge cloud $\phi_a^2$. In our case $\phi_a^2$ is spherically symmetrical. Now (as in Newton's discussion of the gravitational attraction of a spherically symmetrical mass) the charge lying outside the sphere with centre A and radius AB exerts no force on B. The charge cloud inside this sphere behaves as a point charge at A. Hence

$$\int \frac{\phi_a^2\cos\theta_b}{r_b^2}\,d\tau = \frac{Q}{R^2}\,,$$

where $Q$ is the total charge in the atomic cloud $\phi_a^2$ lying inside $r = R$. Thus

$$Q = \int_0^R 4\pi r_a^2\phi_a^2\,dr_a = 4\zeta^3\int_0^R r^2\exp(-2\zeta r)\,dr$$

and is soon found.

With the formula for $F_{\mathrm{B}}$ above and the numerical values in table 7.2.6a, the results shown in table 7.2.6b are obtained. These lead to estimated values

$$R_e = 1{\cdot}72a_0 \quad (\mathrm{VB})\,, \qquad R_e = 1{\cdot}57a_0 \quad (\mathrm{MO})\,.$$

The more highly-charged MO bond [see solution (b)] can be seen to be shorter.

**Table 7.2.6b.** Charge parameters (a.u.) for MO and VB cases.

| $\frac{R}{a_0}$ | MO | | | VB | | |
|---|---|---|---|---|---|---|
| | $\lambda$ | $\mu$ | $R^2F_B$ | $\lambda$ | $\mu$ | $R^2F_B$ |
| 1·2 | 0·574 | 0·574 | 0·303 | 0·645 | 0·479 | 0·328 |
| 1·3 | 0·586 | 0·586 | 0·215 | 0·666 | 0·472 | 0·251 |
| 1·4 | 0·597 | 0·597 | 0·133 | 0·687 | 0·464 | 0·181 |
| 1·5 | 0·609 | 0·609 | 0·053 | 0·708 | 0·455 | 0·116 |
| 1·6 | 0·622 | 0·622 | −0·024 | 0·730 | 0·444 | 0·058 |
| 1·7 | 0·635 | 0·635 | −0·094 | 0·751 | 0·433 | 0·008 |
| 1·8 | 0·648 | 0·648 | −0·159 | 0·772 | 0·420 | −0·034 |

### 7.2.7 Hybrids, bent bonds, and maximum overlap

Since the bonding power of two atomic orbitals $\phi_a$, $\phi_b$ will depend in large degree upon their overlap integral, a simple approximation is to represent it by the overlap integral itself, viz. $S = \int \phi_a \phi_b \, d\tau$. In earlier days Pauling had suggested that the angular factor in a hybrid be used as a measure of its strength. However, this required the assumption that the radial factors of any atomic orbitals to be hybridized were identical. This is far from being true: the overlap criterion is not subject to this limitation.

In general a bond between two atoms will involve hybrids of s-, p-, and perhaps d-orbitals. We assume that hybrids around any one atom are orthogonal. This relates the hybridization ratios to the valence angles. If the bonds are straight, we suppose that the directions of the hybrids lie along the lines joining the nuclei, but if steric or other conditions prevent this, the bonds become bent (Coulson, 1961, p.215), and the overlap integral will be reduced. The object of this problem is to study this effect, particularly in relation to cyclopropane $C_3H_6$.

(a) Let $s$ denote the s orbital in the valence shell of an atom, and $p_i$, $p_j$ denote p orbitals in the directions represented by $i, j$ (figure 7.2.7a). In the case of a carbon atom these would be 2s, $2p_i$, $2p_j$ atomic orbitals. Consider two hybrids represented, in unnormalized form, by $s + \lambda_i p_i$, $s + \lambda_j p_j$. Show that these are orthogonal if $\lambda_i \lambda_j \cos\theta_{ij} = -1$. Deduce that the greater the p-character in a hybrid, the more nearly does the valence angle tend to 90°, but that for real hybrids $\theta_{ij}$ is never less than 90°.

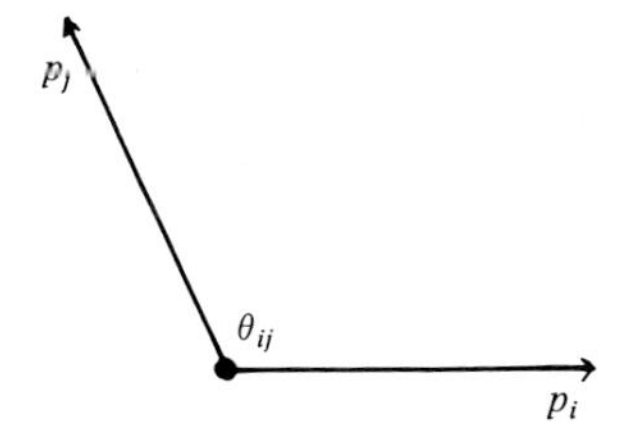


**Figure 7.2.7a.**

(b) Show that if $s, p_x, p_y$ are used to form three hybrids, their directions must all lie in a plane. Moreover, if two are equivalent, and differ from the third, show that they can be taken in the normalized forms:

$$\psi_1 = as + bp_x ,$$

$$\psi_2 = [\tfrac{1}{2}(1-a^2)]^{1/2}s - [\tfrac{1}{2}(1-b^2)]^{1/2}p_x + (\tfrac{1}{2})^{1/2}p_y ,$$

$$\psi_3 = [\tfrac{1}{2}(1-a^2)]^{1/2}s - [\tfrac{1}{2}(1-b^2)]^{1/2}p_x - (\tfrac{1}{2})^{1/2}p_y ,$$

where

$$a^2 + b^2 = 1 ,$$

and

$$\tan\theta_{12} = -\frac{1}{a} , \quad \tan\theta_{23} = -\frac{2a}{b^2} .$$

Verify that if $a = (\tfrac{1}{3})^{1/2}$, $b = (\tfrac{2}{3})^{1/2}$, the three angles are equal to 120°.

It may be assumed that a strong (and therefore short) bond tends to have more s-character than a weak one between two atoms. Would you expect the HCH angle in ethylene to be less than, or greater than, 120°?

(c) Consider (figure 7.2.7b) an atom A at the corner of an equilateral triangle ABC. Let the $y$, $z$ axes lie symmetrically in the plane ABC, so that the $x$ axis is normal to the plane. Show that two pairs of equivalent hybrids may be formed from $s, p_x, p_y$, and $p_z$ of A, and that they may be taken in the form (Coulson and Goodwin, 1962; 1963)

$$\psi_{AB} = as + (\tfrac{1}{2})^{1/2}p_y + bp_z ,$$

$$\psi_{AC} = as - (\tfrac{1}{2})^{1/2}p_y + bp_z ,$$

$$\psi_{AH_1} = bs + (\tfrac{1}{2})^{1/2}p_x - ap_z ,$$

$$\psi_{AH_2} = bs - (\tfrac{1}{2})^{1/2}p_x - ap_z ,$$

where $a^2 + b^2 = \tfrac{1}{2}$.

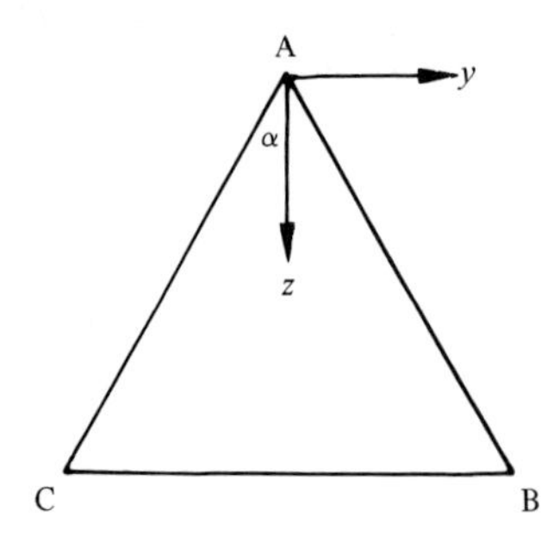

Figure 7.2.7b.

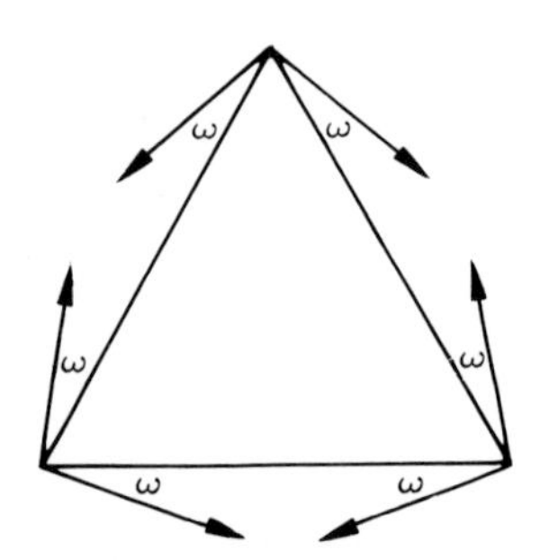

Figure 7.2.7c.

Next suppose that similar hybrids are formed at B and C. We may take A, B, C, to be the three carbon atoms in cyclopropane. Then the carbon–hydrogen bonds will be straight, but the carbon–carbon ones cannot be, since we have already seen that $\theta_{ij}$ can never be less than 90°, whereas the angle CAB is only 60°. Let the directions of the in-plane hybrids be as shown in figure 7.2.7c by the arrows. The C—C bonds are bent by an angle $\omega$ at each end, and the angle between the hybrids is $60° + 2\omega$.

As a rough indication of the degree to which the bonds are bent, suppose that $\omega$ is such that the total overlap in all three C—C and all six C—H bonds is maximum. Express this overlap in terms of $a$, $b$, and the six component overlap integrals, $I_1, ..., I_6$, defined by

$$I_1 = S(s_a, s_b) , \quad I_2 = S(s_a, p\sigma_b) , \quad I_3 = S(p\sigma_a, p\sigma_b) ,$$

$$I_4 = S(p\pi_a, p\pi_b) , \quad I_5 = S(s_a, h_a) , \quad I_6 = S(p\sigma_a, h_a) ,$$

where $p\sigma$, $p\pi$ denote p orbitals of each carbon atom directed along and at right angles to the bond being considered, and $h_a$ is a hydrogenic orbital for one of the hydrogens bonded to A. We neglect all overlap integrals between atoms not directly bonded to each other. Take $I_1 = 0{\cdot}3447$, $I_2 = 0{\cdot}3684$, $I_3 = 0{\cdot}3298$, $I_4 = 0{\cdot}1952$, $I_5 = 0{\cdot}5809$, $I_6 = 0{\cdot}4699$, and show that the maximum overlap occurs when $a = 0{\cdot}4266$, $b = 0{\cdot}5639$, $\omega = 21°26'$. What is the HCH angle?

**Solution**

(a) If $s+\lambda_i p_i$, $s+\lambda_j p_j$ are orthogonal, it follows that

$$\int (s+\lambda_i p_i)(s+\lambda_j p_j)\, d\tau = 0 .$$

Now

$$\int s^2 d\tau = 1 , \quad \int sp_i\, d\tau = \int sp_j\, d\tau = 0 , \quad \int p_i p_j\, d\tau = \cos\theta_{ij}$$

(since a p orbital in direction $j$ can be resolved, like a vector, into the sum of two p orbitals as in figure 7.2.7d by the relation $p_j = p_\xi \cos\theta + p_\eta \sin\theta$).

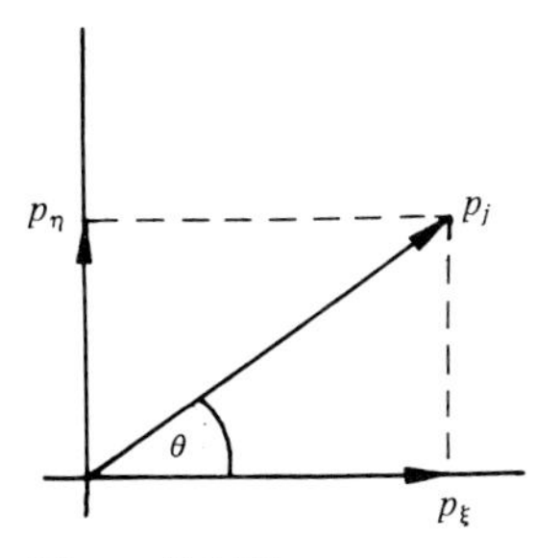


**Figure 7.2.7d.**

Take the $\xi$ direction to be the same as the $i$ direction. Then $\theta \equiv \theta_{ij}$ and $1+\lambda_i\lambda_j \cos\theta_{ij} = 0$.

Greater p character implies larger $\lambda$. Then $\cos\theta$ becomes smaller, tending to zero as $\lambda \to \infty$, so that $\theta \to 90°$ but is always greater than 90° except when one of the hybrids is pure p.

(b) The vector nature of p orbitals discussed in (a) shows that any hybrid of $s$, $p_x$, $p_y$ must lie in the $x$, $y$ plane: hence all three hybrids must lie in this plane.

If the direction of $\psi_1$ is along the $x$ axis, we shall have $\psi_1 = as+bp_x$, where normalization requires that $a^2+b^2 = 1$. If $\psi_2$ and $\psi_3$ are equivalent, then by symmetry they will have the same coefficients of the $s$ and $p_x$ functions, but equal and opposite coefficients for the $p_y$ function. Thus

$$\psi_2 = cs+fp_x+gp_y \ ,$$

$$\psi_3 = cs+fp_x-gp_y \ .$$

Normalization of $\psi_2$ leads to $c^2+f^2+g^2 = 1$, and orthogonality of $\psi_3$ and $\psi_2$ to $c^2+f^2-g^2 = 0$, so that $g^2 = \frac{1}{2} = c^2+f^2$. The only remaining condition is the orthogonality of $\psi_1$ and $\psi_2$, which gives $ac+bf = 0$. This leads to the required form of $\psi_1$, $\psi_2$, and $\psi_3$.

If $a$ and $b$ are positive (why can they be so chosen?) then $\psi_1$ points along the positive $x$ axis. $\psi_2$ is directed at an angle $\theta_{12}$ where

$$\tan\theta_{12} = \frac{g}{f} = -\frac{(\frac{1}{2})^{1/2}}{[\frac{1}{2}(1-b^2)]^{1/2}} = -\frac{1}{a} \ .$$

The value of $\tan\theta_{23}$ follows by elementary trigonometry.

*Check.* If $a = (\frac{1}{3})^{1/2}$ then all three hybrids have the same s character and must therefore be equivalent. But then $\theta_{12} = \tan^{-1}(-1/a) = 120°$, so that all three valence angles are 120°.

In ethylene $\psi_1$ will be used in forming the C—C $\sigma$-bond, $\psi_2$ and $\psi_3$ in forming the C—H bonds. The C—C bond is shortened as a result of being a double bond, so that the s character in $\psi_1$ should exceed $(\frac{1}{3})^{1/2}$. It would then follow that the C—H bonds will have increased p character, and therefore that we expect the HCH angle to be less than 120°. Experimental measurements (Sutton, 1965) give 117·5° ± 0·5° and 115·5° ± 0·6°, in agreement with this conclusion.

(c) The forms of the hybrids $\psi_{AB}$ and $\psi_{AC}$ follow from orthogonality and normalization exactly as $\psi_2$ and $\psi_3$ in (b). $\psi_{AH_1}$ and $\psi_{AH_2}$ must by symmetry be of the form $c_1s \pm c_2p_x + c_3p_z$. Orthonormality soon gives $c_2 = (\frac{1}{2})^{1/2}$, $c_1^2+c_3^2 = \frac{1}{2}$. Orthogonality of $\psi_{AB}$ and $\psi_{AH_1}$ leads to $c_1a+c_3b = 0$, from which the stated forms follow.

Using these forms, we find the overlap integral for the bond C—$H_a$ to be

$$S(\psi_{AH_1}, \psi_{H_1}) = bI_5+(\tfrac{1}{2}+a^2)^{1/2}I_6 \ .$$

We can represent $\psi_{AB}$ in terms of p$\sigma$ and p$\pi$ orbitals on $C_A$ ($\sigma$ and $\pi$ are here referred to the $C_A$–$C_B$ axis) in the form

$$\psi_{AB} = as_a + [b\cos\alpha - (\tfrac{1}{2})^{1/2}\sin\alpha]p\sigma_a + [b\sin\alpha + (\tfrac{1}{2})^{1/2}\cos\alpha]p\pi_a\,, \quad (\alpha = 30°).$$

Thus the C—C overlap integral $S(\psi_{AB}, \psi_{BA})$ is

$$a^2 I_1 + a[b\cos\alpha - (\tfrac{1}{2})^{1/2}\sin\alpha]I_2 + [b\cos\alpha - (\tfrac{1}{2})^{1/2}\sin\alpha]^2 I_3$$
$$+ [b\sin\alpha + (\tfrac{1}{2})^{1/2}\cos\alpha]^2 I_4\,.$$

If now the numerical values are put in, the maximum of $6S(\text{C—H}) + 3S(\text{C—C})$ occurs when $a = 0{\cdot}4266$, $b = 0{\cdot}5639$.

The angle between $\psi_{AB}$ and $\psi_{AC}$ can be found from the formulae derived in (b). Its tangent is equal to $-2^{1/2}b/a^2$, leading to the value $\omega = 21{\cdot}4°$. The HCH angle is $117{\cdot}8°$, in reasonable agreement with the experimental value (Bastiansen *et al.*, 1964) of $115{\cdot}1° \pm 1°$.

A final test of this scheme of bent C—C bonds is in the total electronic density. For a straight bond the overlap charge lies symmetrically around the axis of the bond. But for a bent bond, as obtained with the hybrids of figure 7.2.7c the overlap charge should have its maximum outside the triangle ABC. This is just what has been found experimentally by Hartman and Hirshfeld (1966).

An improvement in this model (Klasinc *et al.*, 1966) consists in weighting the overlap integrals with a factor dependent on the appropriate bond involved. One would then have to find the maximum value of $6k_{CH}S(\text{C—H}) + 3k_{CC}S(\text{C—C})$, where $k_{CH}$ and $k_{CC}$ are distinct known constants.

**References**

Bastiansen, O., Fritsch, F. N., Hedberg, K., 1964, *Acta Crystallogr.,* **17**, 538.
Bayliss, T., 1952, *Q. Rev. Chem. Soc.,* **6**, 319.
Coulson, C. A., 1937, *Trans. Faraday Soc.,* **33**, 1479.
Coulson, C. A., 1961, *Valence,* 2nd edition (Oxford University Press, Oxford). See also McWeeney, R., 1979, *Coulson's Valence*, 3rd edition (Oxford University Press, Oxford).
Coulson, C. A., Goodwin, T. H., 1962, *J. Chem. Soc.,* 2851.
Coulson, C. A., Goodwin, T. H., 1963, *J. Chem. Soc.,* 3161.
Feynman, R. P., 1939, *Phys. Rev.,* **56**, 340.
Greene, E. W., Barnard, J., Duncan, A. B. F., 1971, *J. Chem. Phys.,* **54**, 71.
Griffith, J. S., 1953, *Trans. Faraday Soc.,* **49**, 345.
Hartman, A., Hirshfeld, F. L., 1966, *Acta Crystallogr.,* **20**, 80.
Heitler, H., London, F., 1927, *Z. Phys.,* **44**, 455.
Hellmann, H., 1937, *Einführung in die Quantenchemie* (Deuticke, Leipzig), p.285.
Klasinc, L., Maksić, Z., Randić, M., 1966, *J. Chem. Soc. A,* 755.
Koopmans, T., 1933, *Physica,* **1**, 104.
Lorquet, J. C., 1960, *Rev. Mod. Phys.,* **32**, 312.
Price, W. C., 1935, *Phys. Rev.,* **47**, 444.
Ruedenberg, K., 1958, *J. Chem. Phys.,* **29**, 1232.
Sutton, L. E., 1965, *Interatomic Distances,* Special Publication No.18 (Chemical Society, London).

## 7.3 Arrangements of atoms in solids

G J Bullen, D J Greenslade, University of Essex

### 7.3.1

Show that the cubic close-packed and the hexagonal close-packed arrangements of spherical atoms are equally efficient at filling space.

Calculate the space-filling efficiency of the body-centred cubic arrangement and compare its coordination with that of the two close-packed arrangements.

**Solution**

*Cubic close packing*

The cubic unit cell contains four atoms with their centres at the points (0, 0, 0), $(0, \frac{1}{2}, \frac{1}{2})$, $(\frac{1}{2}, 0, \frac{1}{2})$, and $(\frac{1}{2}, \frac{1}{2}, 0)$ (fractional coordinates).

Let the atom spheres have unit radius. Since the spheres are in contact along a face diagonal, the unit-cell length $a = 2 \times 2^{1/2}$, and the volume of the unit cell is $a^3 = 16 \times 2^{1/2}$. The volume of one sphere is $\frac{4}{3}\pi$. The unit cell contains four spheres, of volume $\frac{16}{3}\pi$. Therefore the fraction of the unit cell filled by the spheres is $\frac{16}{3}\pi/(16 \times 2^{1/2}) = 0{\cdot}741$.

*Hexagonal close packing*

The hexagonal unit cell contains two atoms, at (0, 0, 0) and $(\frac{2}{3}, \frac{1}{3}, \frac{1}{2})$. If the spheres have unit radius, the unit-cell lengths are $a = b = 2$. In order that the layers of spheres parallel to (0001) are in contact, i.e. are close-packed, $c$ must be $(\frac{8}{3})^{1/2}a = 4 \times (\frac{2}{3})^{1/2}$.

The volume of the unit cell is $abc \sin 60° = 16 \times (\frac{2}{3})^{1/2} \times \frac{1}{2} \times 3^{1/2} = 8 \times 2^{1/2}$. The fraction of the unit cell filled by two spheres is thus $\frac{8}{3}\pi/(8 \times 2^{1/2}) = 0{\cdot}741$.

*Body-centred cubic*

The unit cell contains two atoms, at (0, 0, 0) and $(\frac{1}{2}, \frac{1}{2}, \frac{1}{2})$. For spheres of unit radius, the unit-cell length $a = 4/3^{1/2}$, and the volume of the unit cell is $a^3 = 64/(3 \times 3^{1/2})$.

Therefore the fraction of the unit cell filled by two spheres is $\frac{8}{3}\pi/[64/(3 \times 3^{1/2})] = 0{\cdot}680$.

*Coordination*

The two close-packed arrangements have a coordination number of 12, i.e. twelve spheres in contact with a given sphere.

The body-centred cubic arrangement is less efficient at filling space. It has a coordination number of only eight but there are six other atoms at a distance $a$ $(= 4/3^{1/2})$, which is only 15% greater than the closest distance (= 2). Hence the coordination is a less perfect 14 (i.e. 8 + 6).

### 7.3.2

The lattice structures of many ionic solids may be considered as an array of close-packed anions with the (smaller) cations occupying the holes in this array. Deduce the number of holes in a stack of close-packed layers, their positions, and their coordination numbers.

Hence show how the crystal structures of sodium chloride and wurtzite (zinc sulphide) can be described in terms of cation occupation of holes in a close-packed anion lattice.

**Solution**

Two layers of close-packed anions are shown in figure 7.3.2a, the layers being a distance $c$ apart. The rhombus $ABCT_1$ signifies a unit cell of the two-dimensional layer, each unit cell containing the equivalent of one sphere (anion).

There are two types of hole between the two layers (see table 7.3.2). The first type, an *octahedral hole*, occurs at position O. It is surrounded by six spheres (three in each layer) and has its centre halfway between the layers, i.e. at $\frac{1}{2}c$ above the base layer. There is only one such position within the unit cell, so that the number of octahedral holes is equal to the number of anions in one layer. When the layers are in a stack one above the other extending indefinitely, there will be one layer of octahedral holes above each layer of anions and hence a three-dimensional anion array will have as many octahedral holes as there are anions.

The second type, a *tetrahedral hole*, occurs at positions $T_1$ and $T_2$. In each case there are four spheres around the hole, their centres forming a tetrahedron, but the tetrahedra are orientated differently at $T_1$ and $T_2$. The centre of a hole is at the centroid of the tetrahedron and is therefore $\frac{3}{4}c$ above the base layer at $T_1$ and $\frac{1}{4}c$ above the base layer at $T_2$. Since there are equal numbers of holes of each of the types $T_1$, $T_2$, and O in the array, there will be twice as many tetrahedral holes as there are octahedral holes.

**Table 7.3.2.**

| Type of hole | Number per anion | Position | Coordination number |
|---|---|---|---|
| Octahedral | 1 | $\frac{1}{2}c$ above O | 6 |
| Tetrahedral | 2 | $\frac{3}{4}c$ above $T_1$<br>$\frac{1}{4}c$ above $T_2$ | 4 |

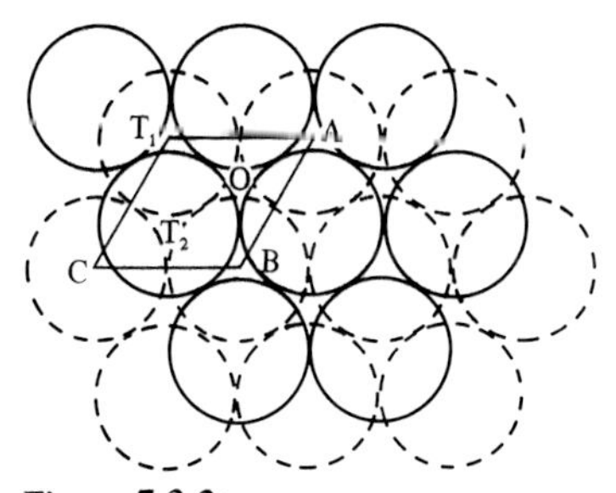


**Figure 7.3.2a.**

The above conclusions apply irrespective of whether the anion array is cubic close-packed or hexagonal close-packed.

*Sodium chloride*
The sodium ions occupy the octahedral holes in a cubic close-packed lattice of chloride ions, there being equal numbers of cations and anions. Two layers of anions are shown in figure 7.3.2b. The cations lying half-way between the layers are at the corners of the cubic unit cell (outlined by dashed lines) and the centres of its faces. The anion layers are parallel to (111) planes.

A single octahedron of chloride ions around a sodium ion is shown in figure 7.3.2c. Although the sodium ions are smaller than the chloride ions, they are in fact too large to fit into the octahedral holes without disturbing the anion lattice which must be expanded slightly.

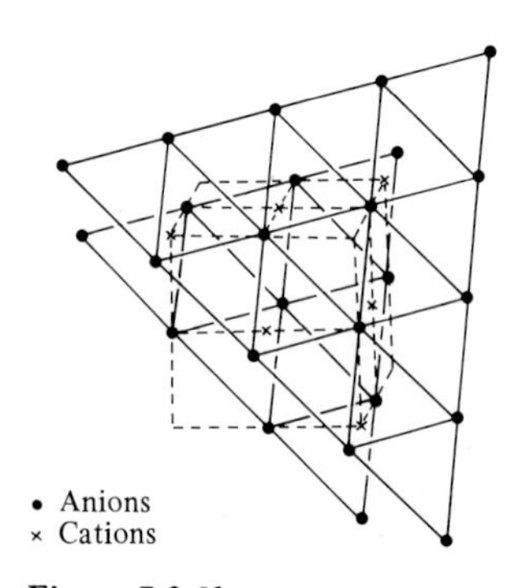


**Figure 7.3.2b.**

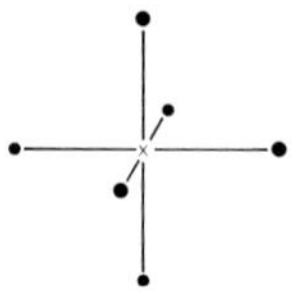

**Figure 7.3.2c.**

*Wurtzite*
In wurtzite, one of the forms of zinc sulphide, the zinc ions occupy *half* of the tetrahedral holes in a hexagonal close-packed lattice of sulphide ions. The holes used are all of the same type, e.g. of type $T_2$ as in figure 7.3.2a. In this case the zinc ions will all be $\frac{1}{4}c$ above the layers of sulphide ions and will all have sulphide ions directly *above* them (figure 7.3.2d; note that the unit cell height $c' = 2c$). Both cation and anion are tetrahedrally coordinated.

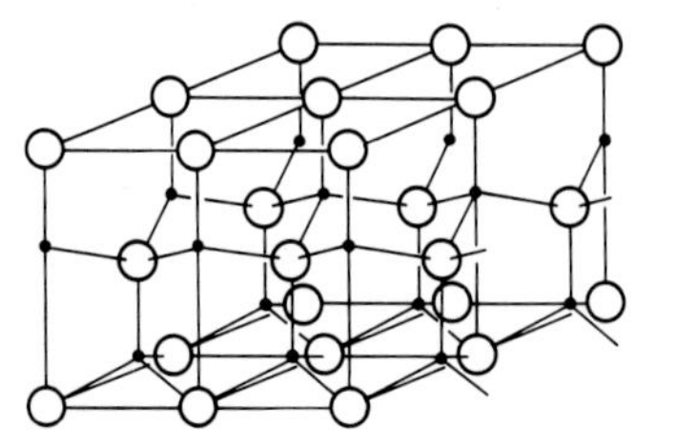

○ Anions
• Cations

**Figure 7.3.2d.**

7.3.3

The crystal structure of FeSe may be viewed as a hexagonal close-packed lattice of selenium atoms with iron atoms occupying the octahedral holes (this is usually termed the nickel arsenide structure). Assuming the $c/a$ ratio to be that of a true hexagonal close-packed lattice, $(\frac{8}{3})^{1/2}$, calculate (a) the closest iron–selenium distance in terms of the lattice constant $a$, and (b) the closest iron–iron distance, also in terms of $a$.

Table 7.3.3a gives the actual $c/a$ ratios and the lattice constants $a$ for a number of iron compounds which have the nickel arsenide crystal structure. Compare the closest Fe–X and Fe–Fe distances in these structures and comment on your results.

**Table 7.3.3a.**

| Compound FeX | FeS | FeSe | FeTe | FeSb | FeSn |
|---|---|---|---|---|---|
| $c/a$ | 1·71 | 1·64 | 1·49 | 1·26 | 1·23 |
| $a$ (Å) | 3·44 | 3·64 | 3·80 | 4·06 | 4·23 |

**Solution**

The hexagonal unit cell contains a layer of selenium atoms at height $z = 0$ (A, B, C in figure 7.3.3a) and another layer at height $z = \frac{1}{2}c$ (D, E, F in figure 7.3.3a). There are two iron atoms in the cell, at P (height $z = \frac{1}{4}c$) and at a point directly above P with $z = \frac{3}{4}c$. The iron atom at P is coordinated by the octahedron of selenium atoms ABCDEF.

(a) AP is one of the closest Fe–Se distances.

Let P′ be the point with $z = 0$ directly below P and let AP′ meet BC at S, its midpoint (see figure 7.3.3b). Then

$$AP' = \tfrac{2}{3}AS = \tfrac{2}{3} \times 3^{1/2} \times \tfrac{1}{2}a = a/3^{1/2},$$

and, since $c^2 = \frac{8}{3}a^2$,

$$(AP)^2 = \tfrac{1}{3}a^2 + \tfrac{1}{16}c^2 = \tfrac{1}{3}a^2 + \tfrac{1}{6}a^2 = \tfrac{1}{2}a^2 .$$

The closest Fe–Se distance is therefore $a/2^{1/2} = 0{\cdot}71a$.

(b) The iron atoms also form layers, so that the atom at P is surrounded by iron atoms in other unit cells at the same height ($z = \frac{1}{4}c$) which are a

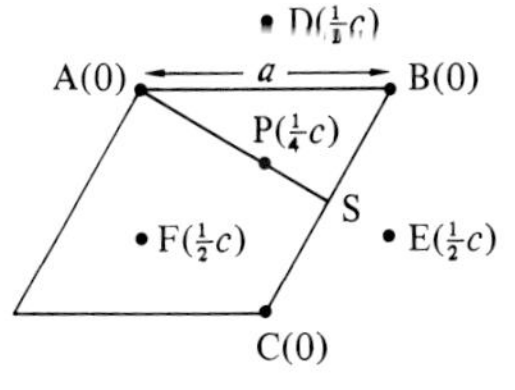


**Figure 7.3.3a.** Numbers in parentheses are the heights of the points in the unit cell.

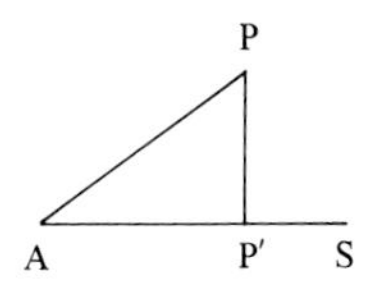


**Figure 7.3.3b.**

distance $a$ away. However, there are also iron atoms directly above and below P, at a distance $\frac{1}{2}c$. With true hexagonal close packing, these two iron atoms are the nearest to P since $\frac{1}{2}c = (\frac{2}{3})^{1/2}a = 0{\cdot}82a$.

An iron atom in this structure is therefore coordinated by six selenium atoms at a distance $0{\cdot}71a$ and by two other iron atoms at $0{\cdot}82a$.

Similar calculations using the actual $c/a$ ratios for FeS, FeSe, FeTe, FeSb, and FeSn give the results summarised in table 7.3.3b. In FeS and FeSe the arrangement of sulphur and selenium atoms is close to true hexagonal close packing but on passing from FeS to FeTe the reduction in $c/a$ results in a progressive shortening of the Fe–Fe distance relative to the Fe–X distance. Thus, as the ionic character of the compound decreases, neighbouring iron atoms assume a greater importance in the coordination polyhedron.

In FeSb and FeSn, which are the least ionic of the five compounds and are more in the nature of alloys, the Fe–Fe distance is actually less than the Fe–Sb or Fe–Sn distance, so that Fe–Fe bonding must be significant.

**Table 7.3.3b.**

| | Fe–X distance (Å) | Fe–Fe distance (Å) |
|---|---|---|
| FeS | $0{\cdot}72a = 2{\cdot}48$ | $0{\cdot}85_5a = 2{\cdot}89$ |
| FeSe | $0{\cdot}71a = 2{\cdot}58$ | $0{\cdot}82a = 2{\cdot}98$ |
| FeTe | $0{\cdot}69a = 2{\cdot}62$ | $0{\cdot}74_5a = 2{\cdot}83$ |
| FeSb | $0{\cdot}66a = 2{\cdot}68$ | $0{\cdot}63a = 2{\cdot}54$ |
| FeSn | $0{\cdot}65a = 2{\cdot}75$ | $0{\cdot}61_5a = 2{\cdot}60$ |

**7.3.4**

The crystal structure type of an ionic compound MX appears, to a first approximation, to depend on the ratio of cation radius to anion radius, $r_M/r_X$. Explain the reason for this and show why the NaCl type of structure is expected when $0{\cdot}73 \geqslant r_M/r_X \geqslant 0{\cdot}41$.

At room temperature and pressure all the alkali-metal halides except CsCl, CsBr, and CsI have the sodium chloride structure. Check, using the ionic radii listed in table 7.3.4a, whether they satisfy the expected criterion. Comment on, and suggest possible reasons for, the occurrence of structures which do not fit the criterion.

**Table 7.3.4a.** Pauling radii of ions.

| Ion | Radius (Å) | Ion | Radius (Å) |
|---|---|---|---|
| $Li^+$ | 0·60 | $F^-$ | 1·36 |
| $Na^+$ | 0·95 | $Cl^-$ | 1·81 |
| $K^+$ | 1·33 | $Br^-$ | 1·95 |
| $Rb^+$ | 1·48 | $I^-$ | 2·16 |
| $Cs^+$ | 1·69 | | |

**Solution**

The lattice energy of an ionic crystal containing ions of charge $+ze$ and $-ze$ is given by

$$U = -\frac{N_A A z^2 e^2}{r_0}\left(1 - \frac{1}{n}\right), \qquad (7.3.4a)$$

where $N_A$ is the Avogadro constant, $A$ is the Madelung constant, $r_0$ is the equilibrium M–X separation ($= r_M + r_X$), and $n$ is the exponent in the repulsive energy term. [Equation (7.3.4a) is based on the simple assumption, made by Born, that the repulsive energy between the ions can be represented by a term $N_A B/r^n$, where $B$ and $n$ are constants. More rigorous expressions for the repulsive energy can be used but equation (7.3.4a), based on Born's simple expression, is adequate for the arguments presented below.]

In general, with few exceptions, anions are larger than cations. The greatest number of anions which can be placed around a cation to form an extended MX lattice is eight, giving the caesium chloride structure. If the cation radius is reduced while the anion radius is kept constant, $r_M/r_X$ falls and so does $r_0$, so making $U$ more negative and the lattice more stable. This fall in $U$ continues as $r_M$ is reduced until the cation becomes so small that the anions around it touch each other. Beyond this point $r_0$ cannot be reduced further by a decrease in $r_M$, and the structure is likely to adopt a different arrangement which will allow the anion and small cation to come closer together and hence give an even lower lattice energy[(1)].

In the CsCl lattice, cation and anion are in contact along the body diagonal of the cubic unit cell so that

$$r_0 = r_M + r_X = 3^{1/2} \times \tfrac{1}{2}a\,,$$

where $a$ is the unit-cell length. When $r_M/r_X$ has fallen to the point where the anions are just in contact, $a = 2r_X$ so that

$$r_M + r_X = 3^{1/2} r_X\,,$$

whence

$$\frac{r_M}{r_X} = 3^{1/2} - 1 = 0{\cdot}73\,.$$

At lower $r_M/r_X$ an octahedral coordination of the cation, as found in the NaCl structure, permits a smaller $r_0$.

In the NaCl lattice, cation and anion are in contact along a face diagonal of the unit cell so that

$$r_0 = r_M + r_X = 2^{1/2} \times \tfrac{1}{2}a.$$

(1) That is, greater stability. Some authors prefer to use the *absolute* value of the lattice energy and hence for them stability implies a high lattice energy.

Again, when the anions are just in contact, $a = 2r_X$ and

$$r_M + r_X = 2^{1/2} r_X$$

so that

$$\frac{r_M}{r_X} = 2^{1/2} - 1 = 0 \cdot 41 \ .$$

Thus the NaCl lattice should be favoured when $r_M/r_X$ lies between 0·73 and 0·41. When $r_M/r_X < 0 \cdot 41$, a smaller $r_0$ can be achieved with tetrahedral coordination of the cation (ZnS type structure).

The $r_M/r_X$ ratios for the alkali-metal halides, calculated from the radii in table 7.3.4a, are given in table 7.3.4b. Only the italicised values are within the expected range. Of the others, the lithium halides, apart from LiF, should adopt tetrahedral coordination and the rubidium halides, apart from RbI, should show eightfold coordination.

A number of reasons for the discrepancies have been put forward (see Greenwood, 1968; Adams, 1974). Greenwood suggests that the balance between NaCl and CsCl type structures is affected by the 3% increase in both cation and anion radii on passing from six-fold to eight-fold coordination. This increase arises because the radius depends on the Madelung constant and on the repulsion constant $B$ (in the term $N_A B/r^n$) which is considered to be proportional to the coordination number. The lattice energies of the NaCl and CsCl type structures are in any case very close and a 3% increase in $r_0$ will close the gap between them or even make the NaCl lattice the more stable even when $r_M/r_X > 0 \cdot 73$. This could explain why the rubidium halides have the NaCl structure.

The problem of the lithium halides is affected by the difference in lattice energy between the NaCl and ZnS type structures. This difference is sufficiently large that, although the lattice energy of the NaCl structure remains constant when $r_M/r_X$ falls below 0·41, it is still lower than the lattice energy of the ZnS structure and continues so until $r_M/r_X$ reaches 0·3 (figure 7.3.4). This could explain the preference of LiCl and LiBr for the NaCl structure though it still leaves LiI as anomalous.

**Table 7.3.4b.** $r_M/r_X$ for NaCl type structures.

| | Li | Na | K | Rb | Cs |
|---|---|---|---|---|---|
| F | *0·44* | *0·70* | 0·98 | 0·92[a] | 0·80[a] |
| Cl | 0·33 | *0·52* | *0·73* | 0·82 | |
| Br | 0·31 | *0·49* | *0·68* | 0·76 | |
| I | 0·28 | *0·44* | *0·62* | *0·69* | |

[a] For RbF and CsF the cation is larger than the anion and the reverse argument applies. The ratio quoted is therefore $r_X/r_M$.

Adams makes the more fundamental criticism that the Pauling ionic radii, which are the most commonly used, are not necessarily the best, and he compares the $r_M/r_X$ values derived from Pauling radii with those derived from the 'corrected' experimental radii of Gourary and Adrian (1960) which are obtained from electron density maps. Unfortunately the correlation between structure and $r_M/r_X$ values based on these 'corrected' radii is no better. The new $r_M/r_X$ values are in general considerably higher so that, while the lithium halides now have radius ratios within the correct range, the rubidium halides are more in error than before and even NaF, KCl, and KBr have $r_M/r_X > 0{\cdot}73$. The preference of the rubidium halides for the NaCl structure is attributed by Adams to a small covalent bond contribution. The NaCl structure can accommodate covalent bonding better than the CsCl structure because its octahedral coordination is suitable for overlap involving the $p_x$, $p_y$, and $p_z$ orbitals.

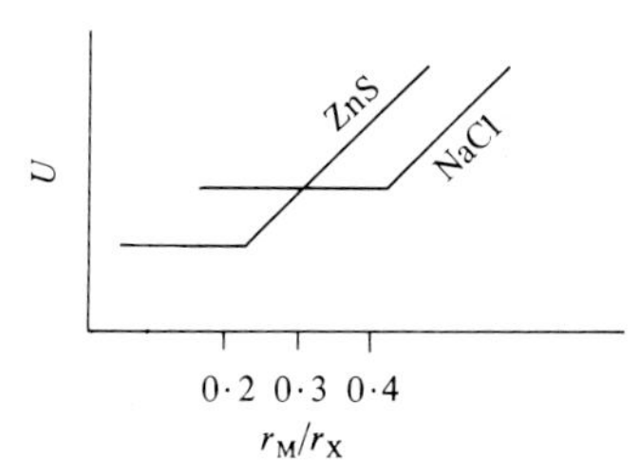


**Figure 7.3.4.**

7.3.5

Hush and Pryce (1957) supposed that the variation of the ionic radii of the first-row transition elements could be partly explained as a crystal-field effect. There is a gain of crystal-field stabilisation as the surrounding ions or dipoles move inwards; counteracting this effect are the forces arising from van der Waals interactions and so forth. These forces are measured by the $a_{1g}$ vibration force constant of the ligands—the metal does not move in this vibration. Considering the weak crystal field for $O_h$ symmetry:

(a) Show that the decrease in ionic radius due to crystal-field effects is $E_s^0/kR$, where $k$ is the force constant, $E_s^0$ is the crystal-field stabilisation energy and $R$ is the ion–ligand separation.

[Hint: $E_s^0 \propto R^{-6}$, see (c) below.]

(b) Show that $E_s^0 = N\Delta$ where $N$ is a number dependent on the electronic configuration of the transition ion and $\Delta$ is the crystal-field splitting.

(c) Given that the crystal-field potential is

$$V_4 = D(x^4+y^4+z^4-\tfrac{3}{5}r^4)\,,$$

(see problem 1.2.5) show that $D = 175e\mu/4R^6$. The coordinates of the electron are $x$, $y$, and $z$; $x^2+y^2+z^2 = r^2$; $\mu$ is the dipole moment of the

dipolar ligands, which are placed on the $x, y, z$ axes at a distance $R$ from the metal ion. The electron charge is $e$.
(d) Correct the radii of the hexahydrate complexes of the ions given in table 7.3.5a using the given data.

**Table 7.3.5a.** Crystal-field splitting parameters, ionic radii, and vibration frequencies of hexahydrate complexes.

| Bivalent ion | Ground-state populations in weak field | | Crystal-field splitting, $\Delta$ (cm$^{-1}$) | Ionic radius, $r$ (Å) | Symmetric stretch, $\nu$ (cm$^{-1}$) |
|---|---|---|---|---|---|
| | $t_{2g}$ | $e_g$ | | | |
| Mn | 3 | 2 | - | 0·90 | - |
| Fe | 4 | 2 | 10000 | 0·85 | 324 |
| Co | 24/5 | 11/5 | 9500 | 0·80 | 248 |
| Ni | 6 | 2 | 8500 | 0·76 | 264 |
| Cu | 6 | 3 | 11200 | 0·80 | 280 |
| Zn | - | - | - | 0·83 | - |

**Solution**

(a) In the harmonic oscillator equation the potential energy term is $\frac{1}{2}kx^2$ ($x$ being the distance of the particle from the mean position). For six molecules at equilibrium distance $R$, a change $\delta R$ produces an energy change $3k(\delta R)^2$. The change in crystal-field stabilisation being $\delta E_s^0$, the total energy change is

$$\delta E \approx \delta E_s^0 + 3k(\delta R)^2 \ .$$

But

$$\frac{dE_s^0}{dR} = -\frac{6E_s^0}{R} \ ,$$

since $E_s^0 = \text{const} \times R^{-6}$. Thus

$$\delta E \approx -\frac{6(\delta R)E_s^0}{R} + 3k(\delta R)^2 \ .$$

At minimum energy and equilibrium $d(\delta E)/d(\delta R) = 0$, so

$$(\delta R)_{\text{min}} = \frac{E_s^0}{kR} \ .$$

(b) For an octahedral field the d orbitals split into $t_{2g}$ and $e_g$ orbitals with energy (Ballhausen, 1962, p.64)

$$E(\phi) = \epsilon_0 + \int \phi^* V \phi \, d\tau \ ,$$

where $\phi$ is either a $t_{2g}$ orbital function or an $e_g$ function. Conventionally,

$E(e_g) - E(t_{2g}) = \Delta$ or $10Dq$. In the weak-field scheme, electrostatic repulsion between the electrons causes the electrons to fill Russell–Saunders states which are then split by the crystal field. The lowest state can then be described by a sum of microstates or Slater determinants, that is, antisymmetrised products of one-electron wave functions, each of which is an eigenfunction of $l^2$ and $l_z$. The energy is then calculated by working out the matrix element (or integral) of the crystal-field potential for this state. Since the potential is a one-electron operator, the result is a sum of one-electron integrals, each of which is proportional to $\Delta$. For example, the $^3F$ state is split to give a lowest state $^3A_{2g}$, which when written out as a sum of the microstates is, for its $M_s = 1$ component,

$$\Psi(^3A_{2g}) = 2^{-1/2}[\{\overset{+}{2}\ \overset{+}{0}\} - \{\overset{+}{0}\ \overset{+}{-2}\}] .$$

Putting this into the integral leads to one electron integrals, each of which is proportional to $\Delta$:

$$E(^3A_{2g}) = \langle 2|V_4|2\rangle + \langle 0|V_4|0\rangle + \langle 2|V_4|-2\rangle = 1\cdot 2\Delta .$$

(See Ballhausen, 1962, p.69 *et seq.* Note that $\{\overset{+}{a}\ \overset{-}{b}\}$ means an antisymmetrised product function, with the first electron in an orbital with $m_l = a$ and spin $+\frac{1}{2}$, the second electron with $m_l = b$ and spin $-\frac{1}{2}$. The parentheses imply a normalised sum of such products, which is antisymmetric with respect to interchange of any two electrons.)

The significance of this result is that one can assign a 'configuration' to the weak-field ground state, in terms of the strong-field functions $t_{2g}$ and $e_g$. The population of the strong-field states is not necessarily integral as is seen in table 7.3.5. The crystal-field stabilisation is readily calculated from such configurations; this is really the reverse of the 'calculation' of the configuration. Thus

$$E(^4A_2,\ t_{2g}^3) = 3E(t_{2g}) = -\tfrac{6}{5}\Delta .$$

(c) The general method for the solution of this type of potential problem is to expand the potential in terms of spherical harmonics, and then $D$ would be deduced by comparison of the expansion with the form of $V$ given (which is in fact a sum of fourth-order Legendre polynomials). The general principle is discussed by Killingbeck and Cole (1971) and is applied to a similar case by Griffith (1961). In this case it is simpler to consider a special position. At $x = r$ and $y = z = 0$, the potential of an electron of charge $e$, due to six point charges $+q$ at a distance $R$ along the six semi-axes is

$$V = -eq[(R-r)^{-1} + (R+r)^{-1} + 4(R^2+r^2)^{-1/2}]$$

$$= -\frac{eq}{R}\left[\left(1-\frac{r}{R}\right)^{-1} + \left(1+\frac{r}{R}\right)^{-1} + 4\left(1+\frac{r^2}{R^2}\right)^{-1/2}\right] .$$

The binomial expansion is then used, and the term in $r^4$ picked out:

$$V_4 = -\frac{eq}{R}\left[2\left(\frac{r}{R}\right)^4 + 4 \times (-\tfrac{1}{2}) \times (-\tfrac{3}{2}) \times \tfrac{1}{2}\left(\frac{r}{R}\right)^4\right] = -\frac{7eqr^4}{2R^5} .$$

But from our original expression

$$V_4 = D(r^4 - \tfrac{3}{5}r^4) = \tfrac{2}{5}Dr^4 .$$

Now we let the point charges become dipoles by superposing the field of six negative charges at $(R-d)$ and that of six positive charges at $(R+d)$:

$$V_4 = \tfrac{35}{4}eq\left[\frac{1}{(R-d)^5} - \frac{1}{(R+d)^5}\right](x^4+y^4+z^4-\tfrac{3}{5}r^4)$$

$$= \tfrac{35}{4}e(x^4+y^4+z^4-\tfrac{3}{5}r^4)\frac{(10R^4dq + ...)}{(R+d)^5(R-d)^5} .$$

Put $\mu = 2dq$, and assume $d \ll R$. Then, dropping small terms, we obtain

$$V_4 = \frac{175e\mu}{4R^6}(x^4+y^4+z^4-\tfrac{3}{5}r^4) .$$

(d) Consider the case of cobalt, for which $E_s^0 = 5700\ \text{cm}^{-1}$; we have $h = 6{\cdot}63 \times 10^{-34}\ \text{J s}^{-1}$, and $c = 3{\cdot}00 \times 10^{10}\ \text{cm s}^{-1}$, whence

$$E = 5{\cdot}7 \times 3 \times 6{\cdot}63 \times 10^{-21}\ \text{J} .$$

$k = m\omega^2$; here $m$ is the mass of $H_2O = 18{\cdot}0 \times 1{\cdot}66 \times 10^{-27}$ kg, and $\omega$ is the angular frequency of the $a_{1g}$ vibration. We have $\tilde{\nu}(a_{1g}) = 248\ \text{cm}^{-1} \equiv 248 \times 3 \times 10^{10}$ Hz, whence

$$k = 18 \times 1{\cdot}66 \times 10^{-27} \times (2\pi \times 248 \times 3 \times 10^{10})^2\ \text{N m}^{-1} .$$

$R$ for hydrates is about 2 Å, being the sum of the ionic radius of the cation and the van der Waals radius of oxygen in water: the dipole moment is, of course, situated more towards the lone pair. We obtain

$$\delta R = \frac{5{\cdot}7 \times 3 \times 6{\cdot}63 \times 10^{16}}{18 \times 1{\cdot}66 \times 4\pi^2 \times (248 \times 3 \times 10^{10})^2 \times 2} = 8{\cdot}8\ \text{pm} .$$

**Table 7.3.5b.** Crystal-field radius corrections.

| Bivalent ion | $a_{1g}$ force constant ($\text{N m}^{-1}$) | $10^{19}E_s^0$ (J) | Radius and correction (pm) |
|---|---|---|---|
| Mn | - | - | 90 |
| Fe | 112 | 7·95 | 85+3·2 |
| Co | 66 | 1·13 | 80+7·8 |
| Ni | 75 | 2·03 | 76+12 |
| Cu | 84 | 1·34 | 80+7·3 |
| Zn | - | - | 83 |

Clearly the accuracy is not high, perhaps not better than the first figure, so that we can say that the ionic radius of divalent cobalt in hydrates would be 0·88 Å, but for crystal-field effects. The corrected radii are listed in table 7.3.5b.

### 7.3.6

In the spinels, of general formula $AB_2O_4$, the oxygen atoms lie in a close-packed cubic lattice. In the normal structure the $A^{2+}$ ions occupy one-eighth of the tetrahedral holes in this lattice with the $B^{3+}$ ions in half the octahedral holes. In the inverse structure half the $B^{3+}$ ions have exchanged places with the $A^{2+}$ ions. Use the data in table 7.3.6a to explain why $CoFe_2O_4$ is an inverse structure whereas $CoCr_2O_4$ is normal.

**Table 7.3.6a.**

| Metal ion | Free ion state | Octahedral state | Tetrahedral state | Octahedral $Dq$ ($cm^{-1}$) |
|---|---|---|---|---|
| $Cr^{3+}$ | $^4F$ | $^4A_{2g}$ | $^4T_{1g}$ | 1760 |
| $Fe^{3+}$ | $^6S$ | $^6A_{1g}$ | $^6A_{1g}$ | 1400 |
| $Co^{2+}$ | $^4F$ | $^4T_{1g}$ | $^4A_{2g}$ | 1000 |

**Solution**

An atomic F state splits in a cubic crystal field to give $A_2$, $T_1$, and $T_2$ states. By symmetry (see problem 1.2.6) it is readily seen that these states are

$$\psi(A_2) = 2^{-\frac{1}{2}}(\Psi_2 - \Psi_{-2})\,, \qquad \psi(T_1) = \Psi_0\,, \qquad \text{etc.,}$$

where the subscript on $\Psi$ refers to the $M_L$ value of the F wave function. In order to calculate the energy of these states in terms of $Dq$, we express them in terms of Slater determinants of one-electron, d, wave functions. The $M_L = 2$ function, with $M_S = \frac{3}{2}$, of a $^4F$ state (of $d^3$ configuration) is simply $\{\overset{+}{2}\ \overset{+}{1}\ -\overset{+}{1}\}$, and $\Psi_{-2} = \{\overset{+}{1}\ -\overset{+}{1}\ -\overset{+}{2}\}$. Thus

$$\begin{aligned} E(^4A_2) &= \tfrac{1}{2}\langle\{\overset{+}{2}\ \overset{+}{1}\ -\overset{+}{1}\}-\{\overset{+}{1}\ -\overset{+}{1}\ -\overset{+}{2}\}|V|\{\overset{+}{2}\ \overset{+}{1}\ -\overset{+}{1}\}-\{\overset{+}{1}\ -\overset{+}{1}\ -\overset{+}{2}\}\rangle \\ &= \tfrac{1}{2}\langle 2|V|2\rangle + \langle 1|V|1\rangle + \langle -1|V|-1\rangle + \tfrac{1}{2}\langle -2|V|-2\rangle - \langle 2|V|-2\rangle \\ &= \tfrac{1}{2}Dq - 4Dq - 4Dq - 5Dq + \tfrac{1}{2}Dq = -12Dq\,, \end{aligned}$$

for octahedral symmetry.

In order to find $\Psi_0$ (with $M_S = \frac{3}{2}$) we operate twice on $\Psi_2$ with $L_-$ (see Ballhausen, 1962, p.11)

$$L_-|^4F, M_L = 2\rangle = (l_{1-} + l_{2-} + l_{3-})\{\overset{+}{2}\ \overset{+}{1}\ -\overset{+}{1}\}$$

$$[3(3+1) - 2(2-1)]^{\frac{1}{2}}|^4F, M_L = 1\rangle$$

$$= [2(2+1) - 1(1-1)]^{\frac{1}{2}}\{\overset{+}{2}\ \overset{+}{0}\ -\overset{+}{1}\} + [6 + 1(-1-1)]^{\frac{1}{2}}\{\overset{+}{2}\ \overset{+}{1}\ -\overset{+}{2}\}$$

$$5^{\frac{1}{2}}|^4F, M_L = 1\rangle = 3^{\frac{1}{2}}\{\overset{+}{2}\ \overset{+}{0}\ -\overset{+}{1}\} + 2^{\frac{1}{2}}\{\overset{+}{2}\ \overset{+}{1}\ -\overset{+}{2}\}\,.$$

Then

$$5^{1/2} \times 12^{1/2} |{}^4\mathrm{F}, M_L = 0\rangle = 3^{1/2}[2\{\overset{+}{1}\ \overset{+}{0}\ -\overset{+}{1}\} + 2\{\overset{+}{2}\ \overset{+}{0}\ -\overset{+}{2}\} + 2^{1/2} \times 6^{1/2}\{\overset{+}{2}\ \overset{+}{0}\ -\overset{+}{2}\}]$$

$$|{}^4\mathrm{F}, M_L = 0\rangle = 5^{-1/2}[\{\overset{+}{1}\ \overset{+}{0}\ -\overset{+}{1}\} + 2\{\overset{+}{2}\ \overset{+}{0}\ -\overset{+}{2}\}] .$$

Using the same procedure as before, we obtain

$$E({}^4\mathrm{T}_1) = 6Dq ,$$

for octahedral symmetry. The $d^7$ states can be considered as three holes in the filled d shell, so the same states occur, but with the sign of $Dq$ reversed, i.e. ${}^4T_1$ is at $-6Dq$ and ${}^4A_2$ at $12Dq$. Furthermore, the $Dq$ value for a tetrahedral field is minus four-ninths that for an octahedral field, if a point ionic model for the crystal-field is assumed. Using these results we can calculate the stabilisation energy, $E_s^0$, of a $Cr^{3+}$ ion placed in an octahedral site in the spinel: $E_s^0 = 12 \times 1760\ \mathrm{cm}^{-1}$. For the tetrahedral case $E_s^0 = 6 \times \frac{4}{9} \times 1760\ \mathrm{cm}^{-1}$, so that for this ion occupation of an octahedral site rather than a tetrahedral site is energetically favourable by 16400 $\mathrm{cm}^{-1}$. This is sometimes called the octahedral site preference energy and is often expressed in macroscopic units such as kJ $\mathrm{mol}^{-1}$ by multiplying by $N_A hc$, $N_A$ being Avogadro's constant, $h$ Planck's constant and $c$ the velocity of light.

Table 7.3.6b lists the crystal-field stabilisation energies and octahedral site preference energies required. Clearly the $Cr^{3+}$ ions have a greater site preference energy than the $Co^{2+}$ ions so that in the cobalt chromium spinel we find the normal structure. In the case of the corresponding iron(III) compound the situation is reversed since there is no crystal-field stabilisation of the $Fe^{3+}$ ion with its half-filled d shell ${}^6S$ state. Although the crystal-field terms are but a few percent of the lattice energy (of the order of $10^6$ J $\mathrm{mol}^{-1}$), they seem to play a critical role in this case. Miller has done careful calculations taking into account other effects; despite this work Katzin (1962), rejects the importance of crystal-field effects. McClure *et al.* (1965) have reviewed this topic.

**Table 7.3.6b.**

| Ion | Octahedral stabilisation energy ($\mathrm{cm}^{-1}$) | Tetrahedral stabilisation energy ($\mathrm{cm}^{-1}$) | Octahedral site preference energy (kJ $\mathrm{mol}^{-1}$) |
|---|---|---|---|
| $Cr^{3+}$ | 21100 | 4700 | 196 |
| $Fe^{3+}$ | 0 | 0 | 0 |
| $Co^{2+}$ | 6000 | 5300 | 8·4 |

### 7.3.7

In table 7.3.7a the electric conductivity of a crystal of sodium chloride doped with $10^{-5}$ mole cadmium chloride per mole sodium chloride is given for various temperatures. In table 7.3.7b the conductivity of a nominally pure crystal of sodium chloride is listed. Assuming that in both of these crystals cation vacancies are largely responsible for the conductivity, calculate the mobility of these vacancies and the number of Schottky defects in a pure crystal, expressing your result as a function of the absolute temperature. What assumptions are made in your calculation?

**Table 7.3.7a.** Conductivity of a sodium chloride crystal doped with cadmium chloride.

| Temperature (°C) | Conductivity ($\Omega^{-1}$ cm$^{-1}$) | Temperature (°C) | Conductivity ($\Omega^{-1}$ cm$^{-1}$) |
|---|---|---|---|
| 402 | $4{\cdot}7 \times 10^{-7}$ | 295 | $3{\cdot}57 \times 10^{-8}$ |
| 372 | $2{\cdot}5 \times 10^{-7}$ | 273 | $1{\cdot}85 \times 10^{-8}$ |
| 344 | $1{\cdot}3 \times 10^{-7}$ | 253 | $9{\cdot}62 \times 10^{-9}$ |
| 318 | $6{\cdot}73 \times 10^{-8}$ | | |

**Table 7.3.7b.** Conductivity of a nominally pure crystal of sodium chloride.

| Temperature (°C) | Conductivity ($\Omega^{-1}$ cm$^{-1}$) | Temperature (°C) | Conductivity ($\Omega^{-1}$ cm$^{-1}$) |
|---|---|---|---|
| 727 | $1{\cdot}55 \times 10^{-4}$ | 527 | $1{\cdot}5 \times 10^{-6}$ |
| 717 | $1{\cdot}45 \times 10^{-4}$ | 464 | $8{\cdot}5 \times 10^{-7}$ |
| 687 | $6{\cdot}89 \times 10^{-5}$ | 436 | $3{\cdot}6 \times 10^{-7}$ |
| 647 | $2{\cdot}4 \times 10^{-5}$ | 402 | $2{\cdot}0 \times 10^{-7}$ |
| 607 | $8{\cdot}6 \times 10^{-6}$ | 376 | $1{\cdot}1 \times 10^{-7}$ |
| 567 | $3{\cdot}5 \times 10^{-6}$ | 344 | $5{\cdot}8 \times 10^{-8}$ |
| 553 | $2{\cdot}2 \times 10^{-6}$ | | |

**Solution**

It is shown in textbooks (Dekker, 1960; Barr and Lidiard, 1970) that the conductivity, $\sigma$, and the mobility, $\mu$, are related by the expression

$$\sigma = ne\mu \ ,$$

where $n$ is the number of charge carriers, only one species contributing appreciably to the conductivity, and $e$ the charge carried by that species. Further, it is shown that the mobility depends on the temperature:

$$\mu = \frac{A}{T}\exp\left(-\frac{B}{T}\right) .$$

In the first case of the impure crystal, the number of cation vacancies, due to the presence of $Cd^{2+}$ ions substituting for $Na^{+}$ ions in the lattice, vastly exceeds the number present in a pure crystal at the same temperature.

For the charges to balance, then, the number of defect sites must equal the number of impurity sites. Thus

$$n = \frac{2\cdot165}{58\cdot44} \times 10^{-5} \times 6\cdot022 \times 10^{23} = 2\cdot231 \times 10^{17}\ \text{cm}^{-3}\ .$$

The conductivity is given in terms of a 1 $\text{cm}^3$ cube weighing 2·165 g; one mole weighs 58·44 g. For the impure crystal we calculate that

$$ne = 1\cdot602 \times 10^{-19} \times 2\cdot231 \times 10^{17} = 3\cdot574 \times 10^{-2}\ \text{C cm}^{-3}\ .$$

We have the relation

$$\lg\sigma T = \lg ne + \lg\mu T = \lg neA - \frac{B}{T}\lg e\ .$$

Thus a plot of $\lg\sigma T$ against $1/T$ will be a straight line of slope $-0\cdot4343B$ and intercept $\lg(0\cdot03574A)$. By drawing this graph, or by means of the least-squares method (Moroney, 1971) we find that

$$\mu = \frac{19\,600}{T}\exp\left(-\frac{9860}{T}\right).$$

In this analysis we have assumed that the cation vacancies are not associated with the bivalent impurity ions (cf. the following question). If we assume that at higher temperatures the conductivity in the pure crystal is largely due to cation mobility, we expect once again a linear relation between $\lg\sigma T$ and $1/T$. Figure 7.3.7 shows that this plot breaks into the so-called extrinsic region, in which impurities in the crystal provide the cation vacancies, and, at higher temperatures, the intrinsic region where most of the vacancies are due to entropy effects.

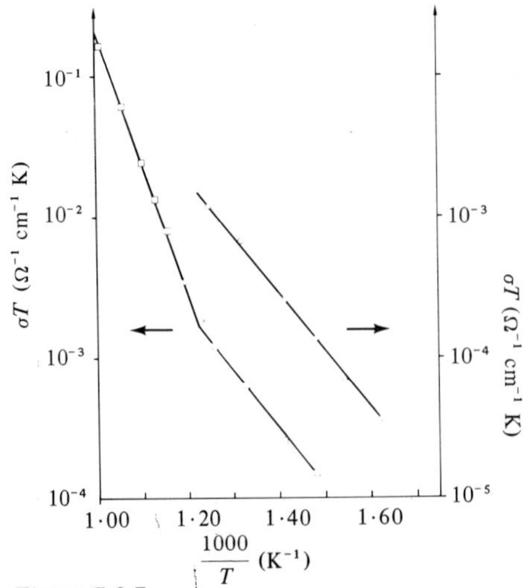


**Figure 7.3.7.**

Extrapolation of the extrinsic conductivity into the intrinsic region allows the calculation of the true intrinsic conductivity. This may then be plotted (as $\lg \sigma T$) against $1/T$ to yield

$$\sigma = \frac{3\cdot 72 \times 10^8 \text{ K}}{T} \exp\left(\frac{21\,600}{T}\right) \Omega^{-1} \text{ cm}^{-1} .$$

Using the previous expression for the mobility, we obtain

$$n = 1\cdot 2 \times 10^{23} \exp\left(\frac{11\,700}{T}\right) \text{ cm}^{-3} .$$

It should be noted that small errors in the determination of the exponents (from the gradient of the logarithmic plots) give rise to large errors in the pre-exponential factors.

The enthalpy of formation, $h$, of Schottky defects is obtained from the relation $n$ = constant $\times \exp(-h/2kT)$. $h = 11\,700 \times 2k$ is in fact rather lower than values now accepted for this quantity. This problem is based on the work of Etzel and Maurer (1950).

7.3.8

The electron spin resonance of sodium chloride crystals doped with manganese (II) chloride has been analysed (Watkins, 1959). At high concentrations of $Mn^{2+}$, a line of width 130 G (13 mT) peak-to-peak is found at $g = 2\cdot 004$. At lower concentrations and high temperature a six-component line at a similar $g$ value is found and is almost isotropic, the multiplets being separated by about 88 G (8·8 mT) at all orientations of the crystal. At lower temperatures the spectrum becomes more complicated, two major components having similar hyperfine splittings, but requiring $DS_z^2$ and $E(S_x^2 - S_y^2)$ terms in the spin Hamiltonian. Both have principal axes of the type [110]. $D$ is about $135 \times 10^{-4}$ and $E$ about $41 \times 10^{-4}$ cm$^{-1}$. A third centre has a similar value of $D$ but $E$ is nearly zero; it is axially symmetric about [100] type axes. In tables 7.3.8a and 7.3.8b the conductivity of such crystals is given for various temperatures and concentrations of $Mn^{2+}$. Discuss the origin of the e.s.r. spectra and analyse the conductivity data. Calculate the number of Schottky defects in a pure crystal at 500°C.

**Table 7.3.8a.** Conductivity, $\sigma$, of NaCl crystals containing various amounts of $MnCl_2$ impurity at a temperature of 400°C.

| | Impurity content (molar p.p.m.) | | | | |
|---|---|---|---|---|---|
| | 150 | 265 | 325 | 500 | 686 |
| $10^5\,\sigma\,(\Omega^{-1}\text{ cm}^{-1})$ | 1·12 | 1·52 | 1·98 | 3·2 | 4·5 |

**Table 7.3.8b.** Conductivity, $\sigma$, at different temperatures, $t$, of NaCl crystals containing various amounts of $MnCl_2$.

| Impurity content (molar p.p.m.) | | | | | |
|---|---|---|---|---|---|
| 686 | | 325 | | nominally pure | |
| $t$ (°C) | $10^5\sigma$ ($\Omega^{-1}$ cm$^{-1}$) | $t$ (°C) | $10^5\sigma$ ($\Omega^{-1}$ cm$^{-1}$) | $t$ (°C) | $10^5\sigma$ ($\Omega^{-1}$ cm$^{-1}$) |
| 657 | 108 | 693 | 62 | 698 | 34 |
| 632 | 88·5 | 636 | 40 | 670 | 12 |
| 612 | 68 | 589 | 25 | 627 | 45 |
| 592 | 58 | 527 | 12·5 | 581 | 1·17 |
| 566 | 48 | 473 | 6·7 | 567 | 0·74 |
| 547 | 36·5 | 445 | 4·5 | 547 | 0·44 |
| 462 | 13·5 | 394 | 1·95 | 533 | 0·285 |
| 445 | 9·7 | | | 527 | 0·22 |
| 423 | 6·2 | | | 502 | 0·124 |
| 403 | 4·5 | | | 479 | 0·073 |
| | | | | 467 | 0·054 |
| | | | | 452 | 0·048 |
| | | | | 426 | 0·033 |
| | | | | 403 | 0·0222 |
| | | | | 385 | 0·0167 |

**Solution**

The $Mn^{2+}$ ion has five d electrons and is in $^6S$ state, which is not split by crystal fields to first order. In the presence of crystal fields of low symmetry, and through the effects of spin–orbit coupling, small splittings of the ground state do occur and are manifested in anisotropy and even splitting of the electron spin resonance spectrum. The manganese nucleus possesses a magnetic moment with $I = \frac{5}{2}$ and this gives rise to the quite large splitting of the isotropic spectrum found at higher temperatures in this case. This spectrum is then assigned to an $Mn^{2+}$ ion on a normal cation site. At high concentrations and low temperature these ions will associate; the e.s.r. spectrum will then be exchange broadened and this is almost certainly the origin of the first spectrum. The spectra requiring the zero-field terms in their spin Hamiltonian must be due to some extra low-symmetry crystal field acting on the $Mn^{2+}$ ion concerned. Since these spectra arise at lower temperatures it seems likely that they are caused by association of a cation vacancy with an $Mn^{2+}$ ion. In the first case the symmetry and the fact that the $E$ term is large suggest that the associated vacancy is in a near-neighbour position. In the latter case the symmetry axis and a low or negligible value of $E$ suggest that the vacancy is at the next-nearest-neighbour position.

The mobility of the vacancies is constant at a given temperature. A plot of $\sigma$ ($= ne\mu$) against concentration of impurity—and thus the number of defects—should be a straight line at constant temperature. Figure 7.3.8a shows that this is so, but that not all the defects contribute to the conductivity, since the line does not pass through the origin. This is in accord with the e.s.r. results which show association of an impurity ion and a cation vacancy. The intercept gives the concentration of impurity bound to vacancies.

As in the previous problem, the variation of mobility with temperature is obtained by plotting $\lg \sigma T$ against $1/T$ (see figure 7.3.8b). The increase in the slope of line I is again evidence for defect and impurity association. The plot of $\lg \sigma T$ against $1/T$ for the nominally pure crystal (figure 7.3.8c) shows the usual intrinsic and extrinsic regions, and from the former it is deduced that

$$\sigma T = \text{const} \times \exp\left(\frac{22700}{T}\right)\ \Omega^{-1}\ \text{cm}^{-1}\ \text{K}\ .$$

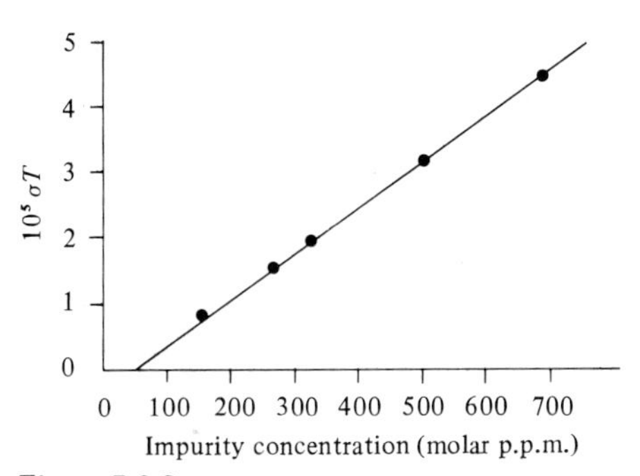


Figure 7.3.8a.

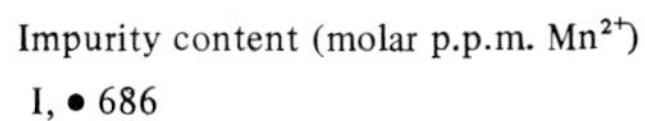


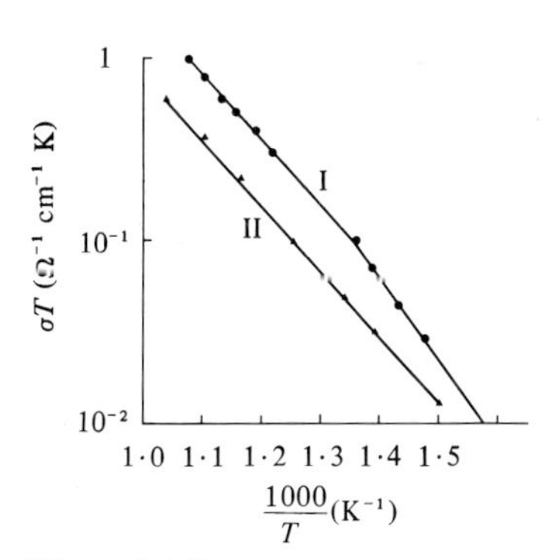


Figure 7.3.8b.

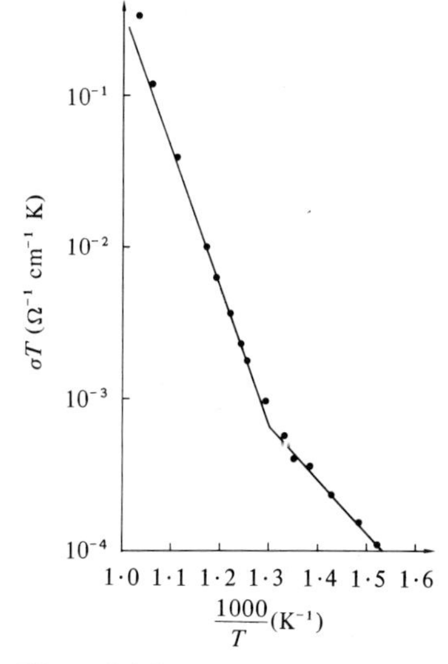


Figure 7.3.8c.

It is to be noted that at the highest temperatures the conductivity is higher than the theory used here would predict. This can be ascribed to the anion vacancies contributing to the conductivity; at even higher temperatures Frenkel defects may be involved (Kirk and Pratt, 1967).

From the relation $n = \sigma/e\mu$, and the expressions for the variation of $\sigma$ and $\mu$ with temperature, we conclude that

$$n = 2 \cdot 82 \times 10^{24} \exp\left(\frac{-14\,500}{T}\right) \text{ cm}^{-3} .$$

Thus at 500°C the number of cation vacancies or Schottky defects in a pure crystal is $2 \times 10^{16}$ cm$^{-3}$. This value can only really be regarded as an order of magnitude. Not only do different workers give different values of the constants in the expressions for $n$, $\mu$, and $\sigma$, indicating the limitations in the accuracy of the experimental method, but a given crystal of sodium chloride will have been subject to a particular thermal history and thus possess a particular set of inbuilt strains which will produce extra defects.

7.3.9

A potassium chloride crystal containing silver chloride is grown from a melt. After x-irradiation at 77 K the electron spin resonance spectrum of the crystal is obtained and found to comprise two components: one is isotropic and is shown in figure 7.3.9a; the other is simplest when the external static magnetic field is directed along a ⟨100⟩ direction as shown in figure 7.3.9b. On warming to room temperature the crystal luminesces,

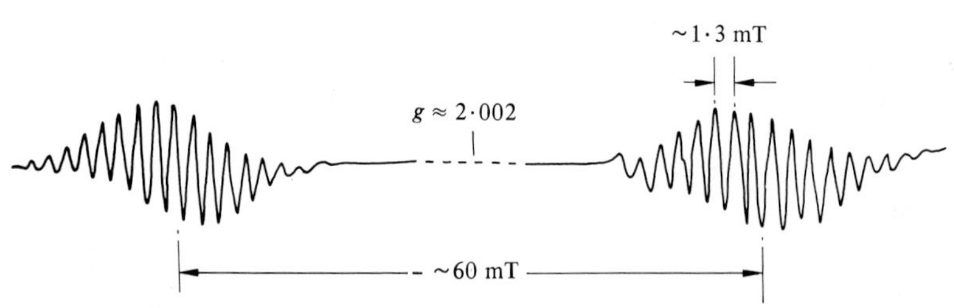


**Figure 7.3.9a.**

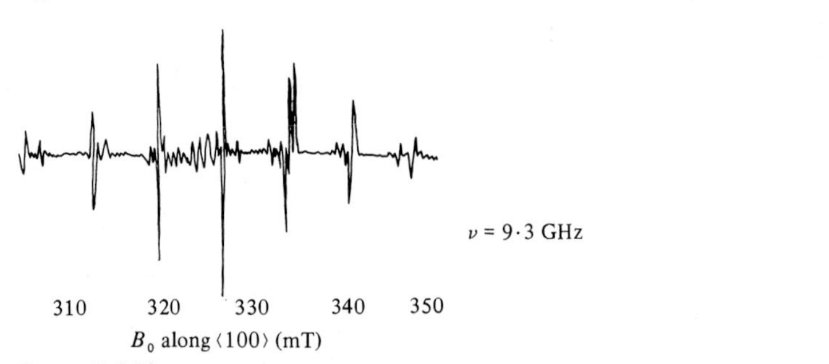


**Figure 7.3.9b.**

the latter spectrum disappears, the former decreases slightly in intensity, and a new spectrum is found. This is axially symmetric about a ⟨100⟩ direction, shows a complex hyperfine splitting pattern about a $g$ value of close to two and is unobservable unless the crystal is cooled to a low temperature. The original irradiated crystal shows an optical absorption at 425 nm, and the latter crystal shows this band at reduced intensity and additional bands at 338 nm and 460 nm.

Discuss the nature of the defects in these crystals. Suggest further experiments to confirm your conclusions and further systems for investigation.

**Solution**

The ionic radius of silver (1·26 Å) is similar to that of potassium (1·33 Å —both on the Pauling scale). It is likely, then, that the silver ions will enter the lattice substitutionally for the potassium. Irradiation with ionising radiation such as x rays will lead to the formation of free holes and electrons in the crystal, and these will move through the crystal until trapped. The pure alkali-metal halide crystals possess a number of Schottky defects which act as traps. A chloride vacancy could trap a hole for example. In this case, however, an electron spin resonance would indicate interaction with six chlorine nuclei. An electron can be trapped by a cation vacancy, leading to an F-centre. This centre shows a very broad e.s.r. line, since the electron is not strongly localised, but wanders out over many shells of ions surrounding the vacancy, and thus the e.s.r. spectrum contains many overlapping, unresolved hyperfine components. Such a line is absent in the present case and we conclude that the electrons are trapped by silver ions. The e.s.r. spectrum in figure 7.3.9a supports this since the electron would be bound into a 5s orbital and would therefore show a large isotropic hyperfine splitting from the silver nucleus and $g \approx 2\cdot002$. There are two silver isotopes abundant in nature: $^{107}$Ag is 51% and $^{109}$Ag is 49% abundant. Each has a nuclear spin $I = \frac{1}{2}$, with free-atom hyperfine splittings of 61·2 and 70·7 mT respectively. In this case, no separation of the splittings due to the two isotopes is possible since each doublet is further split, presumably, by the six surrounding chlorine nuclei at each silver ion site. It seems likely that the splitting these produce is a simple multiple or fraction of the separation of the silver multiplets. The intensity pattern of six nuclei of spin $\frac{3}{2}$ interacting equally with an electron is found from a Pascal diagram:

| | |
|---|---|
| 1 1 1 1 | 1 nucleus splits |
| 1 2 3 4 3 2 1 | 2 nuclei split |
| 1 3 6 10 12 12 10 6 3 1 | 3 nuclei split |
| 1 4 10 20 31 40 44 40 31 20 10 4 1 | 4 nuclei split |
| 1 5 15 35 65 101 135 155 155 135 101 65 35 15 5 1 | 5 nuclei split |

Thus a nineteen-line spectrum is expected from each silver multiplet with intensity ratios 1 : 6 : 21 : 56 : 120 : 216 : 336 : 456 : 546 : 580 .... If the crystal is aligned so that the magnetic field is along the [111] direction then two such overlapping patterns are expected. If the silver isotope multiplets are separated by about twice the chlorine hyperfine splitting, then there will be a central line of relative intensity 1092, flanked by lines of relative intensity 1036, 882, etc. This is approximately the case and suggests that the first crystal contains $Ag^+$ ions at normal cation sites, i.e. surrounded by six chloride ions; on irradiation these trap electrons to become silver atoms or, perhaps more correctly, a hexachloro-silver complex (Symons, 1964).

The holes are trapped by a chloride ion: the resultant chlorine atom is highly reactive and attacks an adjacent chloride ion to give a $Cl_2^-$ radical ion—this is the explanation of the e.s.r. spectrum in figure 7.3.9b which arises from two $^{37}Cl$ nuclei, giving an intensity distribution 1 : 2 : 3 : 4 : 3 : 2 : 1.

On warming the crystal the holes are released and migrate until a more stable trap is found or an electron for recombination. The latter event is less frequent but leads to a slight diminution in the intensity of the e.s.r. of the silver atoms. The former case is the explanation of the new e.s.r. found on warming. Since this spectrum is only seen at low temperature, it must be due to a centre which can relax its energy readily—this suggests the possible presence of an unpaired electron in a d-orbital. Silver is known to be bivalent in some compounds, and it seems natural to suppose that the holes are trapped at $Ag^+$ ions to form $Ag^{2+}$ ions. These are comparable to $Cu^{2+}$ and the complex with surrounding chloride ions is expected to show a Jahn–Teller distortion to give an axially symmetric structure. This is in accord with the electron spin resonance. The absorption spectrum at 338 and 460 nm is understandable since such a complex will have many excited states in this region: a detailed understanding is not readily obtained. Irradiation at room temperature of a potassium chloride crystal doped with silver chloride causes a solid state disproportionation reaction (Greenslade, 1965).

$$2Ag^+ = Ag^0 + Ag^{2+} .$$

The spectrum of the first irradiated crystal is in accord with the assignment, since a silver atom will have a $^2S$ to $^2P$ transition as found in the gas phase at about 330 nm (split by spin–orbit coupling). It is also possible to describe the transition in terms of the $d^{10}s$ molecular complex, and this gives important information on the ordering of molecular orbitals in such a complex (Symons, 1964).

Clearly, the spectra as presented here leave much to be desired in terms of resolution. For this reason crystals containing isotopically pure silver chloride should be studied. Further, by correlating the e.s.r. intensities with the optical intensities during the thermal annealing and during optical bleaching experiments, more definite conclusions can be reached. Finally, the use of electron nuclear double resonance enables the resolution of the hyperfine splittings with great accuracy. Such experiments are described by Delbecq *et al.* (1963) and by Seidel (1963).

A natural extension of this work would be the study of other alkali-metal halides doped with silver ions and the investigation of alkali-metal halides doped with copper.

**7.3.10**

In table 7.3.10 the hyperfine splitting constants, $A$, are listed for the cations which are near neighbours to an F-centre in the corresponding alkali-metal halide, and also the splittings for free alkali-metal atoms. Correlate the density of unpaired s-electrons on the first shell of cations (calculated from the given data) with the anion radius, $r_-$, which to a first approximation is the radius of the F-centre cavity. Also correlate the density with the cube of the ratio of alkali-metal atomic radius, $r_0$, to the lattice parameter, $d$. From your correlations deduce the hyperfine splitting expected from the first shell of cations surrounding an F-centre in sodium cyanide and in rubidium bromide. Mieher (1962) found that the $F_A$-centre in KCl (an F-centre with one of the surrounding cations replaced by a lithium impurity) had hyperfine splitting constants for the potassium ions of 23·71 and 24·59 MHz, and for the lithium ion 8·03 MHz. Use the correlations to calculate the geometry of the centre.

**Table 7.3.10.**

| Metal | Halogen | $A$ (MHz) | $r_-$ (Å) | $r_0^a$ (Å) | $d$ (Å) |
|---|---|---|---|---|---|
| $^7$Li | - | 402 | - | 1·45 | - |
| | F | 39·1 | 1·16 | - | 4·03 |
| | Cl | 19·1 | 1·64 | - | 5·14 |
| | Br | - | 1·8 | - | 5·50 |
| $^{23}$Na | - | 886 | - | 1·80 | - |
| | F | 105·6 | - | - | 4·62 |
| | Cl | 61·5 | - | - | 5·64 |
| | Br | - | - | - | 5·97 |
| | (CN) | - | - | - | (5·83) |
| $^{39}$K | - | 231 | - | 2·20 | - |
| | F | 34·3 | - | - | 5·34 |
| | Cl | 20·6 | - | - | 6·29 |
| | Br | 18·8 | - | - | 6·60 |
| $^{85}$Rb | - | 1012 | - | 2·35 | - |
| | Br | - | - | - | 6·89 |

[a] Source: Slater (1964).

**Solution**

The hyperfine splittings are due to the Fermi contact interaction of the F-centre electron with the nuclei of the alkali-metal cations. This interaction gives rise to a term $A\boldsymbol{I}\cdot\boldsymbol{S}$ in the Hamiltonian with

$$A = \tfrac{8}{3}\pi g\mu_B g_N \mu_N |\psi(0)|^2 ,$$

where $|\psi(0)|^2$ is the density of the electron at the nucleus and the other symbols have their usual meaning. In a molecular orbital picture the electrons reside partly in the cavity and partly in the s orbitals of the six surrounding cations:

$$\psi_F = a\psi_{\text{cavity}} + bs_1 + \ldots + bs_6 ,$$

where $s$ is the one-electron wave function of a cation s orbital. Thus the density of the electron on the s-orbitals is $6b^2$, neglecting overlap effects. $b^2$ is proportional to $|\psi(0)|^2$ and to the hyperfine splitting. If the electron were solely in one of the s orbitals we would have an alkali-metal atom; thus

$$b^2 = \frac{|\psi(0)|_F^2}{|\psi(0)|_{\text{atom}}^2} = \frac{A_F}{A_{\text{atom}}} .$$

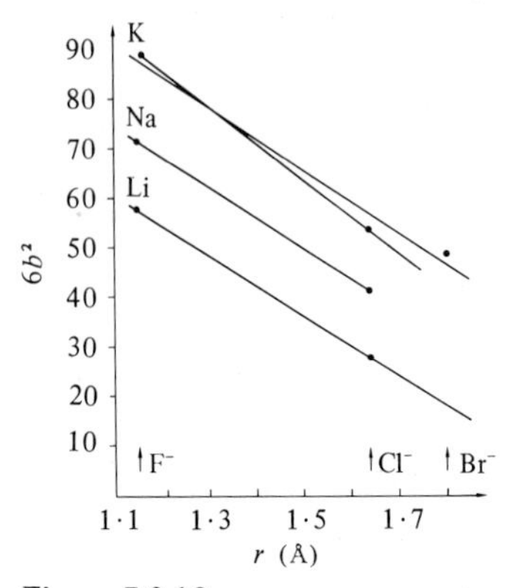


Figure 7.3.10a.

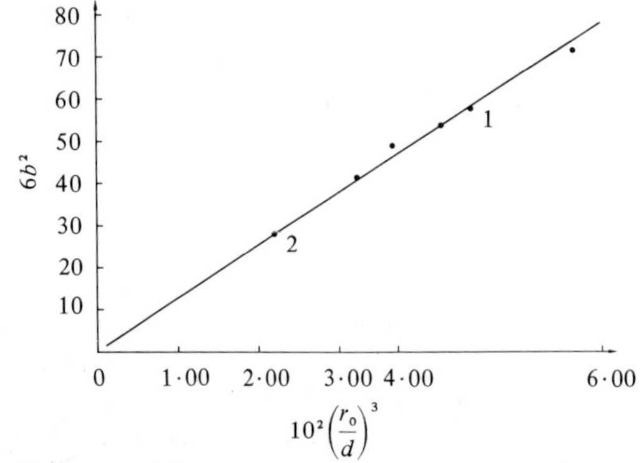


Figure 7.3.10b.

This relation has been used to calculate the required electron densities, $6b^2$, which are plotted against anion radii in figure 7.3.10a and against $(r_0/d)^3$ in figure 7.3.10b. From the latter graph, we deduce that the F-centre in sodium cyanide would have a hyperfine splitting constant for the first shell of sodium ions of 54 MHz, and in rubidium bromide a constant of 84 MHz. From the first graph we conclude that the lithium ion is 2·84 Å from the centre of the cavity and the potassium ions are 3·06 Å (opposite to the lithium ion) and 3·03 Å from the centre of the cavity (Greenslade, 1965; Claxton *et al.*, 1966).

**7.3.11**

Assuming a point electron of $g = 3{\cdot}443$ (as found for $Yb^{3+}$ substituted for $Ca^{2+}$ in $CaF_2$) calculate its distance from a fluorine nucleus ($g = 2{\cdot}6273$) showing a hyperfine splitting (measured by ENDOR) of 1·387 MHz with the external magnetic field along the electron–nucleus vector, and assuming the splitting is dipolar in origin. Given that the lattice parameter of calcium fluoride is 5·4465 Å, assign the given ENDOR line to a fluorine nucleus and calculate the lattice distortion caused by introducing a ytterbium ion into the position of a calcium ion.

**Solution**

The field at $\boldsymbol{r}$ due to a point dipole $\boldsymbol{\mu}$ at the origin is given by

$$H = \frac{\boldsymbol{\mu}}{r^3} - \frac{3(\boldsymbol{\mu} \cdot \boldsymbol{r})\boldsymbol{r}}{r^5} .$$

If the external field lies along $\boldsymbol{r}$, then so does $\mu$ and the expression reduces to $2\mu/r^3$ in magnitude. At the fluorine nucleus this produces a shift in the nuclear magnetic resonance, which is the hyperfine splitting in the electron spin resonance, given by

$$h\nu = g_N \mu_N H$$

where $h$ is Planck's constant, and $\mu_N$ is the nuclear magneton. (In conformity with magnetic resonance literature we write the energy of interaction as $\boldsymbol{\mu} \cdot \boldsymbol{H}$ not as $\boldsymbol{\mu} \cdot \boldsymbol{B}$). Remembering that $\mu = g\mu_B$ for the electron, where $\mu_B$ is the Bohr magneton, we readily combine the previous two formulae to obtain

$$r^3 = \frac{2g\mu_B g_N \mu_N}{hA} ,$$

where $A$ is the hyperfine splitting. Substituting for the values in the question we obtain $r = 4{\cdot}5178$ Å. The ytterbium ion is in the centre of a cube of fluoride ions. The small size of the hyperfine splitting observed indicates that it is not due to nearest neighbours. The next-nearest neighbours are at the corners of the opposite face of the adjacent cube of fluoride ions at a distance $\frac{1}{4} \times 11^{1/2} a_0$, where $a_0$ is the length of the unit cell or the lattice parameter. This distance is thus 4·5160 Å and so there is little if any distortion of the lattice at this point (see Baker *et al.*, 1972). This is not surprising since the radii of ytterbium (0·94 Å) and calcium (0·99 Å) are similar.

**References**


Adams, D. M., 1974, *Inorganic Solids* (John Wiley, New York), pp.98-101.

Baker, J. M., Davies, E. R., Reddy, T.R., 1972, *Contemp. Phys.*, **13**, 45.

Ballhausen, C. J., 1962, *Introduction to Ligand Field Theory* (McGraw-Hill, New York).

Barr, L. W., Lidiard, A. B., 1970, *Physical Chemistry. An Advanced Treatise,* Eds H. Eyring, D. Henderson, W. Jost (Academic Press, New York), Volume X, p.163.

Claxton, T. A., Greenslade, D. J., Root, K. D. J., Symons, M. C. R., 1966, *Trans. Faraday Soc.*, **62**, 2050.

Dekker, A. J., 1960, *Solid State Physics* (Macmillan, London), p.160.

Delbecq, C. J., Hayes, W., O'Brien, M. C. M., Yuster, P. H., 1963, *Proc. R. Soc. London, Ser.A.*, **271**, 243.

Etzel, H. W., Maurer, R. J., 1950, *J. Chem. Phys.*, **18**, 1003.

Gourary, B. S., Adrian, F. J., 1960, *Solid State Phys.*, **10**, 127.

Greenslade, D. J., 1965, Ph D Thesis, Leicester University.

Greenwood, N. N., 1968, *Ionic Crystals, Lattice Defects and Nonstoichiometry* (Butterworths, London), pp.45-46.

Griffith, J. S., 1961, *The Theory of Transition Metal Ions* (Cambridge University Press, Cambridge), p.199.

Hush, N. S., Pryce, M. H. L., 1957, *J. Chem. Phys.*, **28**, 244.

Katzin, L., 1962, *J. Chem. Phys.*, **36**, 3034.

Killingbeck, J., Cole, G. H. A., 1971, *Mathematical Techniques and Physical Applications* (Academic Press, New York), p.472.

Kirk, D. L., Pratt, P. L., 1967, *Proc. Br. Ceram. Soc.*, **9**, 215.

McClure, D. S., Dunn, T. M., Pearson, R. G., 1965, *Crystal Field Theory* (Harper and Row, New York), p.77.

Mieher, R. L., 1962, *Phys. Rev. Lett.*, **8**, 501.

Moroney, M. J., 1971, *Facts from Figures* (Penguin, Harmondsworth), p.277.

Seidel, H., 1963, *Phys. Lett.*, **6**, 150.

Slater, J. C., 1964, *J. Chem. Phys.*, **41**, 3199.

Symons, M. C. R., 1964, *J. Chem. Soc.*, 1482.

Watkins, G. D., 1959, *Phys. Rev.*, **113**, 79.

## Quantities and Units

### Systems of units

Units are important, but if they become an obsession then something is wrong. The metric system and the International System of Units (SI) are to some extent a product of French Rationalism. Indeed, Napoleon, the pragmatic military man, rejected the metric system; and the pragmatic Anglo-Saxon mind is always suspicious of such rationalism. The virtue of this rationalism is order and unity; the evil is unnecessary constraint of the individual worker, to whom reason may suggest a better local choice. Because of this we have not insisted on SI, but have asked authors to give SI equivalents of any other units. In any case there are good reasons for the student to learn other units. As L. F. Bates (1970) said with regard to magnetism, but as is true of most fields, "... Workers specializing in magnetism undoubtedly need flexibility in choice of units, and for many years to come a knowledge of CGS will be a necessity, if only to read the very large published literature ...".

The mechanical base units of SI are the metre, kilogram, and second. Conversion from the corresponding CGS base units to these is straightforward, and this is also true of the conversion of thermochemical calories to joules in thermal problems. The difficulties mostly appear where new names occur (Newton, Pascal, and so on), or with electric and magnetic quantities, which are rationalised in SI. A further difficulty appears in the case of magnetic quantities, since two approaches exist—one due to Sommerfeld, the other to Kennelly. In the former, the magnetic dipole moment, $\mu$, is defined as

$$\boldsymbol{\mu} = I\boldsymbol{A} \ ;$$

in the latter, as

$$\boldsymbol{\mu} = \mu_0 I\boldsymbol{A} \ ,$$

where $I$ is the current in a loop of area and direction $\boldsymbol{A}$, and $\mu_0$ the permeability of free space. Thus the potential energy of the dipole in a magnetic field of strength $\boldsymbol{H}$ and flux density $\boldsymbol{B}$ is $-\boldsymbol{\mu}\cdot\boldsymbol{B}$ in the former system, whereas in the latter it is $-\boldsymbol{\mu}\cdot\boldsymbol{H}$. In the CGS system, of course, $\mu_0 = 1$, and so the distinction is pedantic: papers on electron paramagnetic resonance gaily speak of magnetic field strength and measure it in gauss! Some authors now use the terms '$B$-field' and '$H$-field'. Magnetic susceptibility is dimensionless, but rationalisation requires division of the SI unit by $4\pi$ to obtain the CGS unit, which still seems to be used in papers quoting new measurements. The table below lists conversion factors for the non-SI units used in this book.

Magnetic dipole moments are often quoted in Bohr magnetons, $\mu_B$; electric dipole moments are likewise quoted in Debye units, D. Quantum theorists are keen on atomic units, which lead to a dimensionless

Hamiltonian. The unit of length is the Bohr radius, $a_0$; the unit of energy is the Hartree energy, $E_H$. The latter is simply related to the Rydberg constant, $R$:

$$E_H = 2hcR ,$$

where $h$ is the Planck constant and $c$ the velocity of light. A simple conversion may be useful where reduced mass rather than electron mass is involved, such as in the case of the hydrogen atom. This is discussed by Steiner (1976). The electron volt (eV) is also used occasionally as a unit of energy. All these units are defined numerically in the table below.

The works listed in the bibliography may be found helpful in dealing with units and their conversion.

**Bibliography**

Bates, L. F., 1970, *Contemp. Phys.*, **11**, 306.
Duffin, W. J., 1980, *Electricity and Magnetism,* 3rd edition (McGraw Hill, London).
Smith, D. H., 1970, *Contemp. Phys.*, **11**, 287.
Steiner, E., 1976, *The Determination and Interpretation of Molecular Wave Functions* (Cambridge University Press, Cambridge).

**Table of physical quantities and conversion factors**

| | | |
|---|---|---|
| Avogadro constant | $N_A$ | $6{\cdot}022 \times 10^{23}$ mol$^{-1}$ |
| Bohr magneton | $\mu_B$ | $9{\cdot}274 \times 10^{-24}$ J T$^{-1}$ |
| Bohr radius | $a_0$ | $5{\cdot}292 \times 10^{-11}$ m |
| Boltzmann constant | $k$ | $1{\cdot}381 \times 10^{-23}$ J K$^{-1}$ |
| Elementary charge | $e$ | $1{\cdot}602 \times 10^{-19}$ C |
| Gas constant | $R$ | $8{\cdot}314$ J K$^{-1}$ mol$^{-1}$ |
| Gyromagnetic ratio | | |
| of proton | $\gamma_H$ | $2{\cdot}675 \times 10^8$ s$^{-1}$ T$^{-1}$ |
| of deuteron | $\gamma_D$ | $4{\cdot}106 \times 10^7$ s$^{-1}$ T$^{-1}$ |
| of $^{19}$F | $\gamma_F$ | $2{\cdot}516 \times 10^8$ s$^{-1}$ T$^{-1}$ |
| Hartree energy | $E_H$ | $4{\cdot}360 \times 10^{-18}$ J |
| Mass of electron | $m_e$, $m_0$ | $9{\cdot}110 \times 10^{-31}$ kg |
| Mass of neutron | $m_n$ | $1{\cdot}675 \times 10^{-27}$ kg |
| Nuclear magneton | $\mu_N$ | $5{\cdot}051 \times 10^{-27}$ J T$^{-1}$ |
| Permeability of vacuum | $\mu_0$ | $4\pi \times 10^{-7}$ H m$^{-1}$ |
| Permittivity of vacuum | $\epsilon_0$ | $8{\cdot}854 \times 10^{-12}$ F m$^{-1}$ |
| Planck constant | $h$ | $6{\cdot}626 \times 10^{-34}$ J s |
| Planck constant/$2\pi$ | $\hbar$ | $1{\cdot}055 \times 10^{-34}$ J s |
| Speed of light in vacuum | $c$ | $2{\cdot}998 \times 10^8$ m s$^{-1}$ |

| | | |
|---|---|---|
| Angstrom | Å | $10^{-10}$ m |
| Atomic mass unit | a.m.u. | $1{\cdot}661 \times 10^{-27}$ kg |
| Calorie (thermochemical) | cal | $4{\cdot}184$ J |
| Debye unit | D | $3{\cdot}336 \times 10^{-30}$ C m |
| Dyne | dyn | $10^{-5}$ N |
| Electron volt | eV | $1{\cdot}602 \times 10^{-19}$ J; 8065 cm$^{-1}$; 96484·5 J mol$^{-1}$ |
| Electrostatic unit (charge) | e.s.u. | $3{\cdot}336 \times 10^{-10}$ C |
| Erg | erg | $10^{-7}$ J |
| Fermi (femtometre) | fm | $10^{-15}$ m |
| Gauss | G | $10^{-4}$ T |
| Reciprocal centimetre (spectroscopic) | cm$^{-1}$ | $1{\cdot}986 \times 10^{-23}$ J; $1{\cdot}240 \times 10^{-4}$ eV; 11·962 J mol$^{-1}$ |
| Torr ($\approx$ mmHg) | Torr | 133·3 Pa |

Useful atomic masses ($^{12}$C = 12·0000)

| | | | |
|---|---|---|---|
| C | 12·011 | O | 15·999 |
| H | 1·008 | N | 14·007 |
| D | 2·014 | | |

**Notations used in symmetry**

There are a number of notations used in the classification of molecules into the point groups, the most common being those known as the Schönflies and international systems. The latter, based on the earlier Hermann–Mauguin system, is most readily adapted to extended symmetry and is therefore used by crystallographers and many solid-state physicists. Following Heine (1960), we have adopted it for our character tables in this appendix. The Schönflies system is used by most molecular spectroscopists. The most important difference between the systems is that the international system uses the improper rotation $\bar{n}$ (rotation through an angle $2\pi/n$ radians followed by inversion through the origin of the coordinate system), whereas the Schönflies system uses the rotation reflection $S_n$ (rotation as before, but followed by reflection through the plane perpendicular to the rotation axis and containing the origin). It is, of course, possible to establish the equivalence of the two systems. Thus Cracknell (1968, p.37) defines $S_{4z}^{-}$, a fourfold clockwise rotation about the $z$ axis followed by reflection in the $xy$ plane, as $IC_{4z}^{+}$, a counter-clockwise rotation about $0z$, followed by inversion. This is not strictly in the spirit of the Schönflies system, although formally correct.

A conversion table for the two notations, covering the point groups used in this book, is given in table A2.1. The differences in notation may be summarised in the following pairs of symbols, the first referring to the international system (Henry and Lonsdale, 1952, p.22), the second to the Schönflies system.

Rotation of $2\pi/n$ radians about an axis: $n$, $C_n$. Subscripts may be added to show the orientation of the axis: $2_z$, $C_{4z}$ are about $0z$; $2_d$, $C_{2d}$ imply axes diagonal to the Cartesian axes. Superscripts imply direction of rotation (minus is clockwise) or number of operations. Cracknell (1968 and sections 1.1 and 1.3 of this book) defines $C_{2a}$, $C_{2b}$, $C_{2c}$, $C_{2d}$, $C_{2e}$ and $C_{2f}$ to be rotations about [110], [$\bar{1}$10], [101], [011], [$\bar{1}$01], and [0$\bar{1}$1].

**Table A2.1.** Notation for the point groups.

| International symbol | Full symmetry symbol | Schönflies symbol |
|---|---|---|
| 1 | 1 | $C_1$ |
| $\bar{1}$ | $\bar{1}$ | $S_2(C_i)$ |
| 2 | 2 | $C_2$ |
| $m$ | $m$ | $C_{1h}(C_s)$ |
| $2/m$ | $\frac{2}{m}$ | $C_{2h}$ |
| $mm2$ | $mm2$ | $C_{2v}$ |
| 222 | 222 | $D_2(V)$ |
| $mmm$ | $\frac{2}{m}\frac{2}{m}\frac{2}{m}$ | $D_{2h}(V_h)$ |

**Table A2.1** (continued).

| International symbol | Full symmetry symbol | Schönflies symbol |
|---|---|---|
| $4$ | $4$ | $C_4$ |
| $\bar{4}$ | $\bar{4}$ | $S_4$ |
| $4/m$ | $\frac{4}{m}$ | $C_{4h}$ |
| $4mm$ | $4mm$ | $C_{4v}$ |
| $\bar{4}2m$ | $\bar{4}2m$ | $D_{2d}(V_d)$ |
| $422$ | $422$ | $D_4$ |
| $4/mmm$ | $\frac{4}{m}\frac{2}{m}\frac{2}{m}$ | $D_{4h}$ |
| $3$ | $3$ | $C_3$ |
| $\bar{3}$ | $\bar{3}$ | $S_6(C_{3i})$ |
| $3m$ | $3m$ | $C_{3v}$ |
| $\bar{3}m$ | $\bar{3}\frac{2}{m}$ | $D_{3d}$ |
| $32$ | $32$ | $D_3$ |
| $6$ | $6$ | $C_6$ |
| $\bar{6}$ | $\bar{6}$ | $C_{3h}$ |
| $6/m$ | $\frac{6}{m}$ | $C_{6h}$ |
| $6mm$ | $6mm$ | $C_{6v}$ |
| $\bar{6}m2$ | $\bar{6}m2$ | $D_{3h}$ |
| $622$ | $622$ | $D_6$ |
| $6/mmm$ | $\frac{6}{m}\frac{2}{m}\frac{2}{m}$ | $D_{6h}$ |
| $5$ | $5$ | $C_5$ |
| $\bar{5}$ | $\bar{5}$ | $S_{10}(C_{5i})$ |
| $52$ | $52$ | $D_5$ |
| $5m$ | $5m$ | $C_{5v}$ |
| $\bar{5}m$ | $\bar{5}\frac{2}{m}$ | $D_{5d}$ |
| $23$ | $23$ | $T$ |
| $m3$ | $\frac{2}{m}\bar{3}$ | $T_h$ |
| $\bar{4}3m$ | $\bar{4}3m$ | $T_d$ |
| $432$ | $432$ | $O$ |
| $m3m$ | $\frac{4}{m}\bar{3}\frac{2}{m}$ | $O_h$ |
| $532$ | $532$ | $I$ |
| $\bar{5}\bar{3}m$ | $\bar{5}\bar{3}\frac{2}{m}$ | $I_h$ |
| $\bar{8}2m$ | $\bar{8}2m$ | $D_{4d}$ |

Reflection in a mirror plane: $m$, $\sigma$. Again subscripts are used to designate the orientation of the plane, by means of its normal axis. In the Schönflies system $\sigma_h$ and $\sigma_v$ are used to denote, respectively, planes perpendicular to and containing the principal axis; $\sigma_d$ is a diagonal plane.

Rotation reflection axis: $\tilde{n}$, $S_n$.

Rotation inversion: $\bar{n}$, not used.

Inversion through the origin: $\bar{1}$, $C_i$ (also $i$, $I$, and $S_2$).

Identity operation: 1, $C_1$, but $E$ (and sometimes $I$, although then there is confusion with inversion) is usually used in group theory.

Because of the equivalence that exists between the international and Schönflies systems, the appearance of the character tables for equivalent groups will be the same, apart from the labels of the symmetry operations.

When we come to the notation for representations there is no agreed system, although there are two main ones in use. The first, due to Bethe (1929), simply labels the symmetric representation $\Gamma_1$; increasingly complicated and dimensioned representations, ending with the so-called double-valued representations, are then labelled $\Gamma_2$, $\Gamma_3$, ... . Some workers use slightly different orders from those used by Bethe. The second system is rather more descriptive and is due to Placzek (1934). Rather unfairly, most workers ascribe it to Mulliken (1933), although they cannot have read his original paper, where he freely admits his debt to Placzek. One-dimensional representations are denoted in this notation by $A$ or $B$, two-dimensional and pairs of conjugate single-dimensional representations by $E$, three-, four-, five-, and six-dimensional representations by $T$, $U$, $V$, and $W$. Subscripts are used to distinguish representations of the same dimensionality ($A_1$, $A_2$, $E_g$, $E_u$ for example). The use of subscripts g and u is explained in the introduction to the character tables. Primes ($A'$, $E''$) are sometimes used, especially where the character of a reflection in an important plane changes sign from one representation to another: for example $E'$, $E''$ of $\bar{6}$ have characters for $m_z$ of plus one and minus one respectively. Following the notation of Griffith (1961), we use primes to distinguish the extra (double-valued) representations in the table for the double group 432. Some workers use $F$, $G$, $H$, and $I$ instead of $T$, $U$, $V$, and $W$, but these may be confused with atomic term symbols. Lax (1974) gives a good review of the notation of symmetry.

We have used $\otimes$ to represent the product of representations. Strictly, this symbol means the outer or direct product for groups or for matrices. As discussed by Bradley and Cracknell (1972), the 'product of representations' referred to here is an *inner* direct product, and the sum an *inner* direct sum. For this product, the latter authors use a cross within a rectangle—unfortunately not available to us (cf. Streitwolf, 1971).

The references given here would not be complete without Koster et al (1963) who give Clebsch–Gordan coefficients, and Altmann (1977) who gives a sophisticated mathematical treatment including nonrigid molecules.

**References**

Altmann, S. L., 1977, *Induced Representations in Crystals and Molecules* (Academic Press, London).

Bethe, H. A., 1929, *Ann. Physik,* **3**, 133.

Bradley, C. J., Cracknell, A. P., 1972, *The Mathematical Theory of Symmetry in Solids: Representation Theory for Point Groups and Space Groups* (Oxford University Press, London).

Cracknell, A. P., 1968, *Applied Group Theory* (Pergamon, Oxford).

Griffith, J. S., 1961, *The Theory of Transition-metal Ions* (Cambridge University Press, Cambridge).

Heine, V., 1960, *Group Theory in Quantum Mechanics* (Pergamon, Oxford).

Henry, N. F. M., Lonsdale, K., 1952, *International Tables for X-Ray Crystallography*, Volume 1 (Kynoch Press, Birmingham).

Koster, G. F., Dimmock, J. O., Wheeler, R. G., Statz, H., 1963, *Properties of the Thirty-two Point Groups* (M.I.T. Press, Cambridge, Mass.).

Lax, M., 1974, *Symmetry Principles in Solid State and Molecular Physics* (John Wiley, New York), p.433.

Mulliken, R. S., 1933, *Phys. Rev.,* **43**, 279.

Placzek, G., 1934, *Handb. Radiol.,* **6**, 205.

Streitwolf, H.-W., 1971, *Group Theory in Solid-state Physics* (Macdonald Phoebus, London), p.37.

**Character tables for molecular symmetry**

The following tables are of a compact form and in the international system of notation. Their use is indicated by the diagram below (figure A2.1). The tables for groups containing the inversion operation are not always given, since they are readily derived from one of the other groups: thus the group $4/mmm$ is obtained by adding the inversion $(\bar{1})$ to the elements of 422. Every representation of the group without inversion has two representations in the extended group: $A_2$ of 422 becomes $A_{2g}$ and $A_{2u}$ in $4/mmm$. The first representation is even (*gerade*) under inversion, having a character for inversion equal to plus one, the second is odd (*ungerade*), having the corresponding character of minus one. The formal notation $\mathbf{G}_2 = \mathbf{G}_1 \otimes \bar{1}$ has been used to indicate the group obtained from combining inversion with the group $\mathbf{G}_1$.

Typical element of class of group 1 elements → $E$ $2_z(3)$ ← Number of elements in that class

| | Group 1 | $E$ $2_z(3)$ |
|---|---|---|
| Group 2 | | $E$ $m_z(4)$ ← This line refers to group 2 |
| Irreducible representations of group 2 | Irreducible representations of group 1 | Characters |

**Figure A2.1.** Layout of character tables in this appendix.

**Triclinic, monoclinic, and orthorhombic groups**

| | | $\bar{1}$ | $E$ | $\bar{1}$ |
|---|---|---|---|---|
| | 2 | | $E$ | $2_z$ |
| $m$ | | | $E$ | $m_z$ |
| $A'$ | $A$ | $A_g$ | 1 | 1 |
| $A''$ | $B$ | $A_u$ | 1 | −1 |

| | | 222 | $E$ | $2_x$ | $2_y$ | $2_z$ |
|---|---|---|---|---|---|---|
| | $mm2$ | | $E$ | $m_x$ | $m_y$ | $2_z$ |
| $2/m$ | | | $E$ | $m_z$ | $\bar{1}$ | $2_z$ |
| $A_g$ | $A_1$ | $A$ | 1 | 1 | 1 | 1 |
| $A_u$ | $A_2$ | $B_1$ | 1 | −1 | −1 | 1 |
| $B_g$ | $B_1$ | $B_2$ | 1 | −1 | 1 | −1 |
| $B_u$ | $B_2$ | $B_3$ | 1 | 1 | −1 | −1 |

$mmm = 222 \otimes \bar{1}$

**Tetragonal point groups**

| $\bar{4}$ | $E$ | $2_z$ | $\bar{4}_z$ | $\bar{4}_z^3$ |
|---|---|---|---|---|
| 4 | $E$ | $2_z$ | $4_z$ | $4_z^3$ |
| $A$ | 1 | 1 | 1 | 1 |
| $B$ | 1 | 1 | −1 | −1 |
| $E$ { | 1 | −1 | i | −i |
| | 1 | −1 | −i | i |

$\mathrm{i} = (-1)^{1/2}$

$4/m = 4 \otimes \bar{1}$

| 422 | $E$ | $2_z$ | $4_z(2)$ | $2_x(2)$ | $2_d(2)$ |
|---|---|---|---|---|---|
| $\bar{4}2m$ | $E$ | $2_z$ | $\bar{4}_z(2)$ | $2_x(2)$ | $m_d(2)$ |
| $4mm$ | $E$ | $2_z$ | $4_z(2)$ | $m_x(2)$ | $m_d(2)$ |
| $A_1$ | 1 | 1 | 1 | 1 | 1 |
| $A_2$ | 1 | 1 | 1 | −1 | −1 |
| $B_1$ | 1 | 1 | −1 | 1 | −1 |
| $B_2$ | 1 | 1 | −1 | −1 | 1 |
| $E$ | 2 | −2 | 0 | 0 | 0 |

$4/mmm = 422 \otimes \bar{1}$

**Trigonal and hexagonal point groups**

| 3 | $E$ | $3_z$ | $3_z^2$ |
|---|---|---|---|
| $A$ | 1 | 1 | 1 |
| $E$ | 1 | $\omega$ | $\omega^2$ |
| | 1 | $\omega^2$ | $\omega$ |

$\omega = \exp(\frac{2}{3}\pi i)$
$\bar{3} = 3 \otimes \bar{1}$

| 32 | $E$ | $3_z(2)$ | $2_y(3)$ |
|---|---|---|---|
| $3m$ | $E$ | $3_z(2)$ | $m_x(3)$ |
| $A_1$ | 1 | 1 | 1 |
| $A_2$ | 1 | 1 | −1 |
| $E$ | 2 | −1 | 0 |

$\bar{3}m = 32 \otimes \bar{1}$

| | $\bar{6}$ | $E$ | $\bar{6}_z$ | $3_z$ | $m_z$ | $3_z^2$ | $\bar{6}_z^5$ |
|---|---|---|---|---|---|---|---|
| 6 | | $E$ | $6_z$ | $3_z$ | $2_z$ | $3_z^2$ | $6_z^5$ |
| $A$ | $A'$ | 1 | 1 | 1 | 1 | 1 | 1 |
| $B$ | $A''$ | 1 | −1 | 1 | −1 | 1 | −1 |
| $E_1$ | | 1 | $-\omega^2$ | $\omega$ | −1 | $\omega^2$ | $-\omega$ |
| | | 1 | $-\omega$ | $\omega^2$ | −1 | $\omega$ | $-\omega^2$ |
| | $E'$ | 1 | $\omega$ | $\omega$ | 1 | $\omega^2$ | $\omega^2$ |
| | | 1 | $\omega^2$ | $\omega^2$ | 1 | $\omega$ | $\omega$ |
| $E_2$ | | 1 | $\omega$ | $\omega^2$ | 1 | $\omega$ | $\omega^2$ |
| | | 1 | $\omega^2$ | $\omega$ | 1 | $\omega^2$ | $\omega$ |
| | $E''$ | 1 | $-\omega$ | $\omega$ | −1 | $\omega^2$ | $-\omega^2$ |
| | | 1 | $-\omega^2$ | $\omega^2$ | −1 | $\omega$ | $-\omega$ |

$\omega = \exp(\frac{2}{3}\pi i)$
$6/m = 6 \otimes \bar{1}$

| | | 622 | $E$ | $2_z$ | $3_z(2)$ | $6_z(2)$ | $2_y(3)$ | $2_x(3)$ |
|---|---|---|---|---|---|---|---|---|
| | $6mm$ | | $E$ | $2_z$ | $3_z(2)$ | $6_z(2)$ | $m_y(3)$ | $m_x(3)$ |
| $\bar{6}m2$ | | | $E$ | $m_z$ | $3_z(2)$ | $\bar{6}_z(2)$ | $2_y(3)$ | $m_x(3)$ |
| $A_1'$ | $A_1$ | $A_1$ | 1 | 1 | 1 | 1 | 1 | 1 |
| $A_2'$ | $A_2$ | $A_2$ | 1 | 1 | 1 | 1 | −1 | −1 |
| $A_1''$ | $B_2$ | $B_1$ | 1 | −1 | 1 | −1 | 1 | −1 |
| $A_2''$ | $B_1$ | $B_2$ | 1 | −1 | 1 | −1 | −1 | 1 |
| $E''$ | $E_1$ | $E_1$ | 2 | −2 | −1 | 1 | 0 | 0 |
| $E'$ | $E_2$ | $E_2$ | 2 | 2 | −1 | −1 | 0 | 0 |

$6/mmm = 622 \otimes \bar{1}$

**Fivefold rotation groups**

| 5 | $E$ | $5_z$ | $5_z^2$ | $5_z^3$ | $5_z^4$ |
|---|---|---|---|---|---|
| $A$ | 1 | 1 | 1 | 1 | 1 |
| $E'$ | 1 | $\omega$ | $\omega^2$ | $\omega^3$ | $\omega^4$ |
| | 1 | $\omega^4$ | $\omega^3$ | $\omega^2$ | $\omega$ |
| $E''$ | 1 | $\omega^2$ | $\omega^4$ | $\omega$ | $\omega^3$ |
| | 1 | $\omega^3$ | $\omega$ | $\omega^4$ | $\omega^2$ |

$\omega = \exp(\frac{2}{5}\pi i)$
$\bar{5} = 5 \otimes \bar{1}$

| 52 | $E$ | $5_z(2)$ | $5_z^2(2)$ | $2_y(5)$ |
|---|---|---|---|---|
| $5m$ | $E$ | $5_z(2)$ | $5_z^2(2)$ | $m_x(5)$ |
| $A_1$ | 1 | 1 | 1 | 1 |
| $A_2$ | 1 | 1 | 1 | −1 |
| $E_1$ | 2 | $2\cos x$ | $2\cos 2x$ | 0 |
| $E_2$ | 2 | $2\cos 2x$ | $2\cos x$ | 0 |

$x = \frac{2}{5}\pi$
$\bar{5}m = 52 \otimes \bar{1}$

**Cubic point groups**

| 23 | $E$ | $2_z(3)$ | $3(4)$ | $3^2(4)$ |
|---|---|---|---|---|
| $A$ | 1 | 1 | 1 | 1 |
| $E$ | 1 | 1 | $\omega$ | $\omega^2$ |
| | 1 | 1 | $\omega^2$ | $\omega$ |
| $T$ | 3 | −1 | 0 | 0 |

$\omega = \exp(\frac{2}{3}\pi i)$
$m3 = 23 \otimes \bar{1}$

| $\bar{4}3m$ | $E$ | $3(8)$ | $2_z(3)$ | $m_d(6)$ | $\bar{4}_z(6)$ |
|---|---|---|---|---|---|
| 432 | $E$ | $3(8)$ | $2_z(3)$ | $2_d(6)$ | $4_z(6)$ |
| $A_1$ | 1 | 1 | 1 | 1 | 1 |
| $A_2$ | 1 | 1 | 1 | −1 | −1 |
| $E$ | 2 | −1 | 2 | 0 | 0 |
| $T_1$ | 3 | 0 | −1 | −1 | 1 |
| $T_2$ | 3 | 0 | −1 | 1 | −1 |

$m3m = 432 \otimes \bar{1}$

**Cubic double group**

| $432'$ | $E$ | $\bar{E}$ | $3(8)$ | $3'(8)$ | $2_z/2'_z(6)$ | $2_d/2'_d(12)$ | $4_z(6)$ | $4'_z(6)$ |
|---|---|---|---|---|---|---|---|---|
| $A_1/\Gamma_1$ | 1 | 1 | 1 | 1 | 1 | 1 | 1 | 1 |
| $A_2/\Gamma_2$ | 1 | 1 | 1 | 1 | 1 | −1 | −1 | −1 |
| $E_1/\Gamma_3$ | 2 | 2 | −1 | −1 | 2 | 0 | 0 | 0 |
| $T_1/\Gamma_4$ | 3 | 3 | 0 | 0 | −1 | −1 | 1 | 1 |
| $T_2/\Gamma_5$ | 3 | 3 | 0 | 0 | −1 | −1 | −1 | −1 |
| $E'/\Gamma_6$ | 2 | −2 | 1 | −1 | 0 | 0 | $2^{1/2}$ | $-(2^{1/2})$ |
| $E''/\Gamma_7$ | 2 | −2 | 1 | −1 | 0 | 0 | $-(2^{1/2})$ | $2^{1/2}$ |
| $U'/\Gamma_8$ | 4 | −4 | −1 | 1 | 0 | 0 | 0 | 0 |

$3'$ is a threefold rotation coupled with $\bar{E}$.
$\Gamma_i$ are Bethe's notation for the irreducible representations (see page 458).

**Icosahedral groups**

| 532 | $E$ | $5_z(12)$ | $5_z^2(12)$ | $3(20)$ | $2(15)$ |
|---|---|---|---|---|---|
| $A$ | 1 | 1 | 1 | 1 | 1 |
| $T_1$ | 3 | $\frac{1}{2}(1+5^{1/2})$ | $\frac{1}{2}(1-5^{1/2})$ | 0 | −1 |
| $T_2$ | 3 | $\frac{1}{2}(1-5^{1/2})$ | $\frac{1}{2}(1+5^{1/2})$ | 0 | −1 |
| $U$ | 4 | −1 | −1 | 1 | 0 |
| $V$ | 5 | 0 | 0 | −1 | 0 |

$\bar{5}\bar{3}m = 532 \otimes \bar{1}$

**Eightfold group**

| $\bar{8}2m$ | $E$ | $4_z(2)$ | $2_z$ | $2_d(4)$ | $\bar{8}_z(2)$ | $\bar{8}_z^3(2)$ | $m_d(4)$ |
|---|---|---|---|---|---|---|---|
| $A_1$ | 1 | 1 | 1 | 1 | 1 | 1 | 1 |
| $A_2$ | 1 | 1 | 1 | −1 | 1 | 1 | −1 |
| $B_1$ | 1 | 1 | 1 | 1 | −1 | −1 | −1 |
| $B_2$ | 1 | 1 | 1 | −1 | −1 | −1 | 1 |
| $E_1$ | 2 | 0 | −2 | 0 | $-(2^{1/2})$ | $2^{1/2}$ | 0 |
| $E_2$ | 2 | −2 | 2 | 0 | 0 | 0 | 0 |
| $E_3$ | 2 | 0 | −2 | 0 | $2^{1/2}$ | $-(2^{1/2})$ | 0 |

# Index

The numbers refer to problems; references to whole chapters are denoted by bold type.


Abelian group 1.1.3
Absolute configuration 2.1.9, 4.5.5, 4.5.6, 4.5.7
Acetylene 3.1.1, 3.1.5, 7.2.1
Ammonia 1.4.4, 4.4.2
Angular momentum, orbital 1.1.8, 1.2.11, 4.4.5
Anharmonic potential 3.3.1
Anomalous scattering (x-ray) 2.1.9
Antibonding 4.6.4
Antiferromagnetism 1.3.7
Antisymmetrized cube and square 1.1.7, 1.2.7
Antisymmetrized product function, see Slater determinant
Appearance energy (potential) 6.1.3, 7.1.4
Asymmetric top molecule 3.2.4, 3.3.4
Asymmetry parameter 5.2.0, 5.2.3

Base (mass) peak 6.1.0
Bending vibration 4.6.3
Benzene 3.2.3, 5.1.1, 5.1.2
  $\pi$ orbitals 1.2.4, 7.2.2
Bond
  bent 5.2.6, 7.2.7
  charge-displacement 4.5.7
  dipole 4.4.12
  dissociation energy 7.1.1
  ionic character of 5.2.1, 5.2.2, 5.2.3
  length determination 2.1.4, 2.1.10, 2.3.1, 2.3.5, 2.3.6, 3.2.1, 3.2.3, 3.3.1, 3.3.2, 3.3.5, 5.1.1
Born approximation 2.3.3, 2.3.7
Born-Haber cycle 7.1.9
Bragg angle 2.1.1, 2.1.2, 2.3.1
Bravais lattice 1.1.9, 1.3.9, 2.1.1, 2.1.4, 2.3.1
Brillouin zone 1.3.9, 1.3.11

Carbon dioxide 1.4.2, 3.1.1, 3.1.2
Casimir function 3.3.5
Character, group representation 1.1.3, 1.1.4, 1.2.4, 1.2.5, 1.2.7, 1.3.5, 1.4.1, 4.4.5, Appendix 2
Charge transfer 4.1.8, 4.1.10
Chemical shift 5.1.4, 5.1.5, 5.1.6, 5.1.7, 5.1.8, 5.1.9
  as tensor 5.1.3
Chiral structures 4.5.0, 4.5.4
*Cis-trans* isomers 4.1.3

Class (of group) 1.1.2
  multiplication 1.1.4
Close packing 7.3.1, 7.3.2, 7.3.3, 7.3.6
Configuration, molecular 2.3.4, 3.2.6, 4.5.0, 4.5.2
  (see also Absolute configuration)
Conformation 4.4.1, 4.5.8
Coordinate frame 4.4.4
Coordinate transformation 1.2.2, 1.2.3, 2.1.10
Cotton effect 4.5.0, 4.5.1, 4.5.2
Coupling constant
  quadrupole **5.2**, 5.3.9, 5.3.11
  rotational 3.2.2
  spin-spin 5.1.8, 5.1.9, 5.1.10, 5.1.11
Crystal field 1.2.5, 1.3.4, 4.1.1, 7.3.5, 7.3.6
Crystal texture 2.4.3, 2.4.4
Cubic symmetry 1.2.5, 1.2.6, 1.2.9, 2.1.1, 2.1.4, 2.4.4

Davydov splitting 1.4.6
de Broglie relation 2.2.1, 2.3.2
Debye equation 4.4.2
Debye temperature 5.3.6
Defect lattice 7.3.7, 7.3.8, 7.3.9
Degenerate states 4.4.3, 4.6.6
Degenerate vibrations 3.1.1, 3.1.4, 3.1.5, 3.1.6
Deslandres table 4.1.9
Deuterated crystal 2.2.3, 5.1.1
Diamond 4.2.4
Difference map (x-ray) 2.2.4
Dipole moment
  electric **4.4**, 7.1.6
  magnetic **4.3**
  transition 3.3.3, 3.3.4, 4.5.0
Disorder 2.2.5
  (see also Defect lattice)
Dissociation energy 7.1.1, 7.1.4
Donor-acceptor complex 4.1.8, 4.1.10
Double group 1.1.10, 1.2.9, 4.3.4
Double resonance 5.1.11
  (see also ENDOR)

Effective atomic charge 4.6.9
Einstein photoelectric equation 4.6.0
Electric conductivity 7.3.7, 7.3.8
Electric field gradient 3.3.5, 4.4.11, 5.2.0, 5.3.9
Electric polarisability tensor 1.3.3

Electron affinity 7.1.9
Electron spin resonance **4.2**, 7.3.8, 7.3.9
Electronegativity 4.6.9, 5.2.1, 7.1.8
ENDOR 4.2.1, 7.3.11
Epitaxy 2.4.4
Equivalent positions 1.3.6, 1.3.7, 2.1.3, 2.1.5, 2.1.6, 2.1.7
Ethylene 3.2.4, 7.2.2
Exchange parameter 4.3.7, 4.3.8
Exciton 4.5.0, 4.5.4, 4.5.5
Exclusion rule, see Mutual exclusion rule

F-centre 7.3.9, 7.3.10
Face-centred lattice 1.3.11, 2.1.4
Fine structure 4.1.6
Fluorescence 4.1.9, 4.1.13
Force constant 2.3.6, 3.1.2, 3.1.3, 3.3.1
Fourier transform 2.3.3
Franck-Condon principle 4.6.0, 4.6.1
Free-electron model 1.3.11, 7.2.2
Free radical 4.1.9, 4.2.1
Frenkel defect 7.3.8

*g*-factor 1.2.12, 4.2.1
Gouy method 4.3.1
Graphite 2.4.3
Group multiplication 1.1.2, 1.1.3

Harker section 2.1.6
Harmonic oscillator 3.3.1, 7.3.5
Hartree-Fock wave functions 7.2.5, 7.2.6
Heat of formation 7.1.1
Heavy-atom technique 2.1.5
Heitler-London method 7.2.4
Helix 4.5.0, 4.5.8
Hellmann-Feynmann theorem 7.2.6
Homonuclear diatomic molecule 3.2.1
Host matrix 5.3.3
Hückel theory 1.2.4, 4.2.3, 4.5.7
Hybrid orbital 1.2.2, 5.2.4, 5.2.6, 7.2.7
Hydrocarbons 1.2.1, 1.2.4, 2.3.6, 3.1.1, 3.1.5, 3.2.3 et seq., 3.3.3, 4.1.13, 4.1.14, 4.2.2, 4.5.0, 4.5.4, 4.5.7, 5.1.1, 5.1.2, 6.1.1, 6.1.3, 7.1.2, 7.1.3, 7.1.7, 7.1.8, 7.2.2, 7.2.7
Hydrogen 4.6.0, 4.6.2
  scattering of neutrons and x rays 2.2.0
Hydrogen bond 2.2.4, 2.2.5
Hydrogen chloride 2.3.1, 3.1.1, 4.6.5
Hyperfine splitting (or coupling) 4.2.1, 5.3.10, 7.3.10

Infrared absorption
  active modes 1.4.1, 3.1.4, 3.1.6
  parallel and perpendicular bands 3.2.5
Intensity (of absorption band) 3.2.5, 4.1.5
International symmetry notation, see Symmetry
Ion pair 7.1.6
Ionic character, see Bond
Ionization energy (potential) 4.6.0, 6.1.3, 7.2.1, 7.2.5
Irreducible representation 1.1.3, 1.1.5, 1.1.6, 1.1.7, 1.4.1, 4.4.3
Isomer shift 5.3.5, 5.3.7, 5.3.8
Isomorphism 1.1.3, 1.1.4
Isomorphous (crystal) replacement method 2.1.8
Isotopic abundance 6.1.0
Isotopic species, rotational constants 3.3.2

Jahn-Teller effect 1.4.5, 4.1.1, 4.1.2, 4.2.4, 7.3.9
Jones symbols 1.3.9

Kinetic shift 6.1.3
Kramers' theorem 4.2.5
Koopmans' theorem 7.2.5

Langevin equation 4.3.0
Laplace equation 4.4.8
Laporte rule 4.1.1
Lattice energy 7.1.9, 7.3.4
Laue condition 2.1.2
Layer lines 2.1.2
Lifetime of excited state 5.3.2
Ligand field 4.1.2, 4.1.3, 4.1.4, 4.1.5, 4.3.3 (see also Crystal field)

Madelung constant 7.1.9, 7.3.4
Magnetic anisotropy 4.3.6
Matthieu equation 3.2.4
Maximum overlap 7.2.7
Maxwell-Boltzmann distribution 2.2.1
McLafferty rearrangement 4.1.14
Metal-metal bond 4.1.12
Metastable (mass) peak 6.1.2
Miller indices 2.1.1
Modes of vibration, see Normal modes of vibration
Molecular intensity curve 2.3.3, 2.3.7
Molecular ion 6.1.0
Molecular orbital 7.2.3, 7.2.4, 7.2.6
Moment of inertia 3.2.3, 3.3.1, 3.3.2

Moment, second, of n.m.r. line 5.1.1, 5.1.2, 5.1.3
Monochromatic neutron beam 2.2.1
Morse curve 3.3.1
Mössbauer effect **5.3**
in various phases 5.3.4
Multipole moments 4.4.12
Mutual exclusion rule (infrared-Raman) 1.4.1, 3.1.5, 3.2.5, 3.2.6

Neutron flux 2.2.0
Nonbonding orbital 4.6.5
Non-rigid group 1.2.10
Normal modes of vibration 1.4.1, 3.1.1, 3.1.4, 3.1.5, 3.1.6
assignment 1.4.2, 3.1.6, 3.2.5
Nuclear magnetic resonance 1.4.7, **5.1**
Nuclear quadrupole coupling 3.3.5
Null matrix (neutron diffraction) 2.2.3

*O* branch 3.2.2
Octant rule 4.5.1
Octopole, electric 4.4.5
Operator
orbital angular momentum, see Angular momentum
Operator equivalent 1.2.11
Optical activity 4.5.0
Orbital populations 5.2.3, 5.2.4, 5.2.5, 5.2.6
Orbitals, $\pi$ 1.2.4, 4.5.7, 7.2.2
Oxygen 3.2.1, 3.2.2, 4.6.4

*P* branch 1.4.2, 3.1.4, 3.2.5
Parity 4.4.3
Pascal's constants 4.3.1
Patterson function 2.1.6, 2.1.7
Pauli principle 1.2.7, 1.2.8
Permutation group, see Symmetric group
Photodissociation 4.1.14
Pi orbitals, see Orbitals, $\pi$
Platt diagrams 4.5.5, 4.5.6
Point group
assignment, flow chart for 1.2.1
black-and-white 1.1.9
crystallographic 1.3.4, 2.1.3
noncrystallographic 1.1.1, 1.1.3, 1.2.1
of x-ray diffraction pattern 2.1.3
Polarizability, electric 3.2.3
tensor 1.3.3, 3.1.4, 3.1.6
Polarization spectra 4.1.3
Polymorphs 2.4.3

Powder, diffraction pattern of 2.1.1, 2.4.2, 2.4.4
Preferred orientation 2.4.4
Projection operator 1.2.4, 1.2.5
Pseudoscalar function 4.5.3

*Q* branch 1.4.2, 3.1.4, 3.2.2, 3.2.5
Quadrant rule 4.5.0, 4.5.1
Quadrupole splitting, see Coupling constant, quadrupole
Quantum defect 4.6.8
Quenching 4.1.9

*R* branch 1.4.2, 3.1.4, 3.2.5
Racah parameter 4.1.2, 4.1.7
Radial distribution function 2.3.4, 2.3.5, 2.3.7
Radius ratio (ionic) 7.3.4
Raman spectrum 1.4.1, 1.4.2, 1.4.3, 3.1.4, 3.1.6, **3.2**
polarized and depolarized bands 3.2.5
Reciprocal lattice 1.3.9, 1.3.11, 2.4.3, 2.4.5
Recoil-free fraction 5.3.6
Refractive index anisotropy 4.4.11
Representation
double-valued 1.1.10
wave vector 1.3.10
Resonance energy 7.1.8
Rotation photograph (x-ray) 2.1.2
Rotational constant 3.2.2, 3.2.3, 3.2.4, 3.3.2
Rotational spectrum 3.3.1
Rydberg series 4.6.8, 7.2.1

*S* branch 3.2.2, 3.2.4
Scattering factor, atomic
electron 2.3.1, 2.3.7
neutron 2.2.0
x-ray 2.1.4, 2.1.8, 2.1.9, 2.2.0, 2.3.3, 2.3.7
Schönflies notation, see Symmetry
Schottky defect 7.3.7, 7.3.8
Screw axis 2.1.3, 2.1.5, 2.1.6, 2.1.7, 4.5.0
Seitz symbol 1.3.7, 1.3.8
Selection rules 3.2.6, 4.1.1
Shpolskii spectra 4.1.13
Simple harmonic motion 3.1.3, 4.6.1, 5.2.10
Slater determinant 1.2.5, 1.2.8, 7.3.5
method (for atomic terms) 1.2.8
Space group 1.3.6, 1.3.7, 1.3.8, 1.3.9, 1.3.10, 2.1.3, 2.1.5 et seq.
black-and-white 1.3.7
determination 2.1.3, 2.1.5

Space group (continued)
nonsymmorphic 1.3.10
symmorphic 1.3.9
Spin-orbit coupling 4.3.2, 4.3.4
Spin systems (n.m.r.) 5.1.7 et seq.
Spinels 7.3.6
Stark effect 3.3.3, 3.3.4, 4.4.2, 4.4.3, 4.4.10
Stereographic projection 1.3.1, 1.3.2
Strain energy 7.1.3
Stretching frequency 3.1.3, 3.1.7
Structure factor 2.1.4, 2.1.5, 2.1.8, 2.2.0
Subgroup 1.1.2, 1.1.9
Subspectra (n.m.r.) 5.1.9, 5.1.10
Symmetric group 1.2.9, 1.2.10, 1.4.8
Symmetric top molecule 3.2.3, 3.2.6, 3.3.3, 4.4.6
Symmetrized cube and square 1.1.7, 1.2.7, 1.4.2
Symmetry-adapted orbital 1.2.4, 1.2.7
Symmetry
element 1.1.1, 1.2.1, 1.4.1
operation 1.1.2, 1.2.3, 1.2.11, 1.4.1
international notation 1.2.1, 2.1.3, Appendix 2
Schönflies notation 1.1.5, 1.2.1, 1.4.1, Appendix 2
Systematic absences 2.1.1, 2.1.3, 2.1.4, 2.1.5

Tanabe-Sugano diagram 4.1.2, 4.1.4, 4.1.7
Temperature dependence of spectrum 4.1.7
Temperature factor (x-ray diffraction) 2.1.4, 2.1.8, 2.1.10
Temperature-independent paramagnetism 4.3.2, 4.3.3
Term (atomic) 1.2.6, 1.3.4, 1.3.5
Thermal vibration tensor 2.1.10
Torsional-rotational energy level 1.2.10
Transition-metal compounds (ions) 1.2.1, 1.2.9, 1.3.7, 1.4.1, 1.4.5, 2.1.5, 2.4.4, 4.1.1, 4.1.2, 4.1.3, 4.1.4, 4.1.5, 4.1.6, 4.1.7, 4.1.8, 4.1.12, 4.2.5, **4.3**, 4.5.0, 4.5.3, 4.5.6, 5.2.11, 5.3.1, 5.3.7, 5.3.11, 7.1.5, 7.3.3, 7.3.5, 7.3.6, 7.3.8, 7.3.9, 7.3.11
Transition moment dipole 4.5.0, 4.5.4
Triplet state 4.2.2
Tunnelling 4.4.2

Unit cell 2.1.2, 2.1.3, 2.1.4, 2.1.7, 2.1.8, 2.4.1, 2.4.5

Valence bond theory 4.6.9, 7.2.6
Valence force field 1.4.2, 3.1.2
Variation theorem 4.1.9
Vibration frequency 3.1.2
Vibrational modes, see Normal modes of vibration
Vibronic coupling 1.4.5, 4.1.5, 4.1.6, 4.1.7
Vib-rotational selection rule 1.4.4

Wave properties
electrons 2.3.2
neutrons 2.2.0, 2.2.1
Width of spectral band 4.1.4
Wigner-Eckart theorem 1.2.11, 1.2.12

Young tableaux 1.2.8

Zeeman splitting 1.2.12
Zero-field splitting 4.2.2, 4.2.5
Zero-point motion 3.3.2